Advances in Transboundary Aquifer Assessment

Advances in Transboundary Aquifer Assessment

Editors

Sharon B. Megdal
Anne-Marie Matherne

MDPI • Basel • Beijing • Wuhan • Barcelona • Belgrade • Manchester • Tokyo • Cluj • Tianjin

Editors

Sharon B. Megdal
Water Resources
Research Center
University of Arizona
Tucson
United States

Anne-Marie Matherne
U.S. Geological Survey
New Mexico Water
Science Center
Albuquerque
United States

Editorial Office
MDPI
St. Alban-Anlage 66
4052 Basel, Switzerland

This is a reprint of articles from the Special Issue published online in the open access journal *Water* (ISSN 2073-4441) (available at: www.mdpi.com/journal/water/special_issues/transboundary_aquifer).

For citation purposes, cite each article independently as indicated on the article page online and as indicated below:

LastName, A.A.; LastName, B.B.; LastName, C.C. Article Title. *Journal Name* **Year**, *Volume Number*, Page Range.

ISBN 978-3-0365-8451-5 (Hbk)
ISBN 978-3-0365-8450-8 (PDF)

Cover image courtesy of Andrew Robertson

Contents

About the Editors

Anne-Marie Matherne

Anne-Marie Matherne, Ph.D., is a hydrologist with the U.S. Geological Survey. Her research interests include geomorphology and sediment transport and groundwater–surface water interactions. She has worked on various research projects, including landscape erosion on Navajo Nation and in the Four Corners area of the United States, effects of municipal pumping on stream flow, and post-wildfire debris flow hazard assessments. She has worked internationally in Honduras, El Salvador, and Ethiopia. She is currently Chief of the Environmental Geosciences Unit of the USGS New Mexico Water Science Center and Project Manager of the Transboundary Aquifer Assessment Program. Dr. Matherne holds a PhD in Geotechnical Engineering from the University of Illinois, Chicago and is a registered professional geologist.

Sharon B. Megdal

Sharon B. Megdal, Ph.D., is Director of The University of Arizona Water Resources Research Center (WRRC). Her other primary titles include Professor and Specialist in the Department of Environmental Science, C.W. & Modene Neely Endowed Professor, and Distinguished Outreach Professor. She aims to bridge the academic, practitioner, and civil society communities through water policy and management research, education, and engagement programs. The geographic scope of Dr. Megdal's work ranges from local to international. Her applied research projects include analysis of water management, policy, and governance in water-scarce regions, groundwater recharge, and transboundary aquifer assessment. She is the lead editor of the book *Shared Borders, Shared Waters: Israeli-Palestinian and Colorado River Basin Water Challenges*, and she has guest edited several Special Issues of the journal *Water*. Dr. Megdal teaches the graduate course "Water Policy in Arizona and Semi-arid Regions". In 2020, she was awarded the Warren A. Hall Medal for lifetime achievement in water resources research and education by the Universities Council on Water Resources. Sharon Megdal holds a Ph.D. degree in Economics from Princeton University.

Preface to "Advances in Transboundary Aquifer Assessment"

Groundwater as a resource is being increasingly relied on globally to serve emerging and developing water needs. Characterizing groundwater aquifer systems in terms of geology and groundwater quality, quantity, and sustainability is critical to understanding the physical and socioeconomic implications of its use. Of special consideration is aquifer assessment in a transboundary setting, where the cooperation of multiple jurisdictions, sometimes with different languages and cultures, is required. This Special Issue features studies of transboundary aquifers, particularly at the border shared by Mexico and the United States, a semi-arid to arid region experiencing significant population growth and changing climate conditions. The papers represent a broad array of investigations of complex physical aquifer systems and related institutional settings, including the identification and prioritization of the needs and strategies for sustainable groundwater development and use; the characterization of the physical framework of the aquifer, stressors on the aquifer system, and how those stressors influence the availability of groundwater in terms of its quantity and quality; and the incorporation of stakeholder input and prioritization directly into the process of aquifer assessment and model building. We hope that this collection of diverse papers provides useful information and encourages additional research that contributes to understanding groundwater and aquifer systems.

Sharon B. Megdal and Anne-Marie Matherne
Editors

water

Editorial
Advances in Transboundary Aquifer Assessment

Anne-Marie Matherne [1,*] and Sharon B. Megdal [2]

1 U.S. Geological Survey, New Mexico Water Science Center, Albuquerque, NM 87113, USA
2 Water Resources Research Center, The University of Arizona, Tucson, AZ 85719, USA; smegdal@arizona.edu
* Correspondence: matherne@usgs.gov

Abstract: This Special Issue is intended to highlight both recent work to advance the physical understanding of transboundary aquifers and factors relevant in successful collaboration on transboundary groundwater resource use. The collected papers address: (1) the identification and prioritization of the needs and strategies for sustainable groundwater development and use, along with the complexities introduced by working across borders with differing governance frameworks, institutions, cultures, and sometimes languages; (2) the characterization of the physical framework of the aquifer, stressors on the aquifer system, and how those stressors influence the availability of groundwater in terms of its quantity and quality; and (3) the incorporation of stakeholder input and prioritization directly into the process of aquifer assessment and model building. The papers provide insights into the state of knowledge regarding the physical characterization of important transboundary aquifers, primarily along the U.S.–Mexico border and the opportunities for greater stakeholder involvement in resource evaluation and prioritization. They point the way towards a future focus that combines both of these aspects of transboundary aquifer assessment for informing groundwater management discussions by policymakers.

Keywords: transboundary aquifers; aquifer assessment; groundwater; stakeholder involvement; United States–Mexico border; United States–Canada border

Citation: Matherne, A.-M.; Megdal, S.B. Advances in Transboundary Aquifer Assessment. *Water* **2023**, *15*, 1208. https://doi.org/10.3390/w15061208

Received: 3 March 2023
Accepted: 9 March 2023
Published: 20 March 2023

1. Introduction

Groundwater serves the drinking water needs of about 50% of the global population and contributes to over 40% of the global production of irrigated crops. Over 40% of the world's water is transboundary in nature, crossing a binational border [1]. Management of the joint resource between countries involves the cooperation of multiple jurisdictions, sometimes with different languages and cultures. Management decisions about use of the groundwater resources require a physical understanding of the aquifer [2], including groundwater availability, stressors on the system, and the potential for sustainable groundwater use. Information about the physical system can support informed decisions by governments and managers regarding the shared resource. This Special Issue, "Advances in Transboundary Aquifer Assessment", is intended to highlight both recent work to advance the physical understanding of transboundary aquifers and factors relevant in successful collaboration on transboundary groundwater resource use.

Three themes emerged in the papers that comprise this Special Issue. The first theme "Transboundary governance and stakeholder engagement" (see Section 2.1) includes identifying and prioritizing needs and strategies for sustainable development and use, along with the complexities introduced by working across borders with differing governance frameworks, institutions, cultures, and sometimes languages. Papers in this section focus on the U.S.–Mexico border, with one paper addressing issues along the U.S.–Canada border. The papers focusing on "Aquifer characterization and assessment" (Section 2.2) involve the physical framework of the aquifer, stressors on the aquifer system, and how those stressors influence the availability and quality of groundwater. The papers in Section 2.3 "Integration of stakeholder input into model development" move beyond the reliance on physical data and expert opinion in aquifer assessment and model development to formally include stakeholder participation in the process of assessment and model building; these represent an effort to make models more responsive to current and developing issues and priorities in the aquifers being modeled.

2. Contributions

Much of the work described in the papers for this Special Issue was conducted under the umbrella of the Transboundary Aquifer Assessment Program (TAAP). Initiated through U.S. Congressional legislation in 2006 (U.S. Public Law 109–448, TAA-Act), the 2009 Joint Report of the Principal Engineers of the International Boundary and Water Commission (IBWC/CILA) [3], referred to as the TAAP Cooperative Framework, established the ability of the United States and Mexico to work together to study transboundary aquifers. The two countries agreed to focus on four aquifers: the San Pedro and Santa Cruz River aquifers along the border shared by the states of Arizona (United States) and Sonora (Mexico); and the Mesilla/Conejos-Médanos and Hueco Bolson aquifers along the border shared by New Mexico and Texas (United States) and Chihuahua (Mexico) (Figure 1). The choice of aquifers was based on the location of population centers, industry, and environmental concerns. Much of the work under TAAP has focused on these aquifers, and that focus is reflected in the topics covered in many of the papers in this collection.

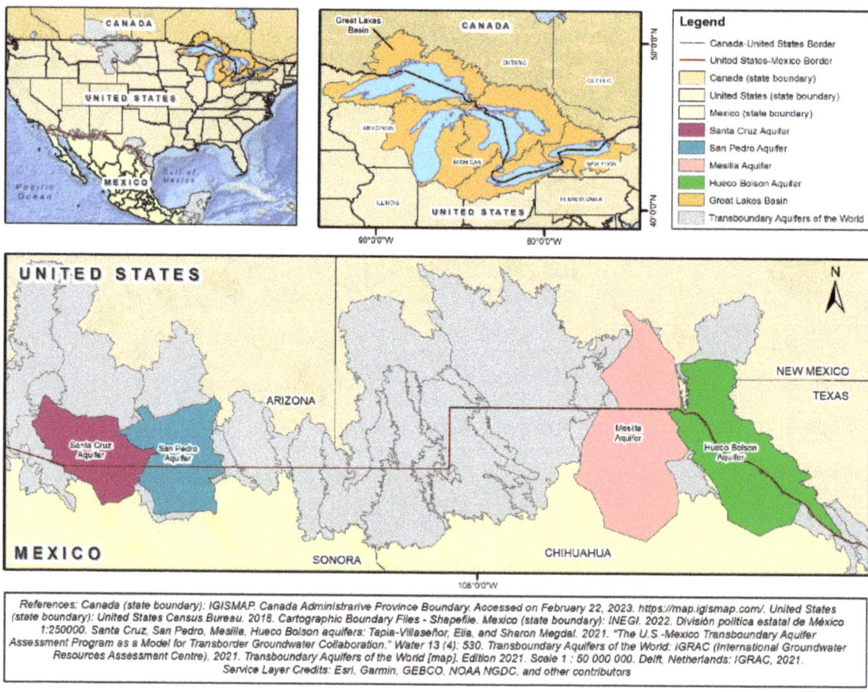

Figure 1. Location of aquifers discussed in the papers of this Special Issue, including the Transboundary Aquifer Assessment Program aquifers of focus and the Laurentian Great Lakes Basin (figure produced by Elia Tapia-Villaseñor; used with permission).

2.1. *Transboundary Governance and Stakeholder Engagement*

In some regions, water use and associated water governance have generally focused on more readily available surface-water resources, with laws and agreements governing groundwater storage and use lagging behind the ability to assess and use groundwater resources. This is the case for the Laurentian Great Lakes Basin along the U.S.–Canada border (Figure 1), the focus of Weekes and Krantzberg [4]. In their paper, "Twenty-first century science calls for twenty-first century groundwater use law: A retrospective analysis of transboundary governance weaknesses and future implications in the Laurentian Great Lakes Basin", they trace the development of water use and its regulation in the transboundary Laurentian Great Lakes Basin. Increasing population, with associated increases in water demand and land-use changes, has resulted in increased groundwater use. Coupled with climate change, increased groundwater use is driving a groundwater storage (GWS) decline. The Great Lakes are net groundwater receivers, and over-pumping aquifers can also reduce groundwater fluxes to surface-water systems. The GWS Governance framework, that is, policies and decision-making standards impacting GWS, are contained in binational-to-municipal-level statutes, voluntary agreements/regulations, common law, and treaties. Weeks and Krantzberg examine the history and development of GWS governance at the binational and at the province (state) levels. Although, in recent decades, groundwater specific policies have been developed, they note the prevalence of policies originally intended to safeguard surface water quantities interpreted to govern groundwater use and to maintain groundwater storage. Weekes and Krantzberg argue for the need to update groundwater policies and regulations to reflect current science and water use in the basin.

Focusing on processes that facilitate and support the integration of science and policymaking, Petersen-Perlman et al. [5], in "Science and binational cooperation: Bidirectionality in the Transboundary Aquifer Assessment Program in the Arizona-Sonora border region", observe that the use of scientific information in management and policymaking depends on salience, credibility, and legitimacy of scientific information; iterative information production; and sociocultural factors. Petersen-Perlman et al. look at six transboundary agreements globally, including TAAP, and note that the production of scientific information and governance, in the form of transboundary water cooperation over use of a shared resource, is iterative. Data production informs governance and policy, which in turn informs further data production. The process is bidirectional, in what the authors term "reciprocal synchronicity". A case-study analysis of TAAP finds that information sharing between the United States and Mexico was only possible after agreeing on and establishing the TAAP Cooperative Framework for data sharing and scientific collaboration between the countries. It has yet to be seen whether the assessments will aid transboundary water governance between the two countries.

Development of transboundary policies and governance between countries relies on collaborative processes that are articulated in some transboundary agreements. Tapia-Villaseñor and Megdal [6], in "The U.S.-Mexico Transboundary Aquifer Assessment Program as a model for transborder groundwater collaboration", note that TAAP was established as a program for the physical characterization of aquifers and is focused on binational information production. This knowledge–improvement phase is an element of the six global transboundary aquifer agreements examined in comparison to TAAP. Although not expressly stated, the binational nature of the TAAP Cooperative Framework, which establishes the ability of the United States and Mexico to perform transboundary assessments, implies and necessitates development of collaborative elements consistent with the principles of other transboundary groundwater management agreements around the world.

Tapia-Villaseñor and Megdal note that the principles of the TAAP Cooperative Framework include elements that promote trust between the United States and Mexico such as data sharing, development of binational aquifer assessment activities, the establishment of technical advisory committees, and the establishment of technical groups. In "Trust, risk, and power in transboundary aquifer assessment collaborations" [7], Brause examines the issue of trust in binational interactions in the Mesilla/Conejos-Médanos Basin, one of the TAAP designated priority transboundary aquifers, and the need to manage asymmetrical relationships of power and unequal levels of risk inherent in collaborating across the border. Brause observes that the TAAP Cooperative Framework does well to manage power inequalities at personal and interpersonal levels and in the context of organizing and managing collaborative exchange, but it cannot mitigate differences in structural power. Structural power differences are a greater issue at times of increased risk to a nation-state's ability to maintain sovereign control over its borderland water resources, such as an ongoing (2022) domestic water lawsuit in the United States that could affect water resources critical to Mexico (Texas v. New Mexico and Colorado, No. 141 Original, Eighth Circuit, United States Court of Appeals [https://www.ca8.uscourts.gov/texas-v-new-mexico-and-colorado-no-141-original]; accessed 15 October 2022).

2.2. Aquifer Characterization and Assessment

Papers in this Special Issue dealing with aquifer characterization examine the four TAAP aquifers of focus. In the San Pedro River aquifer, earlier work produced a hydrogeologic framework model with datasets such as geology, soils, and landcover, harmonized across the U.S.–Mexico border [8]. In "Hydrogeomorphologic mapping of the transboundary San Pedro Aquifer: A tool for groundwater characterization" [9], Minjarez Sosa et al. use datasets from the hydrogeologic framework model to develop a hydrogeomorphologic map of the San Pedro River Basin. Groundwater deficit in the aquifer is attributed to competing use from mining, military, domestic, and agricultural users. Mapping identifies potential areas of recharge in the highland and groundwater discharge in the lowland areas of the basin. This hydrogeomorphologic map can potentially serve as a tool for modeling and the development of strategies for sustainable water resource management.

Studies of the Santa Cruz River aquifer focus on the effects of climate variability and uncertainty on groundwater availability in the region. Shamir et al. [10], in "A Review of climate change impacts on the USA-Mexico Transboundary Santa Cruz River Basin" note current trends of year-round warming and a decline in precipitation and streamflow, especially in the winter months. A review of studies on climate uncertainty in the region in the mid-21st century identifies and describes a continuation of the current warming trend and a projected mid-21st century decline in precipitation events. These projected trends are important considerations in the development of strategies for sustainable water resources management of the Santa Cruz River aquifer. The findings of Shamir et al. are supported by the paper "Assessing groundwater withdrawal sustainability in the Mexican portion of the transboundary Santa Cruz River aquifer" [11]. Tapia-Villaseñor et al. develop a water-budget model for the Mexican portion of the Santa Cruz River aquifer to assess annual water withdrawal. Model results indicate a sharp decline in sustainable groundwater withdrawal for this part of the aquifer, from a maximum of 36.4 million cubic meters (MCM)/year in 1993 to less than 8 MCM/year in 2020, coincident with the drying period also identified in [10]. Based on their analysis, they point to a need to adjust water resource management criteria to respond to the large interannual climate variability in the region.

Because of their importance as regional water sources, there is a long history of research focused on the Mesilla/Conejos-Médanos and Hueco Bolson aquifers [12,13]. Four Special Issue papers focus on the physical assessment of these aquifers, expanding understanding of groundwater/surface-water interactions and of deep and interbasin groundwater circulation, and include a synthesis of Mesilla/Conejos-Médanos research and an updated hydrologic conceptual model. The Rio Grande/Río Bravo del Norte is the primary source of recharge to the Mesilla Basin/Conejos-Médanos aquifer system. Ikard et al., in "Gradient self-potential logging in the Rio Grande to identify gaining and losing reaches across the Mesilla Valley" [14], use gradient self-potential logging to survey an approximately 72 km reach of the Rio Grande from Leasburg Dam near the northern terminus of the Mesilla Valley downstream to Canutillo, Texas. By interpreting an estimate of the streaming-potential component of the electrostatic field in the river, they identify reaches where surface-water gains and losses were occurring and, therefore, areas of aquifer recharge and discharge along this portion of the Rio Grande.

Salinity contributions to the shallow Mesilla/Conejos-Médanos aquifer system and the Rio Grande come from several sources, including upwelling of geothermal groundwater. Pepin et al., in "Salinity contributions from geothermal waters to the Rio Grande and shallow aquifer system in the transboundary Mesilla (United States)/Conejos-Médanos (Mexico) Basin" [15], examine the potential contributions of deep saline groundwater from geothermal sources and demonstrate the use of heat as a groundwater tracer to identify salinity sources. Historical temperature data and groundwater flux estimates indicate that the region's known geothermal systems could account for 22% of Rio Grande salinity leaving the basin each year. Regional water level mapping indicates that upwelling brackish waters flow toward the Rio Grande and the southern part of the Mesilla portion of the basin.

In "Investigation of the origin of Hueco Bolson and Mesilla Basin Aquifers (US and Mexico) with isotopic data analysis" [16], Garcia-Vasquez et al. use the isotopic tracers δO^{18} and tritium to validate an interconnection between the Mesilla (U.S. portion) and Hueco Bolson aquifers. They combine new data from the Mexican portion of the Mesilla/Conejos-Médanos aquifer with results from the U.S. side of the aquifer [17]. Analyzing isotopic data from the Mesilla/Conejos-Médanos together with data from the U.S.–Mexico Hueco Bolson aquifer [18], Garcia-Vasquez et al. find evidence, as stated in [17] and [18], that the groundwater is old (recharged thousands of years ago). Their regional analysis supports groundwater exchange between the Mesilla and Hueco Bolson aquifers. These findings support an earlier geologic study [19] stating that the Mesilla/Conejos-Médanos and Hueco Bolson aquifers were originally part of a single aquifer system.

These more focused studies [14–16] contributed to a synthesis and refinement of the water budget and hydrogeologic framework model for the Mesilla/Conejos-Médanos aquifer [12]. In "Mesilla/Conejos-Médanos Basin: U.S.-Mexico transboundary water resources", Robertson et al. use an updated hydrogeologic framework, a binational water-level map, and previously reported aquifer property assumptions to estimate potentially recoverable fresh to slightly brackish groundwater in the Mesilla portion of the Basin at about 82,600 cubic hectometers (hm^3), largely in agreement with previous estimates. Storage for the Conejos-Médanos portion of the Basin is estimated at 69,100 hm^3. Based on evidence presented in this paper, the Rio Grande alluvium is the only unit currently receiving substantial amounts of recharge from the Rio Grande; the amount of groundwater in the Rio Grande alluvium represents a little less than 0.6% of the entire regional aquifer. The majority of groundwater stored in this basin is thousands to tens of thousands of years old. This water is very slowly being displaced at the boundaries by mountain-front recharge and near pumping centers, where vertical gradients are increased by large groundwater pumping withdrawals.

Work by Sanchez and Rodriguez [20], "Transboundary aquifers between Baja California, Sonora and Chihuahua, Mexico, and California, Arizona and New Mexico, United States: Identification and categorization" completes the western segment of a border-wide assessment of transboundary aquifers [21,22], using datasets and nomenclature harmonized across the U.S.–Mexico border. The combined border-wide assessment identified 72 transboundary hydrogeologic units, of which 50–55% were reported to have good to moderate aquifer potential and good to regular water quality. This combined work provides a high-level assessment to aid in identifying and prioritizing transboundary aquifers for further characterization and evaluation with respect to suitability for resource development.

2.3. Integration of Stakeholder Input into Model Development

Demonstrating a further development for these transboundary studies, we begin to see movement beyond reliance on physical data and expert opinion in aquifer assessment and model development to formally include stakeholder participation in the process of assessing, prioritizing issues of concern, and model building, with two papers focused on the Hueco Bolson and one on the Mesilla/Conejos-Médanos aquifer.

Hydraulic gradients and flow directions in the Hueco Bolson aquifer have changed because of high groundwater withdrawal rates in the two major cities, El Paso, United States and Ciudad Juarez, Mexico, raising questions about long-term aquifer sustainability [13]. Talchabhadel et al., in "Current status and future directions in modeling a transboundary aquifer: A case study of Hueco Bolson" [13], present an overview of the

Hueco Bolson aquifer modeling history and describe a coupled groundwater–watershed model currently (2021) under development. Given the complex set of stressors acting on this transboundary aquifer, they make the point that any sustainable and acceptable management solution will need all stakeholders' buy-in and knowledge co-production. They propose the development of a graphical quantitative modeling framework (e.g., system model and Bayesian belief network) to include expert opinions and enhance stakeholder participation in the model.

Focusing on stakeholder-driven assessment in the Hueco Bolson, Mayer et al., in "Investigating management of transboundary waters through cooperation: A serious games case study of the Hueco Bolson Aquifer in Chihuahua, Mexico and Texas, United States" [23], used a binational, multisector, serious-games workshop to explore collaborative solutions in extending the life of a shared aquifer. The workshop led to increased knowledge building on the part of the participants as well as an agreement on the importance of both binational action and informal binational collaboration in extending the life of the aquifer.

Finally, in a study that addresses the processes used to move between information creation and management decisions, Atkins et al., "Modeling as a Tool for Transboundary Aquifer Assessment Prioritization" [24], use a system dynamics model to quantitatively assess the dynamics of transboundary aquifer assessment information reporting and perception delays in the Mesilla/Conejos–Médanos Basin. The results show that the timing and content of reporting can change the dynamic behavior of natural, human, and technical components of transboundary aquifer systems. Atkins et al. demonstrate the potential for modeling to assist with prioritization efforts during the stakeholder data collection and exchange phases to ensure that transboundary aquifer assessments achieve their intended outcomes.

3. Conclusions

These papers provide insight into the state of knowledge regarding the physical characterization of important transboundary aquifers, primarily along the U.S.–Mexico border, and stakeholder inclusion in resource evaluation and prioritization, while pointing the way towards a future focus that combines both of these aspects of transboundary aquifer assessment. The papers in this Special Issue build on prior TAAP work and other studies. Physical assessment is informed by and can inform questions about binational groundwater management, which is the purview of policy makers. Binational assessment enables the parties to develop a common scientific framework and understanding about groundwater and aquifer conditions, while fostering binational relationships. Methodologies are proposed for incorporating expert opinions and stakeholder participation directly in model and scenario development. These efforts suggest that characterization of the complexities of the physical systems and consideration of binational stakeholders and governance can inform development of sustainable management strategies.

Author Contributions: A.-M.M. led the organization and writing of this paper. S.B.M. assisted the writing of the paper. All authors have read and agreed to the published version of the manuscript.

Funding: S.B.M.'s contribution to this paper, along with her efforts as lead editor of this Special Issue were funded in part by the U.S. Geological Survey (funding authorized by P.L. 109–448) Award Number G17AC00439 for the Transboundary Aquifer Assessment Program.

Conflicts of Interest: The authors declare no conflict of interest.

References

1. IGRAC (International Groundwater Resources Assessment Centre). *Transboundary Aquifers of the World [Map]*; Edition 2021, Scale 1: 50 000 000; IGRAC: Delft, The Netherlands, 2021.
2. Alley, W.M.; Reilly, T.E.; Franke, O.L. *Sustainability of Ground-Water Resources*; Circular 1186; U.S. Geological Survey: Reston, VA, USA, 1999. [CrossRef]

3. IBWC, Joint Report of the Principal Engineers Regarding Joint Cooperative Process. United States-Mexico for the Transboundary Aquifer Assessment Program. 2009. Available online: https://www.ibwc.gov/Files/Minutes/Joint_Report_TAAP_081909.pdf (accessed on 26 August 2022).
4. Weekes, K.; Krantzberg, G. Twenty-first century science calls for twenty-first century groundwater use law: A retrospective analysis of transboundary governance weaknesses and future implications in the Laurentian Great Lakes Basin. *Water* **2021**, *13*, 1768. [CrossRef]
5. Petersen-Perlman, J.D.; Albrecht, T.R.; Tapia-Villaseñor, E.M.; Varady, R.G.; Megdal, S.B. Science and binational cooperation: Bidirectionality in the Transboundary Aquifer Assessment Program in the Arizona-Sonora Border Region. *Water* **2021**, *13*, 2364. [CrossRef]
6. Tapia-Villaseñor, E.M.; Megdal, S.B. The U.S.-Mexico Transboundary Aquifer Assessment Program as a model for transborder groundwater collaboration. *Water* **2021**, *13*, 530. [CrossRef]
7. Brause, H. Trust, risk, and power in transboundary aquifer assessment collaborations. *Water* **2021**, *13*, 3350. [CrossRef]
8. Callegary, J.B.; Minjárez Sosa, I.; Tapia Villaseñor, E.M.; dos Santos, P.; Monreal Saavedra, R.; Grijalva Noriega, F.J.; Huth, A.K.; Gray, F.; Scott, C.A.; Megdal, S.B.; et al. *San Pedro River Aquifer Binational Report*; International Boundary and Water Commission: El Paso, TX, USA, 2016; p. 173. Available online: https://www.ibwc.gov/files/binational_study_transboundary_san_pedro_aquifer.pdf (accessed on 20 October 2022).
9. Minjarez Sosa, J.I.; Gutiérrez, G.A.; Monreal, R.; Grijalva Noriega, F.J.; Tapia-Villaseñor, E.M. Hydrogeomorphologic mapping of the Transboundary San Pedro Aquifer: A tool for groundwater characterization. *Water* **2022**, *14*, 906. [CrossRef]
10. Shamir, E.; Tapia-Villaseñor, E.M.; Cruz-Ayala, M.B.; Megdal, S.B. A Review of climate change impacts on the USA-Mexico Transboundary Santa Cruz River Basin. *Water* **2021**, *13*, 1390. [CrossRef]
11. Tapia-Villaseñor, E.M.; Shamir, E.; Cruz-Ayala, M.-B.; Megdal, S.B. Assessing groundwater withdrawal sustainability in the Mexican portion of the Transboundary Santa Cruz River Aquifer. *Water* **2022**, *14*, 233. [CrossRef]
12. Robertson, A.J.; Matherne, A.-M.; Pepin, J.D.; Ritchie, A.B.; Sweetkind, D.S.; Teeple, A.P.; Granados-Olivas, A.; García-Vásquez, A.C.; Carroll, K.C.; Fuchs, E.H.; et al. Mesilla/Conejos-Médanos Basin: U.S.-Mexico transboundary water resources. *Water* **2022**, *14*, 134. [CrossRef]
13. Talchabhadel, R.; McMillan, H.; Palmate, S.S.; Sanchez, R.; Sheng, Z.; Kumar, S. Current status and future directions in modeling a transboundary aquifer: A case study of Hueco Bolson. *Water* **2021**, *13*, 3178. [CrossRef]
14. Ikard, S.; Teeple, A.; Humberson, D. Gradient self-potential logging in the Rio Grande to identify gaining and losing reaches across the Mesilla Valley. *Water* **2021**, *13*, 1331. [CrossRef]
15. Pepin, J.D.; Robertson, A.J.; Kelley, S.A. Salinity contributions from geothermal waters to the Rio Grande and shallow aquifer system in the Transboundary Mesilla (United States)/Conejos-Médanos (Mexico) Basin. *Water* **2022**, *14*, 33. [CrossRef]
16. Garcia-Vasquez, A.C.; Granados-Olivas, A.; Samani, Z.; Fernald, A. Investigation of the origin of Hueco Bolson and Mesilla Basin Aquifers (US and Mexico) with isotopic data analysis. *Water* **2022**, *14*, 526. [CrossRef]
17. Teeple, A.P. *Geophysics- and Geochemistry-Based Assessment of the Geochemical Characteristics and Groundwater-Flow System of the U.S. Part of the Mesilla Basin/Conejos-Médanos Aquifer System in Doña Ana County, New Mexico, and El Paso County, Texas, 2010–2012*; Scientific Investigations Report; U.S. Geological Survey: Reston, VA, USA, 2017.
18. Eastoe, C.J.; Hibbs, B.J.; Olivas, A.G.; Hogan, J.F.; Hawley, J.; Hutchison, W.R. Isotopes in the Hueco Bolson Aquifer, Texas (USA) and Chihuahua (Mexico): Local and General Implications for Recharge Sources in Alluvial Basins. *Hydrogeol. J.* **2007**, *16*, 737–747. [CrossRef]
19. Hawley, J.W.; Kottlowski, F.E.; LeMone, D.V. Quaternary geology of the south-central New Mexico border region. In Proceedings of the Border Stratigraphy Symposium, El Paso, TX, USA, 30 April 1969; New Mexico Bureau of Mines and Mineral Resources: Socorro, NM, USA, 1969; Volume 104, pp. 89–115.
20. Sanchez, R.; Rodriguez, L. Transboundary aquifers between Baja California, Sonora and Chihuahua, Mexico, and California, Arizona and New Mexico, United States: Identification and categorization. *Water* **2021**, *13*, 2878. [CrossRef]
21. Sanchez, R.; Rodriguez, L.; Tortajada, C. Transboundary aquifers between Chihuahua, Coahuila, Nuevo Leon and Tamaulipas, Mexico, and Texas, USA: Identification and categorization. *J. Hydrol. Reg. Stud.* **2018**, *20*, 74–102. [CrossRef]
22. Sanchez, R.; Lopez, V.; and Eckstein, G. Identifying and characterizing transboundary aquifers along the Mexico–U.S. border: An initial assessment. *J. Hydrol.* **2016**, *535*, 101–119. [CrossRef]
23. Mayer, A.; Heyman, J.; Granados-Olivas, A.; Hargrove, W.; Sanderson, M.; Martinez, E.; Vazquez-Galvez, A.; Alatorre-Cejudo, L.C. Investigating management of transboundary waters through cooperation: A serious games case study of the Hueco Bolson Aquifer in Chihuahua, Mexico and Texas, United States. *Water* **2021**, *13*, 2001. [CrossRef]
24. Atkins, A.; Saeed, E.P.; Langarudi, P.; Fernald, A.G. Modeling as a tool for transboundary aquifer assessment prioritization. *Water* **2021**, *13*, 2685. [CrossRef]

water

MDPI

Article

Twenty-First Century Science Calls for Twenty-First Century Groundwater Use Law: A Retrospective Analysis of Transboundary Governance Weaknesses and Future Implications in the Laurentian Great Lakes Basin

Khafi Weekes [1,*] and Gail Krantzberg [2]

[1] Faculty of Science, School of Earth, Environment and Society, McMaster University, Hamilton, ON L8S 4K1, Canada

[2] Faculty of Engineering, W Booth School of Engineering Practice and Technology, McMaster University, Hamilton, ON L8S 4K1, Canada; krantz@mcmaster.ca

* Correspondence: khafi.weekes@gmail.com

Abstract: How has groundwater use been historically governed by the binational to municipal government levels across the Laurentian Great Lakes Basin (GLB)? To what extent have they contemplated the physical–environmental requirements to maintain aquifer storage in devising policies and making decisions governing groundwater use? Although it is amongst the largest freshwater stores in the globe, cases of groundwater shortages are increasingly being reported across GLB communities, raising questions on the fitness of governance approaches to maintain groundwater storage (GWS) with growing climate and human pressures. Applying retrospective analytical methods to assess the century-old collaboration of the United States and Canada to maintain GLB water quantities, we characterize long-term trends and undertake systematic diagnosis to gain insight into causal mechanisms that have persisted over the years resulting in current GWS governance gaps. We reveal the surprising prominence of policies originally intended to safeguard surface water quantities being used to govern groundwater use and thereby maintain GWS. We also connect these, based on sustainable aquifer yield theory, to growing groundwater insecurity in the Basin's drought-prone and/or groundwater-dependent communities. Based on deep understanding of long-standing policy pathologies, findings inform transboundary GWS governance reform proposals that can be highly useful to multiple levels of government policymakers.

Keywords: groundwater storage; groundwater use; multilevel governance; agreement; transboundary basins; retrospective analysis; United States; Canada

Citation: Weekes, K.; Krantzberg, G. Twenty-First Century Science Calls for Twenty-First Century Groundwater Use Law: A Retrospective Analysis of Transboundary Governance Weaknesses and Future Implications in the Laurentian Great Lakes Basin. *Water* **2021**, *13*, 1768. https://doi.org/10.3390/w13131768

Academic Editors: Sharon B. Megdal and Anne-Marie Matherne

Received: 16 March 2021
Accepted: 21 June 2021
Published: 26 June 2021

1. Introduction

With estimates ranging from 5585 km^3 to 4000 km^3 [1], groundwater accounts for roughly 20% of water stored in the Laurentian Great Lakes Basin (GLB). Groundwater fluxes maintain habitats and baseflows to tributaries of the five (5) Great Lakes [2]. It has also become increasingly vital for society, supporting the USD 6 trillion regional economy [3] of the eight (8) US states and the Canadian province of Ontario that are within the GLB's hydrological boundaries (Figure 1).

Figure 1. States and provinces within the Laurentian Great Lakes Basin hydrological boundary [4].

Rising populations with their attendant water demand and land use changes, coupled with climate change [5], are driving an emerging problem of persistent groundwater storage (GWS) decline. At the Basin scale, long-term satellite monitoring estimates an average GWS loss of 3.8 ± 2.3 km^3/year [6]. Though this rate of decline pales in comparison to the overall water-richness of the GLB, the globe's largest surface freshwater store, much of it occurs in drought-prone and/or groundwater-dependent communities. Located further inland, these locales are without ready access to Great Lakes' waters, and are becoming increasingly water insecure [7]. These trends are emerging as GWS—the volume of water that an aquifer holds at any given time within its voids and interstices—is fundamentally limited by an aquifer's storage capacity, which is based on its unique geometry and geophysical attributes [8]. While the quantity of GWS can fluctuate seasonally, as it is a derivative of a predefined rate of inflow from artificial recharge and/or precipitation, and outflow via natural discharge to surface water bodies and/or pumping, it can be permanently drawn down if subject to long-term overuse and reduction of recharge with climate change and land uses that increase impermeable surfaces [8].

GWS governance, involving planning, coordinating, policy making, implementation, and monitoring of policy outcomes [9], provides the means by which groundwater use may be managed, and socio-environmental stressors on GWS addressed. Normatively, long-term GWS decline indicates that governance may be ill-suited to the physical–environmental sustainability needs to maintain GWS. When governance effectuates actions resulting in increased and/or long-term stability of GWS and optimal economic development, it can be considered sustainable [10,11]. In these cases, consideration is placed on maintaining sustainable aquifer yield—the volume of groundwater that can be withdrawn from aquifer systems that avoids unacceptable environmental, socio-economic, and legal consequences [12]. Determining sustainable yield requires strong science–policy alignment as policymakers must consider the water balance of the overall hydrological system, uncertainties in quantifying GWS with spatial and temporal variation, and how human uses can impact GWS over time [12].

Given the GLB's transboundary basin settings, policies and decision-making standards impacting GWS (also known as the "GWS governance framework") are contained in binational-to-municipal-level statutes, voluntary agreements/regulations, common law, and treaties [13]. Per North American institutional historicism, the most important to maintaining GWS are those directly controlling groundwater use: out-of-basin diversions, pumping rates, allocation, conservation, consumption, and withdrawals [14]. Economic policies are also key, creating fiscal deterrents and/or incentives under which groundwater use decisions are made [15]. Environmental safeguards are another aspect, with requisites for data collection and monitoring as well as technical/environmental standards for well construction and pumping [16].

Researchers have long posited that the GWS governance framework may be unfit for purpose in high-groundwater-stress contexts of the GLB [17–22]. They concur on its inadequate consideration of sustainable yield, in particular its insufficient science-based guidelines and incentives promoting conservation and efficient uses that reflect the unique physical–environmental requirements of aquifers to maintain GWS. Growing cases of GWS decline across the basin highlight the need for binational-to-municipal levels of government within the hydrological boundaries of the GLB to provide policies and decision-making standards guiding management actions [23] that address human and climate drivers of GWS depletion [20]. It also presents an opportunity for the establishment of proactive multilevel governance measures designed to halt further proliferation of this problem. Retrospective analysis of historical governance characteristics has proven useful to deepen understanding of present-day policy gaps, and confirm inferences of why policies have led to current environmental outcomes [24]. Using this analytical approach, we deconstruct the historical evolution of GWS governance, deducing features and inferring causal linkages that are likely to have culminated in growing cases of GWS decline and gaps in the current GWS governance framework. Findings are used to proffer recommendations of governance reforms addressing the growing specter of groundwater insecurity deepening in vulnerable locales.

2. Materials and Methods

We applied causal process tracing (CPT)—a qualitative, retrospective analytical technique useful for deducing change and causation within a temporal sequence of events [25]. Per Figure 2, CPT operates by characterizing the intervening causal mechanism (n1 => n2 => etc.) between the cause(s) (X) and the outcome(s) (Y). The causal mechanism is a chain of events or "empirical manifestations" (n_x) linking causes (X) with their long-term effects and eventual outcomes at the end of the study period (Y). It describes "not simply a relationship that has been found, but one that has been found repeatedly." [26]. As such, the more empirical manifestations that are observed within the study period, the more confident researchers can be of the causal mechanism [27]. CPT depends on detailed descriptions of empirical manifestations as well as the concepts linking and/or used to diagnose them, which are based on the overall hypothesis and theories of how X impacts Y.

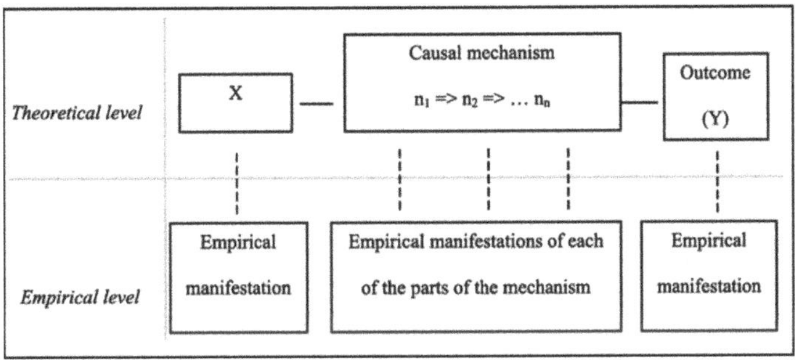

Figure 2. Elements of the causal process tracing method [27].

At its core, our research is a historical process narrative explaining how GWS governance gaps are likely to have persisted over time to feature in current governance and lead to groundwater insecurity. In this context, CPT was applied to design our analysis as outlined in Table 1.

Table 1. Causal process tracing application in research design.

CPT ELEMENT	APPLICATION IN RESEARCH
Causes (X)	Foundational policies and decision-making standards of the current GWS governance framework.
Outcomes (Y)	• Persistent GWS decline in drought-prone and/or groundwater-dependent GLB communities. • Weaknesses and gaps in the present-day GWS governance.
Causal mechanism	Multilevel governance processes, which have evolved over time, defining groundwater uses and environmental safeguards relevant to maintaining GWS.
Empirical manifestation/events (n_x)	Milestones and/or changes in policies and decision-making standards over the timeframe of the evolution of the GWS governance framework, e.g., successive binational treaties, statute amendments, major court decisions, and other governance mechanisms influencing GWS.
Causal linkages (=>)	Established by interpretation and detailed descriptions of policies and decision-making standards over time based on the hypothesis and sustainable aquifer yield theory.

We first characterize the outcomes, providing an overview of GWS governance weaknesses and the emerging problem of groundwater insecurity. In so doing, we describe the human and climate pressures driving GWS vulnerabilities, drawing from official government reports and published literature. We then characterize the emerging GWS decline problem, documenting cases at the sub-watershed scale, using a wide range of indicators including (i) deteriorating water quality with oxygen exposure to lithology [28] and/or upwelling of deeper brines [29]; (ii) collapsing cavities in evaporates (e.g. gypsum) due to dissolution as pumping increases water velocity [30]; (iii) land subsidence due to over-pumping that reduces pore water pressure causing gradual lowering of land [30]; (iv) waning stream levels as baseflow declines [31]; (v) loss of groundwater-dependent ecosystems [32]; (vi) sustained decline of water table levels, defined as the upper limit of the underground where all interstices and voids are saturated with water [33]. Data on these indicators were sourced from desk studies of publicly available reports from peer-reviewed journals, GWS monitoring and governance institutions, and responses to our survey distributed from December 2018 to February 2019 to managers in these institutions. We received a 100% response rate.

To deduce the cause and causal mechanism, the evolution of groundwater use policies and environmental safeguards impacting GWS were studied over the introduction of common law principles in the 19th century, up to the adoption of the 2005 Great Lakes–St Lawrence Basin Sustainable Water Resources Agreement (2005 GLSWRA), the most recent binational agreement controlling groundwater use. Economic policies impacting GWS were reviewed up to the 2020 US Mexico Canada Agreement (2020 USMCA). This study period is sufficient as legal concepts foundational to current GWS governance are drawn from 19th century common law (judge-made, case law long applied by appellate courts to resolve legal disputes related to groundwater use and conservation) [34]. From this, multilevel treaties, rules, and statutes (laws made by legislative bodies of governments at multiple levels) have evolved over the years [24]. As policies and standards component to the present-day GWS governance framework have not changed significantly since the 2020 USMCA and 2005 GLSWRA [22], the dates of the adoption of these binational agreements were considered appropriate for delimiting the study period. Data on the historical policies and standards component to the cause and causal mechanism were sourced from peer-reviewed publications, expert interviews, as well as publicly available government repositories and archives. Policies made by municipalities were not considered as they are not involved in GWS policy and decision making [14,22].

Identification of empirical manifestations and causal linkages was made considering sustainable aquifer yield theory. Aimed at avoiding undesirable social, environmental, and legal outcomes from aquifer pumping, the theory posits a balanced compromise between the contrasting strategies of either little or no pumping of aquifers and the total uptake of natural discharge [8,12,35]. Balancing these opposing governance strategies is largely science-based, as it considers the physical–environmental requirements of aquifers for maintaining GWS [35,36]. We applied the main concepts of sustainable aquifer yield theory as evaluative indicators to identify and assess scopes of policies and standards, pinpointing their changes over the study period, and determining the extent to which they considered (i) the finite volume of groundwater that aquifers can store that is innately limited by their geophysical parameters; (ii) natural recharge of aquifers that are controlled by precipitation and climate; (iii) fluxes required to maintain vital environmental functions; (iv) whether allowed and/or economically incentivized human uses disturbed the equilibrium required to sufficiently maintain GWS while avoiding unwanted outcomes. These evaluations were contextualized by the contemporaneous state of hydrogeological science at key governance milestones, given that the understanding of physical–environmental parameters to maintain GWS evolved over the study period.

To conclude, we synthesized findings, diagnosing the extent to which historical policy gaps have carried over to the current GWS governance framework, and the governance processes by which weaknesses have persisted over time. Insights were then used to link historical governance to emerging GWS decline cases, as well as to provide recommendations to address governance gaps.

3. Results

3.1. Outcomes (Y): The Emerging Problem of Groundwater Insecurity and Linked Governance Gaps

- Characterizing Sub-Watershed-Scale GWS Decline

GLB groundwater is mainly pumped from five principal aquifer systems: the Cambrian–Ordovician, Silurian–Devonian, Mississippian, and Pennsylvanian bedrock aquifers that are composed mainly of carbonates and sandstone, as well as the overlying, surficial aquifer system that is dominated by alluvium and glacial deposits [37]. Due to high permeability and effective porosity, the most productive aquifers are hosted in the unconsolidated sands and gravels of the surficial aquifer system [38] within which most wells are located. As GWS in surficial aquifers is prone to seasonal and climate fluctuations due to their relative shallowness, growing pumping rates often result in long-term groundwater decline, with occurrences being particularly reported in communities located in drought-prone locales and/or are heavily reliant on groundwater [39].

Occurrences of indicators of persistent GWS decline resulting from groundwater over-pumping have not yet been comprehensively documented in the GLB [32]. Based on available information, the impacts of over-pumping on GLB stream baseflow and groundwater-dependent ecosystems are poorly understood [13]. However, one well-documented case in Wisconsin linked excessive pumping to the drying of wetlands causing native habitat loss and invasive species spread [40].

Better documented are cases of long-term pumping reducing riverine baseflow given the interconnectedness of the Basin's surface water bodies with groundwater flow systems [41,42]. The Great Lakes are net groundwater receivers, with their tributaries gaining substantial volumes of water fluxes directly from the Basin's groundwater flow systems [42]. Groundwater contributes from 48% of streamflow in the Lake Erie Sub-Basin up to 79% in the Lake Michigan Sub-Basin [43]. Therefore, as over-pumping aquifers can reduce groundwater fluxes to surface water systems, it can diminish stream baseflow or, in extreme cases, reverse the normal flow of groundwater to surface water bodies. One of the most acute examples is occurring in aquifers supplying residents of the Chicago–Milwaukee metropolitan area and the Green Bay, Wisconsin, and Toledo, Ohio area. Here, long-term pumping has not only reduced stream baseflow but has reversed water flow from surface waters to aquifers [38].

As the Basin's hydrogeologic settings contain a substantial amount of glacial, unconsolidated deposits, some areas are susceptible to land subsidence due to groundwater decline caused by over-pumping [30]. Though not as prevalent in more drought-prone North American states/provinces, localized reports of land subsidence have been reported in Indiana, Wisconsin [44], and Michigan [45]. In regions where aquifers are hosted in karstic rock, sinkholes and cavity collapse can occur due to carbonate dissolution with pumping [46]. To illustrate, municipalities having high risks of gypsum cavity collapse linked to mining dewatering have been documented in Ontonagon, Houghton, Iosco, Keweenaw, Kent, Barry, Eaton, Calhoun, and Jackson counties in Michigan [47].

Upwelling of brines due to excessive mine dewatering has been reported in wells in the townships of Windsor and Romney, Ontario [13]. In Michigan, upwelling of brines due to long-term pumping for drinking water and agriculture has been well documented in Michigan's Lower Peninsula [48], as well as in Ottawa County that abuts northern Lake Michigan [49]. Arsenic concentrations exceeding the US Environmental Protection Agency's maximum contaminant level of 10 μg/L are often reported in well water in Southeast Michigan [50] in the counties of Huron, Tuscola, Sanilac, Lapeer, Genesee, Shiawassee, Livingston, Oakland, Macomb, and Washtenaw. These wells pump the Marshall Sandstone, hosted in the Mississippian basement aquifer system [37]. Relatedly, long-term pumping has caused drinking water of the straddling community of Waukesha, Wisconsin to be contaminated with radium, prompting its successful application for access to GLB water resources [51].

Responses to our survey indicated that persistent groundwater table decline occurs in aquifers supplying roughly 10% of GLB municipalities. Widespread groundwater table decline risks have been modelled in Michigan including the Grand Rapids and the metropolitan area of Detroit and its eight (8) suburban counties including Genesee, Oakland, Macomb, Washtenaw, Wayne, St. Clair, Lapeer, and Monroe (communication from the Department of Environmental Quality on 5 December 2018). This has also been extensively documented in aquifers supplying Milwaukee and Chicago, including its eight (8) eastern suburban counties, as intense pumping beginning in 1864 caused groundwater table levels to decline by as much as 275 m by 1980 [52]. In the Ontario Sub-Basin, aquifers supplying municipalities in the Grand River Watershed, including Kitchener, Waterloo, Cambridge, the City of Guelph, and surrounding townships, have a moderate risk of developing GWS shortages [53]. These risks are particularly in droughts, the summer agricultural growing season, and periods of high municipal water demand used to supply the residential, industrial, and commercial sectors [54].

- Characterizing Present-Day GWS Governance Weaknesses

Incorporated into current federal and state/provincial laws, many of the current policies and decision-making standards governing groundwater use are from the 2005 GLSWRA. The binational agreement, aimed at sustaining the quantity of all GLB waters, generally prohibits withdrawals over 379,000 L/day " ... in any 30-day period (including Consumptive Uses) from all sources ... " (defined as bulk water) or diverting any volume of water from the Basin, except when in containers 20 L or less, without a regional review decision-making process by Great Lakes governors/premiers. Parties are urged to promote efficient water use and to record water uses by sector in a regional data base. Water uses below bulk water definitions are considered " ... reasonable uses ... " for which GLB states/provinces can set their own regulations. The Great Lakes states passed a series of Great Lakes–St Lawrence Basin Sustainable Resources Compact Acts into law between 2007 and 2008, and Ontario brought these policies into effect in Ontario Regulation 225/14 in 2014. These laws limited the scope of the 2005 GLSWRA regional review process to deciding on large water diversions from the GLB, and gave the states/provinces responsibilities to regulate bulk water use; the most common regulation being Permit to Take Water (PTTW) programs.

Relevant economic policies include the 2020 USMCA, state/provincial PTTW and/or well license fees, and municipal water supply tariffs. As the newest North American free trade treaty, the 2020 USMCA allows export of GLB groundwater when embedded in products. It furthers the scope of past trade agreements, including large, medium, and small enterprises, and removes tariffs on a wider range of agricultural products. It is the only binational agreement impacting GWS with legally binding recourse should enterprises perceive unfair barriers to free trade [55].

With identical policies guiding groundwater and surface water use, with high volumetric water use thresholds for bulk water definitions, binational-to-municipal levels of government often overlook fundamental physical–environmental differences between groundwater and surface water [7]. Sustainable aquifer yield considerations also appear to be largely ignored in federal and state/provincial governance of smaller volumes of GLB groundwater use [56]. Some examples are that policies generally do not include volumetric limits controlling groundwater pumped for agricultural purposes or from smaller-capacity wells on private land for domestic use. Policies guiding aquifer pumping in federal lands are also largely absent [34]. Instead, governmental oversight is typically limited to data-recording requirements and technical specifications for commissioning wells [14].

Economic policy tools generally encourage groundwater overuse, furthering groundwater insecurity risks in vulnerable locations [57]. The 2020 USMCA increases competition for groundwater resources by opening up free trade provisions to a greater pool of enterprises. The removal of trade tariffs on a wider set of agricultural products increases pressure on aquifers given that agriculture is the most intense water-consuming sector within the GLB. At the state/provincial level, higher-capacity wells requiring PTTWs attract low permit fees [14], and groundwater used for agriculture and firefighting are exempt from permits [58]. Finally, graduated block rates of municipal water supply tariffs can incentivize water wastage, as rates become progressively cheaper the more water is used [59].

3.2. Causal Mechanisms: Linking Historical GWS Governance to Current Outcomes

- Fundamental Legal and Scientific Principles Underpinning the Evolution of GWS Governance

In North America, controlling who has access to groundwater has historically been tied to land ownership and property rights [24]. This has its origins in the Absolute Ownership Rule of English common law [60] that allowed landowners to use groundwater below their property without limits or obligations to conserve the resource for neighbors or for future uses [61]. Court deliberations in the earliest documented application of the Absolute Ownership Rule—1843 Chasemore vs. Richards (1843-60 All E.R. 77, 81-82 H.L. 1859)—show that the court did not think it could limit the use of "water percolating through underground strata, which has no certain course and no defined limit, but oozes through the soil in every direction in which the rain penetrates." It is apparent that the Absolute Ownership Rule was originally devised based on the idea that groundwater quantity, flow rates, and flow directions were "unknowable", given the embryonic state of hydrogeological science at the time [62]. Later adopted in early North American governments, the Absolute Ownership Rule was modified to the Reasonable Use Rule, limiting groundwater uses to those done without waste or inhibiting the rights of adjacent property owners to access groundwater within their properties [63].

In multiple levels of GLB government, applying the Reasonable Use Rule to govern groundwater use has been nuanced by the Underground Stream Doctrine and the Public Trust Doctrine. The Underground Stream Doctrine interrelates surface water and groundwater rights of use, resulting in groundwater wells traditionally being treated as surface water diversions and groundwater flow considered "tributary" to GLB surface water [14]. Adding to this is the Public Trust Doctrine that originated from sixth century Roman civil law or "Institutes of Justinian", obliging governments to protect in perpetuity "things common to mankind—the air, running water, the sea, and consequently the shores of the sea." Used as the basis for environmental and natural resource protection laws, when the Public Trust Doctrine was adopted in the constitutions of newly formed North American states/provinces, governmental responsibilities to protect water resources originally extended only to surface water [63].

In this context, policies, standards, and court decisions governing GLB water use have traditionally prioritized safeguarding surface water quantities. Unless the purpose of groundwater protection has been closely tied to safeguarding surface water quantity for the greater public good, governmental oversight of groundwater use has been lacking, with groundwater use being traditionally treated as a private property rights issue [24]. Remaining largely unchanged over the years, these legal principles have carried through multilevel GWS governance, despite advances in scientific understanding of groundwater's physical–environmental sustainability requirements and its role in providing a range of vital environmental flows beyond baseflow to surface water bodies.

- The Evolution of Binational GWS Governance

As far back as the 1794 Jay Treaty, aimed at maintaining Great Lakes' levels for international navigation during the Napoleonic wars, binational governance of GLB water uses prioritized maintaining surface water quantities [63]. Modern governance began with the 1909 Boundary Waters Treaty (1909 BWT) that banned large diversions of surface waters straddling the international border. Aiming to ensure equitable "domestic and sanitary uses, navigation uses, and uses for power and irrigation", it established the International Joint Commission (IJC). The IJC did not have a major GWS governance role until the 1988 Cabin Creek Coal Mine case, when its Water Use Reference was updated allowing investigation of GWS issues as a matter of practice [64].

The next significant binational agreement was the 1956 Great Lakes Basin Compact that created the Great Lakes Commission (GLC) to promote "orderly, integrated, and comprehensive development, use, and conservation" of GLB water resources. It was the first agreement to adopt a whole-of-basin approach to governance, explicitly considering the range of water uses: "industrial, commercial, agricultural, water supply, residential, recreational, and other." However, its mandate was limited to the Great Lakes and all connected "rivers, ponds, lakes, streams and other watercourses.", reflecting the original interpretation of the Public Trust Doctrine by excluding groundwater from its purview.

Another important update was the 1985 Great Lakes Charter (1985 Charter). Established as a good faith agreement between the GLB governors and premiers, it is significant as it introduced many of the key standards and policies for the uses of "all GLB waters" still in place in today. It expanded GLC membership to include Canadian premiers, introduced the regional review process for making decisions on bulk water use and diversions, and most significantly, was the first binational agreement to include groundwater in its purview as a public trust responsibility [65]. Improving science–policy alignment, the 1985 Charter introduced volumetric limits to GLB water use to safeguard "nonrenewable" GLB water resources. It required regional review of bulk water uses, defined as any withdrawal exceeding 380,000 L/day in any 30-day average, and any new or increased diversion or consumption of GLB water exceeding 19 million liters per day in any 30-day period. It also initiated the Great Lakes St. Lawrence River Regional Water Use Database that was eventually established in 1988.

Despite these milestones, the 1985 Charter did not appear to consider sustainable aquifer yield in recommending policies and standards to govern groundwater use. Notwithstanding well-documented knowledge that aquifers can be depleted due to over-pumping since 1910 [62], the 1985 Charter did not reflect on groundwater's relative scarcity and lower replenishment rates compared with surface waters as it provided identical volumetric definitions and controls for bulk groundwater and surface water use. By stating its overall aim was to safeguard GLB surface waters, it invoked the Underground Stream Doctrine, considering groundwater flow systems as merely tributaries to surface water bodies, and interrelating the policies governing the uses of both resources. In so doing, it failed to keep pace with groundwater science that had advanced considerably from the 19th century in North America. By 1903, key hydrogeological concepts relevant to sustainable aquifer yield had been developed, including that environmental flows provided by groundwater were not just limited to surface water bodies, as well as the relationship between groundwater budgets and sustainable limits for consumptive uses and aquifer geometry and geological media [62]. Though the Regional Water Use Database has been providing yearly reports on GLB water withdrawals, consumption, and diversions, since its inception it has not had a specific data field for tracking water use from aquifers. This has made it difficult to garner consistent groundwater use data, an essential input for determining sustainable aquifer yield.

Since the 1985 Charter's original policy prescriptions remained largely unchanged in the intervening years, many of its original GWS governance gaps have carried through to the present day. As the 1985 Charter was set up as a non-legally-binding agreement, it did not include enforcing mechanisms. Thus, the GLC later agreed to the 2001 Great Lakes Charter Annex, committing the GLB states/provinces to agree on policies to be included in laws within the next three (3) years. This was fulfilled when the 2005 GLSWRA was passed and subsequently integrated into current state/provincial laws governing GLB water use.

Sustainable aquifer yield considerations have also been absent from binational economic policies affecting GWS. Though they can be traced back to the 1855 Reciprocity Treaty, it was not until 1987 that the first such policy was established that had direct impact on maintaining GWS when both countries established the Canada–United States Free Trade Agreement. Superseded by the 1994 North American Free Trade Agreement that admitted Mexico to the free trade zone, these agreements followed the General Agreement on Tariffs and Trade of the World Trade Organization. Herein, GLB groundwater and surface water were allowed to be exported when "captured whether in bottles, tankers or pipelines." Successive trade agreements have ignored the cumulative impacts the virtual groundwater trade can have over time on source aquifers and the environmental safeguards for maintaining GWS. Instead, these agreements have always included settlement mechanisms for trade disputes, opening the door to growing competition and conflicts between conservationists and industries drawn to the Basin by its cheap, clean, and abundant groundwater supply [66].

- The Evolution of Federal GWS Governance

Per the 1867 Canadian Constitution, the Canadian federal government has had a historically limited role controlling groundwater use, restricted to aquifers within international borders and those underlying railways, federal, and First Nations lands. It has been most involved in geological mapping and tracking GWS levels, founding the Geological Survey of Canada (GSC) in 1947 and expanding its groundwater research commitments in the 1987 Federal Water Policy [67]. The US federal government has also long facilitated similar hydrogeological research, founding the United States Geological Survey (USGS) in 1879. However, it has had a more central GWS governance role, with the Commerce Clause of the 1787 United States Constitution and the 1986 Water Resources Development Act (1986 WRDA) prohibiting diversions of all US waters without Congressional consent. A 2000 amendment to the 1986 WRDA banned all diversions of GLB water unless approved by Great Lakes governors, thus conferring the GLB states' GWS governance role [68]. As such, most GWS governance roles rest with the eight (8) GLB states and Ontario.

Despite the federal governments' long-standing facilitation of hydrogeological research, there is little evidence to suggest that sustainable aquifer yield considerations have been taken into account in successive court rulings or state/provincial laws and decision-making standards impacting GWS. Remaining mostly unchanged from its original 19th century legal doctrines that were based on 19th century scientific understanding of groundwater flow systems, the evolution of state/provincial GWS governance is evaluated below.

- The Evolution of State/Provincial GWS Governance

After agreeing on the 1956 Great Lakes Compact, GLB states/provinces adopted bulk water use and diversion counsels of successive binational agreements. In so doing, they followed the historical trend of overlooking sustainable aquifer yield requirements and favoring surface water preservation objectives. As such, the focus of this assessment is on the governance of smaller volumes of GLB groundwater use within the study period. This is because, despite much of the theoretical foundation and rudimentary groundwater quantification and modelling methodologies being established by 1940 [62], there has been considerable variation in the degree to which these policies and decision-making standards kept pace with these scientific advances and took sustainable yield considerations into account [69]. We also evaluate court rulings to resolve groundwater use conflicts during the study period. As the only state wholly within the Basin's boundaries, we focus analysis on Michigan's court decisions as its many landmark rulings demonstrate well how groundwater conflict resolution has been historically treated in case law.

I. Ontario

Ontario had some of the earliest policies in place impacting GWS in the study period. Its Ontario Water Resources Act (OWRA) mandated licensing and pumping rate data collection since 1961 [67]. A 1990 OWRA amendment introduced more stringent requirements for bulk water use than those of the 1985 Charter, requiring permits for taking over 50,000 L per day, environmental impact assessments (EIAs), and a graduated approach to PTTW fees. Reflecting consideration of lower quantities and replenishment rates of GWS, fees ranged from none for taking water from low-environmental-impact sources, to USD 3000 for groundwater PTTWs issued in high-use regions and/or for water-bottling purposes (Section 34). The 2001 Ontario Municipal Act was the only GLB policy within the study period mandating inclusion of municipalities in PTTW decision making. On regulating pumping from both small- and high-capacity wells, a 2002 Safe Drinking Water Act amendment mandated tracking of pumping rates to avoid uptake of brines, thus reducing aquifer over-pumping risks. The 2002 Ontario Low Water Response Act considered temporal aspects impacting groundwater availability, setting progressive restrictions on water pumping corresponding to reducing levels of streamflow and/or precipitation in times of drought.

II. Pennsylvania

Far stricter than most GLB states/provinces, in Pennsylvania there has been long-standing consideration of the cumulative impacts of smaller water takings (even from aquifers underlying private property), temporal limits to groundwater use, and focus on EIAs before granting bulk groundwater permits. The earliest Pennsylvania statute impacting the Basin's GWS was the 1956 Water Well Drillers License Act (32 P.S. §645.1 et seq), which required users to request and renew annual licenses for small- and large-capacity wells and reporting of water table levels. The 1978 Emergency Management Services Code (35 Pa.C.S. §7101 et seq.) was the first GLB policy to mandate reduced groundwater use during droughts. The 1984 Safe Drinking Water Act appeared to consider sustainable aquifer yield by empowering municipalities to issue permits, at an annual fee capped at USD 500 for persons taking groundwater from publicly owned aquifers. It also required EIAs on aquifers as part of groundwater permit requests. Finally, the 2002 Water Resources Planning Act 220 (27 Pa.C.S. Chapter 31) made it compulsory to report groundwater withdrawals for domestic use from aquifers within private land when exceeding 10,000 gallons per day.

III. Minnesota

Unlike Pennsylvania and Ontario, Minnesota has had far less consideration of sustainable aquifer yield requirements in its statutes and regulations impacting GWS. Instead, the state has had a tradition of having little to no regulations for the use of groundwater within private land, rather focusing on the protection of water within publicly owned lands. In 1897, Minnesota Law first adopted the term public waters (Minnesota Water Law Section 103). However, groundwater was excluded in the original definition of public waters, instead limiting public waters to large lakes and streams that were capable of beneficial public uses such as water supply, fishing, and boating. All other waters were deemed private and beyond the regulation of the state. The catastrophic drought of the mid-1930s demonstrated the need for more stringent water protections, which for the first time included groundwater, as the Minnesota Water Law was amended empowering the state to issue permits to protect the public's interest in the amount of water available for use. Permits were required for large-quantity uses of public waters as well as for the appropriation of public waters for agricultural, industrial, and commercial sectors. Yet, the permit fee structure remained the same for groundwater and surface water, thereby disregarding the differences in availability and recharge rates.

In 1976, the Public Waters Inventory Program was introduced to track water levels (Laws of Minnesota 1976, Chapter 83 and Laws of Minnesota 1979, Chapter 199), reiterating the definition of public waters as those serving "beneficial public purpose" and for the first time including aquifer recharge as public waters. A 1979 amendment confirmed the location of public waters as those within lands to which the State of Minnesota or the federal government hold title. It also made it mandatory for all 87 counties of Minnesota, including the ones to the north east within the GLB, to participate in the public waters inventory. The 1990 Allocating and Controlling Waters of the State (Laws of Minnesota 1990. 103G.255) amended several previous laws to provide further clarity on the state's role in conserving sufficient water resources for public use; however, it did not include specific hydrogeological science-based actions for conserving groundwater. Aiding the protection of groundwater within private and public lands, in response to the 1987–1989 drought, in 1990 the Minnesota Department of Natural Resources was mandated to develop a drought plan (Minnesota Statutes Section 103G.293). Still in use today, the resulting Minnesota Statewide Drought Plan consists of a set of prescribed local action responses to five different conditions/phases of climate (normal to extreme drought) [70].

IV. Wisconsin

Though Wisconsin has not had a long track record of laws reflecting sustainable aquifer yield considerations, and did not have regulations mandating reduced groundwater use during droughts over the study period, it has more recently developed one of the more comprehensive water use and aquifer protection policies of all GLB states/provinces. Its 1983 Comprehensive Groundwater Protection Act 410 (Chapter 160, Wisconsin Statutes) established the Groundwater Coordinating Council to assist state agencies' coordination of water conservation and provision of GWS scientific data. On smaller-capacity wells, it empowered municipalities to regulate—under Wisconsin Department of Natural Resources (DNR) supervision—construction and pump installation for some private wells. The 2003 Groundwater Protection Act (Wisconsin Act 310) mandated EIAs before granting PTTWs for high-capacity wells. The Act also defined the spatial extent of Groundwater Management Areas, mandated pumping rate reporting, and established a decision-making standard for addressing water quantity issues in rapidly growing areas of the state. However, with annual PTTW fees set at USD 100 for both surface water and groundwater, economic incentives did not appear to consider their relative quantity and recharge disparities [71].

V. Indiana

Indiana's approach to GWS governance featured some of the least physical—environmental considerations for protecting GWS of all GLB states/provinces within the study period. Since 1860, Indiana has applied the "Reasonable/Beneficial Use system" to govern both surface water and groundwater uses [72]. Like Minnesota, its application of the Reasonable Use Rule in the Indiana Code (IND. CODE § 14-25-7-6.) permits " … the use of water for a beneficial use in such quantity and manner that is (1) necessary for economic and efficient utilization, and (2) is both reasonable and consistent with the public interest." The first statute to provide some GWS protections was the 1985 Emergency Regulation of Ground Water Rights Act (IC 14-25-4). However, the law was concerned with protecting property rights to groundwater as it protected owners of small-capacity wells from the impacts of high-capacity wells if they significantly lower GWS levels within their properties. Still in use today, this law has been further reinforced in Indiana case law that has held landowners liable for all types of damages caused by the excessive removal of groundwater, including subsidence damage. This is illustrated in the 1998 Indiana Court of Appeals ruling against the GLB City of Valparaiso. Damages were awarded to the plaintiff for land subsidence caused by the City's over-pumping of GLB groundwater (City of Valparaiso vs. Defler, 694 N.E.2d 1177, 1180-82). The Court of Appeals stated that reasonable and beneficial use of groundwater must be maintained to avoid harming the rights of adjacent landowners. The 2003 Water Rights and Resources Act (Indiana Code 14-25-1(1)) furthered this approach to GWS governance. While it defined the types of water subject to government protection for the public welfare, it did not include groundwater. Similar to Minnesota, Indiana provided some recommendations to protect GWS in times of drought. Its 1994 Water Shortage Plan included environmental indicators of water shortages with corresponding groundwater use and management responses.

VI. Michigan

Prior to the passage of the 2005 GLSWRA, Michigan's statutes largely omitted standards to control groundwater use that reflected sustainable aquifer yield considerations [73]. In addition, most controls on groundwater use were set by the courts in settling groundwater use disputes, and rulings were primarily concerned with ensuring equitable access rights to groundwater within property limits. The earliest of these rulings was from the Michigan Supreme Court in the 1917 Schenk vs. City of Ann Arbor case (196 Mich 75, 163 NW 109), where it was found that the City of Ann Arbor did not have greater rights to withdraw groundwater for the provision of public water supply than a private landowner did. The court also ruled on another landmark case, Bernard vs. City of St. Louis in 1922 (220 Mich 159, 189 NW2d 891), in favor of the plaintiff, requiring the City of St. Louis to reduce groundwater withdrawals to maintain adequate water for the plaintiff's use, and awarding compensation for pumping equipment that the plaintiff had to install. In 1982, the Michigan Court of Appeals reaffirmed the outcome of Bernard vs. City of St. Louis, ruling in the Maerz vs. U.S. Steel Corporation case (116 Mich App 710).

Statutes that did cover GWS were first established in the late 1970s. Reflecting the Absolute Ownership Rule in stating that municipal governments had no authority to curb groundwater uses within private land, the 1978 Michigan Public Health Code (PA 368, MCL 333.1101 to 333.25211) indicated that "a local unit of government shall not enact or enforce an ordinance that regulates a large-quantity withdrawal." Another was the 1981 Michigan Right to Farm Act (P.A. 93 Sec. 3 (3)) that listed conditions that offered farmers protection from nuisance suits. Noting that it cannot be applied to resolve water use conflicts, the Act precluded installation of new irrigation equipment or new technologies as grounds for groundwater use complaint suits, paving the way for installation of higher-capacity pumps adding pressure on aquifers. Michigan took its first steps towards conserving GWS based on sustainable aquifer yield considerations when it passed the 1994 Natural Resources and Environmental Protection Act 451 (Mich. Comp. Laws § 324.30106), requiring EIAs before granting permits to take groundwater. The 2003 Aquifer Protection and Dispute Resolution Act added further protections by setting withdrawal thresholds based on a regional groundwater model that can assess the degree to which aquifers are overexploited.

On economic policies impacting GWS during the study period, Michigan's court rulings have had implications on the extent to which free trade treaties could be applied to access groundwater prior to the 2020 USMCA. The Michigan Court of Appeals 2005 ruling on the Michigan Citizens for Water Conservation (MCWC) vs. Nestle Waters North America Incorporated (269 Mich. App. 25, 709 N.W.2d 174) is one of the most significant cases. Nestle previously purchased groundwater rights to a Sanctuary Springs property in Mecosta County, within which it established four high-capacity wells that pumped groundwater at a rate of 400 gallons per minute (576,000 gallons per day). The 1994 Natural Resources and Environmental Protection Act 451 was considered by the court in ruling for the MCWC, preventing Nestle from continuing operations. Considering the MCWC as riparian property owners negatively affected by Nestle's wells, the court found that Nestle's withdrawals unreasonably interfered with MCWC's rights. The court also noted the harmful impacts that Nestle's groundwater extraction was having on the ability of wetlands and watercourses to provide ecosystem services, including the reduction of their ability to provide fisheries habitat, water filtration, and to prevent erosion and flooding. The court ordered Nestle to cease operations pending determination of more sustainable groundwater withdrawal rate, allowing consideration of sustainable aquifer yield factors. It was not until after the study period, in the 2006 amendment to the 1994 Natural Resources and Environmental Protection Act 451, that any statutes were passed that regulated the removal of any quantity of GLB groundwater from an aquifer for free trade purposes [74].

VII. New York

Prior to the 2005 GLSWRA, New York statutes impacting GWS had minimal guidance that reflected sustainable aquifer yield considerations [75,76]. The first was the 1972 New York Environmental Conservation Law (Chapter 43-B) that set standards to reduce over-pumping to prevent upwelling of brines to maintain water quality. The other significant measure during the study period was the 1988 Great Lakes Water Conservation and Management Act (NYS ECL § 15-1501 et seq.) that imposed EIA requirements on public water suppliers that withdrew large amounts of GLB water.

VIII. Illinois

In Illinois, groundwater uses have for the most part proceeded without reasonable use limits, volumetric controls, or policies restricting groundwater use in times of drought. Additionally, Illinois is one of two GLB states initially using the Absolute Ownership Rule in case law applied to resolve groundwater use conflicts, applying it well into the 1980s [77]. The Edwards vs. Haegar (180 Ill. 99) ruling in 1899 allowed for landowners to use groundwater without concern for impacts on neighboring users until the passage of the 1983 Water Use Act. In this Act, the applicability of the Reasonable Use Rule to govern the State's groundwater withdrawals was confirmed. This was reaffirmed in the Bridgman vs. Sanitary District of Decatur (164 Ill. App. 3d 287 4th Dist.) ruling, which stated, "By using the terms 'natural wants' and 'artificial wants' in the definition of reasonable use ... the legislature has adopted the same standards for groundwater withdrawals as that which applies to surface water withdrawals." Another step towards protecting GWS was the adoption of the 1987 Illinois Groundwater Protection Act, which enacted a series of technical programs and procedures to monitor statewide well levels. Though the 1980 Supreme Court Ruling (Wisconsin vs. Illinois, 449 U.S. 48) established the Chicago Diversion, precluding the state from any 2005 GLSWRA obligations, Illinois permitted bulk groundwater pumping for domestic uses following its 1996 Rules and Regulations for the Allocation of Water from Lake Michigan.

IX. Ohio

As per court rulings dating from 1861, Ohio initially applied the Absolute Ownership Rule in regulating how much groundwater landowners could use, joining Illinois as the second state to do so in the GLB [78]. Courts provided no legal remedy for complaints of excessive use until a 1984 Ohio Supreme Court decision in Cline vs. American Aggregates Corporation, which adopted the Reasonable Use Rule in its ruling. The court placed a duty on landowners to make sensible use of groundwater to avoid harm to the groundwater rights of nearby landowners. The next significant step to safeguarding GWS was the 2003 amendment to the Groundwater Rules and Regulations (Ohio Administrative Code Reg. 3745-34) which required groundwater use permits to withdraw over 100,000 gallons per day, the same volumetric limit set for surface water.

4. Discussion

Results of our CPT analysis pinpoint the causes of gaps in the present day GWS governance framework that have led to emerging groundwater supply vulnerabilities in drought-prone and/or groundwater-dependent GLB communities. Referring to key aspects of CPT theory, below we outline the empirical manifestations and causal mechanisms—successive governance milestones/amendments and court rulings within the study period—that comprise the causal chain linking historical causes to present-day outcomes.

- Causes and Outcomes

The greatest strength of GWS governance over the years has been its facilitation of scientific research and data collection, which if applied, would have been relevant to devising groundwater use policies and decision-making standards based on sustainable aquifer yield. This is evidenced with federal governments' early establishment of the USGS and GSC, and their long-term collaboration with the states/provinces in aquifer mapping and monitoring GWS levels [67,68]. GLB states/provinces have fairly consistently required GWS-level data collection, and in some cases, have long required pumping-rate reporting, such as in Ontario, Pennsylvania, Minnesota, and Wisconsin. At the binational level, the 1956 Great Lakes Basin Compact was a key milestone as it initiated the whole-of-basin approach to GWS governance that has come to characterize successive agreements, catalyzing binational hydrological research and data sharing on GLB water resource use [65].

Data and science on aquifer geophysical parameters, flow rates, as well as technologies and methodologies for quantifying and simulating groundwater flow have been steadily improving over the study period. However, our findings suggest that although governments at multiple levels have facilitated much of this science that is relevant to sustainable aquifer yield, they insufficiently leveraged it to develop groundwater use and conservation rules over the study period. This feature is the root cause of present-day GWS governance weaknesses and groundwater decline outcomes.

CPT points to legal principles originating from 19th century court decisions and scientific understanding of groundwater flow systems as the fundamental cause of these outcomes. The oldest of these is the Absolute Ownership Rule. Court deliberations in its earliest documented application, the 1843 Chasemore vs. Richards ruling (1843-60 All E.R. 77, 81-82 H.L. 1859), shed light on the incipient state of hydrogeological science at the time that supported the creation of the legal concept that groundwater use could not be governed because its quantities and flow directions were "unknowable". From 1776 to 1865, the science of hydrogeology was characterized by slow growth in the understanding of underlying principles, especially in the Great Lakes region where springs were plentiful, not requiring early settlers to develop wells to access groundwater, or consider impacts of overuse [62]. Therefore, when the Reasonable Use Rule was later set to govern groundwater use, it was largely oriented towards protecting the rights of adjacent landowners to access groundwater within their property limits. Additionally, for much of the study period, GWS conservation for the greater public good was not prioritized, given the Public Trust Doctrine's incorporation into early state/provincial constitutions that regarded only surface water as a common, public resource. The Underground Stream Doctrine further entrenched this paradigm, as governments typically only stepped in to protect GWS for the purpose of safeguarding surface water.

It is from this scientific and legal basis that during the study period, successive GLB governments and courts at multiple levels largely failed to devise policies and standards for groundwater use and conservation based on sustainable aquifer yield. Carried through to the present day GWS governance framework, we outline the causal mechanisms culminating in current governance weaknesses and GWS decline outcomes.

- Empirical Manifestations and Causal Mechanisms

Focusing first on policies and decision-making standards controlling groundwater pumping, successive, multilevel statutes generally omitted specific, science-based measures based on sustainable aquifer yield during the study period. Echoing Underground Stream Doctrine paradigms, governments typically interrelated surface water and groundwater rights of use, not containing evidence of appreciation that there is five times more surface water than groundwater stored in the Basin [1]. With most governmental controls for groundwater use limited to bulk quantities, it is noteworthy that since being introduced in the 1985 Charter, the same high volumetric water use thresholds were used to define bulk surface water and groundwater, seeming to be better suited to surface water's greater availability and quicker recharge rates [79].

Another significant governance blind spot was the paucity of regulation of smaller quantities of groundwater uses. This is seen in groundwater exports from the Basin in containers 20 L or less being allowed in successive binational agreements leading to the 2005 GLSWRA; these agreements being purportedly aimed at preserving the quantity of all GLB waters. More evidence is that most GLB states, except for Ontario and Pennsylvania, did not have controls on smaller volumes of groundwater pumped within private land. Reflecting the legal principles of the Absolute Ownership Rule, governance focused rather on standards for well construction and pump installation. Groundwater pumping for firefighting and agriculture, regardless of the quantity, was also unregulated, despite the latter being the largest groundwater consuming sector in the Basin [57,59].

Past court rulings also provide more empirical manifestations that reflect 19th century scientific and legal principles. Rulings resolving groundwater use disputes seem to have been rather focused on ensuring equitable groundwater rights of landowners rather than preserving aquifer storage [80]. Courts have generally ruled in favor of those with the deepest wells and highest-capacity pumps, such as in the Bralts and Leighty (no date) Michigan court ruling that "if a neighbor complains that your irrigation pumping is causing their well to go dry, a prudent response would be to offer to deepen their well and consider it an irrigation expense." In other instances, courts typically enforced the Reasonable Use Rule and/or Underground Stream Doctrine and applied jurisprudence on surface water, given its longer track record of case law, to resolve other groundwater use conflicts [72]. Notable examples are deliberations in the City of Valparaiso vs. Defler (694 N.E.2d 1177, 1180-82) ruling in Indiana and the Bernard vs. City of St. Louis ruling (220 Mich 159, 189 NW2d 891) in Michigan.

On GWS conservation, multilevel policies and decision-making standards had a mixed record on considering sustainable aquifer yield during the study period. Positive developments were the introduction of EIA requirements in PTTW decision-making standards for high-capacity wells beginning in the 1980s in some states/provinces, as well as their protection of waters needed for aquifer recharge. Moreover, with regard to the temporal dimension of safeguarding GWS, states/provinces introduced voluntary, judicious water use policies to protect GWS during droughts, with Pennsylvania and Ontario being the only jurisdictions where this was made mandatory during the study period.

Economic policy tools also reflected 19th century legal and scientific concepts, and typically appeared to disregard sustainable aquifer yield. There is little evidence to suggest that they considered quantitative evaluation of the trade-offs between future and current groundwater withdrawals that would be required for dealing with growing groundwater insecurity [63]. To illustrate, historically, fees for municipal water supply and state/provincial well permits and PTTWs have been low or nonexistent. Moreover, as these fees generally were not differentiated from the pricing structure for surface water, multiple levels of governments did not consider groundwater's relative scarcity and lower recharge rates, providing little economic incentives for reducing groundwater use.

- Causal Linkages

Legal principles originating from 19th century groundwater science do not appear to have persisted in successive court decisions, policies, and decision-making standards due to a lack of competence or understanding of hydrogeological science. Much of the theoretical, engineering, and methodological underpinnings needed to quantify groundwater and simulate its flow directions and rates were established since the 1940s [62], and GLB governments have demonstrated their ability to leverage these scientific advances towards safeguarding groundwater quality. GLB governments at multiple levels have had laws protecting groundwater quality since the 1970s that explicitly considered modern science on geophysical and environmental parameters [17]. Kickstarted with major environmental disasters such as the Love Canal catastrophe that leached hazardous chemicals into underlying groundwater in the Niagara escarpment, to widespread eutrophication of Lake Erie, general awareness of GLB water quality crises shifted public opinion, leading to sweeping policy changes [81]. Since then, consecutive amendments of groundwater quality regulations have mostly kept pace with innovations in science, with there being some successes in improving groundwater quality across the GLB [82].

The above suggests that path dependency may be the likely rationale for GWS governance and groundwater quality governance having such contrasting outcomes. The phenomenon of governments starting down a particular track, making the costs of reversal or change extremely high to overcome [83], path dependency is the likely causal link through which GWS governance weaknesses were able to persist well into the present-day governance framework, inexorably contributing to growing water insecurity in high-stress locales. Hansen [84] contends that "path dependence is established only when it can be shown that policy change was considered and rejected for reasons that cannot be explained without reference to the structure of costs and incentives created by the original policy choice.". As such, policies are inherently challenging to reform [85], even when suboptimal to address problems [86]. Often, policymakers typically must wait for critical junctures or exceptional opportunities to enact governance reform [87].

In this context, the evidence conveys that successive GLB governments have had little inducements to amend GWS governance as growing groundwater insecurities have been largely localized and location-specific problems [88]. This is compounded by GLB residents generally having low water risk literacy, lulled into the "myth of water abundance", relatively unaware of risks posed by droughts and rising uses [89,90]. With growing groundwater vulnerabilities not yet garnering widespread public attention, or becoming a Basin-scale problem, public pressure or significant inflection points have not yet demanded GWS governance reforms considering sustainable aquifer yield.

5. Conclusions and Recommendations

Projected increases in climate and human pressures will continue to undermine groundwater security in a "do nothing" policy scenario. Climate change will increase precipitation in the Great Lakes region. However, its pattern will be progressively altered, concentrating more precipitation within winter months when the ground is frozen, and infiltration is reduced. In these conditions, aquifer recharge is expected to decrease by up to 20% [5]. Currently, 10% of the US population and 40% of the Canadian population reside within the GLB [91], with some of the fastest growth in inland peri-urban communities. For many communities, groundwater is often the sole source of public water supply: e.g., almost half of Michigan residents and a third of Ohio residents depend on GLB groundwater for public water supply [92]. With industry increasingly being attracted to the Basin, drawn by clean waters and cheap water prices, these trends have already contributed a thirtyfold increase in regional groundwater withdrawal, currently estimated at 160,000 L/day [2]; as well as an overall 15% increase in groundwater consumption across the Basin, while surface water consumption decreased within the study period [93]. If left unchecked, these trends are likely to proliferate groundwater overuse, particularly in population growth and industrialized hotspots, raising the specter of groundwater insecurity deepening in high-stress locales.

To contend with rising GWS threats, our findings argue strongly in favor of reforms of policies and standards regulating groundwater pumping, use, and conservation. As demonstrated with improvements made with water quality governance due to public pressure, inflection points can make fundamental governance reforms possible [87]. Considering this, our first recommendation is to raise awareness of the true availability and vulnerability of GWS in the Basin. As a water-rich region, these location-specific vulnerabilities are often overlooked. Therefore, raising awareness on the increasing cases and socio-environmental drivers of GWS vulnerabilities across the Basin is key.

Secondly, we urge for groundwater use governance to keep pace with scientific findings of the twenty-first century. It is clear that the Absolute Ownership Rule that underpins the evolution of GWS governance was based on legal concepts predicated on 19th century science, as governments avoided the establishment of specific rules to govern the use of a resource they could not quantify or trace. Through path dependency, they instead applied rules originally devised and better suited to maintain surface water quantity, despite advances in science that increasingly recognized groundwater as quantifiable resources supporting vital environmental functions and economically valuable human uses. In so doing, GLB governments at multiple levels have not recognized that the original interpretation and establishment of these rules were very much a product of their time.

Since then, a great deal more knowledge and data on groundwater flow rates, directions, and quantities have been accrued as the hydrogeological scientific discipline matured. Twenty-first century innovations such as Big Data, GIS, remote sensing, and machine learning technologies to estimate aquifer geometry, quantify GWS, and model groundwater flow directions [94] carry the promise of faster, cheaper, and increasingly accurate estimations of the physical–environmental parameters of sustainable yield [95]. While the significant natural variation in aquifer physical–environmental settings would evidently impact planning needs and options to address highly localized to regional-scale GWS sustainability issues, by leveraging these innovations, more sustainable policies and decision-making standards to better sustain GWS may be created [96].

At the heart of GWS governance are its foundational legal doctrines and scientific assumptions. Courts and governments at multiple levels would need to make a definitive update of the Reasonable Use Rule relevant to the situational contexts of their GWS governance mandates. These considerations imply the abandoning the Underground Stream Doctrine in order to determine reasonable groundwater uses based on sustainable aquifer yield concepts. This contemplates (i) specification of volumetric thresholds for groundwater uses that avoid undesirable consequences on surface water bodies, aquifers, and dependent ecosystems in legal definitions; (ii) adding a temporal dimension to determining reasonable groundwater use, lowering use rates during droughts; (iii) considering the cumulative impacts of smaller-capacity wells over time [97], and (iv) differentiation of bulk water definitions for groundwater and surface water, with lower volumes set for the former given its relative scarcity and differing physical–environmental requirements of aquifers to maintain groundwater.

Restricting what is now considered "reasonable uses" of groundwater will likely require expansion of the Public Trust Doctrine [72]. However, applying public trust principles to govern groundwater use in the GLB has been rejected in the past due to fears over violating private property rights [98]. A 1983 California Supreme Court ruling (National Audubon Society vs. Superior Court 33 Cal.3d 419) provides a practical example for addressing this issue through sharing of public trust responsibilities with private landowners. To resolve a complaint by the National Audubon Society on the lowering Lake Meno's water level due to long-term pumping, the Court ruled that the public trust must be balanced between the Los Angeles Department of Water and Power and land proprietors. In so doing, it rationalized prior appropriation groundwater rights of landowners with the public lands and trust responsibilities of the government for conserving groundwater. By according public trust responsibilities to landowners, it effectively placed a duty on them to conserve groundwater below their lands.

Our third recommendation is to update economic policy tools to incentivize groundwater use efficiency. The structure of costs created by past GWS governance policies has resulted in groundwater being cheap and freely available to well owners, and insufficiently covering the cost of extraction and distribution of municipal water supply [59]. Regarding free trade agreements, these features have been embedded in the business models of industries attracted to the region [99]. While most GLB states/provinces have had voluntary guidelines for water use efficiency, mandatory standards and/or economic incentives should be considered to curtain groundwater overuse. Such incentives can include rebates for installation of efficient plumbing, promotion of judicious irrigation methods, and removing reducing block rates in municipal water supply tariff structures. Economic disincentives may also be considered, as illustrated in Ontario, who since the 1990s has set higher PTTW pricing for withdrawing bulk groundwater than for surface water, and progressively increases costs for PTTWs for higher groundwater volumes.

Looking back at the century-old arc of water resource governance in the GLB, there has been a tradition of collaboration and cooperation across political jurisdictions and government levels. The region's governments have established enduring institutions and more recently, taken steps to enshrine policies into law, suggesting growing political will to have stronger water resource safeguards. Multilevel institutions have also a long tradition of funding and conducting important scientific studies on the current state of the Basin's groundwater resources. With this trajectory, there can be some confidence in GLB continuing its transboundary governance evolution towards better science–policy alignment, to sustain "all waters" of the Basin, rising to the challenges of growing climate and growing human use stressors on vulnerable aquifers.

Author Contributions: Conceptualization, K.W.; methodology, K.W.; validation, K.W. and G.K.; formal analysis, K.W.; investigation, K.W.; data curation, K.W.; writing—original draft preparation, K.W.; writing—review and editing, K.W. and G.K.; visualization, K.W.; supervision, G.K.; project administration, K.W. and G.K.; funding acquisition, G.K. All authors have read and agreed to the published version of the manuscript.

Funding: The research is supported by the Natural Sciences and Engineering Research Council of Canada.

Institutional Review Board Statement: Not applicable.

Informed Consent Statement: Not applicable.

Data Availability Statement: Not applicable.

Conflicts of Interest: The authors declare no conflict of interest. The funders had no role in the design of the study; in the collection, analyses, or interpretation of data; in the writing of the manuscript, or in the decision to publish the results.

References

1. Coon, W.F.; Sheets, R.A. *Estimate of Groundwater in Storage in the Great Lakes Basin, United States, 2006;* USGS Scientific Investigations Report 2006–5180; USGS: Reston, VA, USA, 2006; 19p.
2. Great Lakes Commission. Annual report of the Great Lakes regional water use data base-representing 2016 water use data. In Proceedings of the Great Lakes and St Lawrence Governors and Premiers, Ann Arbor, MI, USA, 29 January 2016.
3. Kavcic, R. A Special report on the Great Lakes and St Lawrence Regional Economy. In Proceedings of the Great Lakes and St Lawrence Governors and Premiers, Ann Arbor, MI, USA, 29 January 2016.
4. National Oceanic and Atmospheric Agency (NOAA). *Basin Map of Great Lakes;* Great Lakes Environmental Research Laboratory: Ann Arbor, MI, USA, 2004. Available online: http://www.flickr.com/photos/noaaglerl/4037600466/ (accessed on 2 January 2019).
5. Lofgren, B.; Hunter, T.; Wilbarger, J. Effects of using air temperature as a proxy for potential evapotranspiration in climate change scenarios of Great Lakes Basin hydrology. *J. Great Lakes Res.* **2011**, *27*, 744–752. [CrossRef]
6. Huang, J.; Halpenny, J.; van der Wal, W.; Klatt, C.; James, T.; Rivera, A. Detecting Groundwater Storage Change within the Great Lakes Water Basin using GRACE. *J. Geophys. Res. Atmos.* **2012**, *117*, B08401. [CrossRef]
7. Howard, K.; Gerber, R. Impacts of urban areas and urban growth on groundwater in the Great Lakes Basin of North America. *J. Great Lakes Res.* **2018**, *44*, 1–13. [CrossRef]
8. Freeze, R.A.; Cherry, J.A. *Groundwater;* Prentice-Hall: Englewood Cliffs, NJ, USA, 1979.
9. Newell, P.; Pattberg, P.; Schroeder, H. Multi-Actor Governance and the Environment. *Annu. Rev. Environ. Resour.* **2012**, *37*, 365–387. [CrossRef]
10. Foster, S.; Loucks, D. *Non-Renewable Groundwater Resources: A Guidebook on Socially-Sustainable Management for Policy Makers;* IHP-VI, Series on Groundwater No. 10; UNESCO: Paris, France, 2006. Available online: https://unesdoc.unesco.org/ark:/48223/pf0000146997 (accessed on 25 October 2019).
11. Majidipour, F.; Mohammad, S.; Taheri, K.; Fathollahi, J.; Missimer, T. Index-Based Groundwater Sustainability Assessment in the Socio Economic Context: A Case Study in the Western Iran. *Environ. Manag.* **2021**, *67*, 648–666. [CrossRef]
12. Maimone, M. Defining and Managing Sustainable Yield. *Groundwater* **2004**, *42*, 809–814. [CrossRef]
13. Granneman, N.G.; Van Stempvoort, D. Groundwater Science Relevant to the Great Lakes Water Quality Agreement: A Status Report for the Great Lakes Executive Committee. 2016. Available online: https://binational.net//wpcontent/uploads/2016/05/GW-Report-final-EN.pdf (accessed on 12 January 2021).
14. Kreutzwiser, R.; Lo, R.; Durley, J.; Priddle, C. Water Allocation and the Permit to Take Water Program in Ontario: Challenges and Opportunities. *Can. Water Resour. J.* **2013**, *29*, 135–146. [CrossRef]

15. Mayer, A.; Mubako, S.; Ruddel, B. Developing the greatest blue economy: Water productivity, freshwater depletion and virtual water trade in the Great Lakes Basin. *Earths Future* **2016**, *4*, 282–297. [CrossRef]
16. Kemper, K.E. Instruments and institutions for groundwater management. In *the Agricultural Groundwater Revolution: Opportunities and Threats to Development*; Giordano, M., Villholth, K.G., Eds.; CABI: Wallingford, UK, 2007; pp. 153–171.
17. Hodge, R.A. Groundwater in the Great Lakes Basin: The Natural System, Use and Abuse, and Policy Implications. *Chi. Kent L. Rev.* **1989**, *65*, 439.
18. Saunders, J. Law and Management of the Great Lakes Basin. *Can. J. Water Resour.* **2000**, *25*, 209–242. [CrossRef]
19. Karkkainen, B. The Great Lakes Water Resources Compact and Agreement: Transboundary Normativity without International Law. *William Mitchell Law Rev.* **2013**, *39*, 1001–1124.
20. Dellapenna, J. Emerging Challenges to Good Governance in the Great Lakes: Changing Legal Regimes: Changing State Water Allocations Laws to Protect the Great Lakes. *Ind. Int. Comp. L. Rev.* **2014**, *9*, 51.
21. Rivera, A. Transboundary aquifers along the Canada-USA Border: Science, Policy and Social Issues. *J. Hydrol. Reg. Stud.* **2015**, *4*, 623–643. [CrossRef]
22. Weekes, K.; Krantzberg, G.; Vizeu, M. Identifying Groundwater Sustainability Implications of Water Policy in High-Use Situations in the Laurentian Great Lakes Basin. *Can. Water Resour. J.* **2019**, *44*, 337–349. [CrossRef]
23. Megdal, S.; Gerlak, A.; Varady, R.; Huang, L. Groundwater Governance in the United States: Common Priorities and Challenges. *Groundwater* **2014**, *53*, 677–684. [CrossRef]
24. Nelson, R.; Quevauviller, P. Groundwater law. In *Integrated Groundwater Management*; Jakeman, A.J., Barreteau, O., Hunt, R.J., Rinaudo, J.D., Ross, A., Eds.; Springer: Cham, Switzerland, 2016. [CrossRef]
25. Beach, D. It's all about mechanisms—What process-tracing case studies should be tracing. *New Political Econ. J.* **2016**, *21*. [CrossRef]
26. Waltz, K.; Bull, H.; Butterfield, H. *Theory of International Politics. Saltzman Institute of War and Peace, Columbia University*; Waveland Press Inc.: Long Grove, IL, USA, 1979.
27. Beach, P.; Pedersen, R. *Process-Tracing Methods: Foundations and Guidelines*; University of Michigan Press: Ann Arbor, MI, USA, 2019. [CrossRef]
28. Farid, H.; Ahmad, I.; Anjum, M.; Khan, Z.; Iqbal, M.; Shakoor, A.; Mubeen, M. Assessing seasonal and long-term changes in groundwater quality due to over-abstraction using geostatistical techniques. *Environ. Earth Sci. J.* **2019**, *78*. [CrossRef]
29. Pophare, A.; Lamsoge, B.; Katpatal, Y.; Nawale, V. Impact of over-exploitation on groundwater quality: A case study from WR-2 Watershed, India. *Earth Syst. Sci.* **2014**, *123*, 1541–1566. [CrossRef]
30. Galloway, D.; Jones, D.; Ingebritsen, S. *Land Subsidence in the United States*; USGS Circular 1182; USGS: Reston, VA, USA, 1999. [CrossRef]
31. Barlow, P.; Leake, S. *Streamflow Depletion by Wells: Understanding and Managing the Effects of Groundwater Pumping on Streamflow*; USGS Circular 1376; USGS: Reston, VA, USA, 2012. Available online: https://pubs.usgs.gov/circ/1376/ (accessed on 12 July 2020).
32. Custodio, E.; Kretsinger, V.; Llamas, M. Intensive development of groundwater: Concept, facts and suggestions. *Water Policy* **2005**, *7*, 151–162. [CrossRef]
33. Scanlon, B.; Faunt, C.; Longuevergne, L.; Reedy, R.; Alley, W.; McGuire, V.; McMahon, P. Groundwater depletion and sustainability of irrigation in the US High Plains and Central Valley. *Proc. Natl. Acad. Sci. USA* **2012**, *109*, 9320–9325. [CrossRef]
34. Expert Panel on Groundwater. *The Sustainable Management of Groundwater in Canada*; The Council of Canadian Academies Report: Ottawa, ON, Canada, 2010.
35. Walton, W.; McLane, C. Aspects of Groundwater Supply Sustainable Yield. *Groundw. Tech. Comment.* **2013**, *51*, 158–159. [CrossRef]
36. Lin, L.; Lin, H. Determination of groundwater sustainable yield using a numerical modelling approach for the Table Mountain Group sandstone aquifer, Rawsonville, South Africa. *Hydrogeol. J.* **2019**, *27*, 841–855. [CrossRef]
37. Sheets, R.A.; Simonson, L.A. *Compilation of Regional Ground-Water Divides for Principal Aquifers Corresponding to the Great Lakes Basin*; U.S. Geological Survey Scientific Investigations Report 2006–5102; USGS: Reston, VA, USA, 2006; 23p.
38. Granneman, N.G.; Hunt, R.J.; Nicholas, J.R.; Reilly, T.E.; Winter, T.C. *The Importance of Groundwater in the Great Lakes Region*; U.S. Geological Survey Water-Resources Investigations Report 00-4008; USGS: Reston, VA, USA, 2000; 14p.
39. Luczaj, J.; Masarik, K. Groundwater Quantity and Quality Issues in a Water-Rich Region: Examples from Wisconsin, USA. *Resources* **2015**, *4*, 323–357. [CrossRef]
40. Great Lakes Science Advisory Board to the International Joint Commission. Groundwater in the Great Lakes Basin. *International Joint Commission.* 2010. Available online: https://legacyfiles.ijc.org/publications/E43.pdf (accessed on 20 November 2020).
41. Great Lakes Executive Committee. Progress Report of the Parties for 2019. *Pursuant to the 2012 Canada-United States Great Lakes Water Quality Agreement.* 2019. Available online: https://binational.net/wp-content/uploads/2019/06/Final-2019-PROP-English-June-7.pdf (accessed on 19 August 2020).
42. Xu, S.; Frey, S.; Erler, A.; Khader, O.; Berg, S.; Hwang, H.; Callaghan, M.; Davison, J.; Sudicky, E. Investigating groundwater-lake interactions in the Laurentian Great Lakes with a fully-integrated surface water-groundwater model. *J. Hydrol.* **2021**, *594*, 125911. [CrossRef]
43. Holtschlag, D.; Nicholas, J. *Indirect Ground-Water Discharge to the Great Lakes*; Open-File Report 98-579; U.S. Geological Survey: Reston, VA, USA, 1998; 25p. [CrossRef]
44. Helmuth, J.; Johnon, D.; Karklins, S.; Lindorff, D. *Status of Groundwater Quantity in Wisconsin*; PUBL-DG-043-97; Wisconsin Department of Natural Resources: Madison, MI, USA, 1997.

45. National Research Council. *Mitigating Losses from Land Subsidence in the United States*; The National Academies Press: Washington, DC, USA, 1991. [CrossRef]
46. Keqiang, H.; Liu, C.; Wang, S. Karst collapse related to over pumping and a criterion for its stability. *Environ. Geol.* **2003**, *43*, 720–724. [CrossRef]
47. Pereira, A. Final Programmatic Environmental Impact Statement for the Central United States. First Responder Network Authority 6. 2019. Available online: https://firstnet.gov/sites/default/files/FirstNet%20FPEIS%20Central%20Chapter%208%20Michigan%20June%202017.pdf (accessed on 2 March 2021).
48. Curtis, Z.; Li, S.; Sampath, P.; Liao, H. Groundwater Sustainability in the Michigan Lowlands—Understanding the Complex Interplay of Natural Brine Upwelling, Human Activity, and Climate Change. In Proceedings of the Fall Meeting, Warszawa, Poland, 15–18 September 2015; American Geophysical Union: Washington, DC, USA, 2015. abstract id. H33I-1731.
49. Curtis, Z.; Liao, H.; Li, S. Ottawa County Water Resources Study—Phase 2. Final Report. Institute for Water Research. 2018. Available online: https://www.miottawa.org/GroundWater/pdf/phase2_report.pdf (accessed on 29 January 2020).
50. Asher, M.; Cleary, E.; Olawoyin, R. Medical Geology of Arsenic in Groundwater and Well Water in South East Michigan. *Environ. Dis.* **2017**, *2*, 9–21. [CrossRef]
51. Choi, W.; Galasinski, U.; Cho, S.; Hwang, C. A Spatiotemporal Analysis of Groundwater Level Changes in Relation to Urban Growth and Groundwater Recharge Potential for Waukesha County, Wisconsin. *Geogr. Anal.* **2012**, *44*, 219–234. [CrossRef]
52. Reeves, H. *Water Availability and Use Pilot. A Multiscale Assessment of the US Great Lakes*; United States Geological Society: Madison, MI, USA, 2011.
53. Grand River Conservation Authority. Groundwater Resources. 2019. Available online: https://www.grandriver.ca/en/our-watershed/Groundwater-resources.aspx (accessed on 1 October 2020).
54. Bruneau, J.; Dupont, D.; Renzetti, S. Economic Instruments, Innovation, and Efficient Water Use. *Can. Public Policy Anal. Polit.* **2013**, *39*, S11–S22. Available online: http://www.jstor.org/stable/23594768 (accessed on 21 November 2020). [CrossRef]
55. Annin, P. *The Great Lakes Water Wars*; The Center for Resource Economics: Washington, DC, USA, 2018.
56. Sterner, R.; Ostrom, P.; Ostrom, N.; Klump, J.V.; Steinman, A.; Dreelin, E.; Zanden, M.V.; Fisk, A. Grand Challenges for Research in the Laurentian Great Lakes Basin. *Limnol. Oceanogr.* **2017**, *62*, 2510–2523. [CrossRef]
57. Burton, A.; Luoma, D.S.S.; Love, N.; Austin, J. Leveraging the Great Lakes Region's Water Assets for Economic Growth. Metropolitan Policy Program at Brookings. 2010. Available online: https://www.brookings.edu/wp-content/uploads/2016/07/0927_great_lakes_water.pdf (accessed on 6 January 2021).
58. International Joint Commission. *Groundwater in the Great Lakes Basin*; International Joint Commission: Washington, DC, USA, 2010.
59. Canadian Ecofiscal Commission. Only Pipes Should be Hidden: Best Practices for Pricing and Improving Municipal Water and Wastewater Services. 2017. Available online: http://ecofiscal.ca/wp-content/uploads/2017/09/Ecofiscal-Commission-Report-Onlythe-Pipes-Should-be-Hidden-FINAL-Sept-26-2017.pdf (accessed on 22 September 2020).
60. Gardner, A.; Bartlett, R.H.; Gray, J.; Carney, G. *Water Resources Law*; LexisNexis Butterworths: Sydney, Australia, 2009.
61. McKay, J. Groundwater as the Cinderella of water laws, policies and institutions in Australia. *Ecol. Econ. J.* **2007**, *141*, 321–327.
62. Rosenshein, J.S.; Moore, J.E.; Lohman, S.W.; Chase, E.B. *Two Hundred Years of Hydrogeology in the United States*; U.S. Geological Survey Open-File Report 86-480; USGS: Reston, VA, USA, 1986.
63. National Research Council. Legal considerations, valuations and groundwater policy—Chapter 5. In *Valuing Ground Water: Economic Concepts and Approaches*; National Academies Press: Washington, DC, USA, 2007. Available online: https://www.nap.edu/read/5498/chapter/1#xi (accessed on 14 January 2021).
64. IJC. *Impacts of a Proposed Coal Mine in the Flathead River Basin*; International Joint Commission: Washington, DC, USA, 1988.
65. Hammer, C. Standing Under the Great Lakes Compact: A Broad-Based Argument Infused with Public Trust Principles for those with Diversion Aversion. *Mich. State Law Rev.* **2018**, *251*, 252–306.
66. Bakker, K.; Cook, C. Water governance in Canada: Innovation and fragmentation. *Int. J. Water Resour. Dev.* **2011**, *27*, 275–289. [CrossRef]
67. Nowlan, L. Out of sight, out of mind? Taking Canada's groundwater for granted. In *Eau Canada*; Bakker, K., Ed.; UBC Press: Vancouver, BC, Canada, 2007.
68. Leshy, J. Interstate groundwater resources: The federal role. *Hastings North West J. Environ. Law Policy* **2008**, *14*, 1475–1498.
69. Kilbert, K.; Merkle, A.; Miller, F. *An Assessment of the Great Lakes States' Implementation of the Water Management and Conservation Provisions of the Great Lakes—St Lawrence River Basin Water Resources Compact*; University of Toledo College of Laws Legal Institute of the Great Lakes: Toledo, OH, USA, 2019.
70. Minnesota Department of Natural Resources. Minnesota Statewide Drought Plan. 2009. Available online: https://files.dnr.state.mn.us/natural_resources/climate/drought/drought_plan_matrix.pdf (accessed on 21 January 2021).
71. Kent, P.; Dudiak, T. *Wisconsin Water Law: A Guide to Water Rights and Regulations*; University of Wisconsin extension: Madison, WI, USA, 2001.
72. Eckstein, G.; Hardberger, A. Groundwater Laws and Regulations: A Preliminary Survey of Thirteen U.S. States. Texas A&M University School of Law Program in Natural Resources Systems. 2017. Available online: https://law.tamu.edu/docs/default-source/faculty-documents/groundwater-laws-reg-13states.pdf?sfvrsn=0 (accessed on 5 January 2021).
73. Lusch, D. *An Overview of Existing Water Law in Michigan Related to Irrigation Water Use and Riparian Considerations*; Michigan State University: East Lansing, MI, USA, 2011.

74. Miller, J. A Critical Look at Michigan Citizens for Water Conservation v. Nestle Waters North America & the Michigan Supreme Court's Recent Jurisprudence. 2008. Available online: https://digitalcommons.law.msu.edu/king/118 (accessed on 5 January 2021).
75. Negro, S.; Porter, K. Water Stress in New York State: The Regional Imperative? *J. Water Law* **2009**, *20*, 5.
76. Daly, J.E. From divining rods to dams: Creating a comprehensive water resource management strategy for New York. *Commem. Pace Law Rev.* **1995**, *1995*, 105–139.
77. Janasie, C. *An Overview of Water Law in Illinois*; NSGLC-20-04-02; Sea Grant Law Center: Oxford, MS, USA, 2020.
78. Hall, P. *Understanding Water Rights in Ohio*; Law Notes; The Ohio State University: Columbus, OH, USA, 2020.
79. Gosman, S. *The Good, the Bad and the Ugly: Implementation of the Great Lakes Compact*; National Wildlife Association: Reston, VA, USA, 2011.
80. Bishop, P. A Short Review of Pennsylvania Water Law. PA Department of Environmental Protection Presentation. 2006. Available online: http://files.dep.state.pa.us/Water/BSDW/WaterAllocation/water_law_review_022806.pdf (accessed on 12 December 2020).
81. Great Lakes Science Advisory Board. Great Lakes Surface and Groundwater Model Integration Review: Literature Review, Options and Approaches and Preliminary Action Plan for the Great Lakes Basin. International Joint Commission Report. 2018. Available online: https://www.ijc.org/sites/default/files/2019-01/Great_Lakes_Surface_and_Groundwater_Model_Integration_Review_Oct2018.pdf (accessed on 21 January 2021).
82. Burlakova, L.E.; Hincheyb, E.; Karatayeva, A.; Rudstamc, L.G. U.S. EPA Great Lakes National Program Office monitoring of the Laurentian Great Lakes: Insights from 40years of data collection. *J. Great Lakes Res.* **2018**, *44*, 535–538. [CrossRef]
83. Cerna, L. *The Nature of Policy Change and Implementation: A Review of Different Theoretical Approaches*; OECD: Washington, DC, USA, 2013. Available online: https://www.oecd.org/education/ceri/The%20Nature%20of%20Policy%20Change%20and%20Implementation.pdf (accessed on 5 January 2021).
84. Hansen, R. Globalization, embedded realism and path dependence: The other immigrants to Europe. *Comp. Political Stud.* **2002**, *35*, 259–283. [CrossRef]
85. Pierson, P. Increasing returns, path dependence and the study of politics. *Am. Political Sci. Rev.* **2000**, *94*, 251–267. [CrossRef]
86. Greener, I. Understanding NHS reform: The policy-transfer, social learning and path-dependency perspectives. *Governance* **2002**, *15*, 161–183. [CrossRef]
87. Capoccia, G.; Kelemen, D. The study of critical junctures: Theory, narrative and counterfactuals in historical institutionalism. *World Politics* **2007**, *59*, 341–369. [CrossRef]
88. Morris, T.J.; Mohapatra, S.P.; Mitchell, A. Conflicts, costs and environmental degradation—Impacts of antiquated groundwater allocation policies in the Great Lakes Basin. *Water Policy* **2008**, *10*, 459–479. [CrossRef]
89. Watershed Council. Great Lakes Water Use and Diversions. 2020. Available online: https://www.watershedcouncil.org/great-lakes-water-use-and-diversions.html (accessed on 7 January 2021).
90. Kane, K. The Great Lakes-St. Lawrence River basin agreement: What happens in the Great Lakes Won't Stay in the Great Lakes. *Mich. State Int. Law Rev.* **2017**, *25*, 432–453.
91. Chaloux, A.; Paquin, S. Green Paradiplomacy and Water Resource Management in North America: The Case of the Great Lakes-St Lawrence River Basin. *Can. Foreign Policy J.* **2013**, *19*, 308–322. [CrossRef]
92. Wilson, G. Groundwater the Sixth Great Lake. Great Lakes Now. 2018. Available online: https://www.greatlakesnow.org/2018/09/groundwater-the-sixth-great-lake/ (accessed on 7 February 2021).
93. Pentland, R.; Mayer, A. Ten Year Review of the International Joint Commission's Report on "Protection of the Waters of the Great Lakes". Alliance for the Great Lakes. 2015. Available online: http://fbheron.issuelab.org/resource/on_track_ensuring_the_resilience_of_the_great_lakes_compact (accessed on 19 November 2020).
94. Sun, A.; Scanlon, B. How can Big Data and machine learning benefit environment and water management: A survey of methods, applications, and future directions. *Environ. Res. Lett.* **2019**, *14*, 073001. [CrossRef]
95. Marçais, J.; de Dreuzy, J.-R. Prospective interest of deep learning for hydrological inference. *Groundwater* **2017**, *55*, 688–692. [CrossRef]
96. Fienen, M.N.; Nolan, B.T.; Feinstein, D.T.; Starn, J. Metamodels to bridge the gap between modeling and decision support. *Groundwater* **2015**, *53*, 511–512. [CrossRef] [PubMed]
97. Water Systems Council. *Who Owns the Water*; National AG Law Center: Washington, DC, USA, 2016. Available online: http://nationalaglawcenter.org/wp-content/uploads/2017/03/Who-Owns-the-Water-2016-Update-FINAL.pdf (accessed on 10 December 2020).
98. Abrams, R. Legal Convergence of the East and West in contemporary Water Law. *Environ. Law J.* **2012**, *42*, 65–91.
99. Kotkin, J.; Schill, M. A Map of America's Future: Where Growth Will Be over the Next Decade. *New Geogr.* 2013. Available online: http://www.newgeography.com/content/003914-a-map-of-americas-future-where-growth-will-be-over-the-next-decade (accessed on 10 September 2020).

water

MDPI

Article

Science and Binational Cooperation: Bidirectionality in the Transboundary Aquifer Assessment Program in the Arizona-Sonora Border Region

Jacob D. Petersen-Perlman [1,2,*], Tamee R. Albrecht [3,4], Elia M. Tapia-Villaseñor [5], Robert G. Varady [3] and Sharon B. Megdal [2]

1 Department of Geography, Planning, & Environment, East Carolina University, Greenville, NC 27858, USA
2 Water Resources Research Center, The University of Arizona, Tucson, AZ 85719, USA; smegdal@email.arizona.edu
3 Udall Center for Studies in Public Policy, The University of Arizona, Tucson, AZ 85719, USA; talbrecht@email.arizona.edu (T.R.A.); rvarady@arizona.edu (R.G.V.)
4 School of Geography, Development & Environment, University of Arizona, Tucson, AZ 85721, USA
5 Departamento de Geología, Universidad de Sonora, Hermosillo 83000, Mexico; elia.tapia@unison.mx
* Correspondence: petersenperlmanj19@ecu.edu

Abstract: Sharing scientific data and information is often cited within academic literature as an initial step of water cooperation, but the transfer of research findings into policy and practice is often slow and inconsistent. Certain attributes—including salience, credibility, and legitimacy of scientific information; iterative information production; and sociocultural factors—may influence how easily scientific information can be used in management and policymaking. However, transnationality usually complicates these sorts of interactions. Accordingly, we argue that the production of scientific information and transboundary water cooperation build upon each other *bidirectionally*, each informing and enhancing the other. We employ a case-study analysis of the Transboundary Aquifer Assessment Program (TAAP), a binational collaborative effort for scientific assessment of aquifers shared between Mexico and the United States. Here, information sharing was possible only by first completing a formal, jointly agreed-upon cooperative framework in 2009. This framework resulted in a collaborative science production process, suggesting that the relationship between sharing data and information and transboundary groundwater governance is iterative and self-reinforcing. In keeping with the publication of the TAAP's first binational scientific report in 2016, we demonstrate the bidirectional relationship between science production and water governance in the TAAP and explore remaining challenges after scientific assessment.

Keywords: transboundary waters; groundwater; US–Mexico; water governance; science production; bidirectionality

Citation: Petersen-Perlman, J.D.; Albrecht, T.R.; Tapia-Villaseñor, E.M.; Varady, R.G.; Megdal, S.B. Science and Binational Cooperation: Bidirectionality in the Transboundary Aquifer Assessment Program in the Arizona-Sonora Border Region. *Water* **2021**, *13*, 2364. https://doi.org/10.3390/w13172364

Academic Editor: Pankaj Kumar

Received: 24 June 2021
Accepted: 21 August 2021
Published: 28 August 2021

Publisher's Note: MDPI stays neutral with regard to jurisdictional claims in published maps and institutional affiliations.

1. Introduction

The arid to semiarid region of the southwestern United States (US) and northwestern Mexico is water-short in most of its geographical reach. Climate-change predictions indicate rising temperatures and increased variability in precipitation patterns, leading to water supply reductions by the middle of the 21st century [1–3]. This hydrological variability affects groundwater basins; the southwestern US is likely to experience declines in groundwater recharge, including in basins such as the San Pedro [4] and Santa Cruz [5,6].

Mexico and the US share four river basins (Tijuana, Colorado, Yaqui, and Rio Grande/Rio Bravo). The two that are by far the largest, the Colorado and Rio Grande/Rio Bravo basins, encompass almost the entirety of the border region. Additionally, 36 aquifers have been identified along the Mexico–US border; 16 of these can be categorized as transboundary [7]. Yet, while surface-water agreements govern and manage the binational Tijuana,

Colorado, and Rio Grande/Rio Bravo river basins, there exists no formal agreement on management of any of the transboundary aquifers.

The absence of binational/multinational transboundary institutions governing internationally shared groundwater is typical among almost all transboundary aquifers. Globally, for the almost 600 identified aquifers crossing international boundaries [8], only a handful of formal agreements over transboundary groundwater exist [9,10], even in locations that exhibit high levels of cooperation regarding other issues. Developing a shared understanding is a prerequisite to joint management of a resource, most especially an unseen one. The US and Mexico have cooperated scientifically—though not managerially—to *assess* four of their shared aquifers: the San Pedro and Santa Cruz, shared between Sonora (Mexico) and Arizona (US), and the Mesilla–Conejos Medanos and Hueco Bolson, shared among three subfederal entities: the Mexican state of Chihuahua, and the US states of New Mexico and Texas (Figure 1).

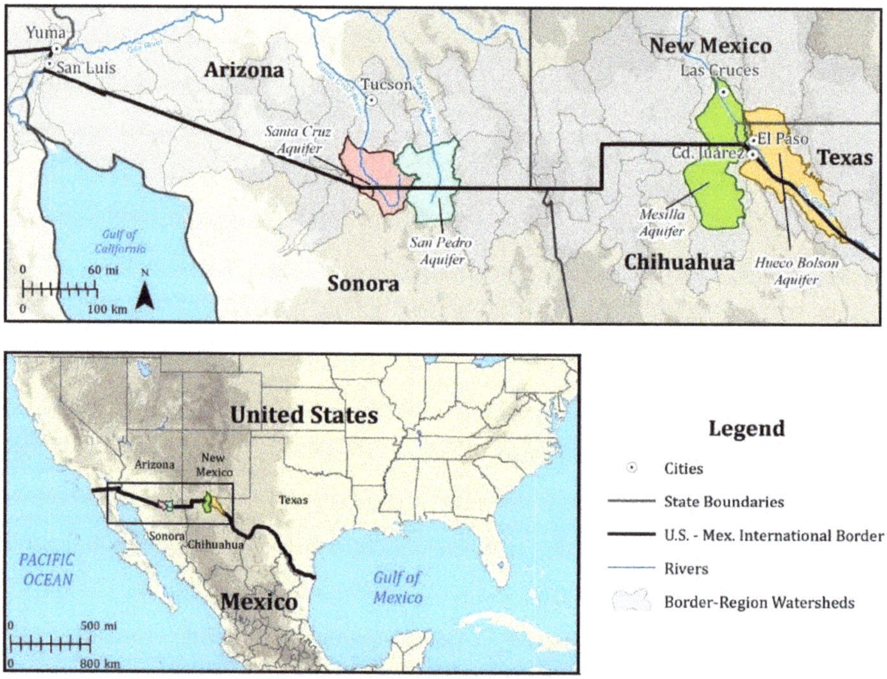

Figure 1. TAAP's aquifers of focus on the Mexico–US border.

Transboundary water resources—whether above or below the surface—span a border and are shared. Here we highlight that the information, interpretation, science, and actions that are needed to manage those resources flow both ways across the border in question. Our aim is to show that the relationship between science production and groundwater governance is bidirectional, with institutions in both countries exerting influence on the outcomes.

The paper seeks to analyze the case study using process tracing ([11]; see, e.g., [12,13]). We argue that analyzing the two-way flow of science production and cooperative governance (e.g., [14]) has yet to be adequately explored in transboundary groundwater governance literature. Specifically, we look at how these processes enhance each other in the Transboundary Aquifer Assessment Program (TAAP) aquifers of focus (Figure 1) shared between Arizona and Sonora—the Santa Cruz and San Pedro aquifers.

First, looking at one direction of the flow, we examine how national and binational policies and cooperative actions, spearheaded by university and government agency research partnerships on both sides of the border, led to advancements in scientific knowledge. The most notable of those advancements was the completion of the first-ever binational scientific aquifer assessments, prepared and released simultaneously in English and Spanish by the (US–Mexico) International Boundary and Water Commission (IBWC).The IBWC/CILA (Comisión Internacional de Límites y Aguas; the Mexican name of the commission) is the binational organization whose mission is "to provide binational solutions to issues that arise during the application of US–Mexico treaties regarding boundary demarcation, national ownership of waters, sanitation, water quality, and flood control in the border region" [15].

Then, viewing the other direction, we discuss how data and information resulting from assessments may contribute to future decision-making for shared groundwater in the border region.

In the following sections, we explore principles of water governance and, specifically, which factors enhance cooperation in groundwater governance. As part of the process, we also examine the role of science in informing policymaking. Next, we use the outlined principles of water governance to analyze how elements of the science–policy interface relate to groundwater governance in the case of the TAAP in Arizona and Sonora.

2. Literature Review

While the science–policy interface has been addressed in water management generally (e.g., [16,17]) and transboundary water management specifically (e.g., [18]), little has been written describing which elements of water governance need to be present for coproduction of knowledge to occur in a transboundary setting (Armitage et al. [19] being a notable exception). Acknowledging that our selection of key principles for the science–policy interface in transboundary groundwater governance is not exhaustive, we review frequently identified principles for analyzing groundwater governance and the science–policy interface in the following paragraphs.

2.1. *Principles for Analyzing Groundwater Governance*

Though governmental entities are more likely to be cooperative than conflictive over shared waters [20], there exist certain factors that make it easier—or more difficult—for cooperation to occur. Among the barriers to transboundary cooperation are spatial and social distance [21–24]; limitations in institutional capacity, financial resources, participation capacity, and data availability [25]; layering and asymmetries of governance structures [26] and intrajurisdictional integration within countries [27,28]; incompatible governance cultures and mandates; and mistrust and/or lack of leadership [25]. Here, social distance refers to disparities in cultural, ethnic, religious, linguistic, political, administrative, legal, and traditional ways of managing and governing water resources. These and other potential asymmetries complicate transboundary resources management. Drivers for transboundary cooperation include leadership, personal relationships, contacts, the existence of binational (or multinational) institutions, and functioning networks [25].

When initiating cooperation on international waters, the chances of such cooperation being successful increase when autonomies of each party are respected, basinwide networks of scientists are established, diverse groups of stakeholders are consulted, and perhaps above all, all parties establish and maintain trust [29]. Such cooperation encourages solving common problems and cultivates interdependence and mutual understanding [30].

Narrowing our focus to transboundary groundwater, scholars have recognized the underdeveloped and/or fragmented structures for resolving critical problems in groundwater governance [10,31–34]. Here, we define groundwater governance as "the overarching framework of groundwater use laws, regulations, and customs, as well as the processes of engaging the public sector, the private sector, and civil society ([35]; p. 678]). Cooperation is a key element in transboundary aquifer governance. Enabling factors include: existing legal mechanisms, functioning regional institutions, funding mechanisms, high institutional capacity, previous water cooperation, scientific research, and strong political will [36].

Several articles have identified elements, or "pillars," of surface water and groundwater governance (e.g., [10,37–41]). Regarding groundwater, principles for management, planning, and assessment can be summarized as follows: *stakeholder engagement and inclusion, proper assessment and data for analysis, management and planning for groundwater use, integrated water management*, and *protection of groundwater resources* [10,38,39].

The reviews of groundwater governance principles and enabling factors presented here feature certain commonalities, including stakeholder engagement, management and planning, integrated water management, protecting groundwater resources, functioning institutional presence and capacity, history of water cooperation, funding, and political will. Other common factors are data sharing and scientific cooperation. Even low levels of scientific research can motivate some degree of cooperation [36], as that research can lead to increased transparency [42].

Complications associated with transboundary aquifers have been recognized for many years by scholars and intergovernmental organizations (e.g., [43,44]). Although few in number, formal international groundwater agreements vary in their legal nature, scope, status, content, duration, and driving motivation, and of course, degree of successful implementation [45]. This suggests that approaches to international groundwater problems are often site-specific and ad hoc in nature [21]. The few existing international groundwater agreements all contain some mechanism for data collection and/or exchange, and most also provide an institutional framework [45]. One of the most prominent such agreements addresses the Guaraní Aquifer in South America. The countries had to overcome asymmetries in political power and water governance structures (e.g., level of centralization) [46]. Despite the original promise of the accord, implementation has been a challenging process [47]. Jordan and Saudi Arabia also signed and ratified a bilateral (though not fully operational) agreement on the Disi Aquifer in 2015 [48] after exploitation from both countries, including withdrawals from a Jordanian pipeline project [49].

2.2. Analyzing the Science–Policy Interface in Transboundary Groundwater Governance

The bidirectional relationship between data-cum-information exchange and groundwater governance is a clear example of the science–policy interface. We define this interface to be how policy actors and scientists interact with each other through processes such as exchanging and joint constructing knowledge [50]. This building of place-based knowledge of groundwater resources is a key step toward effective management and governance [51]. In transboundary contexts, scientific assessment is a needed initial step to determine *whether* and *the extent to which* individual aquifers are cross-border physical systems (see [8,52]). Assessment can also help to catalog existing data sources and identify what data gaps exist. Such lacunae are common for transboundary aquifers because availability of and access to groundwater data are especially constrained [53], of uneven quality and reliability, and sometimes the result of disparate measurement systems and protocols [22].

Long-term planning for sustainable groundwater management also requires both characterization and ongoing monitoring due to the complexity and everchanging conditions of aquifer systems and inherent scientific uncertainty in groundwater evaluation. Management practices need to account for hydrogeological characteristics of transboundary aquifers via such strategies as pollution prevention, integrated land and water management, and context-specific approaches [39,54–56].

Certain attributes—such as salience, credibility, and legitimacy of scientific data and information—may determine how easily scientific information can be utilized in management and decision-making [57]. Salience (the relevance of information to decision-makers and/or the public; [57]) increases when the questions asked are relevant to the actors involved; it can be achieved through cooperative development of project goals via two-way communication [58,59]. Building trust and accountability via long-term relationships also bolsters salience, along with credibility (the creation of information that is believable, trusted, and authoritative; [57]) and legitimacy of the process of knowledge production (the perception of how fair the knowledge production is, whether it includes different perspectives, and appropriate values and concerns; [57,60]). Involving actors perceived as experts in the task at hand tends to enhance credibility. However, if experts do not represent multiple points of view, or if the information is not produced via a transparent process, the data may not be perceived as fully legitimate. Legitimacy can be increased when scientific data are generated via cooperative, inclusive efforts, using mutually agreed-upon protocols. Conversely, lack of consensus on the instruments and methods used for data collection can impede cross-border cooperation [61]. Collaborative knowledge production and institutionalized science–policy processes that engage stakeholders—either via a cross-border organization or established network of stakeholders—can bolster the legitimacy of decision-making processes and knowledge generated in transboundary water contexts [19].

These science–policy processes may not necessarily yield deliberate progress toward some final state, but they do offer a developmental path from an initial state [62]. They may require iteration, building on previous practices, and learning from past successes and failures. Iterative processes are essential to positive science–policy interface outcomes, since capacity, trust building, and adaptability require multiple iterations [63–65]. Multiple iterations may also be necessary when new data or understanding is obtained [62,66].

In some cases, cross-border knowledge generation can provide a foundation for or promote further cooperation, such as formal agreements. Of the existing international agreements on shared groundwater, most were initiated by knowledge-generation efforts and/or from funding and assistance from international organizations [45]. Cooperation can be promoted via joint collection of high-quality data—thereby reducing the potential for data to be contested [67]. Joint monitoring, data collection, and data sharing are recognized as beginning steps within the cooperative process [68]. Joint studies of the Guaraní Aquifer System [47,69], the Nubian Sandstone Aquifer [70], and the Iullemeden Aquifer [69] are examples of this.

Yet in other cases, having prior cooperation or specific agreements in place can also help lead to improved scientific studies. Joint scientific studies may require a foundation of cooperative relations that could include transboundary institutional capacity, a framework for cooperation, or merely the presence of trust and prior working relationships. These studies on transboundary waters are often used to alleviate unidentified gaps in knowledge, missing information, data incompatibility, variation in quality control of data, and lack of scientific understandings [71]. Armitage et al. [19] describe the importance of setting the "conditions for collaboration" early on in transboundary science–policy processes by engaging relevant stakeholders, building relationships and bolstering trust. In both the Danube and the Orange–Senqu basins, for example, establishment of transboundary institutions—river basin organizations—made it possible to conduct cohesive basinwide water quality studies via collaborative studies involving all basin states in knowledge production [19]. Similarly, the US–Mexico agreement on the release of an environmental pulse flow in the Colorado River led to many new scientific discoveries [72].

When science is produced via collaborative processes that engage multiple parties, the information produced is more likely to be accepted by the participating countries. Science–policy processes for transboundary groundwater do not evolve the same way each time—the process is nuanced and involves give-and-take between progress toward scientific investigation and information gathering on the one hand, and political cooperation and agreements enabling people and organizations to work together on the other. Examples of how this critical bidirectional relationship between science and policy manifests in transboundary water governance is demonstrated in Figure 2.

Using the principles of good governance and role of science in decision-making outlined above, we turn now to analyze groundwater governance and processes of science production in one illustrative case: the implementation of the TAAP in Sonora and Arizona.

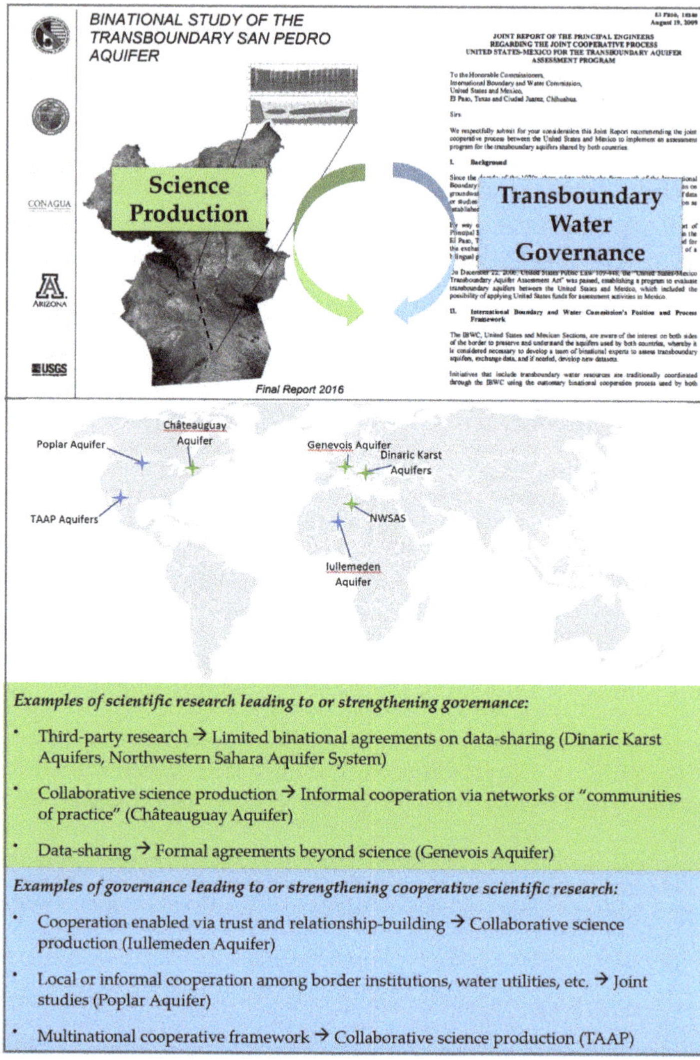

Figure 2. Bidirectionality of science production and transboundary water governance. Case study examples aside from the TAAP are from [69].

3. Materials and Methods

This paper employs a case-study method to investigate the relationship between science production and groundwater governance in the US–Mexico border region. Based on the results of our literature review above, we hypothesize the following:

Hypothesis 1 (H1). *Science contributes to and influences transboundary groundwater governance by informing management with a transboundary scientific understanding on both sides of the international border.*

Hypothesis 2 (H2). *Transboundary groundwater governance, through policies, agreements, and other cooperative efforts, contributes to and influences the course of scientific inquiry by expanding* *cooperative scientific networks across the border, including communities of practice. This in turn helps to generate new scientific knowledge that likely would not be possible without governmental cooperation.*

To test these hypotheses, we employ both process tracing and interviews. First, we analyze elements that have been identified as enabling factors for good groundwater governance, as synthesized from the literature: stakeholder engagement and inclusion, management and planning for groundwater use, integrated water management, protecting groundwater resources, institutional presence and capacity, history of water cooperation, funding, and political will. "Good governance" suggests normative characteristics of being efficient, inclusive, and sustainable. Next, we use key features of scientific information that promote its ease of use in policymaking—salience, credibility, legitimacy [57], and iterative knowledge production [63,64]—to evaluate the bidirectional relationship of science and policy. Finally, we gathered data by conducting 20 interviews of government officials and scientists on both sides of the border (see Supplementary Materials, Table S1), participant observation, and by compiling secondary sources from binational technical meetings, conferences, and stakeholder meetings that took place during the 2010–2019 period.

We asked two sets of interview questions: one for scientists (Table S2), and one for government officials (Table S3). Questions for scientists focused on whether and how transboundary groundwater governance contributes to science by engaging relevant stakeholders, building relationships, and bolstering trust, thereby leading to scientific discoveries. Questions for government officials queried whether and how science (specifically groundwater assessment) contributes to groundwater governance.

4. Results

The adjacent transboundary Santa Cruz and San Pedro aquifers are in southeastern Arizona and northeastern Sonora (Figure 3). Both aquifers support significant populations and economic activities such as mining, agriculture, ranching, tourism, and manufacturing. These aquifers have also seen rapid population growth in recent years [73,74].

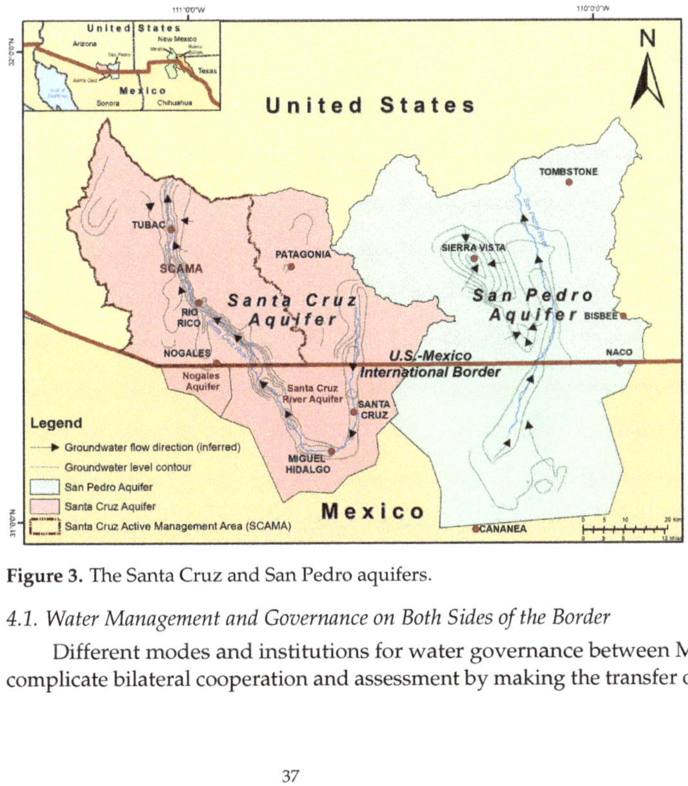

Figure 3. The Santa Cruz and San Pedro aquifers.

4.1. Water Management and Governance on Both Sides of the Border

Different modes and institutions for water governance between Mexico and the US complicate bilateral cooperation and assessment by making the transfer of information and reaching consensus more challenging. Water management and governance in Mexico's nonborder waters tend to be more centralized than in the US [26]. The authority to regulate surface water and groundwater in Mexico resides with Comisión Nacional del Agua (CONAGUA), the national water authority. CONAGUA is responsible for all activities concerning use, management, and protection of national water. At the state and substate levels of government in Mexico, water management is more limited compared to the US; CEA Sonora (Comisión Estatal del Agua Sonora, Sonora's state water commission) assists municipalities in providing water and sanitation services and administers water supply-related programs, and certain municipalities run their own water and wastewater utilities [75].

Water management and governance in the US generally occur at the state and/or substate level. The federal government has built projects for flood control, transportation, hydropower dams, and large-scale water diversions [76] and has set water quality goals through measures such as the 1972 US Clean Water Act and the 1974 US Safe Drinking Water Act. States have authority over implementation of standards, practices, and rules for water use [77]. In Arizona, state law considers groundwater and surface water as distinct water bodies and are regulated as such. While surface water rights in Arizona are regulated under a *prior-appropriation system* ("first in time, first in right"), groundwater use is based on *beneficial use* (e.g., agricultural, industrial, or residential). Groundwater regulations vary across the state. Several regions of the state, including the US portion of the Santa Cruz Aquifer, are designated Active Management Areas (AMAs), where groundwater use is subject to regulations that are meant to be enforced by the Arizona Department of Water Resources [78,79]. As Figure 2 shows, the US portion of the binational San Pedro Aquifer is not part of an AMA.

4.2. Binational Water Management

The 1944 treaty, "Utilization of Waters of the Colorado and Tijuana Rivers and of the Rio Grande," gives the IBWC authority to make rules through adopting minutes (interpretations and clarifications) to the treaty. The IBWC has some authority in water shared between Mexico and the US to ensure compliance with the 1944 treaty, manage joint infrastructure, maintain hydrologic monitoring stations, and communicate information across the border [80]. This is due in part because Mexico's policy requires that all border groundwater (and surface water) issues be handled through the commission [81]. The IBWC comprises two sections, with one section in Mexico (CILA) and one in the US.

In the past century, Mexico and the US have expanded transboundary surface water governance capacity, moved towards inclusion of non-nation-state actors, increased ecological considerations, and have signed agreements related to surface water—most notably the 1944 treaty [82]. There exists no agreement over the management of shared groundwater aside from Minute 242, which addresses groundwater pumping near the US–Mexico border near San Luis, Mexico [83]. Minute 242 authorized the IBWC to begin discussions on a binational groundwater agreement [80], but little progress has been made to date. Issues surrounding water rights on both sides of the border, including those of private parties and concessionaires, remain unresolved [84]. There have also been notable disputes between both countries regarding water management, including over the issue of salinity of the Colorado River as it enters Mexico [85].

4.3. Establishment of the TAAP

Both the US and Mexico have recognized that greater scientific understanding of their shared groundwater resources would be mutually beneficial, particularly within a region where groundwater is a primary component of the water balance and where populations are growing. The two countries signed the La Paz Agreement in 1983, which formally committed the US and Mexico to annual meetings between ministries and reviewing border environmental concerns. The agreement did not include any specific solutions or environmental protections but does provide a mechanism to do so in the future if desired [86]. The US Congress authorized Public Law 109-448 in late 2006, whose formal name is the United States–Mexico Transboundary Aquifer Assessment Act (TAAA) [87]. Though the TAAA signaled US interest in participating in binational studies, Mexican concurrence was needed to proceed with a binational program and identification of aquifers of focus.

From 2007 to 2009, Mexico and the US began the engagement and negotiation process that resulted in approval by the IBWC of the "Joint Report of the Principal Engineers Regarding the Joint Cooperative Process United States–Mexico for the Transboundary Aquifer Assessment Program" (Joint Report [88]). The Joint Report guides the binational study of four transboundary aquifers: the San Pedro, Santa Cruz, Mesilla, and Hueco Bolson. The key word for the cooperation between the two countries is "assessment"—the Joint Report specifies that information that comes from cooperation is "solely for the purpose of expanding knowledge of the aquifers and should not be used by one country to require that the other country modify its water management and use" ([88], p. 3). Further, the Joint Report also states that activities should be beneficial to both countries and cannot limit what each country can do independently within its boundaries.

4.4. Summary of Governance Principles Present

Some of the governance principles for management, planning, and assessment identified in the literature review are present in the TAAP case study. These elements have allowed for successful completion of scientific assessments but have not allowed for a transboundary management regime to manifest at this point.

Stakeholder engagement was one of the keys to the project's success; a broad set of stakeholders and key actors was involved in early efforts that determined the scope of assessment. Stakeholder engagement efforts during the project have included establishing modes of communication through webpages, factsheets, and briefings. The project was most likely aided by the long history of binational stakeholder engagement in the region (see [89]).

There are no binding binational *management and planning* efforts regarding the TAAP aquifers. Aside from Minute 242, which does not involve any of the TAAP aquifers, there were no binding existing legal mechanisms dealing specifically with groundwater prior to the establishment of the TAAP. Elsewhere, the Municipal Water and Sanitation Board of the City of Juárez (Chihuahua, Mexico) and the El Paso Water Utilities Public Service Board (El Paso, TX, USA) signed a legally unenforceable and unofficial memorandum of understanding that calls for cooperation over and information exchange for the Hueco Bolson Aquifer in 1999 [90,91].

In addition to the absence of binding binational management and planning efforts, the two countries also have not engaged in binational *integrated water management*. There have been no binational efforts toward integrated water management elements such as managed aquifer recharge or collaborative modeling (though both countries have expressed interest in building binational models of the aquifers). Neither have there been any specific binational efforts toward *protecting groundwater resources* in terms of IBWC Minutes on water quality or quantity for the TAAP aquifers, though Minutes 261, 276, and 294 do designate impaired water quality resulting from border sanitation as an issue that should be addressed.

Though the IBWC had not previously had a specific focus on groundwater, it was undoubtedly a critical *functioning regional institution* for the function of the TAAP. The IBWC, along with CONAGUA, were key players for the development of the Joint Report. The IBWC's expertise in managing treaty obligations (*previous water cooperation*) and Mexican policies regarding transboundary waters made the binational organization central in formalizing the cooperative framework [22].

Funding was provided by each country, but there were times when the timing of funding availability was asynchronous between the two countries. This resulted in differences between countries in the amount and type of work at a given time.

Political will was evident throughout the TAAP process. The approval of the TAAP had to go through multiple official channels, including the 2006 US TAAA and the IBWC. Stakeholder involvement was also significant in developing the assessment's parameters. Because the parties have limited their efforts to assessment, there has been no test of the political will associated with addressing the institutionally complex matter of joint management.

4.5. Summary of Attributes for Information to Be Used in Decision-Making

Salience of scientific information increases when the questions asked are relevant to actors involved. While binational priorities were developed jointly, some studies focusing on the TAAP aquifers were not binational as each country can conduct work within its own borders without needing to consult the other, in accordance with the Joint Report (e.g., [5]). Binational forums resulted in the development of strategic plans for the two Arizona–Sonora aquifers, outlining priorities and tasks, with annual tactical plans allow for more adaptive research to realities relating to funding, resources, personnel, and new progress [26].

The TAAP team attended and participated at conferences to communicate and exchange information with others in the scientific community. Team members have also disseminated journal articles, theses, and reports to enhance *credibility*. The Sonora–Arizona effort has yielded two reports: the *Binational Study of the Transboundary San Pedro Aquifer*, published in 2016, and the *Binational Study of the Transboundary Santa Cruz Aquifer*, which is undergoing peer review. Both studies address their attention to physical characteristics of the aquifers—the geology, climate, hydrology, landcover, and soils—and integrate these data across the entire geographic extent of the aquifer, evaluating it for the first time as a single physical system. Besides the work published by Pool and Dickinson [92] on the Sierra Vista Subwatershed and Sonoran portions of the Upper San Pedro Basin, binational maps were not common for the Transboundary San Pedro Aquifer. The same is true for the Transboundary Santa Cruz Aquifer, where only a few sources present harmonized binational cartography [93,94]. The San Pedro study produced 42 aquiferwide geographic information system (GIS) layers containing data about the aquifer, which served as the basis for the development of over 34 binational maps that describe multiple aspects of the study area [95]. Each of the two reports represents a one-of-a-kind type of assessment. The studies analyze and harmonize information from two different countries. Analyzing and harmonizing information required overcoming language barriers, institutional asymmetries, mapping and measurement preferences, and review processes.

Identifying each country's team members was one of the initial challenges of the TAAP. In Mexico, for example, some of the members were required by governmental policy to go through CILA for this task, as Mexico requires that all border water issues be handled through CILA. Universidad de Sonora and CONAGUA carried out the studies. US funding was divided among the federally authorized Water Resources Research Institutes in Texas, New Mexico, and Arizona, and the US Geological Survey Water Science centers in those three states, as required by the TAAA. Selection of team members who were bilingual also eased the process of communication.

4.6. Interview Results

Officials interviewed from both countries agreed that the information generated from the reports could be used for recommendations and regulations (though were less certain about whether the information would help regulations), and an informal or formal binational groundwater organization. Interviewees said that the results are potentially useful for, e.g., confirming past conceptual understandings on the other side of the border, some adjudication decisions in Arizona, and providing information for other forms of decision-making. Figure 4 presents a summary of interviewees' scores for salience, engaging stakeholders, bolstering trust and building relationships (interviews of scientists), and salience, credibility, and legitimacy (interviews of government officials). The TAAP's collaborative and iterative process of scientific assessment helped produce information that is more salient, credible and legitimate—regarding the transboundary aquifer—than could have been produced by either country alone (Table 1).

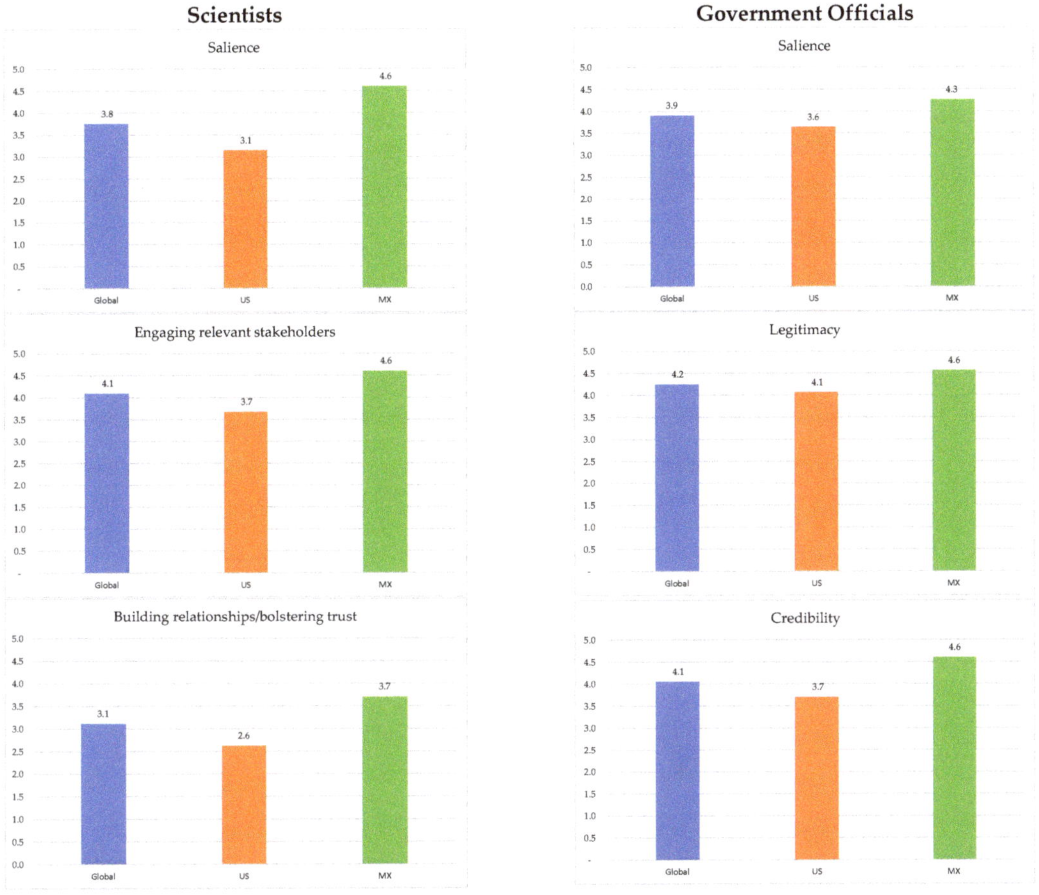

Figure 4. Summary of interview scores.

Table 1. Science production and relevant attributes of science outputs in the TAAP.

Features of Science Production	Relevant Attributes of Science Outputs
Binational development of research aims and focus areas through Binational Technical Group meetings	Legitimacy, salience, iteration
Investment of funding or in-kind investments from both countries	Legitimacy
Involvement of binational experts in knowledge production	Credibility, iteration
Stakeholder involvement in planning	Salience, legitimacy, iteration
Integration and harmonization of data from both nations	Salience, iteration
Bilingual reporting of results (Binational Studies of the San Pedro and Santa Cruz Aquifers)	Legitimacy

Regular and continuous communication and cooperation among the TAAP stakeholders also bolstered the *legitimacy* and acceptability of scientific information produced. The science produced by the Arizona–Sonora TAAP effort achieved legitimacy at the national level, gaining official approval by the governments of both the US and Mexico. Inclusion of social scientists in project design helped address social and institutional aspects of cross-border cooperation, which are critical to the production of legitimate knowledge.

No binational work contributing toward the assessment happened until the 2009 Joint Report. During this time, the Arizona–Sonora team engaged in team- and trust-building. Participation in field trips helped build trust and a shared history among team members. We define trust as an expectation or belief that one group can rely on another's actions and word and/or that the group has good intentions toward others [96]. During the period of the study, the Binational Technical Group meetings established by the Joint Report also helped to build trust: "Through the collaborative work we have learned about the capacity and experience of the other researchers. It is not the same to know a person through what he or she has published as it is to work together with them," one TAAP scientist said. The cooperative process was also aided by previous cooperative work conducted by Mexican and US geologists over the last 50 years. The *multiple iterations* of communication were essential for the process to continue and the work to be completed.

Both groups of Mexican officials and scientists interviewed for this article gave higher scores compared to their US counterparts in all six categories portrayed in Table 1. These higher scores could imply that Mexico perceives more benefits from the transboundary aquifer assessment process through sharing previous studies, data, and technical resources. The program also allows for arguably greater opportunities for Mexican researchers to expand their research networks compared to their US counterparts.

5. Bidirectionality and the Science–Policy Interface

Key elements of "good governance" and of collaborative scientific assessment processes both contributed to fruitful binational cooperation over assessment of these aquifers. We show the bidirectional interaction between groundwater governance and science production in Figure 5.

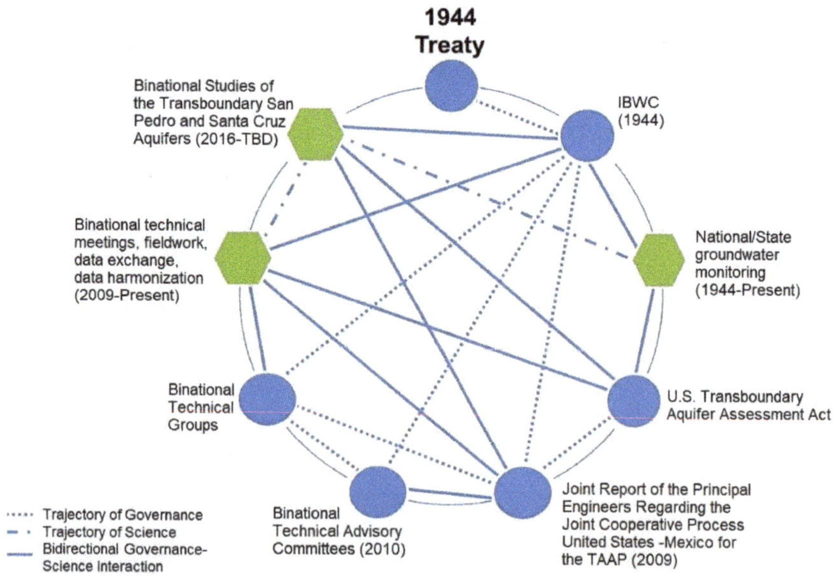

Figure 5. The path to bidirectionality/binationality in the TAAP. Governance elements are represented by blue circles; science elements are represented by green hexagons. Figure by authors.

The TAAP engaged in science production with joint research projects through the binational studies. Transboundary cooperation was established through trust and relationship building among actors. One TAAP scientist interviewed said, "The binational relations have been strengthened in many levels- [on the] local, formal, academic and research level, [we've engaged in] collaboration and [generated] trust, long term projects, medium term, joint collaboration." Trust and relationship-building was aided by the already-established relationship between governments through the IBWC and other governmental avenues of cooperation. This was achieved through formal agreements in data and information sharing through the 1944 agreement and subsequent minutes. The establishment of a framework for a joint or collaborative binational study through the 2009 Joint Report helped to build a formal cooperation channel that was sheltered by the IBWC. Previous collaboration efforts lacked a coordinating body, Binational Technical Groups, and Binational Technical Advisory Committees. This led to ambiguities within studies and barriers associated to information distribution and availability.

In the context of US–Mexico transboundary groundwater, our view is that science can palpably contribute to US–Mexico groundwater governance (Hypothesis 1) in two key ways: (1) by informing management on each side of the border with a transboundary scientific understanding, and (2) by expanding binational cooperative networks—including communities of practice—on local, state, and national levels. These foundational cooperative elements allowed for a collaborative process of science production that goes beyond merely sharing information. From a "science-to-governance" perspective, collaborative scientific assessment of shared aquifers can help to inform water management decisions at the local (e.g., water rights adjudication in Arizona) and national (e.g., determining availability of groundwater in Mexico) levels, provides a shared knowledge-base and strengthened trust among participants. Multiple scientists said that the TAAP has helped scientists in learning to collaborate and promoted making contacts on the other side of the border. All scientists involved with the TAAP said that they have made new contacts thanks to the program: "Through this program we had the fortune to meet several researchers in the field and know what they do and be familiar with their work," one TAAP scientist said. Another TAAP scientist said, "Because of TAAP, now I know who to go to [if I had a question about groundwater across the border]."

From a "governance-to-science" perspective (Hypothesis 2), the previous cooperation over water between Mexico and the US through the 1944 treaty and their subsequent minutes undoubtedly facilitated the establishment of the binational assessment process. Interviewed scientists said that the previous treaty and minutes helped to strengthen communication between countries.

The political will of stakeholders and policy makers played a significant role for the assessments. Funding provided by each country showed investment in the assessment's outcome. The investment of time that it took for the parties to agree upon the 2009 Joint Report was arguably worthwhile, as it allowed them to create a document that struck a balance between independence—where both countries conduct and fund their own research activities on their side of the border—and coordination, including communication of information through sharing data publicly, e.g., through the publication of the assessments. Parties were able to harmonize information and overcome barriers, differences, and preferences through the Binational Technical Group and Binational Technical Advisory Committee meetings. The 2009 Joint Report guided how cross-border scientific efforts were carried out—e.g., via binational teams—which helped reduce the possibility that either nation would object to the knowledge produced. While one scientist pointed out that the 2009 Joint Report allowed for the assessments to happen, another scientist mentioned that it was an obstacle for them due to the level of formality associated with the protocol. The process for sharing data required multiple iterations. Both parties were initially cautious about sharing information. The data sharing process became more efficient as the studies progressed. Over time, scientists interviewed said that participation in multiple, face-to-face meetings built trust and relationships between team members. The meetings led to collaborative science production by helping to resolve issues including jointly determining the delineation of the study area, data needs, and data integration and compatibility: "In both of those watersheds [San Pedro and Santa Cruz], there is a significant amount of data—making the results meaningful to stakeholders needs to be part of the package," one US government official said. However, we should note that the results from scientists involved with the TAAP program may exhibit a more favorable view of the products resulting from the TAAP.

The IBWC was necessary as a coordinating body for the assessments. Its institutional capacity, (manifested through its authority in ensuring compliance with the 1944 treaty), its management of joint infrastructure and maintenance of hydrologic monitoring stations, its protocols for data exchange, its contribution of funding, and its role in transboundary communication, all contributed to helping the assessment process. With suitable adaptation for context, this effort could be replicable for other areas of scientific cooperation between Mexico and the US, and perhaps for other transborder-resource studies [97]. In the TAAP, steps taken on transboundary groundwater governance and production of scientific information built upon each other. This bidirectionality contributed to partially harmonize asymmetries in institutional frameworks between Mexico and the US, particularly because of the central role of the IBWC in its collection of binational data and coordination of the joint studies.

As an example of how governance elements and science production build upon each other bidirectionally, the TAAP process began with a joint decision regarding which aquifers would be assessed first. This consensus-based decision-making helped formulate project aims that are salient for stakeholders on both sides of the border. Datasets produced are comprehensive and harmonized across the border, in turn, allowing for better access to decision-makers and improved legitimacy of the information. Overall, the collaborative process enhanced legitimacy of the information produced through transparency and binational engagement. The resultant cross-border network of scientists and other stakeholders can be leveraged to help guide further efforts toward addressing shared goals. For instance, one TAAP scientist referenced how the TAAP has allowed for advancements in mapping geologic units that were beyond the scope of the original assessments.

There is no agreement to extend cooperation beyond scientific investigation and collaboration. However, if a more formal binational management regime were to come to pass (which appears unlikely according to interviewees), the availability of reliable scientific information would be an initial step [69,70,97–99]. From the outset, building trust and mutual respect have been important components of the assessment's realization. Both countries would need to continue to build trust and navigate jurisdictional overlaps for a development of a binational management agreement, among other things.

It is possible that assessments such as the ones achieved by the TAAP will help to determine the severity of existing challenges and promote joint problem-framing and agenda-setting. Reaching a mutual understanding on aquifer and groundwater conditions is arguably a necessary (though not sufficient) condition for collaborative management. Such a strategy would help direct resources more efficiently to address the problems. That said, assessment can only go so far in leading to resolution of the groundwater management issues in the Santa Cruz and San Pedro aquifers. Rules, regulations, monitoring, enforcement of those rules and regulations, and perhaps most importantly, public acceptance, political will, and financial commitment are needed to resolve management issues. It appears likely that scientific assessments alone will need other factors to generate the political will necessary to create a binational management regime in this case or elsewhere.

There may be some potential for more localized cooperation between subnational jurisdictions within these aquifers. There has also already been informal, local cooperation in sharing water between the cities of Nogales, Arizona, and Nogales, Sonora, in times of serious drought in the Santa Cruz Aquifer [91] or during other specific problems such as fires [100].

6. Conclusions

This article argues, using the case study of the TAAP Sonora–Arizona assessments as an example, that transboundary groundwater governance and the production of scientific information evolve in reciprocal synchronicity—cooperation can enhance science production, and science can lead to advancements in policy. *Both* are needed for transboundary groundwater governance, as they are in nontransboundary situations. Certain elements of governance need to be present for scientific assessments to occur—particularly via collaborative efforts—and for the knowledge gathered through the efforts of the assessments to be potentially usable for future policy- and decision-making. In the case of the TAAP, the establishment of trust, the cooperative framework, and the history of cooperation between the two countries through formal agreements were particularly important to successful assessments. These components helped the binational team overcome challenges of integrating different standards and methods for reporting, peer review, language, measurement units, and technical and financial capacities, among others. While salience, credibility, legitimacy, and iterations of assessment and information-sharing can certainly aid further cooperation, it has yet to be seen whether the assessments will aid transboundary water governance between the two countries. Because of this, it should be noted that the case study is one example of bidirectionality, which may not be present in all cases. More evidence is needed from other cases to prove our argument.

More work lies ahead for policymakers to continue collaborative efforts after the completion of the assessments. A few questions about how momentum can be sustained, how the results of the assessments are being used, identifying the sources for financing, determining if political will exists to continue collaboration and progress to governance, and continuing the trust-building process, are yet to be answered. The politically charged issues surrounding water rights between the two countries also are yet to be solved. Despite the questions listed above, the TAAP case has several elements that enable groundwater collaboration. It is also consistent with some of the principles included in other groundwater management agreements around the world [99].

The TAAP case suggests that the relationship between data-cum-information-sharing and transboundary water governance is iterative and self-reinforcing. All discrete governance and information elements are part of a larger cooperative process. This process could help yield an eventual binational agreement (or agreements) such as those for the Genevois, Guaraní, Iullemeden, and Nubian aquifer systems. Since decision-making ultimately is a political process, we believe that, as elsewhere, science is a necessary condition for forging international groundwater agreements. Along with science, political will, stakeholder engagement, and adequate incentives to cooperate are critical factors for initiating and sustaining transboundary cooperation.

Supplementary Materials: The following are available online at https://www.mdpi.com/article/10.3390/w13172364/s1, Table S1: Data on interviewees, Table S2: Interview questions for scientists, Table S3: Interview questions for government officials.

Author Contributions: J.D.P.-P. and T.R.A. conceived the article; J.D.P.-P., T.R.A. and E.M.T.-V. designed and performed the experiments; J.D.P.-P., T.R.A. and E.M.T.-V. analyzed the data; J.D.P.-P., T.R.A., E.M.T.-V., R.G.V. and S.B.M. wrote the article. All authors have read and agreed to the published version of the manuscript.

Funding: This work was partially funded by the US Geological Survey (funding authorized by P.L. 109–448) Award Number G17AC00439 for the Transboundary Aquifer Assessment Program.

Institutional Review Board Statement: This project has been reviewed and approved by an IRB Chair or designee from East Carolina University (protocol UMCIRB 21-000087).

Informed Consent Statement: Informed consent was obtained from all subjects involved in the study.

Data Availability Statement: Data sharing is not applicable to this study. The information shared during the interview process is confidential and cannot be shared with the public.

Acknowledgments: The authors thank interviewees for their time and their participation. Thanks also to Jaclyn Best for her assistance in data processing and Javier Tadeo Villaseñor for translation and transcription of interviews conducted in Spanish. The authors are grateful for support from the Morris K. Udall and Stewart L. Udall Foundation in Tucson, Arizona; and from Lloyd's Register Foundation in the U.K. Thank you to the reviewers for their helpful comments.

Conflicts of Interest: J.D.P.-P., E.M.T.-V. and S.B.M. were partially funded by the Transboundary Aquifer Assessment Program.

References

1. Pagán, B.R.; Ashfaq, M.; Rastogi, D.; Kendall, D.R.; Kao, S.-C.; Naz, B.S.; Mei, R.; Pal, J.S. Extreme Hydrological Changes in the Southwestern US Drive Reductions in Water Supply to Southern California by Mid Century. *Environ. Res. Lett.* **2016**, *11*, 094026. [CrossRef]
2. Romero-Lankao, P.; Smith, J.B.; Davidson, D.J.; Diffenbaugh, N.S.; Kinney, P.L.; Kirshen, P.; Kovacs, P.; Villers Ruiz, L. North America. In *Climate Change 2014: Impacts, Adaptation, and Vulnerability. Part B: Regional Aspects. Contribution of Working Group II to the Fifth Assessment Report of the Intergovernmental Panel on Climate Change*; Barros, V.R., Field, C.B., Dokken, D.J., Mastrandrea, M.D., Mach, K.J., Eren Billir, T., Chatterjee, M., Ebie, K.L., Otsuki Estrada, Y., Genova, R.C., et al., Eds.; Cambridge University Press: Cambridge, UK; New York, NY, USA, 2014; pp. 1439–1498. ISBN 978-1-107-05816-3.
3. Wilder, M.; Garfin, G.; Ganster, P.; Eakin, H.; Romero-Lankao, P.; Lara-Valencia, F.; Cortez-Lara, A.A.; Mumme, S.P.; Neri, C.; Muñoz-Arriola, F.; et al. Climate change and U.S.-Mexico border communities. In *Assessment of Climate Change in the Southwest United States: A Report Prepared for the National Climate Assessment*; Garfin, G., Jardine, A., Merideth, R., Black, M., LeRoy, S., Eds.; Island Press: Washington, DC, USA, 2013; pp. 340–384. ISBN 978-1-61091-446-8.
4. Meixner, T.; Manning, A.H.; Stonestrom, D.A.; Allen, D.M.; Ajamie, H.; Blasch, K.W.; Brookfield, A.E.; Castro, C.L.; Clark, J.F.; Gochis, D.J.; et al. Implications of projected climate change for groundwater recharge in the western United States. *J. Hydrol.* **2016**, *534*, 124–138. [CrossRef]
5. Shamir, E.; Megdal, S.B.; Carillo, C.; Castro, C.L.; Chang, H.-I.; Chief, K.; Corkhill, F.E.; Eden, S.; Georgakakos, K.P.; Nelson, K.M.; et al. Climate change and water resources management in the Upper Santa Cruz River, Arizona. *J. Hydrol.* **2015**, *521*, 18–33. [CrossRef]
6. Shamir, E.; Tapia-Villaseñor, E.M.; Cruz-Ayala, M.B.; Megdal, S.B. A review of climate change impacts on the USA-Mexico transboundary Santa Cruz River Basin. *Water* **2021**, *13*, 1390. [CrossRef]
7. Sanchez, R.; Eckstein, G. Aquifers shared between Mexico and the United States: Management perspectives and their transboundary nature. *Groundw* **2017**, *55*, 495–505. [CrossRef]
8. International Groundwater Resources Assessment Centre (IGRAC). Transboundary Aquifers of the World. Special Edition for the 7th World Water Forum. 2015. Available online: https://www.un-igrac.org/resource/transboundary-aquifers-world-map-2015 (accessed on 3 August 2021).
9. Burchi, S. Legal frameworks for the governance of international transboundary aquifers: Pre- and post-ISARM experience. *J. Hydrol. Reg. Stud.* **2018**, *20*, 15–20. [CrossRef]
10. Megdal, S.B.; Petersen-Perlman, J.D. Groundwater governance and assessment in a transboundary setting. In *Lake Governance*, 1st ed.; Grover, V., Krantsberg, G., Eds.; CRC Press: Boca Raton, FL, USA, 2018; pp. 38–60. ISBN 9781138633759.
11. Vanhala, L. Process tracing in the study of environmental politics. *Glob. Environ. Politics* **2017**, *17*, 88–105. [CrossRef]
12. Kashwan, P. Forest policy, institutions, and REDD+ in India, Tanzania, and Mexico. *Glob. Environ. Politics* **2015**, *15*, 95–117. [CrossRef]
13. Kauffman, C.; Mann, P. Scaling up Buen Vivir: Globalizing local environmental governance from Ecuador. *Glob. Environ. Politics* **2014**, *14*, 40–58.
14. Milman, A.; MacDonald, A. Focus on interactions between science-policy in groundwater systems. *Environ. Res. Lett.* **2020**, *15*, 090201. [CrossRef]
15. International Boundary and Water Commission-Comision Internacional de Limites y Agua (IBWC). Welcome. Available online: http://www.ibwc.gov/home.html (accessed on 3 August 2021).
16. Dunn, G.; Brown, R.R.; Bos, J.J.; Bakker, K. The role of science-policy interface in sustainable urban water transitions: Lessons from Rotterdam. *Environ. Sci. Policy* **2017**, *73*, 71–79. [CrossRef]
17. Dunn, G.; Bos, J.J.; Brown, R.R. Mediating the science-policy interface: Insights from the urban water sector in Melbourne, Australia. *Environ. Sci. Policy* **2018**, *82*, 143–150. [CrossRef]
18. Bukowski, J. A "new water culture" on the Iberian Peninsula? Evaluating epistemic community impact on water resources management policy. *Environ. Plan. C* **2016**, *35*, 239–264. [CrossRef]
19. Armitage, D.; de Loë, R.C.; Morris, M.; Edwards, T.W.D.; Gerlak, A.K.; Hall, R.I.; Huitema, D.; Ison, R.; Livingstone, D.; MacDonald, G.; et al. Science-policy processes for transboundary water governance. *Ambio* **2015**, *44*, 353–366. [CrossRef]
20. Wolf, A.T.; Yoffe, S.B.; Giordano, M. International waters: Identifying basins at risk. *Water Policy* **2003**, *5*, 29–60. [CrossRef]
21. Blomquist, W.; Ingram, H.M. Boundaries seen and unseen: Resolving transboundary groundwater problems. *Water Int.* **2003**, *28*, 162–169. [CrossRef]

22. Callegary, J.B.; Megdal, S.B.; Tapia-Villaseñor, E.M.; Minjárez-Sosa, I.; Petersen-Perlman, J.D.; Monreal, R.; Gray, F.; Grijalva Noriega, F. Findings and lessons learned from the assessment of the Mexico-United States transboundary San Pedro and Santa Cruz aquifers: The utility of social science in applied hydrologic research. *J. Hydrol. Reg. Stud.* **2018**, *20*, 60–73. [CrossRef]
23. Udall, S.L.; Varady, R.G. Environmental conflict and the world's new international borders. *Transbound Resour Rep.* **1994**, *7*, 5–6.
24. Varady, R.G.; Morehouse, B.J. Moving borders from the periphery to the center: River basins, political boundaries, and water management policy. In *Water: Science, Policy, and Management*; Lawford, R., Fort, D., Hartmann, H., Eden, S., Eds.; American Geophysical Union: Washington, DC, USA, 2003; pp. 143–160. ISBN 978-0875903200.
25. Norman, E.; Bakker, K. Drivers and Barriers of Cooperation in Transboundary Water Governance: A Case Study of Western Canada and the United States; Report to the Walter and Duncan Gordon Foundation. 2005. Available online: https://watergovernance.ca/2017/07/15/drivers-and-barriers-of-cooperation-in-transboundary-water-governance-a-case-study-of-western-canada-and-the-united-states/ (accessed on 4 August 2021).
26. Megdal, S.B.; Scott, C.A. The importance of institutional asymmetries to the development of binational aquifer assessment programs: The Arizona-Sonora experience. *Water* **2011**, *3*, 949–963. [CrossRef]
27. Albrecht, T.R.; Varady, R.G.; Zuniga-Teran, A.A.; Gerlak, A.K.; De Grenade, R.R.; Lutz-Ley, A.; Martín, F.; Megdal, S.B.; Meza, F.; Ocampo Melgar, D.; et al. Unraveling transboundary water security in the arid Americas. *Water Int.* **2018**, *43*, 1075–1113. [CrossRef]
28. Cohen, A.; Norman, E.S. Renegotiating the Columbia River Treaty: Transboundary governance and indigenous rights. *Glob. Environ. Politics* **2018**, *18*, 4–24. [CrossRef]
29. Petersen-Perlman, J.D.; Wolf, A.T. Getting to the first handshake: Enhancing security by initiating cooperation in transboundary river basins. *J. Am. Water Resour. Assoc.* **2015**, *51*, 1688–1707. [CrossRef]
30. Ide, T.; Detges, A. International water cooperation and environmental peacemaking. *Glob. Environ. Politics* **2018**, *18*, 63–84. [CrossRef]
31. De Chaisemartin, M.; Varady, R.G.; Megdal, S.B.; Conti, K.I.; van der Gun, J.; Merla, A.; Nijsten, G.-J.; Scheibler, F. Addressing the groundwater governance challenge. In *Freshwater Governance for the 21st Century*; Karar, E., Ed.; Springer International Publishing: Cham, Switzerland, 2017; pp. 205–227. ISBN 978-3-319-43350-9.
32. Giordano, M. Global groundwater? Issues and solutions. *Ann. Rev. Environ. Resour.* **2009**, *34*, 153–178. [CrossRef]
33. Megdal, S.B. Invisible water: The importance of good groundwater governance and management. *NPJ Clean. Water* **2018**, *1*, 15. [CrossRef]
34. Theesfeld, I. Institutional challenges for national groundwater governance: Policies and issues. *Groundwater* **2010**, *48*, 131–142. [CrossRef]
35. Megdal, S.B.; Gerlak, A.K.; Varady, R.G.; Huang, L.-Y. Groundwater governance in the United States: Common priorities and challenges. *Groundwater* **2015**, *53*, 677–684. [CrossRef]
36. Conti, K.I. Cooperation over transboundary aquifers: Lessons learned from 10 years of experience. In *Free Flow: Reaching Water Security through Cooperation*; Griffiths, J., Lambert, R., Eds.; UNESCO: Paris, France, 2013; pp. 40–44. ISBN 978-92-3-104256-0.
37. Biswas, A.K.; Tortajada, C. Future water governance: Problems and perspectives. In *Improving Water Policy and Governance*; Tortajada, C., Biswas, A., Eds.; Routledge: New York, NY, USA, 2011; pp. 1–13. ISBN 978-0415606288.
38. Akhmouch, A.; Clavreul, D. Assessing and monitoring groundwater governance. In *Advances in Groundwater Governance, Vilholth, K.G., López-Gunn, E., Conti, K., Garrido, A., van der Gun, J., Eds.*; Taylor & Francis Group: London, UK, 2018; pp. 247–265. ISBN 978-1138029804.
39. Albrecht, T.R.; Varady, R.G.; Zuniga-Teran, A.A.; Gerlak, A.K.; Staddon, C. Governing a shared hidden resource: A review of governance mechanisms for transboundary groundwater security. *Water Secur.* **2017**, *2*, 43–56. [CrossRef]
40. Kiparsky, M.; Milman, A.; Owen, D.; Fisher, A.T. The importance of institutional design for distributed local-level governance of groundwater: The case of California's Sustainable Groundwater Management Act. *Water* **2017**, *9*, 755. [CrossRef]
41. Varady, R.G.; Zuniga-Teran, A.A.; Gerlak, A.K.; Megdal, S.B. Modes and approaches of groundwater governance: A survey of lessons learned from selected cases across the globe. *Water* **2016**, *8*, 417. [CrossRef]
42. Ciplet, D.; Adams, K.M.; Weikmans, R.; Timmons Roberts, J. The transformative capability of transparency in global environmental governance. *Glob. Environ. Politics* **2018**, *18*, 130–150. [CrossRef]
43. Puri, S.; Appelgren, B.; Arnold, G.; Aureli, A.; Burchi, S.; Burke, J.; Margat, J.; Pallas, P. *Internationally Shared (Transboundary) Aquifer Resources Management: Their Significance and Sustainable Management*; Non Serial Document, SC-2001/WS/40; Puri, S., Ed.; UNESCO: Paris, France, 2001; pp. 1–71.
44. Puri, S.; Aureli, A. Transboundary aquifers: A global program to assess, evaluate, and develop policy. *Groundwater* **2005**, *43*, 661–668. [CrossRef]
45. Movilla Pateiro, L. Ad hoc legal mechanisms governing transboundary aquifers: Current status and future prospects. *Water Int.* **2016**, *41*, 851–865. [CrossRef]
46. Hussein, H. The Guarani Aquifer System, highly present but not high profile: A hydropolitical analysis of transboundary groundwater governance. *Environ. Sci. Policy* **2018**, *83*, 54–62. [CrossRef]
47. Sindico, F.; Hirata, R.; Manganelli, A. The Guarani Aquifer System: From a beacon of hope to a question mark in the governance of transboundary aquifers. *J. Hydrol. Reg. Stud.* **2018**, *20*, 49–59. [CrossRef]

48. Hussein, H.; Menga, F.; Greco, F. Monitoring transboundary water cooperation in SDG 6.5. 2: How a critical hydropolitics approach can spot inequitable outcomes. *Sustainability* **2018**, *10*, 3640. [CrossRef]
49. Ferragina, E.; Greco, F. The Disi project: An internal/external analysis. *Water Int.* **2008**, *33*, 451–463. [CrossRef]
50. Van den Hove, S. A rationale for science-policy interfaces. *Futures* **2007**, *39*, 807–826. [CrossRef]
51. Ostrom, E. *Governing the Commons: The Evolution of Institutions for Collective Action*; Cambridge University Press: Cambridge, UK, 1990; ISBN 0-521-40599-8.
52. Margat, J.; van der Gun, J. *Groundwater around the World: A Geographic Synopsis*; CRC Press: Boca Raton, FL, USA, 2013; ISBN 9780367576509.
53. Linton, J.; Brooks, D.B. Governance of transboundary aquifers: New challenges and new opportunities. *Water Int.* **2011**, *36*, 606–618. [CrossRef]
54. Foster, S.; Ait-Kadi, M. Integrated Water Resources Management (IWRM): How does groundwater fit in? *Hydrogeol. J.* **2012**, *20*, 415–418. [CrossRef]
55. Giordano, M.; Shah, T. From IWRM back to integrated water resources management. *Int. J. Water Resour. Dev.* **2014**, *30*, 364–376. [CrossRef]
56. Varady, R.G.; van Weert, F.; Megdal, S.B.; Gerlak, A.K.; Abdalla Iskandar, C.; House-Peters, L. GEF-FAO Groundwater Governance Project: A Global Framework for Country Action. *Thematic Paper No. 5: Groundwater Policy and Governance*; With Major Editing by Dellinger McGovern, E. 2013. Available online: https://www.researchgate.net/publication/281824294_Groundwater_Governance_A_Global_Framework_for_Country_Action (accessed on 4 August 2021).
57. Cash, D.; Clark, W.C.; Alcock, F.; Dickson, N.M.; Eckley, N.; Jäger, J. Salience, credibility, legitimacy and boundaries: Linking research, assessment and decision making. In *KSG Working Papers Series RWP02-046*; Harvard University: Cambridge, MA, USA, 2002; p. 25. [CrossRef]
58. Lemos, M.C.; Kirchhoff, C.J.; Ramprasad, V. Narrowing the climate information usability gap. *Nat. Clim. Chang.* **2012**, *2*, 789–794. [CrossRef]
59. Reed, M.S.; Stringer, L.C.; Fazey, I.; Evely, A.C.; Kruijsen, J.H.J. Five principles for the practice of knowledge exchange in environmental management. *J. Environ. Manag.* **2014**, *146*, 337–345. [CrossRef]
60. Kirchhoff, C.J.; Lemos, M.C.; Dessai, S. Actionable knowledge for environmental decision making: Broadening the usability of climate science. *Ann. Rev. Environ. Resour.* **2013**, *38*, 393–414. [CrossRef]
61. Milman, A.; Ray, I. Interpreting the unknown: Uncertainty and the management of transboundary groundwater. *Water Int.* **2011**, *36*, 631–645. [CrossRef]
62. Elshall, A.S.; Tsai, F.T.-C. Constructive epistemic modeling of groundwater flow with geological structure and boundary condition uncertainty under Bayesian paradigm. *J. Hydrol.* **2014**, *517*, 105–119. [CrossRef]
63. Dilling, L.; Lemos, M.C. Creating usable science: Opportunities and constraints for climate knowledge use and their Implications for science policy. *Glob. Environ. Chang.* **2011**, *21*, 680–689. [CrossRef]
64. Sarkki, S.; Tinch, R.; Niemelä, J.; Heink, U.; Waylen, K.; Timaeus, J.; Young, J.; Watt, A.; Neßhöver, C.; van den Hove, S. Adding 'iterativity' to the credibility, relevance, legitimacy: A novel scheme to highlight dynamic aspects of science-policy interfaces. *Environ. Sci. Policy* **2015**, *54*, 505–512. [CrossRef]
65. Sugg, Z.P.; Varady, R.G.; Gerlak, A.K.; de Grenade, R. Transboundary groundwater governance in the Guarani Aquifer System: Reflections from a survey of global and regional experts. *Water Int.* **2015**, *40*, 377–400. [CrossRef]
66. Voss, C.I.; Soliman, S.M. The transboundary non-renewable Nubian Aquifer System of Chad, Egypt, Libya, and Sudan: Classical groundwater questions and parsimonious hydrogeologic analysis and modeling. *Hydrogeol. J.* **2014**, *22*, 441–468. [CrossRef]
67. Kahane, Y. The Turonian-Cenomanian Aquifer. In *Management of Shared Groundwater Resources: The Israeli-Palestinian Case with an International Perspective*; Feitelson, E., Haddad, M., Eds.; Springer: New York, NY, USA, 2001; pp. 83–106. ISBN 978-0-7923-7254-7.
68. Feitelson, E.; Haddad, M. *Identification of Joint Management Structures for Shared Aquifers: A Cooperative Palestinian-Israeli Effort*; World Bank Publications: Washington, DC, USA, 1998; Volume 415, ISBN 0-8213-4307-6.
69. Conti, K.I. *Factors Enabling Transboundary Aquifer Cooperation: A Global Analysis*; UN-IGRAC: Delft, The Netherlands, 2014; p. 108.
70. Grossman, M. Cooperation on Africa's international waterbodies: Information needs and the role of information-sharing. In *Transboundary Water Management in Africa: Challenges for Development Cooperation*; Scheumann, W., Neubert, S., Eds.; Deutsches Institut für Entwicklungspolitick: Bonn, Germany, 2006; pp. 173–236. ISBN 3-88985-326-9.
71. Milman, A.; Gerlak, A.K.; Albrecht, T.R.; Colosimo, M.; Conca, K.; Kittikhoun, A.; Kovács, P.; Moy, R.; Schmeier, S.; Wentling, K.; et al. Addressing knowledge gaps for transboundary environmental governance. *Glob. Environ. Chang.* **2020**, *64*, 102162. [CrossRef]
72. Kendy, E.; Flessa, K.W.; Schlatter, K.J.; de la Parra, C.A.; Hinojosa Huerta, O.M.; Carrillo-Guerrero, Y.K.; Guillen, E. Leveraging environmental flows to reform water management policy: Lessons learned from the 2014 Colorado River Delta pulse flow. *Ecol. Eng.* **2017**, *106*, 683–694. [CrossRef]
73. Prichard, A.H.; Scott, C.A. Interbasin water transfers at the US–Mexico border city of Nogales, Sonora: Implications for aquifers and water security. *Int. J. Water Resour. Dev.* **2014**, *30*, 135–151. [CrossRef]
74. Varady, R.G.; Moote, M.A.; Merideth, R. Water allocation options for the Upper San Pedro basin: Assessing the social and institutional landscape. *Nat. Resour. J.* **2000**, *40*, 223–235.

75. Milman, A.; Scott, C.A. Beneath the surface: Intranational institutions and management of the United States—Mexico transboundary Santa Cruz Aquifer. *Environ. Plan. C Politics Space* **2010**, *28*, 528–551. [CrossRef]
76. McNabb, D.E. Water resource management comes of age. In *Water Resource Management*; Palgrave MacMillan: Cham, Switzerland, 2017; pp. 163–185. ISBN 978-3-319-54816-6.
77. Abrams, R.H. Legal convergence of East and West in contemporary American water law. *Environ. Law* **2012**, *42*, 65–92.
78. Megdal, S.B. Arizona groundwater management. *Water Rep.* **2012**, *104*, 9–15.
79. Arizona Town Hall. Keeping Arizona's Water Glass Gull. 2015. Available online: https://aztownhall.org/resources/Documents/107%20Keeping%20Arizona\T1\textquoterights%20Water%20Glass%20Full%20FINAL%20Report%20web.pdf (accessed on 21 June 2021).
80. Mumme, S.P. Innovation and reform in transboundary resource management: A critical look at the International Boundary and Water Commission, United States and Mexico. *Nat. Resour. J.* **1993**, *33*, 93–120.
81. Mumme, S.P. Minute 242 and beyond: Challenges and opportunities for managing transboundary groundwater on the Mexico-U.S. border. *Nat. Resour. J.* **2000**, *40*, 341–378.
82. Wilder, M.O.; Varady, R.G.; Gerlak, A.K.; Mumme, S.P.; Flessa, K.W.; Zuniga-Teran, A.A.; Scott, C.A.; Pineda Pablos, N.; Megdal, S.B. Hydrodiplomacy and adaptive governance at the U.S.-Mexico border: 75 years of tradition and innovation in transboundary water management. *Environ. Sci. Policy* **2020**, *112*, 189–202. [CrossRef]
83. IBWC. Minute 242: Permanent and Definitive Solution to the International Problem of Salinity of the Colorado River. 1973. Available online: https://www.usbr.gov/lc/region/pao/pdfiles/min242.pdf (accessed on 6 August 2021).
84. Hatch Kuri, G. A joint management of transboundary aquifers: From asymmetries to environmental protection. *Front. Norte* **2018**, *30*, 129–154. [CrossRef]
85. Brownell, H.; Eaton, S.D. The Colorado River salinity problem with Mexico. *Am. J. Int. Law* **1975**, *69*, 255–271. [CrossRef]
86. Mumme, S.P.; Collins, K. The La Paz Agreement 30 years on. *J. Environ. Dev.* **2014**, *23*, 303–330. [CrossRef]
87. Alley, W.M. Five-Year Interim Report of the United States—Mexico Transboundary Aquifer Assessment Program: 2007–2012. Available online: https://pubs.usgs.gov/of/2013/1059/pdf/ofr2013-1059.pdf (accessed on 23 June 2021).
88. IBWC. Joint Report of the Principal Engineers Regarding the Joint Cooperative Process United States-Mexico for the Transboundary Aquifer Assessment Program. 2009. Available online: https://www.ibwc.gov/Files/Minutes/Joint_Report_TAAP_081909.pdf (accessed on 6 August 2021).
89. Rojo, H.A.; Bredehoeft, J.; Lacewell, R.; Price, J.; Stromberg, J.; Thomas, G.A. Sustaining and Enhancing Riparian Migratory Bird Habitat on the Upper San Pedro River: Public Review Draft. 1998. Available online: https://searchworks.stanford.edu/view/4782615 (accessed on 6 August 2021).
90. Eckstein, G.E. Managing buried treasure across frontiers: The International Law of Transboundary Aquifers. *Water Int.* **2011**, *36*, 573–583. [CrossRef]
91. Memorandum of Understanding between City of Juárez, Mexico Utilities and the El Paso Water Utilities Public Services Board and the City of El Paso, Texas (MOU). 1999. Available online: http://www.internationalwaterlaw.org/documents/regionaldocs/Local-GW-Agreements/El-Paso-JuarezMoU.pdf (accessed on 21 June 2021).
92. Pool, D.R.; Dickinson, J.E. *Ground-Water Flow Model of the Sierra Vista Subwatershed and Sonoran Portions of the Upper San Pedro Basin, Southeastern Arizona, United States, and Northern Sonora, Mexico*; No. 2006-5228; U.S. Geological Survey: Reston, VA, USA, 2007.
93. Wallace, C.S.A.; Villarreal, M.L.; Norman, L.M. *Development of a High-Resolution Binational Vegetation Map of the Santa Cruz River Riparian Corridor and Surrounding Watershed, Southern Arizona and Northern Sonora, Mexico*; No. 2011-1143; U.S. Geological Survey: Reston, VA, USA, 2011.
94. Villareal, M.L.; Norman, L.M.; Wallace, C.S.A.; Van Riper, C. *A Multitemporal (1979–2009) Land-Use/Land-Cover Dataset of the Binational Santa Cruz Watershed*; U.S. Geological Survey: Reston, VA, USA, 2011.
95. Callegary, J.B.; Heilman, J.A.; Tapia-Villaseñor, E.M.; Knight, J.E. *San Pedro River Aquifer Data Release—Transboundary Aquifer Assessment Program (TAAP)*; U.S. Geological Survey Data Release: Reston, VA, USA, 2018. [CrossRef]
96. Dirks, K.T. Trust in leadership and team performance: Evidence from NCAA basketball. *J. Appl Psychol.* **2000**, *85*, 1004. [CrossRef]
97. Megdal, S.B. The cooperative framework for the Transboundary Aquifer Assessment Program: A model for collaborative transborder studies. *Water Resour. Impact* **2018**, *20*, 10–11.
98. Mechlem, K. Past, present and future of the International Law of Transboundary Aquifers. *Int. Community Law Rev.* **2011**, *13*, 209–222. [CrossRef]
99. Tapia-Villaseñor, E.M.; Megdal, S.B. The U.S.-Mexico Transboundary Aquifer Assessment Program as a model for transborder groundwater collaboration. *Water* **2021**, *13*, 530. [CrossRef]
100. Coppola, M.C. Blaze in Nogales, Sonora Battled from Both Sides of the Border. *Nogales International (Nogales, AZ)*. 10 May 2012. Available online: https://www.nogalesinternational.com/news/blaze-in-nogales-sonora-battled-from-both-sides-of-the/article_3267761e-9b0d-11e1-85e7-001a4bcf887a.html (accessed on 6 August 2021).

water

MDPI

Article

The U.S.-Mexico Transboundary Aquifer Assessment Program as a Model for Transborder Groundwater Collaboration

Elia M. Tapia-Villaseñor [1,*] and Sharon B. Megdal [2]

[1] Departamento de Geología, Universidad de Sonora, Hermosillo 83000, Mexico
[2] Water Resources Research Center, The University of Arizona, Tucson, AZ 85719, USA; smegdal@arizona.edu
* Correspondence: elia.tapia@unison.mx

Abstract: The assessment of transboundary aquifers is essential for the development of groundwater management strategies and the sustainable use of groundwater resources. The Transboundary Aquifer Assessment Program (TAAP) is a joint effort by the United States and Mexico to evaluate shared aquifers. This study examines the TAAP Cooperative Framework as a guide for further transboundary groundwater collaboration. We compared lessons learned from six transboundary aquifers that currently have mechanisms for groundwater collaboration to identify common elements of collaboration. Though the TAAP Cooperative Framework governs an assessment-only program, the elements of collaboration included are consistent with the principles of other institutional agreements around the world. Importantly, all the analyzed agreements included a knowledge-improvement phase, which is the main objective of the TAAP Cooperative Framework. The present study finds evidence of successful outcomes within the TAAP Cooperative Framework consistent with available transboundary groundwater management agreements, demonstrating that this approach is suited to serve as a model for those wishing to engage in transborder aquifer assessments. Furthermore, the TAAP elements of collaboration can help to establish the meaningful and robust binational cooperation necessary for the development of U.S.-Mexico groundwater management agreements at the aquifer level.

Keywords: transboundary aquifers; United States; Mexico; assessment; agreements; groundwater management

Citation: Tapia-Villaseñor, E.M.; Megdal, S.B. The U.S.-Mexico Transboundary Aquifer Assessment Program as a Model for Transborder Groundwater Collaboration. *Water* **2021**, *13*, 530. https://doi.org/10.3390/w13040530

Academic Editors: Frédéric Huneau

Received: 30 December 2020
Accepted: 10 February 2021
Published: 18 February 2021

Publisher's Note: MDPI stays neutral with regard to jurisdictional claims in published maps and institutional affiliations.

1. Introduction

Groundwater is an important source of fresh water for populations and the environment. Fresh water represents only 2.8% of the total water resources in the world, with 70% of fresh water composed of polar ice layers and continental ice, 1% from surface watercourses, and 29% from groundwater [1]. Transboundary rivers, lakes, and aquifers are home to over 70% of the world's population and supply water for around 60% of global food production [2]. Approximately 600 transboundary aquifers have been identified around the world [3]. However, only six of them have formal binational or multinational mechanisms of cooperation: (1) the Guaraní Aquifer System in Brazil, Argentina, Paraguay, and Uruguay; (2) the Franco-Swiss Genevese Aquifer System in France and Switzerland; (3) the Northwestern Sahara Aquifer System in Algeria, Libya, and Tunisia; (4) the Iullemeden Aquifer System in Mali, Niger, and Nigeria; (5) the Nubian Sandstone Aquifer System shared by East Libya, Egypt, Northeast Chad, and North Sudan, and; (6) the Al-Saq/Al-Disi Aquifer System in Jordan and Saudi Arabia (Figure 1).

Although geographically widespread, these aquifers represent only 1% of identified transboundary aquifers, a proportion that is quite different from the proportion of transboundary river basins with international basin agreements. While there are 310 transboundary river basins around the world, a total of 688 transboundary basin agreements have been signed between 1820 and 2007 [4–6]. These agreements apply to 133 river basins, representing 36% of the identified transboundary basins [4]. The reasons for such a disparity between the number of basin agreements and the number of groundwater agreements include the "invisible" nature of groundwater [7,8], limited and dissimilar groundwater data [9], and the lack of institutional capacity for groundwater governance [10].

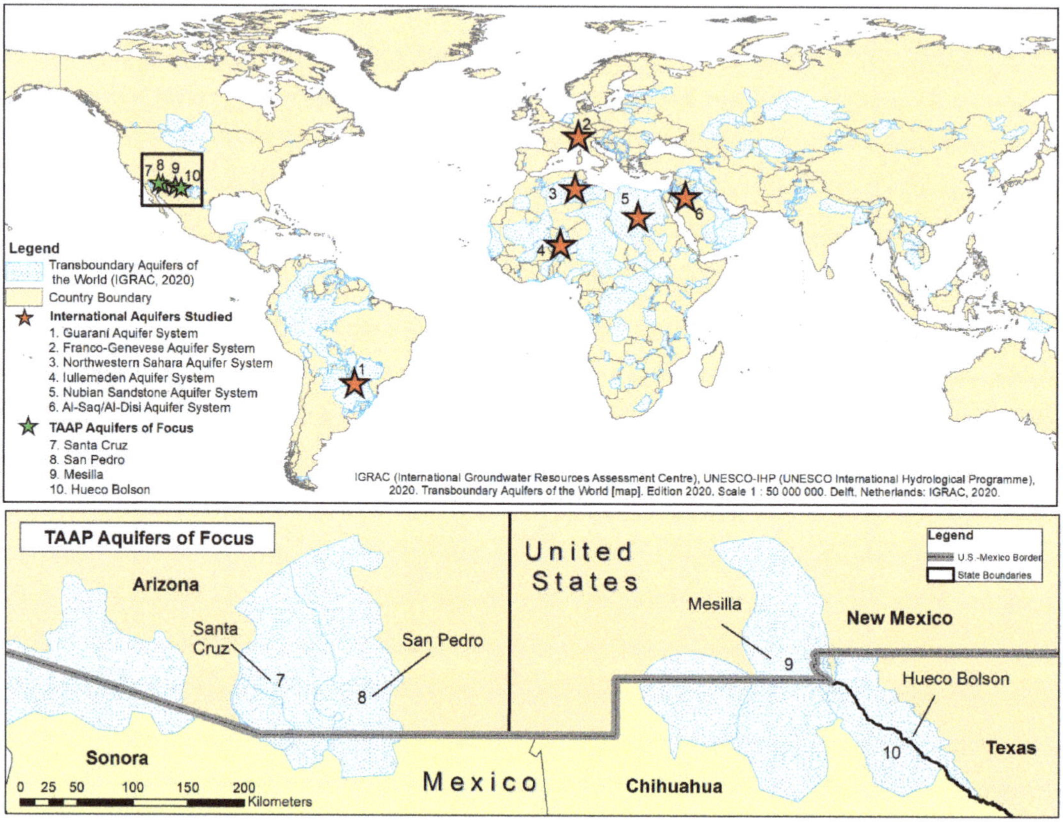

Figure 1. International Aquifers Studied and Transboundary Aquifer Assessment Program (TAAP) Aquifers of Focus, based on Transboundary Aquifers of the World [3].

Groundwater collaboration between the United States and Mexico is similar to other transboundary settings around the world. Efforts by the two countries to understand and manage groundwater resources have been scarce and sporadic [11]. The two countries have a surface water agreement, the 1944 Water Treaty Regarding the Utilization of Waters of the Colorado and Tijuana Rivers and of the Rio Grande (1944 Treaty); however, groundwater was left unmentioned. Only Minute 242 was approved in 1973 by the International Boundary and Water Commission (IBWC), one of many interpretations of the 1944 Treaty includes a provision that is relevant to groundwater. The IBWC is the international body that oversees the application of U.S.-Mexico treaties related to boundary demarcation, national ownership of waters, sanitation, water quality, and flood control in the border region [12]. Challenges in the management of groundwater resources in the U.S.-Mexico border region include rapid urbanization and industrialization, agricultural intensification, contamination of surface and groundwater resources, increase in surface and groundwater demands, and climate uncertainties [13–15]. These challenges indicate the need for binational transboundary collaboration to secure water for populations and the environment.

Such a collaboration could take the form of a binational agreement for the management of groundwater resources. However, scholars have recognized that the assessment of shared aquifer systems is a necessary antecedent to the development of any groundwater management agreement [8,16–19]. For example, Kirstin I. Conti [18] indicated that scientific research is an enabling factor for groundwater cooperation, along with existing legal mechanisms, regional institutions, high institutional capacity, funding mechanisms, strong political will, previous water cooperation, and third-party involvement.

The Joint Report of the Principal Engineers Regarding the Joint Cooperative Process United States-Mexico for the Transboundary Aquifer Assessment Program (TAAP Cooperative Framework) [20], guides the joint effort between the United States and Mexico to improve the knowledge base of transboundary aquifers. The program began in 2006 with the Transboundary Aquifer Assessment Act (U.S. Public Law 109–448, TAA-Act). The TAA-Act authorized the United States Geological Survey (USGS) and the Water Resources Research Institutes (WRRIs) of Arizona, New Mexico, and Texas to work with Mexican counterparts on the development of transboundary aquifer assessments. The TAA-Act authorized U.S. involvement in binational studies of the Santa Cruz and San Pedro Aquifers, shared by the state of Arizona in the United States and the state of Sonora in Mexico, and the Mesilla and Hueco Bolson aquifers, shared by the states of Texas and New Mexico in the United States and the state of Chihuahua in Mexico (Figure 1). These priority aquifers were selected based on their proximity to highly populated areas, increasing groundwater demands, and water quality issues [21]. The binational TAAP was formally initiated in 2009 upon the signing of the TAAP Cooperative Framework by the principal engineers of the U.S. and Mexican sections of the IBWC. The two countries agreed upon the TAAP aquifers of focus consistent with the TAA-Act priority aquifers (Figure 1). According to the TAAP Cooperative Framework, either of the two countries can propose an aquifer of focus, but both countries must agree to develop a joint assessment.

The TAA-Act and the TAAP Cooperative Framework offer a foundation for collaboration to study shared groundwater resources through an effective partnership among federal agencies, academic institutions, and federally established water resources research institutes [21,22]. The TAAP can also be considered a climate and water adaptation initiative for the western U.S.-Mexico border [13], a transboundary regional initiative that has the potential to build adaptive capacity [15], an activity that can support decision-making processes related to groundwater management in each country [23], and a precedent for a binational partnership that can promote and implement a new binational aquifer assessment [9]. However, the relevance of the TAAP Cooperative Framework as a model mechanism for groundwater collaboration has not been fully addressed in the literature.

The TAAP Cooperative Framework is limited to assessment only, with four transboundary aquifers studied to date. The Map of Transboundary Aquifers of the World [3] includes 11 shared transboundary aquifers along the border between the United States and Mexico. Yet, a review of technical studies, reports, and publications on U.S.-Mexico transboundary aquifers suggest that at least 36 transboundary aquifers are shared by the two countries [24]. Clearly, additional study opportunities exist, and the activities undertaken by the TAAP can serve as the basis for assessment that goes beyond the current TAAP aquifers of focus and that can even guide future dialogue regarding groundwater governance and management [8]. The primary objective of the study is to determine whether the elements of the TAAP Cooperative Framework can serve as a model for others wishing to engage in transboundary aquifer assessment. Expert interviews and lessons learned from evaluating six existing international groundwater agreements helped to determine whether the objectives, framework/process, funding, principles, and communication arrangements of the TAAP Cooperative Framework can guide further groundwater cooperation.

2. Materials and Methods

The present work provides an assessment of the TAAP Cooperative Framework as a model for transboundary groundwater collaboration. To achieve this, we compared the elements of collaboration in the TAAP Cooperative Framework with the components of six transboundary groundwater collaboration agreements. The TAAP Cooperative Framework elements of collaboration were taken from the Joint Report of the Principal Engineers Regarding Joint Cooperative Process United States-Mexico for the Transboundary Aquifer Assessment Program (Cooperative Framework) [20]. These elements are presented in Table 1 and were used as a basis for comparison. Selected institutional governance agreements include the existing collaboration mechanisms for the following: (1) the Guaraní Aquifer System (GAS), (2) the Franco-Swiss Genevese Aquifer System, (3) the Northwestern Sahara Aquifer System, (4) the Iullemeden Aquifer System, (5) the Nubian Sandstone Aquifer System (NSAS), and (6) the Al-Saq/Al-Disi Aquifer System. These transboundary aquifers were selected because of their formal mechanisms of groundwater cooperation [17,18,25–27].

Table 1. TAAP Cooperative Framework elements of collaboration [20].

Objectives	Facilitate data exchange Ensure the concurrence for binational aquifer assessment activities Facilitate agreement on the aquifers, which will be evaluated jointly Establish and coordinate binational technical advisory committees Establish an official repository for binational project reports
Framework (Process)	Either of the two countries can propose an aquifer to study The International Boundary and Water Commission (IBWC) will coordinate with agencies from both countries to jointly define the scope of the assessment Binational technical groups will be established and coordinated by the IBWC The IBWC will facilitate concurrence of joint work plans Whoever carries out the joint studies will update the binational technical groups with the project progress The final reports that proceed from the joint studies will be published in English and Spanish and will be made available for publication once they have been approved within the IBWC framework
Funding	Each country will be responsible for any costs on projects conducted in its territory Either country may contribute to the costs of work done in the other country Contributions will be distributed according to the process agreed on through the IBWC All projects and measures considered are subject to the availability of funds
Principles	Activities should be beneficial to both countries Activities should be agreed on within the framework of the IBWC Activities should respect the legal framework and jurisdictional requirements of each country No provision set forth in this agreement will limit what either country can do independently in its own territory No part of this agreement may contravene what has been stipulated in the boundary and water treaties The information generated from these projects is solely for the purpose of expanding knowledge
Communication	The IBWC will be an official repository of records The final joint binational reports will be available to the public in each country and will be posted on the website of each section of the IBWC Information obtained should be considered as official data and should be shared without any restrictions Credit will be given to those who provide information

Stakeholder interviews implemented during 2019 and 2020 served to identify whether TAAP lessons can be generalized to other aquifers along the U.S.-Mexico border and elsewhere. The selection of participants was based on purposive sampling. This nonrandom technique does not need a set number of participants and interviewees are selected according to the qualities or knowledge they possess [28]. Interviewees consist of two IBWC experts (interview 1 and 2), two experts in political sciences (interview 3 and 4), and two researchers/scientists (interview 5 and 6). Selected interviewees were familiar with transboundary aquifer assessment and management and with the principles of the TAAP Cooperative Framework. Interview questions included:

- According to your experience, what factors promote the successful groundwater collaboration between nations that share one or various aquifers?
- Do you think we can generalize the TAAP principles of collaboration to other aquifers within the U.S.-Mexico border?
- Do you think the TAAP Cooperative Framework can serve as a model for the assessment of other transboundary aquifers?
- Do you think the TAAP Cooperative Framework can serve as a basis for the development of future groundwater management agreements in the borderlands of the United States and Mexico?

3. Results

This section describes the history of transboundary groundwater collaboration around the world and the mechanisms of collaboration included in the analyzed groundwater agreements. We also present a comparison of the elements of the TAAP Cooperative Framework and the components of the six analyzed aquifer agreements. Expert interviews regarding the applicability of the TAAP Cooperative Framework to others contemplating transborder collaboration are also reported in this section.

3.1. *Transboundary Groundwater Resources and International Law*

Some of the international guidelines related to transboundary groundwater resources include the 1966 Helsinki Rules, the 1986 Seoul Rules, the 1997 UN Convention on the Law of the Non-Navigational Uses of Transboundary Watercourses (UN Watercourses Convention), the 1999 Convention on the Protection and Use of Transboundary Watercourses and International Lakes, the 2004 Berlin Rules, and the 2008 Draft Articles on the Law of Transboundary Aquifers [29–31]. These guidelines serve as a reference for groundwater management. However, only some of them recognize the connection between surface water and groundwater. For example, the UN Watercourses Convention addresses surface water and groundwater but fails to recognize confined aquifers [30]. The Convention on the Protection and Use of Transboundary Watercourses and International Lakes also fails to recognize confined transboundary aquifers even though it documents the importance of groundwater in the management of drainage basins [29].

The 2008 Draft Articles on the Law of Transboundary Aquifers (UN Draft Articles) do recognize confined aquifers [1,32,33]. Among many provisions, the UN Draft Articles include principles related to the sovereignty of the countries sharing an aquifer (Article 3), provisions for equitable and reasonable utilization of groundwater resources (Article 4), the obligation not to cause significant harm (Article 6), a general obligation to cooperate (Article 7), requirements for the regular exchange of data and information (Article 8), stipulations for the protection and preservation of ecosystems (Article 10), and guidelines for monitoring (Article 13). The United Nations General Assembly (UNGA), however, has not ratified the UN Draft Articles, though the item has been on its agenda several times in it will be again in 2022 [8]. Partly in response to this development in international law, other cases of groundwater collaboration mechanisms have been signed between countries [8]. For example, the Guaraní Aquifer System Agreement and the Bamako Declaration for the Iullemeden Aquifer System both refer to the UN Draft Articles [8,32]. Below, we present a summary of the mechanisms analyzed in this study.

3.2. *Transboundary Groundwater Collaboration around the World*

Diverse cultures, countries, and states connect to groundwater in a hydropolitical matrix that comprises the policies, social exchanges, discussions, and agreements between different nations [34]. Water allocation is a key component of water governance, and in transboundary settings, this process involves a variety of users competing in an unavoidable conflictual process [35]. The principle of equitable and reasonable utilization of water resources should guide groundwater allocation between different countries, yet there is no universal theory of justice to satisfy every water user [36].

According to the United Nations Sustainable Development Goals (SDG 6.5.2), "specific arrangements or agreements between co-riparian countries are a precondition to ensure long-term sustainable cooperation" [37,38]. In this section, we detail the six transboundary aquifers that have agreements or other arrangements established for groundwater collaboration. Table 2 presents a summary of the analyzed transboundary groundwater agreements, the countries involved, dates, and the purpose of each agreement, followed by an explanation of the main characteristics of each aquifer agreement. Focusing on the elements and history of each collaboration, we subsequently compare the agreements with the U.S.-Mexico TAAP Cooperative Framework.

Table 2. Transboundary Groundwater Agreements around the World.

Aquifer System	Agreement	Countries Involved	Date(s)	Agreement Characteristics
Guaraní Aquifer System (GAS)	Guaraní Aquifer Agreement	Argentina Brazil Paraguay Uruguay	Signed in 2010 Ratified in 2018	Promotes the sustainable development of the aquifer system Solves issues arising between countries Aligned to the UN Draft Articles
Franco-Swiss Genevese Aquifer System	Convention on the Protection, Utilization, Recharge, and Monitoring of the Franco-Swiss Genevese Aquifer	France Switzerland	1978–2008 (New convention established in 2008)	Focused on groundwater quality, quantity, and artificial recharge The only treaty to date that allocated volumes of water
Northwestern Sahara Aquifer System	The Permanent Consultation Mechanism for the Northwestern Sahara Aquifer System	Algeria Libya Tunisia	2008	Developed a hydrogeologic database and model Maintains an observation network Analyzes socioeconomic activities Develops joint studies Formulates proposals for optimization and consultation mechanisms
Iullemeden Aquifer System	Bamako Declaration	Mali Niger Nigeria	2009	Serves as a consultative mechanism Improves knowledge and strenghtens regional cooperation
Al-Saq/Al-Disi Aquifer System	Agreement between the Government of the Hashemite Kingdom of Jordan and the Government of the Kingdom of Saudi Arabia for the Management and Utilization of the Ground Waters in the Al-Saq/Al-Disi Layer	Jordan Saudi Arabia	2015	Restricts groundwater extractions in protected areas Governs the digging of observational wells Controls pollution
Nubian Sandstone Aquifer System	Programme for the Development of a Regional Strategy for the Utilization of the Nubian Sandstone Aquifer System and the Terms of Reference for the Monitoring and Exchange of Groundwater Information of the Nubian Sandstone Aquifer System	East Libya Egypt Northeast Chad North Sudan	2000	Focuses on data exchange and monitoring efforts

3.2.1. Guaraní Aquifer System (GAS)

Located across four South American countries, the GAS is one of the largest freshwater reservoirs in the world [26,39]. The GAS covers an area of 1,087,879 square kilometers (km^2), with the largest portion situated in Brazil, followed by Argentina, Paraguay, and Uruguay [26,40]. The breakdown of the reservoir's water resources usage is as follows: municipal water supply (66%), industries (16%), thermal tourism (13%), and irrigation (5%) [40]. The four countries sharing the aquifer are known for their collaboration regarding the La Plata River Basin [8] and have benefited from continuous research and development projects, such as the Environmental Protection and Sustainable Development of the Guaraní Aquifer System Project supported by the Organization of American States and the Common Market of the South [39,41].

The Guaraní Aquifer Agreement was signed in August 2010, but it was not ratified by all four countries until 2018. This document points to the principles described by the UN Draft Articles to promote the sustainable development of the aquifer system and to solve some of the issues that might arise in the aquifer countries. For instance, Articles 2 and 3 of the Guaraní Aquifer Agreement state that each of the parties has the sovereign right to promote the management, utilization, and monitoring of their portion of the aquifer system as long as they follow the principle of reasonable use. Data exchange and knowledge improvement are essential, as expressed in Articles 8 and 12. Finally, a commission oversees compliance by all parties with the principles of agreement [42]. Factors enabling transboundary collaboration among the countries that share the GAS include the existing regional institutions, funding mechanisms, high institutional capacity, previous water cooperation, scientific research, strong political will, and third-party involvement [18].

3.2.2. Franco-Swiss Genevese Aquifer System

The Franco-Swiss Genevese Aquifer System is shared by France and Switzerland and has an approximate areal extent of 19 km^2 [27]. Ten wells on the Swiss side of the aquifer and four wells on the French side supply water to the Swiss Canton of Geneva and the neighboring French Territory (Haute-Savoie). The Convention on the Protection, Utilization, Recharge, and Monitoring of the Franco-Swiss Genevese Aquifer was established in 1978 after a dramatic decrease in groundwater levels associated with groundwater pumping [18,43].

The 1978 convention focused on groundwater quality, quantity, and artificial recharge, and it is the only treaty for transboundary aquifers that allocates specific volumes of water to the involved parties [17,18,44]. Despite the lack of provisions related to sovereignty rights, each of the parties has the right to make decisions around groundwater pumping, equipment, and abstraction margins [8,38]. Recharge from the Arve River is treated and channeled into the aquifer, helping to balance a seven million cubic meters (Mm^3) per year overdraft [43,45]. A joint commission oversees the preparation of groundwater management plans, the monitoring of groundwater, the efforts to gain approval for new infrastructures, and the verification of construction and operation costs of artificial recharge facilities [17]. The commission consists of six members, at least four of whom are experts in water-related issues [17].

When the 1978 convention expired, a new convention came into effect on 1 January, 2008 [46]. This new agreement includes the French communities of Annemasse, the rural districts of Genevois, and the municipality of Viry. The fact that the 1978 convention did not include participation from the federal government of either country adds a local dimension to the arrangement that was essential for the success of the agreement [25]. Technical and scientific studies were also crucial for resolving the overexploitation problems, and they served as the basis for collaboration efforts [43].

3.2.3. Northwestern Sahara Aquifer System

Shared by Algeria, Libya, and Tunisia, the Northwestern Sahara Aquifer System has an areal extent of 1,019,000 km^2. The Permanent Consultation Mechanism for the Northwestern Sahara Aquifer System was signed in July 2008 by the three countries' representatives [26]. The consultation mechanism is composed of a steering committee and a scientific committee [47]. The many goals of the agreement include (1) the development of a hydrogeologic database and groundwater flow model; (2) the setup of an observation network to process, analyze, and validate data; (3) the analysis of the socioeconomic activities of the region; (4) the coordination for the development of joint studies, and (5) the formulation of proposals for optimization and consultation mechanisms [48].

Collaboration among the countries sharing the aquifer has lasted at least 45 years, with activities designed to improve scientific knowledge about the aquifer. Features of collaboration for the Northwestern Sahara Aquifer System include political will, funding, and available institutions, such as the Observatoire du Sahara et du Sahel (Sahara and Sahel Observatory—OSS) [18]. Thus far, collaborative efforts have focused on scientific studies of the aquifer, and transboundary groundwater management has not yet occurred.

3.2.4. Iullemeden Aquifer System

The Sahel region aquifers in West Africa include the Iullemeden-Taoudeni/Tanezrouft Aquifer System. Shared by Mali, Niger, and Nigeria, the Iullemeden Aquifer System has an areal extent of 525,000 km^2. The cooperative and financing mechanisms of the region, along with their regional institutions, have been shaped by 20 years of collaboration that has led to the development of two agreements: (1) the Protocol on Cooperation of the Utilization of the Niger River, signed by Mali and Niger in 1988, and (2) the Joint Commission for Cooperation on Equitable Sharing for Development, Conservation, and Utilization of the Common Water Resources, signed by Niger and Nigeria in 1990 [18]. Later, the Bamako Declaration for the Iullemeden Aquifer, which is a Memorandum of Understanding (MOU) that encourages collaboration between the three countries, was signed by Mali, Niger, and Nigeria in 2009. The MOU recognizes the importance of water resources for alleviating poverty, acknowledges the rights and duties of the countries sharing the aquifer, appreciates the achievements of the improvement of the scientific knowledge associated with the aquifer, and highlights the importance of cooperative management of the Iullemeden Aquifer System in improving the management of shared groundwater resources [49].

Additionally, the countries commit to adopting the principles of the equitable and reasonable use of shared groundwater resources, exchanging information, giving prior notification of planned work, and adopting environmental protection regulations [32]. This collaborative effort evolved into the development of another MOU in 2014, this time, among the countries of Algeria, Benin, Burkina Faso, Mali, Mauritania, Niger, and Nigeria, for the establishment of a Consultative Mechanism for the Iullemeden and Taoudeni/Tanezrouft Aquifer Systems (ITAS). This MOU, however, is not yet in effect, pending the signatures of three of the parties [50]. The consultative mechanisms of the ITAS demonstrate the readiness to develop management strategies, though the agreement has not evolved into additional management actions [1,26].

3.2.5. Nubian Sandstone Aquifer System (NSAS)

The Nubian Sandstone Aquifer System consists of a series of laterally and/or vertically interconnected aquifers that extend across more than 2,000,000 km^2 in East Libya, Egypt, Northeast Chad, and North Sudan [17]. The formal agreements ratified by the countries sharing the aquifer include (1) the Programme for the Development of a Regional Strategy for the Utilization of the Nubian Sandstone Aquifer System and (2) the Terms of Reference for the Monitoring and Exchange of Groundwater Information of the Nubian Sandstone Aquifer System, both signed in October 2000 [47]. The NSAS projects, which are considered initial stages of groundwater collaboration, are widely supported by donors and the scientific community [51].

Though these agreements are relatively recent, the collaboration between Libya and Egypt dates back to 1991, when the Joint Authority of the Nubian Sandstone Aquifer System (JASD-NSAS) was established [47]. The first countries to join were Libya and

Egypt, with Sudan and Chad following in 1996 and 1999, respectively [52]. The information shared under the NSAS agreement includes yearly groundwater extractions, electrical conductivity measurements, chemical analysis, and water-level measurements [48]. In 2012, the Regional Strategic Action Program for the Nubian Sandstone Aquifer System was negotiated through the Action Programme for the Integrated Management of the Shared Nubian Aquifer, which included guidance for future groundwater management agreements [18,53]. The JASD-NSAS has a regional expert group, with offices in each of the countries sharing the aquifer, as well as specific units for public relations, follow-up, finance, technical affairs, information, and administration.

3.2.6. Al-Saq/Al-Disi Aquifer System

The Al-Saq/Al-Disi Aquifer is a reservoir of fossil water shared by Jordan (Al-Saq Aquifer) and Saudi Arabia (Al-Disi Aquifer). Groundwater recharge in the region is minimal, and the two countries seem to be involved in a pumping race that might lead to the inevitable depletion of the groundwater resource [54]. The Agreement between the Government of the Hashemite Kingdom of Jordan and the Government of the Kingdom of Saudi Arabia for the Management and Utilization of the Ground Waters in the Al-Saq/Al-Disi Layer was signed on 4 April, 2015. The Al-Saq/Al-Disi Aquifer area covered by the agreement has an area extent of 308,000 km^2 [55].

The agreement restricts groundwater extractions in protected areas, encourages the drilling of observation wells, and includes pollution control statements. The agreement authorizes the drilling of wells in the management area between Jordan and Saudi Arabia but limits water usage for municipal purposes. A joint Saudi/Jordanian technical committee formed by five members from each country is responsible for supervising the implementation of the terms of the agreement, monitoring groundwater quality and quantity, and exchanging data and information between the involved parties. The agreement calls for members of the joint committee to have one meeting every six months. However, it was reported that as of 2018, the committee has never met [50]. According to the agreement, data exchange with a third party is not allowed unless approved by the two countries. Activities by the joint committee can be completed with the help of experts, technicians, officials, and citizens from the two countries. The agreement will be reviewed every 25 years, and any amendment will be studied by the joint committee and referred to the appropriate authorities. In this case, informal political meetings contributed to the development and signing of the MOU in 2015 [36]. However, the countries have not truly reached a bilateral treaty over the use of their shared groundwater [36].

3.3. Transboundary Groundwater Collaboration between the United States and Mexico

The allocation of shared surface water resources between the United States and Mexico is governed by the 1944 Water Treaty Regarding the Utilization of Waters of the Colorado and Tijuana Rivers and of the Rio Grande (1944 Treaty). The treaty, however, leaves groundwater unmentioned. The IBWC, established in 1889, is the international body that oversees the application of U.S.-Mexico treaties regarding boundary demarcation, water resources, and sanitation in the border region [12]. It received the name of the International Boundary Commission (IBC) before the signing of the 1944 Treaty.

The IBWC is composed of the U.S. and a Mexican Section. The U.S. Section is housed in the U.S. Department of State and has headquarters in El Paso, Texas. The Mexican Section is operated by the Mexican Ministry of Foreign Affairs, with headquarters in Ciudad Juarez, Chihuahua. To implement international treaty provisions, the IBWC requires specific agreements, which have been recorded in the form of Minutes and date back to 1889. A key pillar of the 1944 Treaty is that it allows for interpretations or modifications (Minutes) to adapt to new challenges that emerge between the two countries [56]. These Minutes are considered extensions and applications of the treaty [56]. To date, 324 Minutes act as binding obligations between the United States and Mexico, but only Minute 242 for the "Permanent and Definitive Solution to the International Problem of the Salinity of the Colorado River" specifically includes groundwater management provisions [57]. Resolution 5 of Minute 242 establishes that "pending the conclusion by the Governments of the United States and Mexico of a comprehensive agreement on groundwater in the border areas, each country shall limit pumping of groundwater in its territory within eight kilometers of the Arizona-Sonora boundary near San Luis to 197,358,000 cubic meters annually" [57]. Minute 323, "Extension of Cooperative Measures and Adoption of a Binational Water Scarcity Contingency Plan in the Colorado River Basin," is a relevant example of cooperation for many reasons, including the assessment of desalination impacts [58]. While this Minute does not consider groundwater, it does consider a binational assessment effort within the context of the IBWC and the 1944 Treaty.

Aside from the 1944 Treaty framework, the Bellagio Draft Treaty represents another fine example of the progress being made toward the understanding and management of the U.S.-Mexico transboundary aquifers [16]. The treaty suggests a structure by which the United States and Mexico can work cooperatively, describing the development of a bilateral institution that will allow the United States and Mexico to jointly study and manage their shared groundwater resources [59]. Moreover, it emphasizes the importance of knowledge improvement for the development of joint agreements and the management of groundwater resources. Another collaborative effort between the United States and Mexico, the MOU between Ciudad Juárez Water Utilities and El Paso Water Utilities promotes the exchange of information and the development of binational studies in the region [18,47]. This surface-water and groundwater assessment effort represents a local approach arranged by interested communities [11], indicating the presence of different paths toward scientific groundwater collaboration on a local or regional scale.

Finally, the TAAP Cooperative Framework represents another mechanism of binational collaboration between the United States and Mexico. The knowledge-improvement goals included in the TAAP Cooperative Framework coincide with the data-collection efforts and assessment of shared water resources described in the Bellagio Draft Treaty of 1989. The TAAP Cooperative Framework is described below.

3.4. *The United States-Mexico Transboundary Aquifer Assessment Program*

Recognizing the interest of the United States and Mexico to understand their shared aquifers, and with U.S. Public Law 109–448 as a precedent, the Principal Engineers of the U.S. and Mexican sections of the IBWC signed the Joint Cooperative Process United States-Mexico for the Transboundary Aquifer Assessment Program (TAAP Cooperative Framework) in August 2009. While some scholars argue that TAAP marginalizes issues such as water rights and management [60], others have contended that improving understanding of the U.S.-Mexico transboundary aquifers, which is the objective of the TAAP Cooperative Framework, is a necessary first step toward a binational groundwater management agreement between the United States and Mexico [19].

The TAAP Cooperative Framework promotes the development of binational technical groups to evaluate shared aquifers, advocates for knowledge improvement and data exchange, and states that each country has an obligation to cooperate [24]. It is worth noting that these principles correspond to UN Draft Articles 7 and 8, "General Obligation to Cooperate" and "Regular Exchange of Data and Information" [19]. Additionally, TAAP principle 4 considers the sovereignty of each nation by stating, "no provision set forth [in this agreement] will limit what either country can do independently in its own territory." This principle is consistent with UN Draft Article 3, "Sovereignty of Aquifer States" or countries that share the aquifer. UN Draft Article 13, regarding monitoring, is also consistent with the overall TAAP objective of improving knowledge of transboundary aquifer conditions.

Accomplishments of the TAAP include the development of the Mesilla Valley Hydrologic Model; the completion of the Binational Study of the Transboundary San Pedro Aquifer; the establishment of research projects in Texas, New Mexico, Arizona, Sonora, and Chihuahua; the output of numerous publications and conference presentations; and fieldwork in the U.S. and Mexican portions of the priority aquifers [61]. Additionally, over 50 binational meetings have taken place, many of them were between the technical workgroups established pursuant to the Cooperative Framework.

An existent legal mechanism for collaboration, regional institutions, funding mechanisms, high institutional capacity, previous water cooperation, and scientific research are some of the enabling mechanisms for groundwater collaboration that are present in the TAAP collaboration and that might facilitate future groundwater collaboration between the two countries.

3.4.1. Common Elements of Collaboration between the TAAP Cooperative Framework and International Aquifer Agreements

A comparison of the elements of the TAAP Cooperative Framework and the components of the six international groundwater agreements is presented in Table 3. Five items were particularly relevant as common features of collaboration: (1) the presence of data exchange provisions, which was true for all the agreements but the Al-Saq/Al-Disi Aquifer System; (2) the concurrence for binational aquifer assessment, agreed on and implemented by all the countries sharing an aquifer; (3) the establishment of technical advisory committees, which occurred in all of the countries; (4) the presence of technical groups, discussed in every agreement except the Bamako Declaration, and; (5) respect for the legal framework and jurisdictional requirements of each country, which was inferred from the content of each of the agreements and which apply to all of the analyzed aquifers.

Table 3. Common elements of collaboration between transboundary groundwater agreements (✔ = Component present in the agreement, ✘ = component absent in the agreement, * = inferred present component, ? = unspecified component, not shown in the agreement and cannot be inferred from additional content).

	Elements of Collaboration	TAAP	GAS	Franco-Swiss Genevese	Northwestern Saharan	Iullemeden	NAS	Al-Saq/ Al-Disi
Objectives	Exchange data	✔	✔	✔	✔	✔	✔	✘
	Concur on binational aquifer assessment activities	✔	✔	✔	✔	✔	✔	✔
	Establish and coordinate technical advisory committees	✔	✔	✔	✔	✔*	✔	✔
	Establish an official repository for binational project reports	✔	?	?	✔	?	?	✘
Framework (Process)	Establish technical groups	✔	✔	✔	✔	X	✔	✔
	Develop project progress reports	✔	✔	✔	✔	?	✔	✔
	Publish final reports	✔	?	✔	✔	?	✔	✘
Funding	Arranged between the parties	✔	?	✔	✔	?	?	?
Principles	Activities should be beneficial to both countries	✔	✔*	✔*	✔*	✔*	✔*	✔*
	Activities should be agreed on within the framework of the coordinating agency	✔	?	?	?	?	?	?
	Activities should respect the legal framework and jurisdictional requirements of each country	✔	✔	✔	✔	✔	✔	✔
	No provision set forth in this agreement will limit what either country can do independently in its own territory	✔	✔	✔	✔	?	✔	✘
	No part of this agreement may contravene what has been stipulated in the Boundary and Water Treaties	✔	✔	✔	✔	?	✔	✔
	The information generated from these projects is solely for the purpose of expanding knowledge	✔	✘	✘	✔	✘	✔	✘
Communication	An official repository of records will be present	✔	?	?	✔	?	?	✘
	Reports will be available to the public in each country	✔	?	✔	✔	?	✔	✘
	Information obtained will be considered official data and will be shared without any restrictions	✔	?	✔	✔	?	✔	✘
	Credit will be given to those who provide information	✔	✔*	✔*	✔*	✔*	✔*	?

These features of collaboration, present in all of the agreements except the Al-Saq/Al-Disi Aquifer System and the Iullemeden Aquifer, are in alignment with the main objective of the TAAP Cooperative Framework, which is to improve knowledge of transboundary aquifers. These features also demonstrate that groundwater assessment is necessary for managing transboundary aquifers. A notable difference between the agreements and the TAAP Cooperative Framework is the use of information. While the data generated through the TAAP serve only to improve knowledge, as do the data generated for the NSAS and the NSA, the information and monitoring outcomes from the GAS, Franco-Swiss Genevese Aquifer, Al-Saq/Al-Disi Aquifer System, and Iullemeden Aquifer can be used for decision-making purposes with respect to groundwater management.

Another difference between the agreements and the TAAP Cooperative Framework is the availability of aquifer assessment data to the public, which is not discussed by or does not apply to the rest of the transboundary aquifers. The IBWC is the official repository of the available studies, which are published in both English and Spanish through the organization's official website. The financial arrangements, as described in the TAAP Cooperative Framework, were rather uncommon, being present in the Franco-Swiss Genevese and the Northwestern Saharan Aquifer agreements only.

3.4.2. Expert Interviews on U.S.-Mexico Transboundary Groundwaters

Expert interviews on transboundary groundwater served to determine whether the TAAP principles of agreement could guide transboundary aquifer assessment in areas that have not entered into formal agreements for binational collaborative studies, ultimately leading to the development of groundwater management arrangements or agreements (Table 4).

Table 4. Summary of expert interview responses.

Question	Interviewee 1	Interviewee 2	Interviewee 3	Interviewee 4	Interviewee 5	Interviewee 6
According to your experience, what factors promote the successful groundwater collaboration between nations that share one or various aquifers?	-Interest -Compliance with existing agreements -Respect for cultural differences -Consideration for institutional asymmetries	-Friendly relations between countries -Pre-existing framework for collaboration	-Pre-existing institutional framework -Data-sharing mechanisms -Trust -Interest -Funding	-Trust -Considering imbalances between countries - Processes that promote trust, e.g., data sharing and prior notification -Avoiding water-right discussions in the initial stages	-Trust -Common issues/problems -People—not institutions—promoting cooperation	-Local agreements -Avoiding water-right discussions -Focusing on water quality
Do you think we can generalize the TAAP principles of collaboration and apply them to other aquifers within the U.S.-Mexico border? Do you think the TAAP Cooperative Framework can serve as a model for the assessment of other transboundary aquifers?	-General principles can be utilized -Take into account the uniqueness of each aquifer system	-Principles can help other countries deal with the use of shared groundwater resources -Absolute commitment is needed between countries	-Important model to consider in countries with or without pre-existing frameworks for collaboration -Depends on the circumstances -Require higher-level discussions -Partnerships among universities, federal agencies, and coordinating agencies provide a favorable model	-Follow the basic rules of cooperation and apply them to the specific needs of each aquifer	-	-Elements are good but worked due to the leadership of specific members -Does not have federal strength -Political sensitivity limits data sharing -Formality of relations slows down collaboration

Table 4. *Cont.*

Question	Interviewee 1	Interviewee 2	Interviewee 3	Interviewee 4	Interviewee 5	Interviewee 6
Do you think the TAAP Cooperative Framework can serve as a basis for the development of future groundwater management agreements in the borderlands of the United States and Mexico?	-Principles can serve as a foundation -There has to be a reason/interest that drives the development of a groundwater management agreement	-Smaller, localized agreements are needed -States should be involved	-A framework for talking about groundwater management is needed	-A generic framework for collaboration is needed -You can implement a framework for the whole border stating that each aquifer must have its own regime	-I don't think so -Water management differs greatly between the borderlands	-Not as it stands right now -Does not help to plan, manage, or learn about the border -It is limited to only four aquifers

Interviewees reported that the TAAP principles are general enough to be used as a guide to promoting additional groundwater collaboration for the assessment of other transboundary aquifers in the United States and Mexico. This can be supported by a statement included within the TAAP Cooperative Framework (framework/processes): "Either of the two countries can propose an aquifer to study. Within the IBWC framework, it will be determined whether the proposal is in common interest and, as appropriate, a joint program developed." However, the commitment of the involved countries must be "absolute"; i.e., the time frame, funding, and political support for the analysis should be established between the collaborating parties. Participants also commented that because the TAAP effort involves partnerships among universities, federal agencies, and coordinating agencies, it provides a favorable model for collaboration. Nevertheless, its application would depend on the circumstances of each partner country and also on the leadership of the involved members. Historic or high levels of distrust between the two countries may interfere with the process of collaboration, while a pre-existing foundation, as demonstrated by the 1944 Treaty and the IBWC Framework in the case of Mexico, has facilitated collaboration. On the other hand, it was also expressed that even though the elements of the TAAP Cooperative Framework are effective, it lacks a binding capacity and federal and institutional support from both the United States and Mexico, a fact that sometimes hinders cooperation between the two nations.

Regarding the development of transboundary groundwater management agreements between the United States and Mexico, interviewees stated that the uniqueness of each aquifer system might require the development of aquifer-specific agreements. It was also expressed that the TAAP Cooperative Framework, as it stands right now, cannot serve as the foundation for groundwater management agreements due to the differences in water management between the United States and Mexico. An alternative proposed during the interview process was the possible development of a regional agreement for the use of groundwater resources in the border region. Such an agreement could employ principles of the TAAP Cooperative Framework, and it should be consistent with the UN Draft Articles and existing international groundwater agreements, as well. According to the interviews, the case of aquifer-specific agreements will require the direct involvement of each of the individual states, due to the decentralized way in which water resources are managed in the United States. Each state within U.S. territory has different needs, goals, funds, and management schemes; therefore, each state must play an active role in the development of an agreement.

Interviewees expressed that trust is a key factor in successful groundwater collaboration and that data sharing processes can improve trust between different countries. The interest between involved parties and common issues can also promote collaboration. Respect for cultural differences, institutional asymmetries, and economic imbalance is essential. Finally, additional indicators of successful groundwater collaboration include the presence of a pre-existing institutional framework and the countries' compliance with pre-existing agreements.

3.4.3. The TAAP Cooperative Framework as a Model for Groundwater Collaboration

The TAAP is a binational scientific effort that enabled groundwater data exchange and harmonization, knowledge improvement of the TAAP aquifers of focus, and trust-building among the federal agencies, academic institutions, and water resources research institutes that collaborated in the program. It was enabled by a governance approach, the Joint Report of the Principal Engineers Regarding Joint Cooperative Process United States-Mexico for the Transboundary Aquifer Assessment Program of 2009, and, for the United States, the TAA-Act of 2006 (Figure 2). The TAAP history of collaboration exhibits six out of eight enabling factors for groundwater collaboration described in the current literature [18]: (1) a strong regional institution like the IBWC (U.S. and Mexican Section); (2) existing legal mechanisms, such as the 1944 Treaty and the IBWC Minutes; (3) previous water collaboration for solving water-related issues; (4) third-party involvement from entities that do not belong to the government of each country, such as academic institutions; (5) scientific research on transboundary aquifers, and; (6) funding mechanisms (Figure 2). Although the two remaining features, high institutional capacity, and strong political will are not fully present in the TAAP, they are not absent either. Strong political will is identified when high-ranking officials prioritize transboundary water management [18]. However, the facilitation of diplomatic events and meetings like the ones hosted by the IBWC and TAAP can be considered an early sign for the strengthening of political will. The monitoring and modeling efforts that are present in the TAAP can also be considered factors that strengthen institutional capacity.

The TAAP represents a pre-existing institutional arrangement that promotes trust development between the United States and Mexico, in addition to the development of groundwater assessment studies within the aquifers of focus. These outcomes position the two countries to move forward in one of two ways: (1) implement additional assessment within the transboundary aquifers shared by the United States and Mexico, or (2) initiate dialogue toward the need of developing groundwater management mechanisms for the two countries. While a dialog is needed between the United States and Mexico to determine the need for a possible groundwater management agreement and the scale of the agreement itself, the importance of considering the unique physical, cultural, institutional, and economic characteristics surrounding specific aquifers is essential, as expressed during the expert interviews.

The analysis of common elements of collaboration between existing groundwater cooperation mechanisms and the TAAP indicates that the program itself, which is guided by the TAAP Cooperative Framework, has laid the groundwork for the development of additional aquifer assessment studies along the U.S.-Mexico border. The principles contained within the framework, which already include the majority of the elements described in pre-existing aquifer agreements, allow the two countries to continue studying additional aquifers and help to build trust between the involved parties (Figure 2). Moreover, the TAAP encompasses tenets such as communication and funding principles not previously mentioned in other aquifer agreements. These statements support the fact that the TAAP Cooperative Framework can be used as a model for transborder groundwater collaboration for the assessment of transboundary aquifers between the United States and Mexico and around the world.

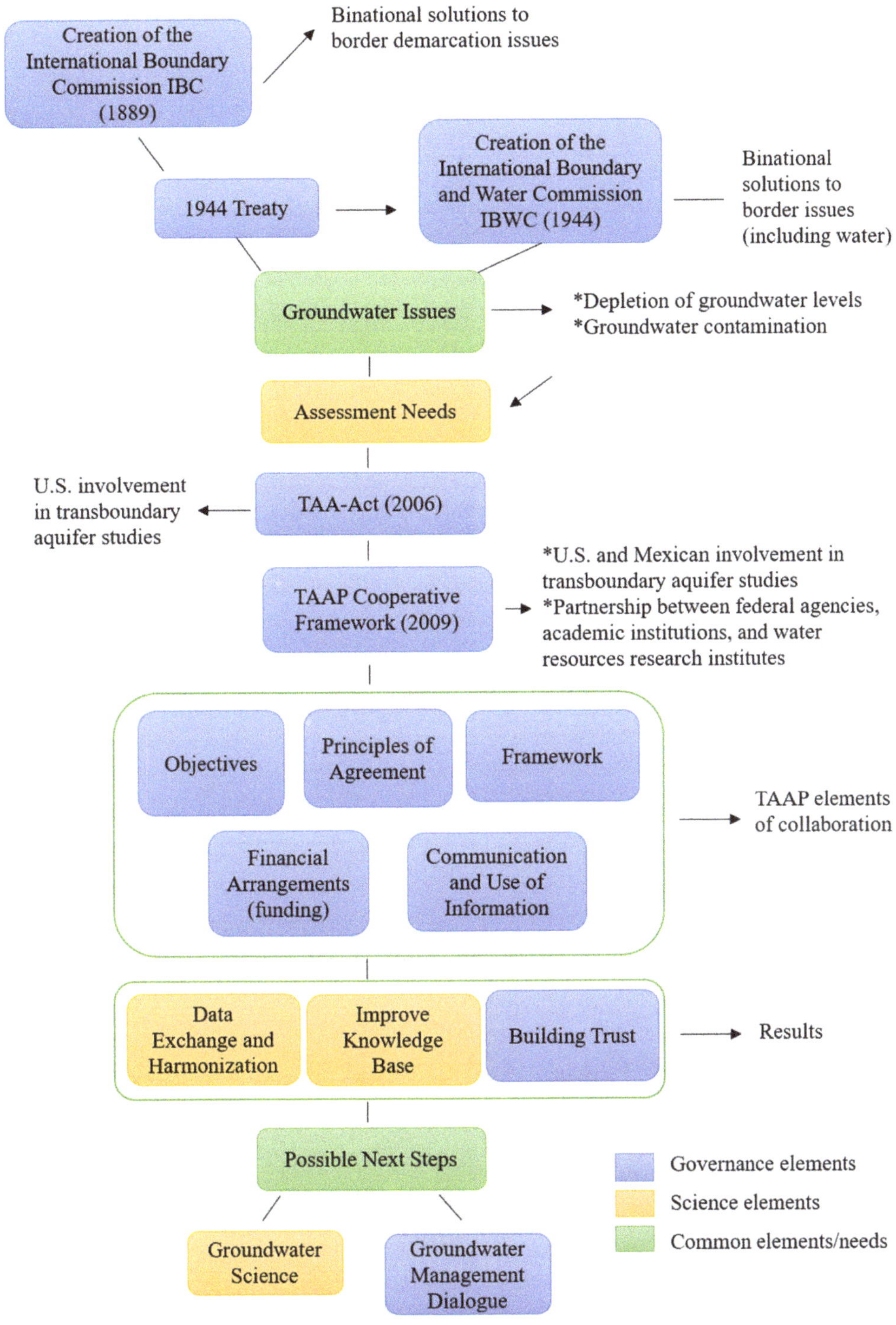

Figure 2. Evolution of the United States-Mexico Transboundary Aquifer Assessment Program.

4. Discussion

Currently, there is no groundwater treaty between the United States and Mexico. The 1944 Water Treaty regarding the Utilization of Waters of the Colorado and Tijuana Rivers and of the Rio Grande (1944 Treaty) is the primary surface-water-allocating mechanism for the two countries. The treaty, however, does not mention groundwater. The Joint Report of the Principal Engineers Regarding the Joint Cooperative Process United States-Mexico for the Transboundary Aquifer Assessment Program (TAAP Cooperative Framework) is a case of groundwater collaboration for the assessment of the U.S.-Mexico transboundary aquifers of focus: the Santa Cruz, San Pedro, Mesilla, and Hueco Bolson aquifers. However, at least 36 transboundary aquifers shared by the United States and Mexico have been identified so far [25].

Relevant studies on U.S.-Mexico groundwater governance have analyzed (1) the intranational institutions for the management of shared groundwater resources [22]; (2) the importance of institutional asymmetries for transboundary aquifer assessment [20]; (3) the institutional assessment of the Transboundary Santa Cruz and San Pedro Aquifers [21], and; (4) the management perspectives for the shared aquifers of the United States and Mexico [25]. While most of these studies discussed the outcomes, advantages, and disadvantages of the TAAP Cooperative Framework and the program itself, the components of the TAAP Cooperative Framework have not been analyzed as a model for groundwater collaboration.

This study analyzed the TAAP Cooperative Framework as a guide for furthering scientific assessment in areas that have not entered into formal agreements for binational collaborative studies. Through literature review and analysis of existing transboundary groundwater management agreements, we found that common elements of collaboration between the TAAP Cooperative Framework and existing groundwater management agreements include provisions for the exchange of data, concurrence for binational aquifer assessment, the establishment of technical advisory committees and technical groups, and respect for the legal framework and jurisdictional requirements of each country.

The TAAP exhibits several features that enable groundwater collaboration: existing legal mechanisms, previous water collaboration, third-party involvement, scientific research, and funding mechanisms. Additionally, the framework is consistent with four UN Draft Articles on the Law of Transboundary Aquifers, findings that may indicate the readiness of the two countries to move on to a next step: to implement additional aquifer assessment along the U.S.-Mexico border or to initiate dialogue toward the development of groundwater management mechanisms.

Some scholars have argued that the TAAP marginalizes issues such as groundwater rights and management [60] and lacks a binding capacity (personal communication, 2020), and we agree with this premise. The information generated through the TAAP is "solely for the purpose of expanding knowledge" [20]. However, the present study has found that scientific assessment is a prior step for the development of groundwater management agreements. In fact, interviews with experts on transboundary waters explored two ways in which the TAAP could guide groundwater management: through local agreements for the management of specific aquifer systems or through a regional agreement that guides the use of groundwater resources in the border region. In any case, lessons from the TAAP Cooperative Framework and the program itself remain as a model of robust binational groundwater collaboration with principles that have the potential to guide future groundwater assessment and management not just along the U.S.-Mexico border, but across the world.

5. Conclusions

The United States and Mexico share rivers, basins, and aquifers. Yet they do not share a water management agreement that suits the needs of the border communities that completely rely on groundwater resources. Challenges for managing shared groundwater in the region include population growth, industrialization, increase in agriculture, contamination, increase in surface water and groundwater demands, and climate uncertainties. These challenges indicate that some sort of binational arrangement is needed to protect and manage the shared groundwater resources. However, the topic has only been mentioned twice since the IBWC was created. Almost five decades after the signing of Minute 242 and three decades after the development of the Bellagio Draft Treaty, there has been no effort to establish a comprehensive groundwater agreement. Meanwhile, efforts toward increasing understanding of the U.S.-Mexico transboundary aquifers have taken place. This study analyzed the TAAP Cooperative Framework as a guide for furthering scientific assessment in areas that have not entered into formal agreements for binational collaborative studies. To achieve this, we compared the elements of collaboration present within the TAAP Cooperative Framework and six transboundary aquifer agreements around the world.

From this analysis, we found that five elements were particularly relevant as common features of collaboration that align with the TAAP Cooperative Framework: (1) the presence of data exchange provisions, (2) the concurrence for binational aquifer assessment, (3) the establishment of technical advisory committees, which occurred with all of the aquifers, (4) the presence of technical groups, and (5) respect for the legal framework and jurisdictional requirements of the involved countries. Expert interviews also served to identify lessons learned from the TAAP and global challenges for groundwater collaboration, which included the importance of trust-building between border communities sharing water resources, groundwater assessment, and a pre-existing framework for collaboration. It was also suggested that the TAAP principles are general enough to be used as a guide to promoting additional groundwater collaboration for the assessment of other transboundary aquifers in Mexico and the United States and around the world. Yet, the applicability of the TAAP Cooperative Framework will depend largely on the unique circumstances of the involved countries.

We conclude for several reasons that the transboundary aquifer assessment efforts following the TAAP Cooperative Framework represent a model for others wishing to engage in transboundary aquifer assessment. The TAAP Cooperative Framework is a concisely written and readily available document that has been successfully approved and signed by two countries. It has promoted productive scientific collaboration between the United States and Mexico in a manner consistent with the Draft Articles on the Law of Transboundary Aquifers (UN Draft Articles). Its elements are also consistent with the information gathering portion of successful groundwater management agreements around the world. It includes funding and communication provisions that are uncommon in existing international agreements but that facilitate groundwater cooperation, as made evident by the collaboration to date. Finally, according to the TAAP Cooperative Framework, either of the two countries can propose an aquifer of focus, meaning that there is no need to develop a new cooperative framework for assessing additional transboundary aquifers shared by the United States and Mexico.

The present study finds evidence of successful outcomes within the TAAP Cooperative Framework consistent with available transboundary groundwater management agreements, demonstrating that the approach is suited to serve as a model for others wishing to engage in transborder aquifer assessments worldwide. Furthermore, the principles of the TAAP Cooperative Framework include elements that promote trust between the United States and Mexico (e.g., data sharing, development of binational aquifer assessment activities, the establishment of technical advisory committees, and establishment of technical groups). These and the rest of the TAAP elements of collaboration can help to establish the meaningful and robust binational cooperation necessary for the development of U.S.-Mexico groundwater management agreements at the aquifer level.

Author Contributions: Conceptualization, E.M.T.-V. and S.B.M.; supervision, S.B.M.; methodology, E.M.T.-V. and S.B.M.; validation, E.M.T.-V. and S.B.M.; formal analysis, E.M.T.-V.; investigation, E.M.T.-V.; resources, E.M.T.-V.; data curation, E.M.T.-V.; writing—original draft preparation, E.M.T.-V.; writing—review and editing, S.B.M.; visualization, E.M.T.-V. and S.B.M.; project administration, S.B.M.; funding acquisition, S.B.M. All authors have read and agreed to the published version of the manuscript.

Funding: This work was partially funded by the U.S. Geological Survey (funding authorized by P.L. 109–448) Award Number G17AC00439 for the Transboundary Aquifer Assessment Program, the Babbitt Dissertation Fellowship Program from the Lincoln Institute of Land and Policy, and the Consejo Nacional de Ciencia y Tecnología (CONACYT).

Institutional Review Board Statement: This project has been reviewed and approved by an IRB Chair or designee from the University of Arizona (protocol number 1809938973, approved on 17 October 2018).

Informed Consent Statement: Informed consent was obtained from all subjects involved in the study.

Data Availability Statement: Data sharing is not applicable to this study. The information shared during the interview process is confidential and cannot be shared with the public.

Acknowledgments: The authors wish to express their thanks to all the people that contributed to the development of this study via interviews, revisions, and through data sharing.

Conflicts of Interest: The authors declare no conflict of interest. The funders had no role in the design of the study; in the collection, analyses, or interpretation of data; in the writing of the manuscript, or in the decision to publish the results.

References

1. Fried, J.; Ganoulis, J. *Transboundary Groundwater Resources: Sustainable Management and Conflict Resolution*; Lambert Academic Publishing: Saarbrücken, Germany, 2016.
2. Earle, A.; Neal, M.J. Inclusive Transboundary Water Governance. In *Freshwater Governance for the 21st Century*; Global Issues in Water Policy; Springer: Cham, Switzerland, 2017; Volume 6, pp. 145–158.
3. IGRAC (International Groundwater Resources Assessment Centre); UNESCO-IHP (UNESCO International Hydrological Programme). *Transboundary Aquifer of the World [Map]: Update 2020: Scale 1: 50 000 000*; IGRAC: Delft, The Netherlands, 2020.
4. Giordano, M.; Drieschova, A.; Duncan, J.A.; Sayama, Y.; de Stefano, L.; Wolf, A.T. A review of the evolution and state of transboundary freshwater treaties. *Int. Environ. Agreem. Polit. Law Econ.* **2014**, *14*, 245–264. [CrossRef]
5. Petersen-Perlman, J.D.; Wolf, A.T. Getting to the First Handshake: Enhancing Security by Initiating Cooperation in Transboundary River Basins. *J. Am. Water Resour. Assoc.* **2015**, *51*, 1688–1707. [CrossRef]
6. McCracken, M.; Wolf, A.T. Updating the Register of International River Basins of the world. *Int. J. Water Resour. Dev.* **2019**, *35*, 732–782. [CrossRef]
7. Lopez-Gunn, E.; Jarvis, W.T. Groundwater governance and the Law of the Hidden Sea. *Water Policy* **2009**, *11*, 742–762. [CrossRef]
8. Megdal, S.B.; Petersen-Perlman, J.D. Groundwater Governance and Assessment in a Transboundary Setting. In *Lake Governance*; Grover, V.I., Ed.; Series: Water: Emerging Issues and Innovative Responses | "A Science Publishers Book"; CRC Press: Boca Raton, FL, USA, 2018; pp. 40–64.
9. Callegary, J.B.; Megdal, S.B.; Tapia Villaseñor, E.M.; Petersen-Perlman, J.D.; Minjarez Sosa, J.I.; Monreal Saavedra, R.; Gray, F.; Grijalva Noriega, F.J. Findings and lessons learned from the assessment of the Mexico-United States transboundary San Pedro and Santa Cruz aquifers: The utility of social science in applied hydrologic research. *J. Hydrol. Reg. Stud.* **2018**, *20*, 60–73. [CrossRef]
10. De Chaisemartin, M.; Varady, R.G.; Megdal, S.B.; Conti, K.I.; van der Gun, J.; Merla, A.; Nijsten, G.; Scheibler, F. Addresing the Groundwater Governance Challenge. In *Freshwater Governance for the 21st Century*; Springer International Publishing: Cham, Switzerland, 2017; pp. 205–227.
11. Sanchez, R.; Eckstein, G. The path towards groundwater management in the borderlands of Mexico and Texas. *Wiley Interdiscip. Rev. Water* **2019**, *7*, e1399. [CrossRef]
12. IBWC. U.S. IBWC. Available online: https://ibwc.gov/home.html (accessed on 25 December 2020).
13. Varady, R.G.; Scott, C.A.; Wilder, M.; Morehouse, B.; Pineda Pablos, N.; Garfin, G.M. Transboundary adaptive management to reduce climate-change vulnerability in the western U.S.–Mexico border region. *Environ. Sci. Policy* **2013**, *26*, 102–112. [CrossRef]
14. Updike, R.G.; Ellis, E.G.; Page, W.R.; Parker, M.J.; Hestbeck, J.B.; Horak, W.F. *United States-Mexican Borderlands: Facing Tomorrow's Challenges through USGS Science*; U.S. Geological Survey Circular 1380; U.S. Department of the Interior, U.S. Geological Survey: Reston, VA, USA, 2013; p. 336.
15. Wilder, M.; Scott, C.A.; Pineda Pablos, N.; Varady, R.G.; Garfin, G.M.; McEvoy, J. Adapting Across Boundaries: Climate Change, Social Learning, and Resilience in the U.S.-Mexico Border Region. *Ann. Assoc. Am. Geogr.* **2010**, *100*, 917–928. [CrossRef]

16. Mumme, S.P. Minute 242 and beyond: Challenges and opportunities for managing transboundary groundwater on the Mexico-US border. *Nat. Resour. J.* **2000**, *40*, 341–378.
17. Puri, S.; Appelgren, B.; Arnold, G.; Aureli, A.; Burchi, S.; Burke, J.; Margat, J.; Pallas, P. *Internationally Shared (Transboundary) Aquifer Resources Management*; United Nations Educational, Scientific and Cultural Organization: Paris, France, 2001.
18. Conti, K.I. *Factors Engabling Transboundary Aquifer Cooperation: A Global Analysis*; IGRAC: Delft, The Netherlands, 2014; p. 108.
19. Megdal, S.B. The Cooperative Framework for the Transboundary Aquifer Assessment Program: A Model for Collaborative Transborder Studies. *Ariz. Water Resour.* **2017**, *25*, 1–2.
20. IBWC. Joint Report of the Principal Engineers Regarding Joint Cooperative Process. United States-Mexico for the Transboundary Aquifer Assessment Program. 2009. Available online: wrrc.arizona.edu/sites/wrrc.arizona.edu/files/images/Joint-Report-TAAP-8-19-09.pdf (accessed on 25 December 2020).
21. Megdal, B.S.; Scott, C.A. The Importance of Institutional Asymmetries to the Development of Binational Aquifer Assessment Programs: The Arizona-Sonora Experience. *Water* **2011**, *3*, 949–963. [CrossRef]
22. Megdal, S.B.; Sención, R.; Scott, C.A.; Díaz, F.; Oroz, L.; Callegary, J.; Varady, R.G. Institutional Assessment of the Transboundary Santa Cruz and San Pedro Aquifers on the United States-Mexico Border. In Proceedings of the ISARM2010 International Conference, Paris, France, 6–8 December 2010.
23. Milman, A.; Scott, C.A. Beneath the surface: Intranational institutions and management of the United States-Mexico transboundary Santa Cruz aquifer. *Environ. Plan. C Gov. Policy* **2010**, *28*, 528–551. [CrossRef]
24. Sanchez, R.; Lopez, V.; Eckstein, G. Identifying and characterizing transboundary aquifers along the Mexico–US border: An. *initial assessment. J. Hydrol.* **2016**, *535*, 101–119.
25. Sanchez, R.; Eckstein, G. Aquifers Shared Between Mexico and the United States: Management Perspectives and Their Transboundary Nature. *Groundwater* **2017**, *55*, 495–505. [CrossRef] [PubMed]
26. Villar, P.C. International cooperation on transboundary aquifers in South. America and the Guarani Aquifer case. *Rev. Bras. De Política Int.* **2016**, *59*. [CrossRef]
27. Movilla Pateiro, L. Ad hoc legal mechanisms governing transboundary aquifers: Current status and future prospects. *Water Int.* **2016**, *41*, 851–865. [CrossRef]
28. Etikan, I. Comparison of Convenience Sampling and Purposive Sampling. *Am. J. Theor. Appl. Stat.* **2016**, *5*, 1–4. [CrossRef]
29. Matsumoto, K. *Transboundary Groundwater and International Law: Past Practices and Current Implications*; Oregon State University: Corvallis, OR, USA, 2004.
30. Jarvis, T.; Giordano, M.; Puri, S.; Matsumoto, K.; Wolf, A. International Borders, Ground Water Flow, and Hydroschizophrenia. *Ground Water* **2005**, *43*, 764–770. [CrossRef]
31. Albrecht, T.R.; Varady, R.G.; Zuniga-Teran, A.A.; Gerlak, A.K.; Staddon, C. Governing a shared hidden resource: A review of governance mechanisms for transboundary groundwater security. *Water Secur.* **2017**, *2*, 43–56. [CrossRef]
32. Eckstein, G.; Sindico, F. The Law of Transboundary Aquifers: Many Ways of Going Forward, but Only One Way of Standing Still: The Law of Transboundary Aquifers. *Rev. Eur. Comp. Int. Environ. Law* **2014**, *23*, 32–42. [CrossRef]
33. UNGA. United Nations General Assembly Resolution 63/124 (2008) on the Law of Transboundary Aquifers. See Also the Subsequent UNGA Resolutions 66/104 (2011), 68/118 (2013), 71/150 (2016), 74/85 (2019). Available online: https://www.internationalwaterlaw.org/documents/intldocs/) (accessed on 25 January 2021).
34. Zeitoun, M. *Power and Water in the Middle East: The Hidden Politics of the Palestinian-Israeli Water Conflict*; I.B. Tauris: London, UK; New York, NY, USA, 2008.
35. Allan, J.A. Water in the environment/socio-economic development discourse: Sustainability, changing management paradigms and policy responses in a global system. *Gov. Oppos.* **2005**, *40*, 181–199. [CrossRef]
36. Neal, M.J.; Greco, F.; Connell, D.; Conrad, J. The social-environmental justice of groundwater governance. In *Integrated GroundWater Management*; Springer: Cham, Switzerland, 2016; pp. 253–272.
37. UN-Water. Step-by-Step Monitoring Methodology for Indicator 6.5.2. 2016. Available online: http://www.unwater.org/publications/step-step-methodology-monitoring-transboundary-cooperation-6--5-2/ (accessed on 17 February 2021).
38. Hussein, H.; Menga, F.; Greco, F. Monitoring Transboundary Water Cooperation in SDG 6.5.2: How a Critical Hydropolitics Approach Can. Spot Inequitable Outcomes. *Sustainability* **2018**, *10*, 3640.
39. Internationally Shared Aquifer Resources Management (ISARM), United Nations Educational Scientific and Cultural Organization (UNESCO), International Hydrological Programme (IHP), Paraguay Ratified the Guaraní Aquifer System Agreement. Available online: https://isarm.org/paraguay-has-ratified-guaran%C3%AD-aquifer-system-agreement (accessed on 25 January 2021).
40. Hussein, H. The Guarani Aquifer System, highly present but not high profile: A hydropolitical analysis of transboundary groundwater governance. *Environ. Sci. Policy* **2018**, *83*, 54–62. [CrossRef]
41. Sugg, Z.P.; Varady, R.G.; Gerlak, A.K.; de Grenade, R. Transboundary groundwater governance in the Guarani Aquifer System: Reflections from a survey of global and regional experts. *Water Int.* **2015**, *40*, 377–400. [CrossRef]
42. Acuerdo sobre el Acuífero Guaraní. *Acuerdo sobre el Acuífero Guaraní Suscipto por la República de Argentina, La República Federativa de Brasil, La República de Paraguay y La República Oriental del Uruguay en San Juan, Argentina, August 2, 2010*; Secretaria. Carpeta no 728 de 2011, Repartido no 486; República Oriental del Uruguay, Cámara de Senadores: Montevideo, Uruguay, 2012; p. 31.
43. De los Cobos, G. The Genevese transboundary aquifer (Switzerland-France): The secret of 40 years of successful management. *J. Hydrol. Reg. Stud.* **2018**, *20*, 116–127. [CrossRef]

44. Eckstein, G. Managing buried treasure across frontiers: The international Law of Transboundary Aquifers. *Water Int.* **2011**, *36*, 573–583. [CrossRef]
45. Garner, E.L. Factors identifying aquifers with a high probability of management success. *Water Int.* **2019**, *44*, 354–362. [CrossRef]
46. De los Cobos, G. A historical overview of Geneva's artificial recharge system and its crisis management plans for future usage. *Environ. Earth Sci.* **2015**, *73*, 7825–7831. [CrossRef]
47. UNESCO. *Atlas of Transboundary Aquifers: Global Maps, Regional Cooperation and Local Inventories*; UNESCO: Paris, France, 2009; p. 322.
48. Burchi, S.; Mechlem, K. *Groundwater in International Law: Compilation of Treaties and Other Legal Instruments*; FAO: Rome, Italy; UNESCO: Paris, France, 2015; Volume 20, pp. 15–20.
49. Republic of Mali, Republic of Niger, Federal Republic of Nigeria. *Bamako Declaration of the Ministers in Charge of Water Resources of the Countries Sharing the Iullemeden Aquifer System (Mali, Niger, Nigeria)*; Republic of Mali: Bamako, Mali, 2009. Available online: internationalwaterlaw.org/documents/regionaldocs/Iullemeden_Bamako_Declaration-2009.pdf (accessed on 25 December 2020).
50. Burchi, S. Legal frameworks for the governance of international transboundary aquifers: Pre- and post-ISARM experience. *J. Hydrol. Reg. Stud.* **2018**, *20*, 15–20. [CrossRef]
51. Cascão, A.E.; Zeitoun, M. Power, hegemony and critical hydropolitics. *Transbound. Water Manag. Princ. Pract.* **2010**, *27*, 42.
52. Gaglio, E. The Nubian Sandstone Aquifer: An. Institutional Approach to Water Management. *Calif. West. Int. Law J.* **2016**, *47*, 32.
53. Grossman, M. Cooperation on Africa's International Waterbodies: Information Needs and the Role of Information-Sharing. In *Transboundary Water Management in Africa: Challenges for Development Cooperation*; Deutsches Institut für Entwicklungspolitick: Bonn, Germany, 2006.
54. Ferragina, E.; Greco, F. The Disi project: An internal/external analysis. *Water Int.* **2008**, *33*, 451–463. [CrossRef]
55. Eckstein, G. The Newest Transboundary Aquifer Agreement: Jordan and Saudi Arabia Cooperate over the Al-Saq /Al-Disi Aquifer. International Water Law Project Blog 2015. Available online: https://www.internationalwaterlaw.org/blog/2015/08/31/the-newest-transboundary-aquifer-agreement-jordan-and-saudi-arabia-cooperate-over-the-al-sag-al-disi-aquifer/ (accessed on 25 December 2020).
56. Mumme, S. *The 1944 U.S.-Mexico Water Treaty as a Constitutional Document*; Rice University's Baker Institute for Public Policy: Houston, TX, USA, 2019; p. 15.
57. IBWC. Minute No. 242 Permanent and Definitive Solution to the International Problem of the Salinity of the Colorado River. 1973. Available online: https://www.ibwc.gov/Files/Minutes/Min242.pdf (accessed on 25 December 2020).
58. IBWC. Minute No. 323 Extension of Cooperative Measures and Adoption of a Binatioal Water Scarcity Contingency Plan in the Colorado River Basin. 2017. Available online: https://www.ibwc.gov/Files/Minutes/Min323.pdf (accessed on 25 December 2020).
59. O'Leary, M.C. The Bellagio Draft Treaty as a Tool for Solving Border Groundwater Issues. *United States-Mex. Law J.* **2003**, *11*, 5.
60. Kuri, G.H.; Rivera, J.J.C.; Huizar, R.Á. Evaluación crítica del acuífero transfronterizo río San Pedro. *Reg. Cohes.* **2019**, *9*, 61–85. [CrossRef]
61. Petersen-Perlman, J.D. *The Transboundary Aquifer Assessment Program: Past, Present, and Future US-Mexico Border Water Summit*; Water Resources Research Center: El Paso, TX, USA, 2018.

water

MDPI

Article

Trust, Risk, and Power in Transboundary Aquifer Assessment Collaborations

Holly Brause

New Mexico Water Resources Research Institute, New Mexico State University, 3170 S Espina St., Las Cruces, NM 88001, USA; hbrause@nmsu.edu

Abstract: In events and discussions about transboundary aquifer assessment, trust is often cited as an essential component of collaborative efforts. However, there is little discussion of what trust is, how it is built, what diminishes trust, and why it is so important. This study uses ethnographic research carried out between 2019 and 2021 with the Transboundary Aquifer Assessment Program (TAAP) to examine the role and significance of trust in U.S./Mexico TAAP collaborations. This study demonstrates that trust is best understood in relationship to power and risk. It examines the strengths and weaknesses of the TAAP program in managing asymmetrical relationships of power and unequal levels of risk in participation. In TAAP collaborations, the insistence on establishing trust should signal participants to consider and address the underlying issues of risk and power.

Keywords: Transboundary Aquifer Assessment Program; collaboration; trust; power; risk; U.S./Mexico border; ethnography

Citation: Brause, H. Trust, Risk, and Power in Transboundary Aquifer Assessment Collaborations. *Water* **2021**, *13*, 3350. https://doi.org/10.3390/w13233350

Academic Editors: Sharon B. Megdal and Anne-Marie Matherne

Received: 1 July 2021
Accepted: 18 November 2021
Published: 25 November 2021

Publisher's Note: MDPI stays neutral with regard to jurisdictional claims in published maps and institutional affiliations.

1. Introduction

In April 2019, the Binational Summit on Groundwater at the U.S./Mexico Border drew a lively crowd to the TecH2O Learning Center in El Paso, Texas. Scholars, governmental officials, water managers, and reporters from both sides of the U.S./Mexico border filed into the modern, angular, cement and glass building, were provided with a glossy program in either English or Spanish and a headset for simultaneous translation, and were directed to the main auditorium. Over the two days of the summit, participants arranged themselves along the curved wooden tables and navy office chairs in the auditorium seating to watch presentations and panel discussions regarding the important themes surrounding binational groundwater: data sharing, salinity, geohydrology, modeling, water law, watershed restoration, collaborative governance, etc. Throughout the groundwater summit, however, an unexpected theme emerged—trust. The Binational Summit in 2019 was my first introduction into binational efforts to understand shared groundwater resources, so I was surprised to hear the word trust repeated in so many panel presentations, keynotes, and discussions. I was not the only one who noticed. In her closing remarks, the then newly appointed Commissioner of the U.S. International Boundary and Water Commission (IBWC), Jayne Harkins, commented that she had also not expected to hear the word "trust" so often at the Binational Summit. To newcomers to the world of binational groundwater collaborations, the emphasis on trust was surprising. Yet, the longer one works on issues of binational groundwater, the more one comes to expect conversations about data, water quality, and aquifer recharge will also include conversations about trust. In this paper, I examine the role of trust in the Transboundary Aquifer Assessment Program (TAAP) and contextualize its importance in relationship to power and risk.

The TAAP program was written into law by the U.S. government in 2006 with the signing of Public Law 109–448, the United States–Mexico Transboundary Aquifer Assessment Act [1]. The law authorized the study of priority transboundary aquifers along the U.S./Mexico border, an effort to be led by the United States Geological Survey (USGS) and the Water Resources Research Institutes of Arizona, New Mexico, and Texas. The law creating the TAAP program, however, was unilaterally developed and passed by the U.S. without agreement from corresponding Mexican institutions. In 2009, after nearly three years of binational negotiation, the TAAP program as a binational effort was established in The Joint Report of the Principal Engineers Regarding the Joint Cooperative Process United States–Mexico for the Transboundary Aquifer Assessment Program, known as the TAAP Cooperative Framework. The TAAP Cooperative Framework serves as a formal agreement to collaborate, establishes a central role for the IBWC and its Mexican counterpart CILA (Comisión Internacional de Límites y Aguas), and defines the rules of engagement agreed upon by both sites.

The TAAP program, at its inception and today, includes participation by members of state agencies such as IBWC/CILA, the USGS, and CONAGUA (Comisión Nacional del Agua), along with university scientists from both sides of the border. Regional TAAP efforts may also include local water authorities, such as the inclusion of JMAS (Junta Municipal de Agua y Saneamiento) and EL Paso Water in the Hueco Bolson and Mesilla/Conejos-Medanos TAAP working group. TAAP provides the foundation for the exchange of data between Mexico and the U.S. about groundwater resources mutually identified as priority aquifers that are shared across the border. Although the data exchange occurs through a legally codified framework supported by both nations, the success of the collaboration is uneven across time and space. Collaboration varies by working group. It may be successful for a stretch of time, then slow, or come to a grinding halt. There are false starts, restarts, and failures along with great successes. These inconsistencies reveal what is obscured by political and legal discussion of the program—that the binational data exchange depends on the social relationships between the participants. Throughout my participation in TAAP and transboundary water activities, the presence of trust in binational social relationships was repeatedly identified as the most important aspect of the binational cooperation. Despite the fact that the word "trust" comes up in nearly every binational waters event, there is little to no discussion about what is meant by the concept. What is trust? How is trust achieved? How is trust broken? Why is trust important?

The act of trusting, scholars have noted, is necessary for social life and essential to the functioning of relationships and institutions [2,3]. The importance of trust appears in multiple scholarly publications about the TAAP program and binational aquifer governance [4–7]. The meaning of trust in collaboration on issues of binational water, however, is not closely examined. Trust is a complex topic, and one that is recently receiving renewed attention in fields from economics and political science to sociology and evolutionary biology [1]. Some authors describe trust as a disposition, affect, or feeling that goes beyond rational, transactional calculations of risk [2,8]. Others argue that trust is informed by past experiences, but is, at its core, an anticipatory orientation towards the future [2,8,9]. One study identified and analyzed 126 definitions of trust, ending on the concise definition of "trust as confidence in the face of risk" [10]. I use Fink et al.'s useful definition of trust as a basis for my own working definition of trust in the context of TAAP collaborations: trust is the willingness to proceed with collaboration despite risks. By operationalizing Fink et al.'s "confidence" as "willingness to proceed with collaboration", trust and mistrust become ethnographically observable as a set of practices that either halt or facilitate collaboration in the TAAP program.

In trust studies, symmetrical relationships and the domestic sphere have been emphasized, while trust in hierarchical relationships is underdeveloped [2]. This means that the relationship between trust and power is undertheorized. Yet, as medical anthropologist Harald Grimen states, "Analyses of trust that neglect power are naïve" [11]. To understand trust in asymmetrical relationships, it is necessary to consider the "nexus of power, trust, and risk" [11].

The concept of risk is present in both the Fink et al. definition of trust and my own working definition of trust. Unequal relationships of power amplify risk. It is important to note that power is not simply a dominating force, but is "an aspect of all relations among people" that operates differently at different scales [12]. Eric Wolf's description of the four modalities of power help clarify how power is manifested at different scales. I condense his longer description here: (1) "the power of potency of capability that is seen to inhere in an individual"; (2) power "manifested in interactions and transactions among people and refers to the ability of an ego to impose its will in social action upon an alter"; (3) "power that controls the contexts in which people exhibit their capabilities and interact with others"; and (4) structural power that "manifest in relationships that not only operates within settings and domains but also organizes and orchestrates the settings themselves, and that specifies the direction and distribution of energy flows" [12]. Wolf further clarifies that structural power is related to Foucault's definition of governance [12]. I will use these different scales of power in my analysis below as I identify sources of risk that are essential to understanding trust in TAAP collaborations.

The object of this article is to interrogate the meaning of trust in the TAAP binational collaborations, and to explicitly link the experience of trust and mistrust to the concept of power. Using ethnographic, interview, and document analysis data, I examine the factors that contribute to trust, and those that are detrimental to trust. I then contextualize the discussion of trust in relationships of power using the Framework for Assessing Power in Collaborative Governance Processes [13]. I show in which ways TAAP successfully negotiates relationships of power, and where it falls short. I then discuss the issue of risk and examine what is at stake in the binational groundwater collaborations. Ultimately, I argue that discussions of trust should cue TAAP participants to consider and discuss the underlying issues of risk and power at multiple scales to strength collaborative binational relationships.

2. Materials and Methods

The data for this article were collected as part of a larger ethnographic research project that examines the social and political context of water use, planning, and management on both sides of the U.S./Mexico border. This is an ongoing research project, so the results presented here are preliminary findings. Ethnographic research methods were used to collect the data presented in this work. Three elements are considered central to ethnographic research practices: participant observation, interviews, and analysis of relevant documents, archives, and scholarly literature [14]. Participant observation prioritized long-term emplacement, face-to-face interactions, taking part in daily activities and special events, and recording such interactions in fieldnotes [15]. Over the course of twenty-seven months between 2019 and 2021, I used participant observation to engage in a large array of events, meetings, and conversations related to local, regional, national, and international water use and management. These events included domestic and international, formal and informal, TAAP meetings. During this time, I managed New Mexico Water Resources Research Institute's TAAP participation and reporting. I helped plan, attended, and presented at binational groundwater conferences and workshops. I was in frequent communication with other TAAP program participants about the program, the current state of cooperation and research, and the future goals of the program. The topics of the meetings, presentations, and conversations included: descriptions of the physical qualities of the aquifers and the quality of the water they contain, the binational exchange of data planned and carried out during this time, discussions of the successes and obstacles to TAAP collaboration, debates about what kind of governance systems might be most appropriate for binational aquifer management in the future, the role of science in water management, and more. I recorded the observations made during fieldwork in ethnographic fieldnotes [16].

Ethnographic fieldnotes provide the bulk of the data used in this article. The process of recording ethnographic fieldnotes includes taking record of the event while in process to the extent possible, then typing up detailed, descriptive notes and reflections after the event. Fieldnotes include record of the words spoken at a meeting, as well as careful observations of non-verbal communication. For example, my notes from the in-person, formal binational TAAP meeting in Juarez in 2019 included relevant information about where the meeting was held, who participated, and what was discussed. It also included notes on which topics of conversation were avoided in this setting, how people arrange themselves in space, notes on the tone of voice used throughout the meeting, non-verbal communicative signaling such as scowls, sideways glances, expressions of surprise, folding arms, etc. From an anthropological perspective, thick description of data on all kinds of social signaling is as important, if not more so, than the spoken transcript of official discussion topics.

I systematically analyzed the data collected through a process of "qualitative analytic coding" [16] using NVivo software. Using NVivo, I assigned each piece of text one or multiple codes and/or subcodes. I began by creating a code structure from the themes that I already knew would be important. For example, I created a code I call Governance. Nested under Governance are subcodes for different policies, including international treaties and minutes, and national policies such as Waters of the U.S. I also used "open coding"—designating descriptive codes for a wide range topics as they emerged in the data [16]—in order to allow unexpected patterns to emerge from the data. The process included "affective coding" [17] to capture emotions, conflict, and values present in the data. Under the code Affect, I use subcodes for Trust, Anger, Optimism, Fear, and more. The data set used for this article currently has 210 codes and subcodes, and will likely continue to grow. The process of coding is not just a matter of sorting by words—the word "power" does not have to be mentioned in a section of data to be coded as having a conceptual relationship to manifestations of power. I used my working definition of trust as "the willingness to proceed with collaboration despite risks" to guide my coding. As such, ethnographically recorded displays of hesitancy to proceed, or obstructing the collaborative process, were coded as relating to trust. Qualitative analytic coding, therefore, is a practice in analysis and interpretation of data, rather than merely sorting. Having analyzed and organized my data through qualitative analytic coding, I was able to pull out the most important themes and patterns relating to the idea of trust, power, and risk in the TAAP program.

I supplemented my ethnographic fieldnotes with four targeted semi-structured interviews [18] carried out in the spring of 2021 for the purpose of cross-checking the results of my ethnographic data analysis. I selected the interviewees using "purposeful sampling", a technique that selects interviewees based on their knowledge and experience about a phenomenon, as opposed to randomized interviews intended to serve as a representative sample of a larger population [19]. The interviewees shared a combined forty-seven years of participation in the TAAP program. Two were government agency representatives and two were scholarly researchers. Two were from Mexico and two from the United States. I asked questions such as: What is your role in the TAAP program and how did you become involved? What is the role of trust in the TAAP program? Can you describe a time when trust was broken in binational collaborative efforts? What is at risk in binational collaborations? As the interviews were semi-structured, I did not strictly adhere to the interview script and instead allowed for the development of a natural conversation with appropriate follow-up questions. The interviews served to confirm or challenge the patterns in my ethnographic data, and provided additional data. I recorded the interviews either on Zoom or using a digital voice recorder, transcribed the interviews, and coded them in NVivo.

I also compiled and analyzed an archive of relevant documents. The compiled archive included foundational TAAP documents such as the 2009 Joint Report of the Principal Engineers Regarding the Joint Cooperative Process United States–Mexico for the Transboundary Aquifer Assessment Program, and the United States–Mexico Transboundary Aquifer Assessment Act of 2006. I also included scholarly articles published by TAAP researchers and popular coverage of TAAP activities from newspapers and online publications. Collecting data from three sources—fieldnotes, interviews, and documents—enabled me to cross-check the findings and validate the data through triangulation [19].

3. Results

In this section, I show the factors that contribute to trust and the factors that diminish trust identified in the data. I then demonstrate the link between trust and relationships of power, and use the Framework for Assessing Power in Collaborative Governance Processes [13] to examine how the TAAP program navigates relationships of power. I then show how the Framework falls short in its ability to identify important sources of power imbalances in the TAAP program.

3.1. *What Promotes and Diminishes Trust*

In the TAAP program, scholars and representatives from government institutions from both sides of the border join together to exchange data and carry out collaborative research on the shared aquifers. There is a common understanding amongst TAAP participants that trust is essential to the collaborative process. Through the process of applying qualitative analytic coding to the ethnographic data, interviews, and documents, patterns emerged that reveal what contributes to trust in TAAP collaborations, and what diminishes trust. Table 1 Shows important factors revealed in the ethnographic data and interviews that can contribute to trust amongst TAAP collaborators. Many people in both formal interviews and informal conversations during fieldwork explained that trust was not automatically present in TAAP interactions from the start. It had to be established through relationship building through long-term engagement. For some TAAP collaborations, building a history of showing up, following through, and working together over the years resulted in a solid baseline of trust. Frequent face-to-face interactions were also cited as useful for building trust. In-person interactions that occur in formal TAAP meetings are important, as are informal face-to-face interactions such as fieldtrips organized on each side, and/or interacting with other participants at conferences. These factors increase confidence in one another by building familiarity at an interpersonal level, diminishing the perception of risk in the collaboration. Participants identified the need for transparency about motivations for carrying out research. Assurances that the information collected would be for the betterment of science and beneficial for both sides help establish confidence in the collaboration. Similarly, working together to establish mutually agreed upon common goals was identified as important for relationships of trust. Establishing clear boundaries about what the program was intended for, and could not be used for, was also essential. Importantly, establishing transparency about motivations, goals, and boundaries in the beginning of the relationship is not sufficient; TAAP participants emphasized the need to communicate and reaffirm these aspects of the collaboration throughout the collaborative process. In the ethnographic data, there were many examples of participants from both sides publicly reaffirming the goals and boundaries of the TAAP program. For instance, in the context of planning for a data exchange, there was a lively discussion regarding whether or not data on the topic of governance was appropriate to exchange within the boundaries of TAAP. In the end, both parties agreed not to include such data during that exchange. Respect for the processes set out in the TAAP Collaborative Framework also works to build trust (described in more detail below). Participants also identified the importance of the personal traits of patience and sensitivity to building trust. Patience in the process of establishing relationships of trust and other participants' timelines is highly valued. Likewise, sensitivity to the needs, interests, and fears of the other side is important. Lastly, successful collaborations, including completed data exchanges, the establishment of work plans, and the publication of joint reports all contributed to feelings of trust and confidence in the group's ability to further collaborate.

Table 1. Factors that contribute to and diminish trust.

Contributes to Trust	Diminishes Trust
Long-term Engagement	Unilateral Decision Making
Relationship Building	Pushing Agenda or Timeline
Face-to-Face Interactions	Controlling Funding
Transparency about Motivations	Unmet Expectations in Exchange
Common Goal	Turn-over of Participants
Clear Boundaries	Absence of Key Participants
Respect for Process	Disagreeable Personalities
Communication	
Patience and Sensitivity	
Successful Collaborations	

Table 1 shows that there are also clear patterns in the data about the factors that detract from trust. One factor that limits trust is unilateral decision making. An interviewee who was present for the very start of the TAAP program explained that this was a problem in the beginning as the United States passed Public Law 109–448, creating the TAAP program without input or collaboration with Mexico. Mexico, in turn, did not recognize this U.S. domestic law and insisted on collaboratively developing a binational framework. The TAAP Cooperative Framework therefore became "the bridging document in order to implement the TAAP in that binational arena." Relatedly, the data also revealed that attempts by one side to push a specific research agenda or timeline harms trust. An example of this occurred at a binational TAAP meeting in 2019. After a long deliberation about next steps to exchange data, and the assertion by the Mexican participants that all other collaborations would be on hold until the completion of the data exchange, a U.S. newcomer to the TAAP project interrupted to insist on simultaneous engagement in another collaborative project. Pushback from both sides was immediately evident in body language—crossing arms, scowling, uncomfortable chuckles, and sideways glances. A long-term U.S.-based TAAP participant stepped in to smooth things over and shelve the suggested endeavor, but another participant later admitted she feared that years of work could have been unwittingly undone by the comment.

One side controlling funding can also damage trust. The U.S. TAAP law includes a provision that money for research on shared aquifers can fund Mexico-based research teams to collect data on the Mexican side of the border. Early attempts to fund studies in Mexico, however, were contentious. Contracts from the U.S. side had specific stipulations for deliverables, reporting, and timely information exchange. As one U.S.-based interviewee explained, "Mexico felt that we were trying to control them through payment of funds." While the early study was successful, disagreements in the process damaged the relationship.

Additionally, unmet expectations in data exchanges can lead to mistrust. In formal TAAP meetings now, a one-time failure of the U.S. to provide what Mexico considered to be equivalent, current, and properly formatted data that occurred a decade ago still emerges in conversation as a source of mistrust and obstacle to moving forward with collaborative projects. The group now takes special care to clarify what data is available and what format it is most useful in before attempting to exchange data.

Other factors that limit trust are related to individual participants and personalities. The turn-over of personnel involved in the TAAP program can be problematic. Changes in administrations or ruling party can lead to changes in who holds important government offices, retirement of collaborators and scholars leaving for positions elsewhere all disrupt the continuity of relationships and necessitate rebuilding trust. Even when membership remains the same, the absence of key members in formal meetings can also cause mistrust by throwing into doubt the commitment of the individual to the collaboration. Lastly, disagreeable personalities can greatly affect the formation and dissolution of trusting relationships. In the data, traits such as being "pushy", "domineering", "forceful", "impatient", and "aggressive" were identified as having a detrimental effect on relationships of trust.

Each of these factors that negatively contribute to trust are distinct mechanisms, but I argue that they contribute to mistrust as each highlight existing power inequalities in ways that amplify the awareness of risk. Even personality traits such as being "pushy" are related to the threat of one side using its power to coerce the other side. It is significant that, in both formal and informal conversations about TAAP, and in my ethnographic field notes, the negative factors listed were generally discussed in relationship to the actions of the U.S., the party with more social, political, and economic power.

3.2. Why Is Trust Important?

The above data are useful for understanding how trust is established and maintained, but less revealing about why it is so important. I argue that trust is important in TAAP collaborations as it helps to mitigate unequal relationships of power. All social relationships are also relationships of power, but in the case of TAAP, the unequal relationship of power is intensified by the fact that the United States is a significantly more powerful player in global politics, is economically dominant, and is backed by the world's most powerful military force. In this section, I use the Framework for Assessing Power in Collaborative Governance Processes [13] to demonstrate how the TAAP program manages unequal relationships of power, and where it falls short of that goal.

The Framework for Assessing Power in Collaborative Governance Processes identifies three arenas for power: formal authority (right to make decisions and take action), discursive legitimacy (ability to represent a discourse of social value), and resources (ability to deploy financial, material, and knowledge resources) (Purdy 2012). In regard to authority, the original U.S. domestic law that created the TAAP program without input from Mexico started the program off with unequal authority. Authority, however, can be negotiated. In the formation of the TAAP Cooperative Framework, representatives from the U.S. and Mexico set stipulations for the sharing of authority within the project after Mexico's refusal to accept the unilateral authority of the United States. Formal authority in the TAAP program is therefore managed through the TAAP Cooperative Framework.

The category of discursive legitimacy is also relatively even on both sides. The U.S. and Mexico both appeal to the shared social values of protecting natural resources, binational cooperation, and scholarly advancement. The use of discourses about trust fits into this category—it is a socially salient idea that can be used to manage relationships of power and increase standing in an unequal relationship.

The category of resource-based power is less effectively managed in the TAAP program. Ideally, under the TAAP Cooperative Framework, both the United States and Mexico would agree on a priority aquifer, approve a particular study, each side would carry out data collection on their own side of the border using their own funds, and then the data would be shared and a joint report would be produced. In practice, the U.S. has significantly more funding, personnel, and institutional infrastructure to devote to binational aquifer assessment than Mexico. As one interviewee explained, there have been times that TAAP participants on both sides have agreed to the importance of collecting a certain kind of data, but then the Mexican side declines to move forward with the joint study due to lack of funding. In meetings and in formal presentations by Mexican officials, the budget cuts to participating agencies and lack of personnel and equipment have at times been acknowledged as an obstacle to participation. Other times, more subtle actions, such as delaying studies, or giving "maybe in the future" responses to requests for collaboration may also stem from the gap in resourced-based power. The TAAP Cooperative Framework does include the following stipulation: "Either country may contribute to costs for work done in the other country" [20]. This rule is intended to compensate for the unequal resources. As discussed above, however, the U.S. control over funding for research completed by Mexico has previously been experienced as coercive. Therefore, the practice of sharing funds meant to equalize participation on each side of the border does not mitigate the gap in resourced-based power in TAAP collaborations.

The Framework for Assessing Power in Collaborative Governance Processes juxtaposes the three arenas of power with three sources of power—participants (who is involved and who leads), content (what issues are addressed), and process design (the where, when, and how of collaborative interaction) [13]. The TAAP Cooperative Framework is largely successful in managing power asymmetries in these three sources. TAAP brings together participants from agencies and scholars on both sides of the border with roughly equivalent standing. Representatives from each side have the power to move the collaboration forward or to delay and/or end particular collaborative efforts. In terms of content, representatives from each country have the ability to propose a priority aquifer, but the other side must agree to move the collaboration forward. Both countries must agree on the topic and scope of work of joint research.

In formal TAAP meetings, unequal relationships of power are mitigated through process design. In-person meetings are held on both sides of the border as to not privilege one site over the other. Each side speaks the language of their own country, regardless of language ability, and simultaneous translation is provided in both languages so that neither Spanish or English is privileged over the other. Formal meetings are arranged and hosted by IBWC/CILA. Meeting organizers balance the agenda between each side, and agendas are distributed in both languages prior to the meeting. The representatives from the U.S. are seated together and at the opposite side of the table from the Mexican representatives. The strict rules of engagement create a "ritualized" meeting space, with a high degree of "formalization" [21]. Formalization is characterized by restricted codes of behavior, rigid schedule of events, and invariance to form, and is often a way of clarifying social hierarchies [21]. In the case of formal TAAP meeting, the formalization serves to symbolically confer equal standing to participants from each side. The rules of engagement for TAAP meetings therefore help to moderate the effects of unequal relationships of power.

When relationships of power are managed, trust is more easily established. Examining the TAAP program with the Framework for Assessing Power in Collaborative Governance Processes is a useful exercise to illuminate the strengths and weaknesses in the TAAP program at mitigating unequal relationships of power and establishing trust. In the analysis above, resource-based power is shown to be the most significant and persistent source of power inequality. However, the Framework was not necessarily intended to examine international attempts at collaborative governance processes and therefore falls short of illuminating important power relationships and corresponding barriers to trust in the TAAP program. Referencing again Wolf's four modalities of power, the Framework provides tools to assess (1) personal power, (2) interpersonal power, and (3) the context in which power is exercised, but is does not address (4) structural power. Below, I examine the importance of structural power in shaping the context of relationships of power in the TAAP program.

3.3. Risk, Power, and Sovereignty

The gap in structural power between the United States and Mexico is particularly important to consider in the TAAP program, an endeavor that engages with two important ways that modern nation-states assert sovereignty: control over borderlands, and control of natural resources. Border scholars show that demonstrating control of national borders, particularly through militarization and policing, have become essential ways of asserting sovereignty in an era of globalization [22–27]. Likewise, control over and efficient use of natural resources is a central component of nation-states' claims of political legitimacy, modernity, and sovereignty [28,29]. Demonstrating appropriate control over natural resources in the U.S./Mexico borderlands is therefore deeply tied to structural power due to its symbolic significance in regard to assertions of sovereignty by each country.

Assertions of sovereignty at the border are material as well as symbolic [25], and the importance of the gap in structural power between the U.S. and Mexico is also both material and symbolic. Collaboration across structural power imbalance results in different levels of symbolic and material risk between sovereign nations with significantly different levels of political, economic, and military power. To understand the level of risk that each side faces, it is important to contextualize the relationships between the United States and Mexico in the history of exercises of power between the two nations. The last military conflict between the U.S. and Mexico, the Mexican–American War, ended in 1848 with the signing of the Treaty of Guadalupe in which Mexico ceded more than half its territory to the U.S. To understand the historical significance of this to the TAAP program, it is important to consider that each of the Water Resources Research Institutes of Arizona, New Mexico, and Texas are all located on land ceded to the U.S. by Mexico through military intervention. While this event occurred more than a century and a half ago, historical legacies haunt the present and expectations about the future. The United States military remains dramatically more powerful than that of Mexico. If the exchange of data about shared groundwater resources resulted in conflict over the resources, Mexico would be at a distinct disadvantage to defend its rights to water. Therefore, Mexico is taking a greater risk in collaborating than the United States, causing an asymmetrical relationship of power that hinders the development of trust.

There is also the perceived risk of losing control over groundwater resources through litigation. For the TAAP participants that I spoke to about risk, legal battles over shared groundwater resources were a more pressing concern than military conflict. One interviewee explained that U.S.-based domestic lawsuits have disrupted TAAP collaborations in the past, as Mexico does not want to be drawn into a legal battle over water rights. Instead, the interviewee explained, Mexico prefers to step back and wait for the U.S. lawsuit to be resolved before continuing binational work. Litigation between U.S. states therefore increases perceived risk of participation in TAAP and decreases trust in the process.

The TAAP Cooperative Framework includes provisions to limit the risk of participation by explicitly prohibiting TAAP data from being used to intervene in the water management practices of either sovereign nation. In the section titled "Principles of the Agreement", the last three of the six principles are aimed at this task, stating:

1. No provision set forth in this agreement will limit what either country can do independently in its own territory.
2. No part of this agreement may contravene what has been stipulated in the Boundary and Water Treaties between the two countries.
3. The information generated from these projects is solely for the purpose of expanding knowledge of the aquifers and should not be used by one country to require that the other country modify its water management and use [20].

The language of the agreement strongly imposes limits to what the data produced and shared through the TAAP program can be used for. Yet, data and research results, once made public, can be taken up in unexpected and undesirable ways, regardless of the intentions of the scientists involved. There will always be a degree of risk in producing and sharing data about shared resources, leading to the question: If risk is inherent in the TAAP data exchange, why trust at all?

TAAP participants choose to trust one another and the collaborative process as the risk of not exchanging knowledge about shared groundwater resources seems even greater. Along the U.S./Mexico border, urban centers, rural communities, industrial and agricultural economies, and ecosystems rely on groundwater resources. Border communities are looking to prolong the life of the aquifers they depend on. However, how can that be accomplished with incomplete knowledge of basic facts about shared aquifers? How many aquifers are shared across the U.S./Mexico border? What is the capacity of each aquifer? How far into each country does a particular aquifer extend? How much water is being extracted from each side? What factors are affecting the quality of the shared groundwater? The sharing of data that addresses these questions is essential to understanding the future of the binational aquifers and the futures of the communities that depend on them. Therefore, extending trust and engaging in collaborative binational groundwater assessments are worth the risk to participants.

There is also the hope amongst many participants that successful collaborative assessment of binational aquifers will eventually lead to successful cooperative management of the shared resources. As one interviewee said:

It really begins with us. If we can demonstrate that we can collect this information, and do it where it's equal and transparent, that's going to hopefully help the next effort, where they do start having discussions on how the resources are managed. People are afraid to take that step because they fear of losing control of their resource. But we're hoping that at some point, they'll think, we're more afraid that we're not going to have a resource if we don't talk about it ... So, I'm hopeful that our work is laying the groundwork for those future discussions.

The risk involved in the collaborative assessment of binational groundwater is magnified in attempts at collaborative governance. The asymmetry of power between the U.S. and Mexico makes Mexico particularly hesitant to engage in negotiations over water [30]. The hope is that the framework for mitigating unequal relationships of power put into practice in the TAAP program, as well as the years of successful binational collaboration and relationship building that has flourished under TAAP, will result in sufficient trust that collaborative binational resource management may become possible.

4. Discussion

In public presentations, official meetings, informal discussions, scholarly publications, and interviews, trust is identified as essential to the success of the TAAP program. There are specific behaviors and occurrences that contribute to trust, and others that detract from trust, as shown above. Yet, the question of why trust is so important is not often discussed. Considering the nexus of power, trust, and risk [11], trust is important due to what is at risk—loss of control by a sovereign nation of shared groundwater resources. The level of risk is high due to the unequal relationship of power that characterizes the United States and Mexico. The TAAP Cooperative Agreement and rules of engagement for formal TAAP meetings help to mitigate the unequal power relationship, but important asymmetries remain.

Consider the factors that contributed to trust listed in Table 1—ethnographic and interview data support that factors such as long-term engagement, relationship building, and face-to-face interactions are important to establishing trust in the TAAP program. However, the data also show that adherence to the listed factors does not always result in trust. I have witnessed TAAP participants who know each other well, have worked together for many years, who respect TAAP processes, and who communicate well about goals and boundaries, demonstrate resistance to further collaboration in the form of delaying action or refusing to commit to a project. Using my working definition of trust, an unwillingness to proceed in collaboration demonstrates a lack of trust. I argue that this is because, considering trust in relationship to risk and power, these factors only address differences in power at the first, second, and third levels of power identified by Wolf as described in the introduction. They do not diminish risk based on unequal relationships of power at the fourth, structural level of power. That same limitation is found in the factors identified in the Framework for Assessing Power in Collaborative Governance Processes. The TAAP Cooperative Framework does well to manage power inequalities at personal and interpersonal levels, and in the context of organizing and managing the exchange, but it cannot mitigate differences in structural power.

I argue that the gap in structural power is more of a hinderance at times of increased risk to a nation-state's ability to maintain sovereign control over its borderland water resources, such as a domestic lawsuit over water in the U.S. While work on other shared aquifers along the U.S./Mexico border continues to progress, work on the Mesilla Bolson/Conejos Medanos and the Hueco Bolson at the Texas/New Mexico/Chihuahua border has slowed and experienced setbacks in the last decade. Ethnographically, this was visible at the conclusion of a virtual Mesilla/Hueco TAAP meeting in 2021 when it came time to discuss next steps. A U.S. representative asked Mexico to commit to a consistent meeting time at agreed upon intervals throughout the year as a way to ensure the collaboration moved forward. A Mexican representative rejected the proposal. The U.S. representative then asked to set a date just for the next meeting. Again, the Mexican representative refused, stating that they needed to take time to review the data that the U.S. provided in the most recent exchange before committing to any further meetings. Earlier in the meeting, discussions of further collaborations beyond data exchanges were tabled by Mexico as contingent upon a successful data exchange.

Discussion by U.S.-based collaborators in regard to these slowdowns often center around trust at an interpersonal level—participants wondering what more they can do to earn the trust of the other side in order to continue. However, I interpret the refusal to progress in collaboration not as a lack of interpersonal trust, but as a reaction to increased risk of collaboration brought about by the still unresolved lawsuit over water resources brought to the Supreme Court in 2013 by the state of Texas against New Mexico. This increased risk highlights the unequal relationship of power between the two countries. In the context of this unequal relationship of power, Mexico's practice of delaying progress, tabling ideas, and rejection of establishing a set meeting schedule can all be understood as forms of negation as described by James Scott—ways for the less powerful party to assert agency without directly confronting the source of power [31]. Insistence on establishing relationships of trust before proceeding further is another way of "pumping the breaks" or slowing down the collaborative process. It can be frustrating for participants who are ready to progress with their scholarship, and for parties hoping to meet funding timelines and produce deliverables. However, using trust as a tactic to slow down, especially when used by the underpowered party, should be understood as an important tool for navigating very real risk and a form of asserting power in an asymmetrical context. The emphasis on trust makes sense when contextualized in the power, trust, and risk nexus.

When I asked participants if issues of power were ever discussed openly between TAAP participants from the U.S. and Mexico, no one could recall such an instance. In my ethnographic fieldwork, I did not observe frank conversations about power. Instead, trust is the acceptable language with which to address issues that I argue are rooted in power. The lack of frank conversations about power indicates that important asymmetrical power relationships between participants remain, despite efforts mediate such differences.

In this research, representatives of government agencies were more able to identify the structural risks of TAAP than were participating scholars, who tended to emphasize interpersonal trust. However, all participants should be encouraged to think critically about trust, risk, and power when engaging in binational collaborations. This is especially true in times of heightened risk, such as ongoing litigation over water resources. If refusals to move forward can be attributed to heightened risk for the Mexican collaborators, both sides may need to adjust in the short-term to keep the long-term goals of the program intact. For example, if both sides could agree to name another priority aquifer (or set of aquifers) that presents less risk, collaboration could continue while tabling stalled efforts on temporarily contentious aquifers. This would require addressing the gap in structural power head-on and working together to find suitable workarounds.

In this article, I use the working defined trust in TAAP collaborations as the willingness to proceed with collaboration despite risks. This definition is useful for analyzing trust in practice in TAAP collaborations; it is also useful for conceptualizing what trust is. Similar to Grimen [11], I analyze trust in relationship to risk and power, but it is important in this context to consider how manifestations of power are scaled hierarchically ask described by Wolf [12]. Trust is a social relationship, but it is not only an aspect of interpersonal relationships. Risk related to structural relationships of power is an essential component to understanding trust and its social function. This is particularly true when considering asymmetrical binational collaborations.

One interviewee stated, "it's critical that you establish trust and it doesn't matter how long it takes, it'll benefit whatever you're doing. I think that's been one of the most valuable lessons that I learned participating in the TAAP." Establishing trust in a context of unequal risk and power requires intentionality, commitment, patience, and a major investment of time. The binational relationships established, and scholarship collaboratively produced, through the TAAP program are therefore a commendable feat. Yet, the TAAP program would benefit from critical discussions of power and risk, particularly in reference to the gap in structural power between the U.S. and Mexico. Such conversations are particularly necessary when progress is stalled, revealing a lack of trust. Once underlying issues of power and risk are identified, creative and workable solutions could be identified to keep the important work of the TAAP program going.

Funding: The author's work has been supported by funding through the U.S. Geological Survey Award #G17AC00441 to New Mexico State University, entitled "Transboundary Aquifer Assessment Program".

Institutional Review Board Statement: The study was conducted according to the guidelines of the Declaration of Helsinki, and approved by the Institutional Review Board of New Mexico State University (protocol code 20982 and date of approval 23 October 2019).

Informed Consent Statement: Informed consent was obtained from all subjects involved in the study.

Data Availability Statement: Data sharing is not applicable to this study. Ethnographic and interview data is confidential.

Acknowledgments: The author thanks all those who helped shape this paper, including those who participated in interviews and informal discussions about the topic, and thanks those who offered advice and revisions.

Conflicts of Interest: The author declares no conflict of interest. The funders had no role in the design of the study; in the collection, analyses, or interpretation of data; in the writing of the manuscript, or in the decision to publish the results.

References

1. United States–Mexico Transboundary Aquifer Assessment Act. Public Law 109–448 2006 United States Congress. Available online: https://www.govinfo.gov/content/pkg/PLAW-109publ448/pdf/PLAW-109publ448.pdf (accessed on 15 June 2021).
2. Broch-Due, V.; Ystanes, M. (Eds.) *Trusting and Its Tribulations: Interdisciplinary Engagements with Intimacy, Sociality and Trust*; Berghahn Books: New York, NY, USA, 2016.
3. Coates, J. Trust and the Other: Recent Directions in Anthropology. *Soc. Anthropol.* **2019**, *27*, 100–105. [CrossRef]
4. Callegary, J.B.; Megdal, S.B.; Villaseñor, E.M.T.; Petersen-Perlman, J.D.; Sosa, I.M.; Monreal, R.; Gray, F.; Noriega, F.G. Findings and Lessons Learned from the Assessment of the Mexico-United States Transboundary San Pedro and Santa Cruz Aquifers: The Utility of Social Science in Applied Hydrologic Research. *J. Hydrol. Reg. Stud.* **2018**, *20*, 60–73. [CrossRef]
5. Eckstein, G.E. Rethinking Transboundary Ground Water Resources Management: A Local Approach along the Mexico-U.S. Border. *Georget. Int. Environ. Law Rev.* **2012**, *25*, 95.
6. Megdal, S.B.; Scott, C.A. The Importance of Institutional Asymmetries to the Development of Binational Aquifer Assessment Programs: The Arizona-Sonora Experience. *Water* **2011**, *3*, 949–963. [CrossRef]
7. Tapia-Villaseñor, E.M.; Megdal, S.B. The U.S.-Mexico Transboundary Aquifer Assessment Program as a Model for Transborder Groundwater Collaboration. *Water* **2021**, *13*, 530. [CrossRef]
8. Pink, S. Sensuous Futures: Re-Thinking the Concept of Trust in Design Anthropology. *Senses Soc.* **2021**, *16*, 193–202. [CrossRef]
9. Pedersen, E.O.; Liisberg, S. Introduction: Trust and Hope. In *Anthropology & Philosophy: Dialogues on Trust and Hope*; Berghahn Books: New York, NY, USA, 2015; pp. 1–19.
10. Fink, M.; Harms, R.; Möllering, G. Introduction: A Strategy for Overcoming the Definitional Struggle. *Int. J. Entrep. Innov.* **2010**, *11*, 101–105. [CrossRef]
11. Grimen, H. Power, Trust, and Risk: Some Reflections on an Absent Issue. *Med. Anthropol. Q.* **2009**, *23*, 16–33. [CrossRef] [PubMed]
12. Wolf, E.R. *Envisioning Power: Ideologies of Dominance and Crisis*; University of California Press: Berkley, CA, USA, 1999.
13. Purdy, J.M. A Framework for Assessing Power in Collaborative Governance Processes. *Public Adm. Rev.* **2012**, *72*, 409–417. [CrossRef]
14. Elliott, D.; Culhane, D. (Eds.) *A Different Kind of Ethnography: Imaginative Practices and Creative Methodologies*; University of Toronto Press: North York, ON, Canada, 2016.

15. Dewalt, K.M.; Dewalt, B.R.; Wayland, C.B. Participant Observation. In *Handbook of Methods in Cultural Anthropology*; Russell, H.B., Ed.; AltaMira Press: Walnut Creek, CA, USA, 1998.
16. Emerson, R.M.; Fretz, R.I.; Shaw, L.L. *Writing Ethnographic Fieldnotes*, 2nd ed.; University of Chicago Press: Chicago, IL, USA, 2011.
17. Saldaña, J. *The Coding Manual for Qualitative Researchers*, 3rd ed.; SAGE Publications Ltd.: Los Angeles, CA, USA; London, UK, 2015.
18. Bernard, R.H. *Research Methods in Anthropology: Qualitative And Quantitative Approaches*, 5th ed.; AltaMira Press: Lanham, MD, USA, 2011.
19. Patton, M.Q. *Qualitative Research & Evaluation Methods*, 3rd ed.; SAGE: Thousand Oaks, CA, USA, 2002.
20. TAAP Cooperative Framework. 2009. Available online: https://wrrc.arizona.edu/sites/wrrc.arizona.edu/files/images/Joint-Report-TAAP-8-19-09.pdf (accessed on 15 June 2021).
21. Bell, C. *Ritual: Perspectives and Dimensions*; Oxford University Press: Oxford, UK, 1997.
22. Alonso, A. Sovereignty, the Spatial Politics of Security, and Gender: Looking North and South from the US-Mexico Border. In *State Formation: Anthropological Perspectives*; Pluto Press: Ann Arbor, MI, USA, 2005.
23. Andreas, P. *Border Games: Policing the U.S.-Mexico Divide*; Cornell University Press: Ithaca, NY, USA, 2000.
24. De Genova, N. Migrant "Illegality" and Deportability in Everyday Life. *Annu. Rev. Anthropol.* **2002**, *31*, 419–447. [CrossRef]
25. Donnan, H.; Wilson, T.M. *Borders: Frontiers of Identity, Nation and State*, 1st ed.; Routledge: New York, NY, USA, 1999.
26. Rosaldo, R. Cultural Citizenship, Inequality, and Multiculturalism. In *Latino Cultural Citizenships*; Beacon Press: Boston, MA, USA, 1997.
27. Rosas, G. *Barrio Libre: Criminalizing States and Delinquent Refusals of the New Frontier*; Duke University Press: Durham, NC, USA, 2012.
28. Ferry, E.E.; Limbert, M.E. Introduction. In *Timely Assets: The Politics of Resources and Their Temporalities*; School for Advanced Research Press: Santa Fe, NM, USA, 2008; pp. 3–24.
29. Richardson, T.; Weszkalnys, G. Introduction: Resource Materialities. *Anthropological Quarterly.* **2014**, *87*, 5–30. [CrossRef]
30. García-Acevedo, M.; Ingram, H. Conflict in the Borderlands. *NACLA Rep. Am.* **2004**, *38*, 19–24. [CrossRef]
31. Scott, J.C. *Weapons of the Weak: Everyday Forms of Peasant Resistance*; Yale University Press: New Haven, CT, USA, 1985.

water

MDPI

Article

Hydrogeomorphologic Mapping of the Transboundary San Pedro Aquifer: A Tool for Groundwater Characterization

José Ismael Minjárez Sosa, Grisel Alejandra Gutiérrez Anguamea, Rogelio Monreal, Francisco Javier Grijalva Noriega and Elia M. Tapia-Villaseñor *

Departamento de Geología, Universidad de Sonora, Hermosillo 83000, Mexico; ismael.minjarez@unison.mx (J.I.M.S.); grisel.gutierrez@unison.mx (G.A.G.A.); rogelio.monreal@unison.mx (R.M.); francisco.grijalva@unison.mx (F.J.G.N.)
* Correspondence: elia.tapia@unison.mx

Abstract: Hydrogeomorphology is an emerging discipline that studies the relationship between landforms and hydrology, focusing on groundwater and surface water interactions. This study presents the methodology for the elaboration of a hydro-geomorphological map oriented to illustrate the relationships between the aquifer components and geomorphological characteristics in the United States-Mexico Transboundary San Pedro Aquifer (TSPA). This information contributes to a further understanding of the TSPA, facilitates the location of groundwater recharge and discharge zones, is useful for the development of sustainable groundwater management strategies, and could be useful in developing conceptual and numerical groundwater models for the region.

Keywords: hydrogeomorphology; transboundary aquifer; recharge; discharge; United States; Mexico

Citation: Sosa, J.I.M.; Gutiérrez Anguamea, G.A.; Monreal, R.; Grijalva Noriega, F.J.; Tapia-Villaseñor, E.M. Hydrogeomorphologic Mapping of the Transboundary San Pedro Aquifer: A Tool for Groundwater Characterization. *Water* **2022**, *14*, 906. https://doi.org/10.3390/w14060906

Academic Editors: Sharon B. Megdal and Anne-Marie Matherne

Received: 1 October 2021
Accepted: 14 February 2022
Published: 14 March 2022

1. Introduction

Granular and fractured aquifers represent an important source of fresh water in arid and semi-arid regions that are highly dependent on groundwater resources. Factors such as climate, topography, geomorphology, and lithology influence groundwater-flow interactions [1–4]. However, groundwater availability ultimately relies on the rainfall rate of the site, and the location and characteristics of the aquifer's recharge and discharge zones [2,3,5]. A better understanding of these areas contributes to the development of groundwater-management plans and strategies that promote water-resources sustainability, which is essential in transboundary settings where water resources are shared by two or more countries. Hydrogeomorphologic studies have proven to be useful for investigating the associations between landforms and hydrological processes that affect surface-water and groundwater flow, identifying the potential impacts of changes in land-use practices, and locating possible groundwater recharge and discharge areas [1,5–11].

A term first introduced in 1972, hydrogeomorphology broadly described the study of landforms produced by different hydrologic processes [10]. Over the years, hydrogeomorphology evolved into "an interdisciplinary science that focuses on the interaction and linkage of hydrologic processes with landforms or earth materials and the interaction of geomorphic processes with surface and subsurface water in temporal and spatial dimensions" [11]. Frequently, geomorphologic and hydrogeomorphologic studies have focused on flood assessment and surface-water controls, landslide assessment, and in atmosphere-hydrosphere-lithosphere interactions [12–15]. On the other hand, hydrogeomorphologic studies have also been associated with the analysis of groundwater resources (e.g., [1,5–7]). For instance, hydrogeomorphologic mapping allowed the identification and classification of hydro-objects in Southern Italy and the modeling of catchment contribution areas [16]. Additionally, scholars in this area of study have defined the connection between landforms and hydrology and expressed the need for a holistic approach that considers the relationships among landscape engineering, agriculture, natural areas, and water-resources management [17].

The importance of using criteria including landform, elevation, lithology, and hydrology for the assessment of groundwater resources has been described by Chaminé et al. (2015) [18]. In the arid Río Asunción Basin of Sonora, México, the correlation between the basin's groundwater-storage capacity and its lithologic units, ability to resist weathering and erosion, and presence of faults was described by Gutiérrez Anguamea (2013) [6]. The mapping methodology presented by [6] was later used for the development of a hydrogeomorphologic map for the state of Sonora, Mexico [7], an approach published in collaboration with the Mexican National Water Commission (CONAGUA) that serves as a guide for water-resources management in the region.

Located in Northwestern Mexico, the state of Sonora is bordered to the north by the state of Arizona in the United States. The United States and Mexico share history, culture, people, and water. A recent transboundary-characterization study indicates that based on geological correlations, there are 72 hydrogeologic units, or aquifers, that cross the U.S.-Mexico border [19]. One of these aquifers is the Transboundary San Pedro Aquifer (TSPA), located in the Arizona-Sonora border region. The TSPA is a Transboundary Aquifer Assessment Program (TAAP) aquifer of focus, which is a joint effort between the United States and Mexico to evaluate shared aquifers [20–23]. A number of studies and technical activities have been carried out through the TAAP in Arizona and Sonora over the last decade (i.e., [22–25]). For example, in 2016 the International Boundary and Water Commission published the Binational Study of the Transboundary San Pedro Aquifer [24]. This study, jointly developed by TAAP partners from the two countries, binationally described the physical geography, geology, hydrology, hydrogeology, and hydro-geochemistry of the TSPA. In this study, we aim to contribute to the TAAP knowledge base by using hydrogeomorphologic mapping as a tool for groundwater characterization, a novel approach that could guide land and water-management decisions in both the United States and Mexico.

2. Study Area

Located in the eastern portion of the Arizona-Sonora border, the TSPA is drained by the San Pedro River (Figure 1). The San Pedro River has its headwaters east of Cananea, Sonora, and flows northward to the United States until its confluence with the Gila River. The San Pedro River sustains hundreds of species—for example, it is an important bird habitat—and the basin contains one of the major unfragmented landscapes in the Southwest [26]. Several authors who have studied the aquifer basin have reported concerns regarding the impact of groundwater pumping on the San Pedro River, an ongoing issue that has captured the attention of scientists and stakeholders within the region [24,26–28].

The TSPA has an approximate area of 5000 km^2 and a population of around 97,235 (Table 1). The climate in this border region is arid to semi-arid with bimodal patterns of precipitation characterized by intense summer rains associated with the North American Monsoon and winter precipitation associated with the presence of Pacific cold fronts [24,27]. The mean annual precipitation in the Mexican portion of the TSPA has been reported to be 553 mm [28]. On the other hand, 330 mm were reported in Tombstone, and 960 mm on the Huachuca mountains in Arizona [24]. Mean annual temperature was reported to range between 12 °C and 18 °C [24]. The TSPA is located in the Basin and Range Province, bordering the Sonoran and Chihuahuan Deserts [24,27].

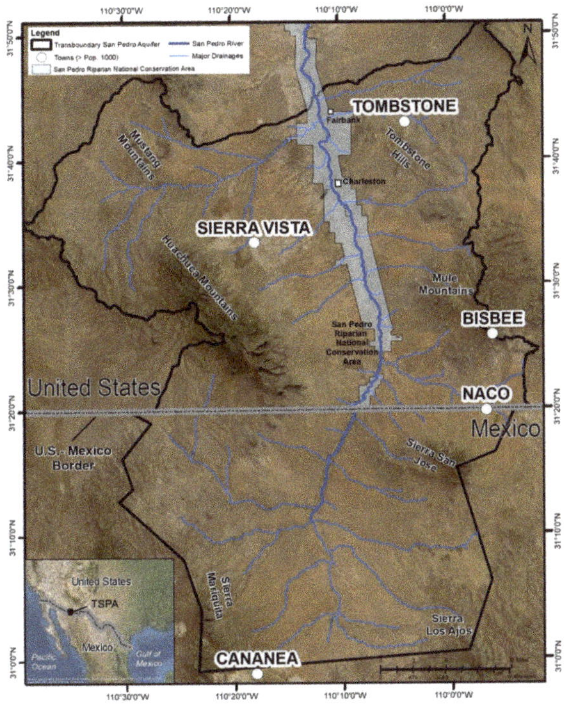

Figure 1. The Transboundary San Pedro Aquifer (TSPA). Source: Authors' development based on Callegary et al. (2016) [24].

Table 1. Population centers located in the TSPA. Source: INEGI, 2010, INEGI, 2020, U.S. Census Bureau 2020 [29–31].

Town	Population
Sierra Vista	45,308
Tombstone	1209
Naco	6064
Bisbee	5203
Cananea	39,451
Total	97,235

The major economic activities within the region include tourism and military operations in the United States, and livestock, agriculture, and mining in Mexico [24]. According to data from the Mexican Public Registry of Water Rights [32], 82 wells are registered in the Mexican portion of the TSPA under the following activities: 41 for livestock activities, two for industrial uses, 14 for agricultural uses, and 25 for public, urban, residential, and miscellaneous uses. Annual groundwater extractions from these wells equals 30.67 million cubic meters per year (MCM/year) [32], and in 2015, the aquifer registered an annual groundwater deficit of −7.49 MCM [28]. In 2014, groundwater demand in the U.S. portion of the TSPA, the Sierra Vista Sub-watershed, was reported to be 38.38 MCM/year (31,119 acre-feet per year) [33]. This region also reported a groundwater deficit of 5.21 MCM (4229 acre-feet per year) during the same year [33].

3. Materials and Methods

In this study, we analyzed the satellite imagery available through the ArcGIS Online Server [34] and combined the geologic and hydrogeologic information [24], and the topographic features (Digital Terrain Model SRTM1N30W109V3, [35]) to identify composition, topographic arrangement, and the presence or the absence of structures (see Figure 2). A visual inspection of satellite imagery allows for the differentiation of rock units based on the identification of textures and tones, i.e., smoothness, roughness, and compaction [36–38].

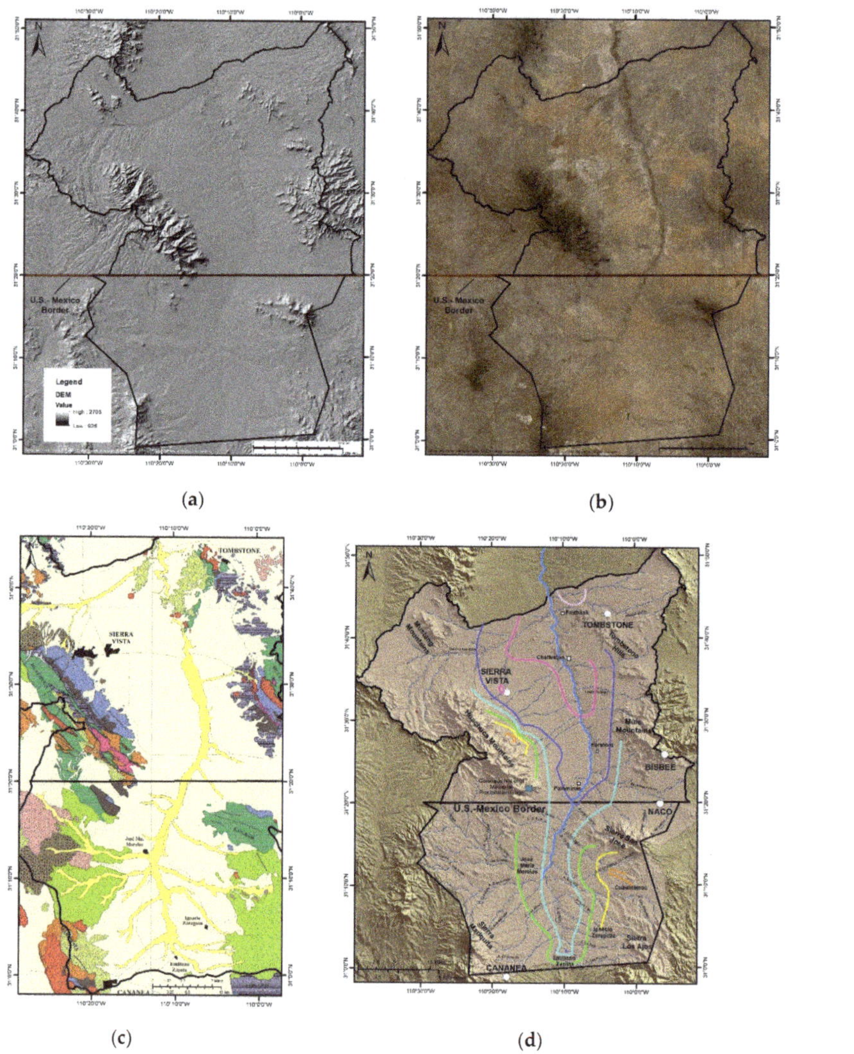

Figure 2. GIS layers used for the development of the hydrogeomorphologic map of the TSPA. (**a**) Digital Elevation Data [35]; (**b**) Satellite Image [34]; (**c**) Geology and Structures [24]; (**d**) Hydrography and Groundwater Levels [24].

3.1. Topographic Characterization (Landform Identification)

A topographic characterization was carried out using the Digital Terrain Model information available for the study site [35]. Elements within the study area were classified into four major types of terrain (described below): mountains, hills, piedmonts, and plains [39].

Mountains: Landforms with a relative height greater than 200 m, associated with endogenous folding processes, magmatism, vulcanism, and the dissection of endogenous formation structures [40]. Relative heights were considered from the base to the top of each formation analyzed in this study.

Hills: Landforms with a relative height less than 200 m. This group originates from the leveling of mountains (endogenous) or the dissection of a sloping plain (erosive exogenous). However, hills may be associated with low-elevation endogenous landforms or the product of quaternary tectonics [41].

Piedmonts: Mountainous margins or transitional zones distinguished by a change of slope and considerably lower height, ranging from 0 to 200 m depending on the behavior of the terrain. Piedmonts are composed of detrital material and present fluvial drainage [6,40].

Plains: Land surfaces with minimal slope and altitude difference. Correspond to the cumulative exogenous terrain of alluvial, wind, and coastal deposits [6,40]. The following factors were considered in the identification of a plain: land use (agricultural and urban), change in slope, and drainage pattern.

3.2. Geologic Characterization

A diverse tectonic evolution has shaped a complex geology in the TSPA with intrusive, metamorphic, volcanic-sedimentary, sedimentary, and volcanic rocks [24,30]. A Precambrian basement covered by sedimentary platform sequences—mainly carbonates—is exposed along southeastern Arizona/northeastern Sonora [24,30]. The oldest Mesozoic rocks within this region are Jurassic volcanic and sedimentary sequences covered by Cretaceous-Tertiary rocks, which are widely distributed throughout the TSPA [24,30]. The lithological units considered in this study, based on Callegary et al. (2016) [24], are shown in Table 2.

Table 2. Lithological units in the TSPA. Source: Authors' development based on Callegary et al. (2016) [24].

Legend	Lithostratigraphic Units	Description
	Precambrian Igneous-Metamorphic Complex	Igneous and metamorphic rocks
	Early Paleozoic Sedimentary Unit	Localized outcrops of detrital-carbonate rocks within the Mexican portion of the TSPA
	Late Paleozoic Sedimentary Unit	Limestone and sandstone exposed in most topographic highs in the TSPA
	Jurassic Felsic Volcano-Sedimentary	Intercalation of volcanic rocks, sandstones, agglomerates, basalt flows, sills, and intermediate composition
	Jurassic Intrusive Complex	Intrusive hypabyssal bodies mainly exposed on the U.S. side of the TSPA
	Late Jurassic–Early Cretaceous Sedimentary Unit	Conglomerate, sandstone, shale, and limestone from the Bisbee Group
	Late Cretaceous Sedimentary Unit (KsVs, Ks)	Sedimentary sequences

Table 2. *Cont.*

Legend	Lithostratigraphic Units	Description
	Cretaceous–Paleocene Volcano-Sedimentary Unit	Rhyolitic clastic and volcanic rocks
	Tertiary–Cretaceous Intrusive Complex	Intrusive felsic rocks
	Tertiary Felsic Volcanic Unit	Rhyolitic rocks from the west-central portion of the TSPA
	Tertiary Volcano-Sedimentary Unit	Continental rocks, mainly conglomerates with intercalations of sandstone and tuff
	Plio–Quaternary Sedimentary Unit	Coarse sediments (gravels and sands) distributed within the center of the TSPA
	Alluvium	Gravel, sands, silts, and clay.

3.3. Hydrological and Hydrogeological Information

The proposed hydrogeomorphologic map includes hydrological and hydrogeological data for a better visualization of the impact groundwater extractions have on the aquifer's distinct units. Information for this study includes a spatial layer with the hydrology, the locations of wells, and the groundwater levels for the year 2011. In addition, we identified the permeability, hydraulic conductivity of the rock units based on Gutiérrez Anguamea (2013) [6] and Freeze and Cherry (1979) [3] (Figure 3).

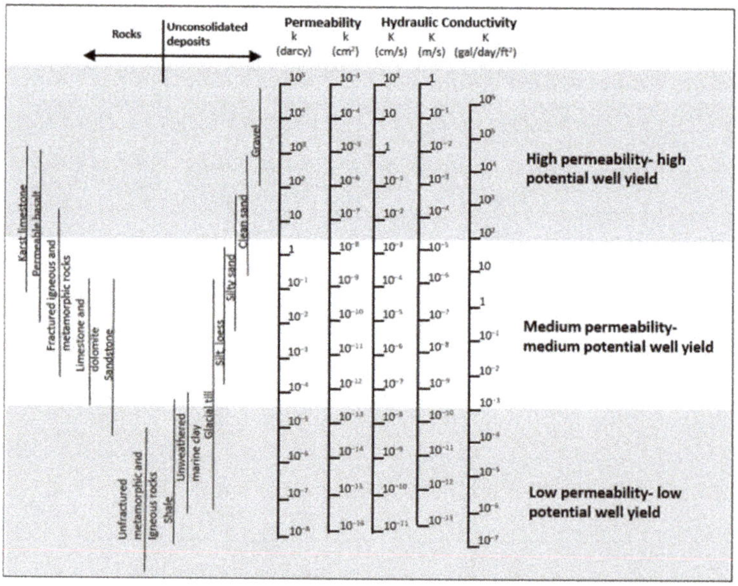

Figure 3. Permeability and hydraulic conductivity of rock units and unconsolidated deposits. Modified from Freeze and Cherry (1979) [3].

The primary permeability is a property directly related to the origin and formation of rock material to allow water to pass through it [42]; likewise, a secondary permeability can be interpreted based on the number and interconnection of structures that are present in a lithological unit [3,43,44]. Although it is true that the hydraulic potential of materials can be defined by direct and indirect methods—such as petrography, stratigraphy, and resistivity estimation [43]—in this study the permeability was determined based on the characteristics inherent to the formation of the rock (i.e., Figures 3 and 4) and its subsequent fracturing by movements of the earth crust.

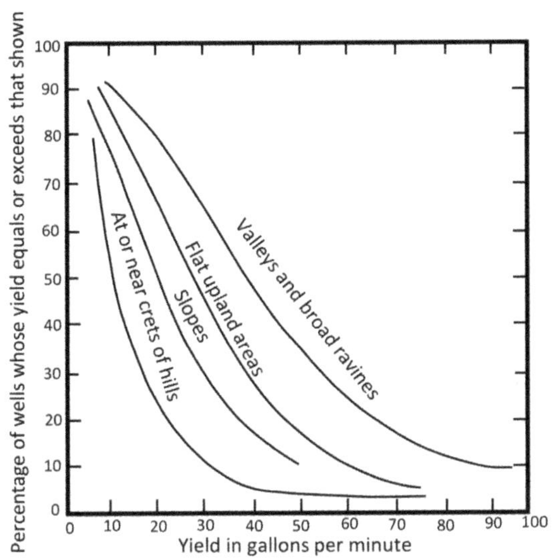

Figure 4. Relation between landform and potential well yield. Source: Freeze and Cherry (1979) [3] and LeGrand (1954) [45].

It is also presumed that the combination of permeability (primary and/or secondary) with the shape of the terrain is directly related to the potential well yield of a lithological unit (Figure 3). In other words, a portion of materials with a significant primary permeability, such as a smoothed conglomerate hill, is likely to allow water to flow through it easily, as it is composed of elements with varied granulometry and flat topography [6]. In contrast to the above, when it comes to more compact and steeper materials where the speed of surface runoff increases and the spaces between the rock crystals are smaller, the potential well yield of groundwater can be reduced [6].

The combination of the aforementioned factors allowed for the assignment of a groundwater permeability and potential well yield category to each of the elements contained in the TSPA. Categories were based on the aquifer materials, permeability, and hydraulic conductivity to describe how much and how quickly water moves within an aquifer [6].

4. Results

Based on the hydrogeomorphological analysis of the TSPA, a total of 22 units of high, medium, low, very low, and very low/null permeability/potential well yields were characterized (Table 3, Figure 5). The legend of the map is based on [46]. Extensive and highly productive (high aquifer well yield) intergranular aquifer units are shown in shades of blue. The green color range represents fissured environments. The brown areas signify the units of local extension and limited resources, as well as those that are considered to have very low well yield. Alluvial plains, cultivated plains, upper divergent plains, and unconsolidated polymictic conglomerate foothills were identified as intergranular mediums with high permeability and potential well yield.

Table 3. Hydrogeomorphologic Units in the Transboundary San Pedro Aquifer.

Recharge/Non-Recharge	Unit	Description
Discharge Zone	High permeability intergranular environment (high potential well yield)	Crop plain
		Upper divergent plain
		Alluvial plain
		Unconsolidated polymictic conglomerate piedmont
		Water
	Medium permeability environment (medium potential well yield)	Unconsolidated polymictic conglomerate hill
	Low permeability environment (low potential yield)	Consolidated polymictic conglomerate hill
		Consolidated polymictic conglomerate and basalt hill
Recharge Zone	Medium permeability fissured environment (medium potential well yield)	Fissured limestone, sandstone, and shale hill
		Fissured sandstone and shale mountain
	Low permeability fissured environment (low potential well yield)	Fissured volcanic mountain
	Very low permeability fissured environment (very low potential well yield)	Fissured polymictic conglomerate and volcanic mountain
		Fissured limestone, sandstone, and shale mountain
		Fissured volcanic and sandstone mountain
		Fissured plutonic mountain
		Fissured metamorphic mountain
		Fissured volcanic hill
		Fissured plutonic hill
		Fissured metamorphic hill
Impervious Areas		Urban zone
		Volcanic hill
		Plutonic hill

Regarding the units of medium permeability/potential well yield, only hills of unconsolidated polymictic conglomerate were identified as such. Conglomerate hills, consolidated polymictic conglomerate hills, consolidated basalts, consolidated polymictic conglomerate foothills, consolidated sands, polymictic conglomerate mountains, and polymictic conglomerates were identified as units of limited potential well yield.

The presence of fractures and faults indicates fissured environmental units with a very low permeability and potential well yield. A second subcategory comprises medium permeability fissured units such as fissured limestone, sandstone and shale hill, polymictic conglomerate mountain and fissured sandstone, sandstone mountain, and fissured shale. A third category, the fissured volcanic mountain, was identified in a fissured medium of low permeability and potential well yield. Finally, those units whose permeability is characterized as very low/null are considered non-aquifer units. According to the flow-direction lines identified for the study area, groundwater flows from south to north and towards the San Pedro River. This information is consistent with Callegary et al. (2016) [24], who also identified cones of depressions near the cities of Sierra Vista, Tombstone, and Cananea. For this study, the hydrological discharge zones were located within the San Pedro River and its tributaries, while the recharge areas were mainly located within mountainous areas: the Huachuca, Mule, and Mustang Mountains (in the United States) and the Sierra Mariquita, Sierra Los Ajos, and Sierra San Jose (in Mexico).

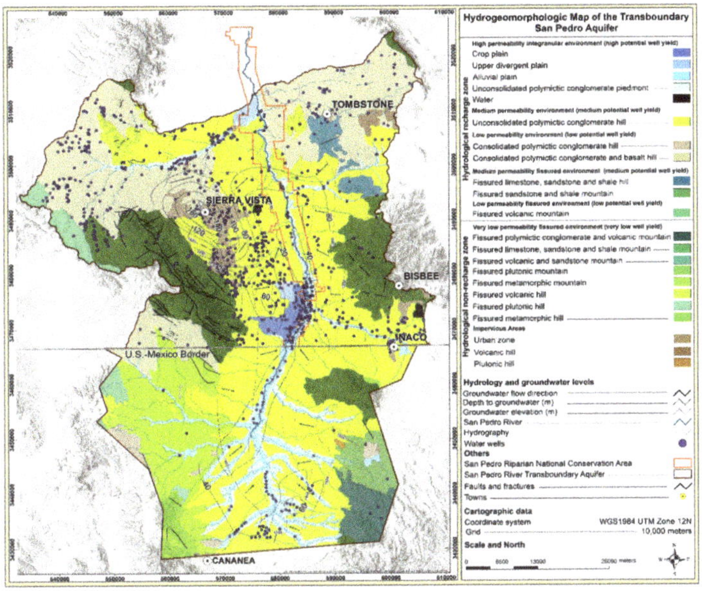

Figure 5. Hydrogeomorphologic map of the San Pedro River Basin.

5. Discussion and Conclusions

The TSPA is an aquifer shared between the United States and Mexico. It is also a TAAP aquifer of focus that has been deeply studied over the last decade. The *Binational Study of the Transboundary San Pedro Aquifer* [24] is one of the most relevant binational studies of the region and includes information regarding the physical geography, geology, hydrology, hydrogeology, and geochemistry of the region. According to Chaminé et al. (2015) [18], groundwater characterization must be approached based on different disciplines. These disciplines might include geology, hydrology, hydrogeology, and geochemistry, but also geomorphology and hydrogeomorphology.

Hydrogeomorphology studies describe the interactions between hydrologic processes, landforms, and lithology, and can be useful for determining potential well yields, along with recharge and recharge zones. In this study, we developed a hydrogeomorphologic map for the TSPA. This aquifer is currently experiencing groundwater deficit, with mining, military, domestic, and agricultural users competing for groundwater resources. Hydrogeomorphologic units are defined based on their formation, composition, and original texture of the different rock formations [6]. According to this study, highlands constitute potential recharge zones, and lowlands serve as groundwater-flow discharge areas.

This map makes it possible to quickly identify the functioning of the aquifer system, with the recharge and discharge zones clearly discernible. The information presented here can be used as the basis for the development of sustainable water-resources strategies that consider the hydrogeomorphic characteristics of a given aquifer region to determine how feasible it is to extract or continue extracting water in that area. Moreover, the assessment of binational aquifers needs to use consistent and harmonized methodologies to identify discharge and recharge areas in need of conservation efforts. The application of this methodology allows the locations of these areas to be identified within the framework of a pragmatic morphogenetic mapping.

Author Contributions: Conceptualization, J.I.M.S. and E.M.T.-V.; supervision, J.I.M.S. and E.M.T.-V.; methodology G.A.G.A. and J.I.M.S., validation, J.I.M.S., formal analysis, G.A.G.A., investigation, J.I.M.S. and G.A.G.A., resources, J.I.M.S., R.M., and F.J.G.N., data curation, G.A.G.A., writing—J.I.M.S., E.M.T.-V., G.A.G.A., and R.M., original draft preparation, J.I.M.S., writing—review and editing, E.M.T.-V., visualization, E.M.T.-V. and G.A.G.A., project administration, E.M.T.-V. All authors have read and agreed to the published version of the manuscript.

Funding: This work was partially funded by the U.S. Geological Survey (funding authorized by P.L. 109–448) Award Number G17AC00439 for the Transboundary Aquifer Assessment Program.

Institutional Review Board Statement: Not applicable.

Informed Consent Statement: Not applicable.

Data Availability Statement: Not applicable.

Acknowledgments: The authors would like to acknowledge the Transboundary Aquifer Assessment Program (TAAP).

Conflicts of Interest: The authors declare no conflict of interest. The funders had no role in the design of the study; in the collection, analyses, or interpretation of data; in the writing of the manuscript; or in the decision to publish the results.

References

1. Teixeira, J.; Chaminé, H.I.; Marques, J.E.; Gomes, A.; Carvalho, J.M.; Albertí, A.P.; Rocha, F.T. Integrated approach of hydrogeomorphology and gis mapping to the evaluation of ground water resources: An example from the hydro-mineral system of caldas da cavaca, nw portugal. In Proceedings of the 33rd International Geological Congress, Oslo, Norway, 6–14 August 2008 .
2. Fetter, C.W. *Applied Hydrogeology*; Pearson: London, UK, 2000.
3. Freeze, R.A.; Cherry, J.A. *Groundwater*; Prentice-Hall: Hoboken, NJ, USA, 1979.
4. Surrette, M.; Allen, D.M.; Journeay, M. Regional evaluation of hydraulic properties in variably fractured rock using a hydrostructural domain approach. *Hydrogeol. J.* **2007**, *16*, 11. [CrossRef]
5. Teixeira, J.; Chaminé, H.I.; Carvalho, J.M.; Pérez-Alberti, A.; Rocha, F.T. Hydrogeomorphological mapping as a tool in groundwater exploration. *J. Maps* **2013**, *9*, 263–273. [CrossRef]
6. Gutiérrez Anguamea, G.A. Metodología para la Elaboración del Mapa Hidrogeomorfológico de la Cuenca del Rio Asunción. Master's Thesis, Universidad de Sonora, Hermosillo, Mexico, 2013.
7. Gutiérrez Anguamea, G.A.; Minjárez Sosa, J.I.; Oroz Ramos, L.A. *Mapa Hidrogeomorfológico del Estado de Sonora*; CONAGUA: Hermosillo, Mexico, 2015.
8. Babar, M. *Hydrogeomorphology: Fundamentals, Applications and Techniques*; New India Publishing Agency: New Delhi, India, 2005.
9. Leopold, L.B. Field data: The interface between hydrology and geomorphology, scientific basis of water-resource management. In *Studies in Geophysics*; The National Academies Press: Cambridge, MA, USA, 1982.
10. Scheidegger, A.E. Hydrogeomorphology: Review. *J. Hydrol.* **1973**, *20*, 193–215. [CrossRef]
11. Sidle, R.C.; Onda, Y. Hydrogeomorphology: Overview of an emerging science. *Hydrol. Processes* **2004**, *18*, 597–602. [CrossRef]
12. Gregory, K.J. Hydrogeomorphology: How applied should we become? *Prog. Phys. Geogr.* **2016**, *3*, 84–101. [CrossRef]
13. Barrera, F.B.; Borja, C.; Núñez, M.F.; Lama, A. Dinámica hidrogeomorfológica e impacto antrópico en la cuenca del arroyo del partido (NW del Parque Nacional de Doñana, Huelva, España): Evaluación de procesos actuales. *Cuatern. Geomorfol. Rev. Soc. Española Geomorfol. Asoc. Española Para Estud. Cuatern.* **2009**, *23*, 45–64.
14. Arnaud-Fassetta, G.; Cossart, E.; Fort, M. Hydro-geomorphic hazards and impact of man-made structures during the catastrophic flood of June 2000 in the Upper Guil catchment (Queyras, Southern French Alps). *Geomorphology* **2005**, *66*, 41–67. [CrossRef]
15. Verstappen, H.T.; van Zuidam, R.A. *The ITC System of Geomorphologic Survey: A Basis for the Evaluation of Natural Resources and Hazards*; ITC Publication: Murarrie, Australia, 1981.
16. Guida, D.; Cuomo, A.; Palmieri, V. Using object-based geomorphometry for hydro-geomorphological analysis in aMediterranean research catchment. *Hydrol. Earth Syst. Sci.* **2016**, *20*, 3493–3509. [CrossRef]
17. Kudrna, K.; Šindelářová, M. Principles of hydrogeomorphology as a basic precondition for solution of territorial structure of unitary system of agricultural, forest and water management zákony hydrogeomorfologie jako základní předpoklad pro řešení teritoriální struktury unitární soustavy zemědělského, lesního a vodního hospodářství. *J. Cent. Eur. Agric.* **2006**, *7*, 669–676.

18. Chaminé, H.I.; Carvalho, J.M.; Teixeira, J.; Freitas, L. Role of hydrogeological mapping in groundwater practice: Back to basics. *Eur. Geol. J.* **2015**, *40*, 34–42.
19. Sanchez, R.; Rodriguez, L. Transboundary Aquifers between Baja California, Sonora and Chihuahua, Mexico, and California, Arizona and New Mexico, United States: Identification and Categorization. *Water* **2021**, *13*, 2878. [CrossRef]
20. IBWC. *Joint Report of the Principal Engineers Regarding the Joint Cooperative Process United States-Mexico for the Transboundary Aquifer Assessment Program*; International Boundary and Water Comission: El Paso, TX, USA, 2009.
21. Tapia-Villaseñor, E.M.; Megdal, S.B. The U.S.-Mexico Transboundary Aquifer Assessment Program as a Model for Transborder Groundwater Collaboration. *Water* **2021**, *13*, 530. [CrossRef]
22. Callegary, J.B.; Megdal, S.B.; Tapia Villaseñor, E.M.; Petersen-Perlman, J.D.; Minjárez Sosa, I.; Monreal, R.; Gray, F.; Grijalva Noriega, F. Findings and lessons learned from the assessment of the Mexico-United States transboundary San Pedro and Santa Cruz aquifers: The utility of social science in applied hydrologic research. *J. Hydrol. Reg. Stud.* **2018**, *20*, 60–73. [CrossRef]
23. Petersen-Perlman, J.D.; Albrecht, T.R.; Tapia-Villaseñor, E.M.; Varady, R.G.; Megdal, S.B. Science and Binational Cooperation: Bidirectionality in the Transboundary Aquifer Assessment Program in the Arizona-Sonora Border Region. *Water* **2021**, *13*, 2364. [CrossRef]
24. Callegary, J.B.; Minjarez, I.; Tapia Villaseñor, E.M.; dos Santos, P.; Monreal Saavedra, R.; Grijalva Noriega, F.J.; Scott, C.A.; Megdal, S.B.; Oroz Ramos, L.A.; Rangel Medina, M.; et al. *Binational Study of the Transboundary San Pedro Aquifer*; International Boundary and Water Comission: El Paso, TX, USA, 2016.
25. Tapia-Villaseñor, E.M.; Shamir, E.; Megdal, S.B.; Petersen-Perlman, J.D. Impacts of Variable Climate and Effluent Flows on the Transboundary Santa Cruz Aquifer. *JAWRA J. Am. Water Resour. Assoc.* **2020**, *56*, 409–430. [CrossRef]
26. The Nature Conservancy. *San Pedro River*; The Nature Conservancy: Arlington, VA, USA, 2022.
27. Minjarez Sosa, I.; Monreal Saavedra, R.; Rangel Medina, M.; Grijalva Noriega, F.; Ochoa Granillo, A.; Tapia Villaseñor, E.M.; Ramírez Días, L.L.; Cirett Galán, S.; Moncada García, G.; Montijo González, A. *Actividades Hidrogeológicas en el Acuífero Río San Pedro, Estado de Sonora, Primera Etapa*; Universidad de Sonora, CILA, CONAGUA: Hermosillo, Mexico, 2011.
28. CONAGUA. *Actualización de la Disponibilidad Media Anual de Agua en el Acuífero Río San Pedro (2616), Estado de Sonora*; Diario Oficial de la Federación: Ciudad de México, Mexico, 2015.
29. Instituto Nacional de Estadística y Geografía (INEGI). *Censo de Población y Vivienda 2010*; INEGI: Ciudad de México, Mexico, 2010; Available online: https://www.inegi.org.mx/programas/ccpv/2010/ (accessed on 30 September 2021).
30. Instituto Nacional de Estadística y Geografía (INEGI). *Censo de Población y Vivienda 2020*; INEGI: Ciudad de México, Mexico, 2020; Available online: https://www.inegi.org.mx/programas/ccpv/2020/ (accessed on 30 September 2021).
31. United States Census Bureau. Census Bureau QuickFacts: Sierra Vista City, Arizona—Tucson City, Arizona. 2022. Available online: https://www.census.gov/quickfacts/fact/table/sierravistacityarizona,tucsoncityarizona/BZA210219 (accessed on 30 September 2021).
32. CONAGUA. Consulta a la Base de Datos del Registro Público de Derechos de Agua (REPDA). Available online: https://www.gob.mx/conagua/acciones-y-programas/consulta-la-base-de-datos-del-repda (accessed on 12 May 2021).
33. Watershed Management Group; City of Sierra, V. A Stormwater Action Plan for Sierra Vista. 2016. Available online: https://verderiver.org/wp-content/uploads/2017/07/stormwater-action-plan-for-sierra-vista.pdf (accessed on 30 September 2021).
34. Esri; Maxar; GeoEye; Earthstar Geographics; CNES/Airbus DS; USDA; USGS; AeroGRID; IGN; GIS User Community. World Map. 2021. Available online: https://services.arcgisonline.com/ArcGIS/rest/services/World_Imagery/MapServer (accessed on 30 September 2021).
35. Farr, T.G.; Rosen, P.A.; Caro, E.; Crippen, R.; Duren, R.; Hensley, S.; Kobrick, M.; Paller, M.; Rodriguez, E.; Roth, L.; et al. The Shuttle Radar Topography Mission. *Rev. Geophys.* **2007**, *45*, RG2004. [CrossRef]
36. Panizza, M. *Environmental Geomorphology. Developments in Earth Surface Processes*; Elsevier: Amsterdam, The Netherlands, 1996; Volume 4.
37. Muñoz Jiménez, J. Paisaje y geosistema, una aproximación desde la geografía física. In *Seminario "Paisaje y Medio Ambiente"*; Universidad de Valladolid, Secretariado de Publicaciones e Intercambio Editorial: Valladolid, Spain, 1996.
38. García-Romero, A. An Evaluation of Forest Deterioration in the Disturbed Mountains of Western Mexico City. *Mt. Res. Dev.* **2002**, *22*, 270–277. [CrossRef]
39. Tapia, V.G.; Lopez, B. Mapeo geomorfológico analítico de la porción central de la Cuenca de México: Unidades morfogenéticas a escala 1:100,000. *Rev. Mex. Cienc. Geológicas* **2002**, *19*, 50–65.
40. Lugo Hubp, J. *Diccionario Geomorfológico*; Instituto de Geografía, Universidad Nacional Autónoma de México: Ciudad de México, Mexico, 1989.
41. Córdova, C. Cartografía Geomorfológica en Escala Pequeña del Occidente de la República Mexicana. Bachelor's Thesis, Facultad de Filosofía y Letras, Universidad Autónoma de México, Ciudad de México, Mexico, 1988.
42. Castany. *Tratado Práctico de las Aguas Subterráneas. Primera Parte. Elementos de Hidrología de Superficie*; Ed. Omega: Barcelona, Spain, 1971.
43. Pedroza, V.; Gutierrez, M.; Le Roux, J.; Hernández, E.; Vasquez, J.P. Estudio Comparativo del Coeficiente de Permeabilidad K. Available online: https://biblioteca.sernageomin.cl/opac/DataFiles/14905_v2_pp_238_240.pdf (accessed on 30 September 2021).
44. Díaz, E.J.; Contreras, N.M.; Pinto, J.E.; Velandia, F.; Morales, C.J.; Hincapie, G. Evaluación hidrogeológica preliminar de las unidades geológicas de la mesa de los santos, santander. *Boletín Geol.* **2009**, *31*, 10.

45. LeGrand, H. *Geology and Ground Water in the Statesville Area*; North Carolina Department Conservation Development, Divison Mineral Resources: Raleigh, NC, USA, 1954.
46. UNESCO. *International Legend for Hidrogeological Maps*; UNESCO: Paris, France, 1983.

water

Review

A Review of Climate Change Impacts on the USA-Mexico Transboundary Santa Cruz River Basin

Eylon Shamir [1,*], Elia M. Tapia-Villaseñor [2], Mary-Belle Cruz-Ayala [3] and Sharon B. Megdal [4]

1 Hydrologic Research Center, 11440 West Bernardo Court, Suite 375, San Diego, CA 92127, USA
2 Departamento de Geología, Universidad de Sonora, Hermosillo 83000, Mexico; elia.tapia@unison.mx
3 Arid Lands Resource Sciences, The University of Arizona, Tucson, AZ 85719, USA; marybelca@email.arizona.edu
4 Water Resources Research Center, The University of Arizona, Tucson, AZ 85719, USA; smegdal@arizona.edu
* Correspondence: eshamir@hrcwater.org

Abstract: In the parched Upper Santa Cruz River Basin (USCRB), a binational USA–Mexico basin, the water resources depend on rainfall-triggered infrequent flow events in ephemeral channels to recharge its storage-limited aquifers. In-situ data from the basin highlight a year-round warming trend since the 1980s and a concerning decline in average precipitation (streamflow) from 1955–2000 to 2001–2020 by 50% (87.6%) and 17% (63%) during the winter and summer, respectively. Binational sustainable management of the basins water resources requires a careful consideration of prospective climatic changes. In this article we review relevant studies with climate projections for the mid-21st century of four weather systems that affect the region's precipitation. First, the North American Monsoon (NAM) weather system accounts for ~60% of the region's annual rainfall. The total NAM precipitation is projected to decline while heavy rainfall events are expected to intensify. Second, the frequency of the pacific cold fronts, the region's prevalent source of winter precipitation, is projected to decline. Third, the frequency and intensity of future atmospheric rivers, a weather system that brings winter rainfall to the region, are projected to increase. Fourth, the frequency and intensity of large eastern pacific tropical cyclones (TC) are expected to increase. On rare occasions, remnants of TC make their way to the USCRB to cause storms with considerable impact on the region's water resources. In contrast to the high confidence projections for the warming trend to persist throughout the mid-21st century, the precipitation projections of these four weather systems affecting the region encompass large uncertainties and studies have often reported contradicting trends. An added source of uncertainty is that the USCRB is located at the periphery of the four rain-bearing weather systems and small mesoscale changes in these weather systems may have accentuated impacts on their edges. Despite the high uncertainty in the projections of future precipitation, the early 21st century drying trend and the projected mid-21st century decline in precipitation events serve as a pressing call for planning and actions to attain sustainable water resources management that reliably satisfies future demands.

Keywords: Santa Cruz River; climate change; water resources; transboundary aquifer; transboundary aquifer assessment; Arizona; Sonora

Citation: Shamir, E.; Tapia-Villaseñor, E.M.; Cruz-Ayala, M.-B.; Megdal, S.B. A Review of Climate Change Impacts on the USA-Mexico Transboundary Santa Cruz River Basin. *Water* **2021**, *13*, 1390. https://doi.org/10.3390/w13101390

Academic Editor: Maria Mimikou

Received: 24 April 2021
Accepted: 10 May 2021
Published: 16 May 2021

1. Introduction

Water supply in the Upper Santa Cruz River Basin (USCRB), a binational United States of America (USA)–Mexico basin, relies on a relatively storage-limited aquifer system that is recharged primarily during occasional rain-triggered flow events in ephemeral desert channels. Because of the region's scarce water resources and the added complexity that requires bridging binational regulatory and policy differences in order to manage the shared aquifer, the region was selected for the USA–Mexico Transboundary Aquifer Assessment Program (TAAP), a program that aims to improve the knowledge base of transboundary aquifers between the United States and Mexico [1–4]. In order to consider a sustainable water resources plan for the USCRB that addresses binational needs, a comprehensive impact assessment of future climatic changes on the region's water resources is needed. However, identifying quantitative future climate projections for the region that should be used for hydrological impact assessments is challenging. This is because of the region's complex and variable climate, which includes two very different rainy seasons (i.e., winter and summer). Winter (November–March) rain is caused by cold fronts from the Pacific Ocean that yield widespread stratiform rainfall events while summer rain (July–September) is triggered by the North American Monsoon (NAM) weather system that brings intense, brief, and local rainfall events.

The Southwest chapter of the USA Fourth National Climate Assessment (hereinafter, Fourth Assessment) [5] provides projections for the mid-21st century that, when being interpreted for the USCRB, are uncertain and point to conflicting trends. For the winter, the Fourth Assessment projects an increase in the frequency of high-pressure weather systems that would trigger longer durations of dry spells, while also projecting an increase in the frequency of rain-bearing atmospheric rivers to hit the Pacific Ocean's eastern shore. For the summer, although the Fourth Assessment projects an increase in extreme daily precipitation due to a warmer atmosphere that can hold larger amount of water vapor, it also states that the projected total summer precipitation is uncertain. The Fourth Assessment's future projections, which cover the entire domain of southwest USA, are clearly too general and insufficient as a quantitative projection to be used for hydrologic impact assessment in the USCRB. In addition, because the USCRB is located at the periphery of both the winter and summer rain-bearing weather systems (as we will further discuss), the region is very sensitive to small changes in these prevailing weather systems. The challenge of climate models to represent the weather at the periphery of these rain-bearing weather systems is an added source of uncertainty to the future climate projections of the USCRB.

In order to conduct a quantitative hydrologic impact assessment, time series of projected future precipitation and surface temperature are needed. These time series are commonly available from dynamic simulations of coupled atmospheric-ocean global climate models (GCMs). The GCMs are often simulated in a spatial resolution that is too coarse to represent the salient regional climatic features and therefore the GCMs' simulations require additional spatial downscaling. The uncertainty in the future rainfall and temperature projections for the USCRB is demonstrated in Figure 1. In this figure, the projected mid-21st century (2040–2069) changes in total winter (DJF) and summer (JJA) precipitation and temperature for the study region are shown. These projections are from 20 statistically downscaled CMIP5 Global Climate Models (GCM) with Representative Concentration Pathway 4.5 and 8.5 (RCP 4.5 and RCP 8.5), which are future greenhouse gas (GHG) concentration trajectories [6]. The simulated projections all agree on the warming trends and the warmer projections are seen for RCP 8.5, which is the higher GHG concentration scenario. Concerning precipitation, the average projected changes for both GHG concentration trajectories suggest a future with a slightly (less than <5%) wetter summer and dryer winter. However, there is a large spread and contradicting trends among the GCM projections, pointing to wetter and/or dryer winters and/or summers. The uncertainty in the projected precipitation is even larger when considering GCMs that are dynamically downscaled using Regional Climate Models (RCM) [7,8]. Furthermore, the projected changes in future precipitation are highly magnified as precipitation transforms to runoff and streamflow that is eventually recharged into the aquifers [7,9,10].

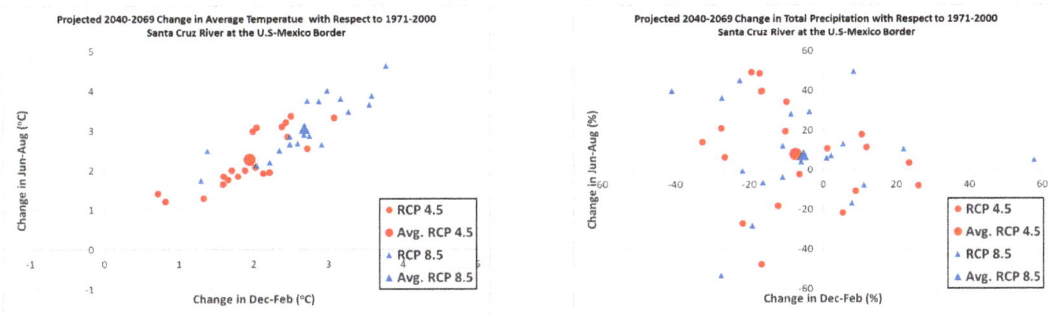

Figure 1. Projected 20402069 changes, as compared with 1971-2000, of average temperature and average precipitation relative changes (percent) in December-February (X-axis) and June-August (Y-axis). The projections are from twenty 4-km statistically downscaled CMIP5 GCMs with RCP 4.5 and RCP 8.5 emission scenarios. The downscaling is based on the Multivariate Adaptive Constructed Analogs procedure [6] using meteorological data and interpolation method as described in Abatzoglou, (2013) [11]. The dataset was retrieved from https://climate.northwestknowledge.net/MACA/index.php (accessed on 13 May 2021).

In view of this large uncertainty, our objective in this paper is to review and synthesize the current knowledge on climate change trends in precipitation and temperature that are relevant to the USCRB.

2. Study Area

The Santa Cruz River flows southward from its origin in the San Raphael Valley in south-central Arizona to cross the international border into the state of Sonora, Mexico. Flowing ~60 km into Mexico, the river initially flows south, bends west, and then bends again to the north and crosses the international border back into the USA, about 8 km east of the city of Nogales, Arizona. From its border crossing, the river flows in a general northward direction to its confluence with the Gila River, which is a tributary of the Colorado River that drains much of southern Arizona and parts of western New Mexico. Our region of interest is the Upper Santa Cruz River Basin (USCRB) (Figure 2), which is composed of the drainage area of the river's headwater in the San Rafael Valley USA (470 km^2), the Mexican portion of the basin (1122 km^2), and the USA Santa Cruz Active Management Area (1791 km^2).

Based on Scott et al., (2012) [12], about 60% of the water in the USCRB is consumed by municipal demand (12.8 and 20.9 MCM/Yr in USA and Mexico, respectively) to primarily supply the twin cities of Nogales, Sonora (~220,000 people) and Nogales, Arizona (~20,000 people), the two largest population centers in the USCRB. The remaining ~40% of the water is consumed by irrigated agriculture (13.2 and 7.4 MCM/Yr in USA and Mexico, respectively). The main source of water supply in the basin comes from relatively shallow alluvial aquifers with limited storage capacity. These aquifers that are also known as the Transboundary Santa Cruz Aquifer system are composed of the Santa Cruz River Aquifer and Nogales Aquifer in Mexico, and the aquifer system of the Santa Cruz Active Management Area and San Rafael Valley in USA. The aquifers are being recharged through infiltration during occasional and highly variable rainfall-driven streamflow events in the ephemeral channels of the Santa Cruz River and its tributaries (e.g., [13,14]). The relatively limited storage of the aquifers and their dependence on streamflow events for recharge makes the region's water resources availability tightly linked to the local prevailing climate and its variability (e.g., [9,15–17]).

Figure 2. Map of the study region.

3. Climate and Historical Trends

The climate at the USCRB is classified as hot semi-arid according to Köppen-Geiger, with an extremely hot summer (average daytime temperature is about 40 °C) and a mild winter (average daytime temperature is about 15 °C). The monthly 1980–2020 average temperature in Figure 3, confirms the reported warming trend that the southwest USA has been experiencing (e.g., [5,18]). This warming is seen for all months with linear trends ranging from 1 °C to 3 °C in 40 years. There is a consensus and high confidence among various climate projections that this historical warming trend will persist in the future (e.g., Figure 1; [5,18]). This near-surface warming leads to an increase in vapor pressure deficit, which is likely to increase the potential evapotranspiration (ET). Potential ET (PET) is the ET that would occur if a sufficient water source were available. However, because the USCRB is already a region with a limited moisture supply in which the PET largely exceeds the Actual ET (AET), changes in the AET due to the prospective future increase in PET may be negligible. In a study conducted at the Mexican part of the USCRB, the annual AET was estimated as 90–95% of the annual precipitation [19]. In this region, most of the rainfall either quickly evaporates from the soil back to the atmosphere or runs off as surface flow. With the absence of open water surfaces in the USCRB, such as lakes, reservoirs and perennial sections of the river channels, the only perennial source of moisture available for ET is the sedimentary aquifer underlining the river channels, which is tapped by the roots of the riparian vegetation. Serrat-Capdevila et al., (2011) [20], who studied the riparian aquifer of the San Pedro River, a river east of the USCRB with similar geographical and climatological features, found that although PET is projected to increase, the riparian AET rates during the growing season will remain unchanged because of transpiration regulation by the stomata. They also suggested that the future warming trend would prolong the duration of the growing season, a change that may slightly increase the total annual AET.

The AET response to the warming atmosphere is often complex and dependent on processes such as changes of the region's land cover (e.g., [21]), physiological adaption of the plants to carbon intake and regulation of transpiration (e.g., [20]), and complex atmospheric mesoscale processes that may even cause for the warming to decrease the AET (e.g., [22]). Another response to the warming that may be important in the USCRB is the anthropogenic change in water demand. In summary, the impact of the existing and projected warming trends on water resources in the USCRB is yet unclear and it is likely secondary in its importance to prospective changes in the precipitation regime.

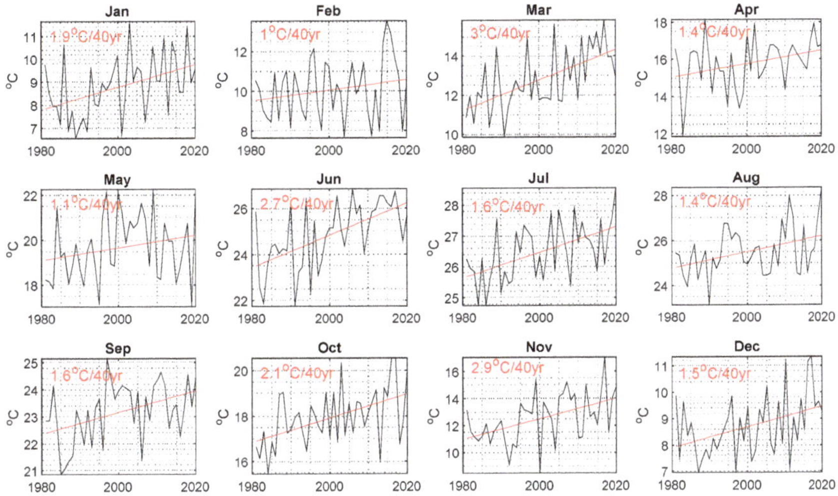

Figure 3. Average 1981–2020 monthly temperature from the Santa Cruz River near the border crossing from Sonora to Arizona. Data are available from the 4 km monthly climatological gridded surface meteorological dataset [11]. The estimated linear trend are shown as red lines and the estimated monthly rate of change over 40 years are indicated at the top left of the panels.

In Figure 4, the summer and winter precipitation from a gauge near the city of Nogales, Arizona and observed streamflow from a hydrometric station on the Santa Cruz River, less than a 1 km north of the border crossing from Sonora to Arizona, are shown for the 1937–2020 and 1955–2020 water years, respectively. While the average summer precipitation during these years (259 mm) was more than twice as much as the winter (115 mm), the average winter streamflow (8.5 MCM) was only about 7% smaller than the summer streamflow (9.15 MCM). Clearly, the inter-annual averages of the seasonal values by themselves are insufficient descriptors of the statistical characteristics of these highly variable and positively skewed seasonal precipitation and streamflow time series. The differences between the winter and summer, as well as changes in the statistical distributions as the rainfall transforms into streamflow, point to the complex and seasonal-dependent hydrological processes of the region, which must be considered for water resources assessment.

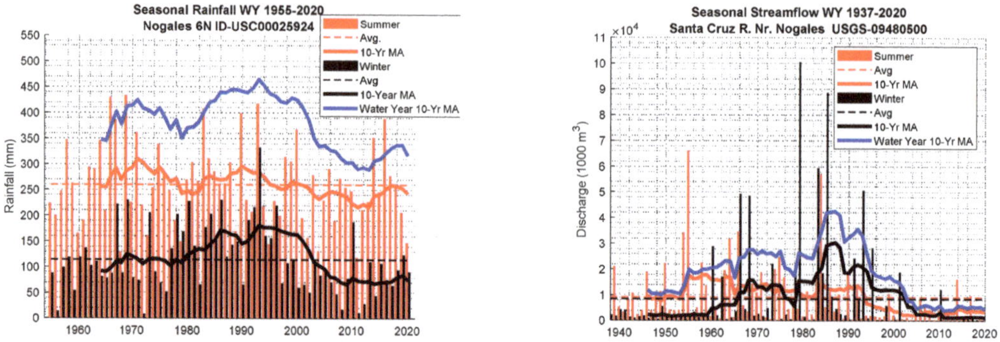

Figure 4. Observed summer (July–September) and winter (November–March) 1955–2020 precipitation (mm) from a rain gauge near Nogales, Arizona (USC00025924) and 1937–2020 streamflow (1000 m^3) on the Santa Cruz River near the border crossing from Sonora to Arizona (USGS-09480500). The inter-annual averages are indicated as horizontal solid lines. The seasonal and the water year (blue) 10-year moving averages are indicated as solid lines.

An alarming drying trend of a sizeable decline in the early 21st century is shown in Figure 4 and Table 1. This drying trend is seen when comparing the early 21st century (2001–2020) to the average historical record from the 20th-century of streamflow (1937–2000) and precipitation (1955–2000) for both winter and summer seasons. The average decline from 1955–2000 to 2001–2020 in summer precipitation is 10%, and the reduction in winter precipitation is 33%. These declines are substantially larger at the streamflow record (65% and 78% for summer and winter flow, respectively). The 10-year moving averages that are shown in Figure 4 shows that both for summer and winter the precipitation and streamflow are below the inter-annual averages since about 2000. Although dry seasons were seen in the historical records, the consequence of having both dry winters and summers is shown when looking at the total annual 10-year moving average (blue), which indicates that the 2001–2020 period is the dryer period in the observed record.

Table 1. Observed average summer and winter precipitation (mm) in USC-00025924 and streamflow (MCM) in USGS-09480500. The square parentheses indicate the percent reduction of the averages for the precipitation and streamflow respectively from 1955–2000 and 1937–2000 to 2001–2020.

	Summer	Winter
	Precipitation (mm)	
1955–2000	259	115
2001–2020	233 (10%)	* 77 (33%)
	Streamflow (MCM)	
1937–2000	9.2	8.5
2001–2020	* 3.2 (65%)	* 1.9 (78%)

* Null hypothesis in which the 2001–2020 is from the same distribution as 1955–2000 and 1937–2000 for the precipitation and the streamflow, respectively was rejected at 1% significance level using the non parametric Kolmogorov-Smirnov test.

We used the agnostic non-parametric Kolmogorov–Smirnov test to examine the null hypothesis, stating that the seasonal precipitation and streamflow during the early 21st-century (2001–2020) and the 20th-century (1955–2000 and 1937–2000 for the precipitation streamflow, respectively) can be considered as statistical samples that were taken from the same seasonal population. Except for summer precipitation, the null hypothesis was rejected at 1% significance level for the summer precipitation and for both winter and summer streamflow. The rejected null hypothesis implies that the statistical distributions of the early 21st century (2001–2020) seasonal hydrometeorological time series are significantly different than the distributions of the historical data that are available from the 20th century. While the reduction in the precipitation is most likely attributed to climatic factors, the intensified reduction of the streamflow during the early 21st century should be further investigated to understand whether this reduction can entirely be attributed to climatic factors, or whether other factors such as changes in land cover, land use, and/or water resources management are responsible for this reduction as well.

The significant change in winter precipitation, as seen in Table 1, are also seen in two other in-situ stations from the region (i.e., USC000026282 in Patagonia (1980–2020) and USC000028865 in Tumacacori (1965–2020)). In these two stations the null hypothesis was rejected in a significant levels of 0.01 and 0.05 for the Patagonia and Tumacacori stations, respectively. A streamflow analysis of the Lochiel hydrologic station (USGS 09480000), a station on the Santa Cruz at the border crossing from Arizona to Sonora, showed a significant decrease in 2001–2020 streamflow when compared to 1950–2000 during both summer and winter seasons. We note that data from the Lochiel hydrologic station were missing during 2014–2019.

4. Summer

Summer rainfall in the USCRB is mainly driven by the North American Monsoon (NAM) climate system that triggers localized convective cells, which often produce thunderstorms and intense short-lived rainfall events. The NAM starts in early June along the western slopes of the Sierra Madre Occidental in southern Mexico and later expands northward to reach the USCRB by early July, lasting until early September. A high-pressure, subtropical ridge forming over northwest Mexico in June causes hot and dry weather, with southwesterly winds in Arizona. As summer progresses, the subtropical ridge normally moves northward until its center of circulation is located over west Texas and New Mexico. This northward movement of the sub-tropical ridge alters the low-level winds in Arizona to a southerly or southeasterly direction. The southerly-southeasterly wind direction combined with the daytime surface low pressure (thermal low) caused by the intense heating of the desert ground, conveys pulses of low-level moist air from the Gulf of California and the eastern Pacific to the region. The Gulf of California moisture surges—which is a major source for the low-level moisture transport into Arizona and Sonora [23]—are triggered by tropical easterly waves and passing tropical cyclones (e.g., [24]). Additional upper-level moisture is transported by easterly winds aloft from the Gulf of Mexico (e.g., [25]). This combination of a seasonally warm land surface and ample atmospheric maritime moisture is conducive to the development of local convective clouds that produce afternoon isolated thunderstorms followed by local and short-lived intense precipitation events. From mid-September the NAM withdraws as the southwestern ridge decays and retreats southward (e.g., [25]).

We present the 1981–2020 climatological spatial extent of the NAM's precipitation in Figure 5a–c. Inspecting these climatological maps, we outline the core of the NAM to be in western and southeast Mexico as the areas with precipitation exceeding 500 mm (a) and a relatively small inter-annual variability (coefficient of variation (CV) smaller than 0.3 (c)). Our study area (demarked with purple dots) that received an average of about ~200 mm per summer with a CV greater than 0.3 is noticeably situated at the northern edge of the NAM weather system. However, as the contribution of the NAM to the total annual precipitation quickly diminishes northward of the study region (Figure 5b), the NAM precipitation accounts, on average, for ~60% of the annual precipitation at the USCRB. The increase of the precipitation inter-annual variability in the NAM's northern boundaries is attributed to the variability in the strength and latitudinal position of the subtropical ridge and the low-level moisture surges from the Gulf of California (e.g., [26,27]).

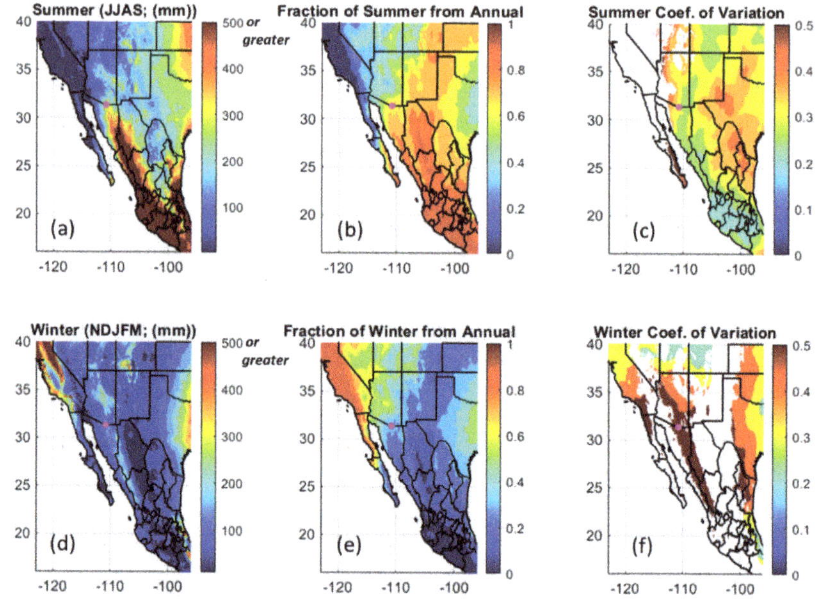

Figure 5. Average 1981–2020 summer and winter precipitation ((**a**) and (**d**) for summer and winter, respectively), their portion from the average annual precipitation ((**b**) and (**e**) for summer and winter, respectively), and their inter-annual variability represented by the coefficient of variation for locations with average seasonal rainfall that exceed 100 mm ((**c**) and (**f**) for summer and winter, respectively). The purple dots demark the Santa Cruz River at its crossing of the international border from Sonora Mexico to Arizona U.S. The dataset is from the global monthly 4 km TerraClimate dataset [28].

The NAM's onset and rainfall amount are influenced by Sea Surface Temperature (SST) anomalies that shape the land-sea thermal contrast. During El Niño–Southern Oscillation's (ENSO) El Niño (La Niña) phase, periods when the SST of the eastern Pacific Ocean near the equator are warmer (colder) than normal, later (earlier) monsoon onset in southwestern USA was observed [29]. Moreover, dry summers in the northern edge of the NAM system tend to follow wet winters that are associated with El Niño phase [29]. However, in the USCRB, ENSO indices were found to have a weak association with summer rainfall [16,30].

4.1. Historical Changes

As seen in Figure 3 and as reported by previous studies, an apparent positive trend in summertime temperatures have been observed in recent decades in the southwestern USA (e.g., [5,18,31]). This warming can potentially trigger two competing processes that may affect the NAM precipitation regime. On the one hand, higher terrestrial temperatures that increase vapor pressure may enhance convective activity to produce more intense rainfall events. On the other hand, the warming SST may decrease the contrast between the land and sea to reduce the maritime moisture transport and increase atmospheric stability that suppresses convection activity [32]. Although we observed a reduction in the total summer rainfall in the USCRB (Figure 4; Table 1), studies on historical changes of NAM precipitation characteristics asserted, at times, contradicting conclusions. These studies are commonly based on regional analyses of either daily to seasonal precipitation observations (e.g., [27]), or high-resolution atmospheric model simulations (e.g., [31,33,34]). The interpretation of the trends from these studies to Northern Mexico–Southern Arizona (NMSA) and in particular to the USCRB, is challenging because of the region's high inter-annual variability within the NAM weather system. Moreover, these analyses are often deficient in representing the small temporal and spatial scale of NAM precipitation events, typical to the study region. A recent study by Demaria et al., (2019) [35] analyzed 1961–2017 sub-daily precipitation in the densely gauged Walnut Gulch Experimental Watershed (149 km^2) in southeastern Arizona (about 150 km northeast of the Santa Cruz river at its border crossing from Sonora to the USA). Attempting to represent the region's rainfall small spatiotemporal scale, they reported intensification of NAM precipitation starting at the mid-1970s. Studies conducted in the USCRB indicated several observed changes in the region's hydrologic regime, such as a reduction in summer streamflow volume, a reduction in the number of summer streamflow occurrences [30,36] and a reduction in the duration of baseflow [17]. These reductions increased the monthly inter-annual variability of streamflow since the 1970s (e.g., [30]). Again, we point to the results shown in Table 1 of ~63% reduction in average summer streamflow from 1937-2000 to 2001–2020. With respect to precipitation, the analysis presented in Table 1 indicates a 17% reduction in total summer precipitation, which agrees with Shamir et al., (2007a) [37], who reported a reduction in the number of summer precipitation events.

4.2. Future Projections

The future of the NAM weather system is an active study area and recent detailed reviews have been presented by Wang et al. (2021) [26] and Pascale et al. (2019) [32]. In the following, we attempt to review the findings that are relevant to the USCRB. Almost all future climate projections agree that the increase in atmospheric moisture associated with the warming of the atmosphere will increase the intensity of the extreme rainfall events (e.g., [5,26,32]). However, future NAM projections are uncertain mainly because the horizontal resolution of the GCMs is often too coarse to adequately represent the complex topography of the NAM's region and the regional (mesoscale) atmospheric processes that control the NAM convective storms. Some important processes for assessing the NAM characteristics and their simulations that are challenged by the GCMs include the location and characteristics of the North American subtropical high pressure, the regional wind patterns, the near-surface onshore flow into NMSA, and the northward low-level jets along the Gulf of California (e.g., [23,38–41]). Furthermore, GCM biased simulations of NAMs' teleconnections, such as SST in the North Atlantic and the Pacific, often leads to unrealistic easterly low-level moisture flux across the Caribbean region, which affect the simulations of the NAMs precipitation (e.g., [42]).

Several studies that analyzed CMIP5 GCMs, reported an expected future seasonality shift of the NAM with no significant change in the total seasonal precipitation. This seasonal shift is expressed as a delay of the NAM's onset at the beginning of the summer (June–July) due to increased atmospheric stability and increased precipitation during the late summer (September–October) [43–45]. Geil et al., (2013) [41] evaluated the skill of 21 CMIP5 GCMs to simulate the NAM system. They identified several GCMs that failed to simulate NAM characteristics. Among the well-performing GCMs, large differences in their future projections with contradicting trends were identified. They also concluded that all the evaluated GCMs, because of simulated SST biases, simulated excessive precipitation in September and failed to effectively simulate the retreat of the seasonal NAM (as also reported by Colorado Ruiz et al., 2018 [46]). Correcting for the GCM's SST biases, Pascale et al., (2017) [42] projected a reduction in NAM rainfall over the Southwestern U.S. that is attributed to an increase in the atmospheric stability due a uniform warming of the SST, which dampens the convective activity. Bukovsky et al., (2013 and 2015) [39,40] evaluated four CMIP3 GCMs that were dynamically downscaled by six different RCMs. When RCMs were used to downscale the historical reanalysis data, various atmospheric patterns of the NAM were simulated with high skill, but all simulations showed a dry bias over Arizona. Bukovsky et al. (2013) [39] concluded that except from one GCM, all other GCMs provided reasonably adequate boundary conditions to the RCMs during the summer season. All the RCM simulations suggested an overall trend of decrease NAM precipitation. However, the RCMs were varied in their skill to simulate the NAM's climatological regional features. Based on the best-performing GCM-RCM, a slight but significant trend for drying summers by mid-21st century across the southwestern USA, with increased frequency and intensity of heavy precipitation events, was projected (Bukovsky et al., 2015) [40].

Castro et al. (2017) [38] implemented a dynamically downscaled high horizontal resolution (3 km) RCM with a convective permitting scheme in order to skillfully simulate the thermodynamic conditions during extreme NAM rain events. These extreme rain events are infrequent late summer events of organized monsoon thunderstorms over relatively large geographic areas, bringing large amounts of rainfall (e.g., [47]). Castro et al., (2017) [38] observed a fewer severe, strong, and organized convective summer events in the southwest USA over the past sixty years. However, this frequency reduction was accompanied by increase in the intensity of these severe events. Their analysis points to a future that is consistent with the trends of the historical record. We note that the organized severe thunderstorms events that are documented in Castro et al., (2017) [38], Maddox et al., (1995) [47] and the 27–31 July 2006 event that brought heavy rainfall to Santa Catalina Mountains and the lower Santa Cruz River [48] had not caused exceptional storm events in the USCRB. As of now, we are unaware of an extreme summer event in the USCRB that was the result of well-organized convective complexes.

In summary, most studies point to a future with warmer temperature during the NAM season. With respect to precipitation, the studies remain uncertain. Some studies reported a seasonal shift but these observed projected shifts may be associated with GCM simulations' biases in SST. An agreed upon projection points to an overall decrease in total seasonal precipitation but an increase in rainfall intensity during large events. A major caveat of these projections is that the results of most studies are dominated by the core region of the NAM system, while the USCRB, which is at the fringe of the NAM, experiences much higher inter-annual variability than in the core.

5. Winter

The 1981–2020 climatological spatial extent of winter precipitation (Figure 5d–f) displays the importance contribution of the winter season (~50%) to the annual precipitation (d). However, as was shown for the summer precipitation, the USCRB is located at the southeast edge of the winter weather system, which is shown by the region's large inter-annual variability (f). This large inter-annual variability, which makes the region sensitive to subtle changes, is likely to increase the uncertainty of future winter precipitation projections.

In recent years, winter storms in the region are broadly classified into two categories: Pacific cold fronts and Atmospheric Rivers. The Pacific cold fronts are large-scale low-pressure cold front systems approaching from the Pacific Ocean. These storms bring high winds and cloudy skies that cause persistent rain over large areas. The frequency of these storms is closely associated with SST in the Pacific Ocean. During El Niño years, the southwest experiences wetter winters because the upper-level, subtropical, westerly jet stream over California and Baja California (Mexico) is displaced southward, a displacement that transports onshore moisture to the southwest. During La Niña years, the northward displacement of the subtropical jet stream brings dryer winters in the southwest (e.g., [49,50]).

Specifically, in the USCRB, Shamir (2017) [16] reported that during 1949–2016 most years with September–October ENSO3.4 anomalies greater than 1.5 °C experienced above-normal winter rainfall. On the other hand, during most of the strong September-October La Niña conditions (ENSO3.4 < -1 °C), excluding one year, the winter rainfall was below normal. Winter rainfall had no clear association with neutral years that experienced September-October ENSO3.4 anomalies between -1 and 1.5 °C. It is interesting to note that during strong El Niño years the streamflow at the Upper Santa Cruz River was not always above normal. However, during all strong La Niña years the annual streamflow was below normal. Although the ENSO phases show a firm association with winter precipitation, the wettest years and the largest inter-annual variability were observed during the neutral ENSO years.

Atmospheric Rivers (ARs) are a middle latitude synoptic phenomenon that transports high water vapor concentration from the tropical ocean to the land through a relatively long and narrow corridor. Although ARs are a year-round phenomenon, summer ARs are prevalent in high latitudes and do not affect the precipitation in NMSA. During 1988–2011, an average of seven ARs-related rainfall events per year penetrated inland to hit Arizona causing about half of the largest winter streamflow events [51]. The region in Arizona that is most impacted by ARs is the Salt and Verde watersheds, draining the Central Highlands and Mogollon Rim (e.g., [52]). In the USCRB, about 10% of the largest annual flow events and 38% of the winter flow events are attributed to ARs [51] and about 10–15% of the cool season precipitation during 1998–2008 are caused by ARs [53]. The ARs that reach the USCRB are ARs with a northeastward trajectory that make landfall along the Baja California Peninsula, south of the U.S.–Mexico border (32.5° N Latitude). These are about 8% of the cool season ARs that hit the eastern Pacific coast [54]. Water vapor concentration near the coast in these southern ARs are typically smaller than those ARs farther to the north. The southern ARs often deplete their moisture over the mountains of Southern California and the Baja California Peninsula while about half maintain their AR properties as they penetrate to the lower Colorado River basin [54]. The association of ARs with ENSO is still inconclusive. Some evidence exists for ARs to be more pronounced during El Niño and neutral ENSO conditions [55].

Future Projections

Most studies projecting increased probability of drier winters attributed to the projected widening of the high-pressure subtropical Hadley cell, which in turn will displace the moisture carrier subtropical jet stream northward (e.g., [18,56]). Some studies also point to an expected increase in the frequency of winter's prolonged dry spells during the 21st century in the Southwest [57]. However, AR related storms in the USCRB are a small fraction of the ARs to hit the Pacific coast, they constitute a substantial portion of the region's winter precipitation. The projected AR changes stated by the Fourth Assessment [5] is that for higher GHG concentration trajectories (RCP 8.5), AR frequency, and intensity in the mid-21st century is projected to increase. This has been the general agreement since the analysis of selected CMIP3 GCMs by Dettinger et al., (2011) [58]. Increase in atmospheric water vapor holding capacity due to increasing temperature is the main driver for the projection of significant increases in AR frequency and magnitude (e.g., [58–61]). The magnitude and characteristics of these projected changes are still a vibrant research topic. Warner et al., (2015) [60] who analyzed 10 CMIP5 GCMs using RCP 8.5, projected an increase in winter-average precipitation along the southern offshore transect of 11%–18% when comparing 1970–1999 to 2070–2099. In their analysis, however, the southernmost sampling site was just offshore the Santa Barbara's coast (35 °N Latitude), about 2 degrees north of the U.S.–Mexico border.

Gershunov et al. (2019) [62] compared 1951–2000 to 2051–2100 of selected five well-performing statistically downscaled CMIP5 GCMs using RCP 8.5 trajectories. In general, an agreement among the GCMs about the ARs hitting the Eastern Pacific Coast, projected positive linear trends in their intensity (~10%), frequency (~20%), and duration (~20%). Taking a spatial perspective, it appears that the projected increase in AR intensity is shown north of 33° latitude while in Baja California Peninsula this increase is less than 5%. The largest increase in AR contribution to annual precipitation is projected for Northern Baja California Peninsula coast (>20%). This increase in potential AR contribution to annual precipitation along the coast is diminished to ~5% in the USCRB region. The large reduction in the impact of the ARs may also be attributed to the relatively large portion of summer contribution to the annual rainfall in the USCRB, which is not the case in the coastal regions.

Since AR s are commonly formed immediately south of the subtropical jet stream, the projected future poleward shift of the jet stream is expected to reduce the AR frequencies in the south and increase them in the north. However, studies of GCM projections point to significant uncertainty with respect to shifts in future ARs. While some support the hypothesis of a northward shift in AR frequencies, which will reduce the frequency of ARs hitting south of the U.S.–Mexico border [63], others do not support this projected shift [59–61].

6. Tropical Storms

Northeastern pacific tropical cyclones (TC), which are prevalent during May-November, often transport surges of moisture from the Gulf of California to enhance the NAM activity in the region. About half of the Gulf of California moisture surges to the southwest are associated with TCs in the eastern Pacific. These surges yield more precipitation than non-TC surges [24]. Ritchie et al. (2011) [64] reported that, during 1992–2005, an average of about three TC per year affected the NAM precipitation in NMSA. In the USCRB, about 10% of the summer precipitation was attributed to TC moisture and in 1992 about 30–40% were attributed to TC.

Although relatively infrequent events, during autumn months (September–October) the eastern north Pacific TC can potentially bring torrential rainfalls with substantial contribution to the USCRB water resources. In the hydrometric station (USGS 09480500) on the Santa Cruz River at the Sonora–Arizona border crossing, at least two of the largest instantaneous flow events recorded since 1930 were associated with remnants of TC events (880 m^3/s on 9 October 1977 from Hurricane Heather; 484 m^3/s on 4 August 1974; and; 459 m^3/s on 2 October 1983 from Hurricane Octave). Most TC in the northeastern Pacific travel in a northwest trajectory, rarely cross the 25° N, to weaken and eventually dissipate over the cold Pacific water. Occasionally, later in the season (September–October), a storm may shift to a northward trajectory towards NMSA or southern California. These storms, upon their landfall, weaken to a tropical depression (winds lower than 35 knots). When accompanied with a stable middle-latitude upper-level trough over the eastern Pacific, these tropical depressions can funnel a large quantity of tropical moisture to NMSA region. These storms that bring torrential precipitation to the region have a recurrence interval of five years.

At present, despite clear evidence for warming oceans and atmosphere, there is no clear evidence for detectable changes in northeast Pacific TC activity (e.g., [65]). Future projections of global and regional TC have been an active research topic that suffers from the limited skill of many GCMs in simulating TC climatological features, mainly because of their relatively coarse horizontal resolution [66]. In a recent review of the expected TC changes in the northeastern Pacific Ocean the following was suggested: the frequency of TC events would decrease (median 5%; with range from 90% of the projections of 30–20%), frequency of large TC (4–5 Saffir-Simpson Scale) would increase (~35%, 10–100%), average TC intensity would increase (6%, 3–10%), and TC rainfall rate would increase (20%, 5–30%). Furthermore, there were no clear indications for projected systematic changes in the TC tracks and locations [66].

Although future changes in TC characteristics may have a substantial impact on the water resources in the USCRB, using the projected regional changes to comprehend the rare TC events that affect the USCRB is clearly thus far problematic.

7. Previous Climate Assessment Studies in the Region

In a study over the state of Sonora Mexico, Magaña et al. (2012) [67], who analyzed 20 CMIP3 statistically downscaled scenarios, projected a 10–25% decrease in annual precipitation for 2040−2099. However, although most models showed a decreasing trend, the magnitude of the average projected precipitation change is smaller than the variability among the GCMs. Several other climate assessments that were conducted in the USCRB and NMSA are based on projected CMIP3 and CMIP5 downscaled simulations of selected GCMs that were found to adequately simulate climatic indicators of southwest USA [68]. Dominguez et al., (2010) [68] graded CMIP3 GCMs with respect to their performance in simulating monthly temperature and precipitation climatology over southwest USA (i.e., the states of Arizona, New Mexico, Utah, and Colorado). In addition, they assessed the

GCMs skill to simulate ENSO teleconnection with winter precipitation and the location of the subtropical jet stream. Among the CMIP3 GCMs, they identified the Max Planck Institute ECHAM5 and the UK Met Office HadCM3 as the best performing GCMs. These GCMs projected amplification of the La Niña conditions that is accompanied with drying trends in winter precipitation.

Following the recommendations of Dominguez et al. (2010) [68], Ajami et al., (2012) [69], in a study at the San Pedro River, reported an average change in annual precipitation of 14–7% projected for 2050–2099. This projected change in precipitation transforms to a decline in the aquifer recharge rate of 15–27%, with higher sensitivity to changes in winter precipitation. Another study from the San Pedro River Basin, which used a multi-model approach of 17 statistically downscaled CMIP3 GCMs with four different emission scenarios, projected 17–30% changes in average annual precipitation [70]. Evaluating the impact of the changes in precipitation on the recharge of the San Pedro aquifer, they estimated a substantial decline in the recharge rate through 2100. Nevertheless, this projected decline in the groundwater recharge includes a large variation among the GCMs that ranged from 100% to 30%. Meixner et al. in their 2016 synthesis of previous studies for the San Pedro River basin, which is mainly based on the two mentioned above studies, projected an average decline in groundwater recharge of 10–20%.

Shamir et al. (2015) [9] used dynamically downscaled 35 km simulations of the two CMIP3 GCMs recommended by Dominguez et al., (2010) [68] using an A2 emission scenario (model configuration is described in Castro et al., 2012 [23]) to assess the future climate impact on water resources in the shallow aquifers on the American side of the Santa Cruz River. Their analysis reveals that a dominant climatic change element with direct impact on water resources is the occurrence rate of three seasonal wetness categories (i.e., wet, medium, or dry) in both the winter and summer rainy seasons. They reported for the mid-21st century (2041–2070) a clear trend for drying summers that is expressed by an expected increase (decrease) in the occurrence frequency of dry (wet) summers. For the winter, they reported an expected increase in the frequency of both dry and wet winter seasons, implying a lower chance of experiencing a moderate winter. They provided a probability assessment given different management schemes of the aquifer. They concluded that the projected changes in precipitation would introduce challenges for the City of Nogales, Arizona to meet water demand, a long-term increase in cumulative water deficit, and a general decrease in the aquifer recharge rate to the aquifer.

The impacts of the CMIP5 version of the GCMs that were recommended by Dominguez et al., (2010) [68] on the USCRB water resources were assessed by Shamir and Halper, (2019) [7] and Tapia et al., (2020) [15]. These studies used dynamically downscaled simulations at 25 km of RCP 8.5 emission scenario [38]. Projected CMIP5 precipitation changes from the two selected models provided contradicting results. The downscaled simulations of the Hadley GCM expected a wetter mid-21st century (2020–2059) in the USCRB, expressed by the frequency of dry and wet winters to increase and the frequency of dry summer to decrease and wet summer to increase. On the other hand, the MPI dynamically-downscaled simulations points to a much dryer mid-21st century with projections for an increase in the frequency of dry winters and summers, and a decrease in frequency of wet winters and summers. The projected changes in the seasonal frequency of the wetness categories were compared with statistically downscaled simulations of the same models to find similar but substantially weaker trends. In their study, Shamir and Halper [7] provided a probabilistic perspective to represent the variability and known sources of uncertainties. Herein we use the distributions' median to report the range of the projected changes. The dry (wet) projections of the MPI (HAD) expect a −15% (7.5%) annual change in precipitation that is transformed to a −22% decline (13% increase) in streamflow and a 15% increase (7.3% decrease) in water supply for the City of Nogales, Arizona expected to be substituted from a different source.

Tapia et al. (2020) [16] used the same precipitation projections to evaluate the impact on water resources to the effluent dominant region of the USCRB. In this region of the

USCRB, the impact of the projected median change on the region's water deficit would increase (decrease) for the MPI (HAD) by about 4.5 MCM/year, which is about 22% of the annual groundwater withdrawal in this region.

8. Summary

In this review, we summarize the current state-of-knowledge of future climatic changes in the binational USCRB that are thought to be relevant for the region's water resources. In the USCRB water resources replenishment is primarily depends on rainfall-triggered streamflow in ephemeral channels that recharge the alluvial aquifers. This aquifer recharge in the region is contingent on highly variable and different winter and summer rainy seasons. Since the USCRB is located at the geographical edge of these weather systems, small future changes in their cores may prompt unforeseen consequences in the USCRB. Observations of local in-situ data from the basin for 1980–2020 reveals a warming trend for all months, since the 1980s. While the warming trend is in agreement with other previous reports (e.g., [5,18]), we also observed an alarming declining trend in summer and winter precipitation in the USCRB that is likely not explained by the observed inter-annual variability. The declines in average precipitation from 1955–2000 to 2001–2020 of 50% in the winter and 17% in the summer were accentuated by reductions in streamflow on the Santa Cruz River of 87.6% and 63%, respectively.

Our review of projected climate in the region points to an overall agreement that the historical warming trend will highly-likely to continue in the future. Nevertheless, the impact of this future warming on the region's water resources is yet unclear. To synthesize the future projected precipitation for the region, we reviewed climate studies that addressed four rain-bearing weather systems that influence the USCRB (i.e., summer North American Monsoon, winter cold fronts, atmospheric rivers, and tropical cyclones). Although highly uncertain, most studies reported that for the mid-21st century the total summer NAM precipitation is expected to decline with the expectation for intensification of the large events. For the winter, the Pacific cold fronts' frequency and intensity are projected to decline due to SST warming that will displace the subtropical jet poleward to resemble a La Niña like conditions. On the other hand, the frequency and the intensity of future ARs that will likely to hit the USCRB in the winter are projected to increase. Although rare, TC remnants can cause very significant rain events for replenishing the region's water resources. While the frequency of TC in the eastern Pacific are expected to decrease, the rain rate and the frequency of large TC are expected to increase by mid-21st century.

We note that the general projection statements listed above are highly uncertain and many studies of projected future precipitation reported ranges of projected changes that often have contradicting trends.

Although large uncertainty is associated with the projected future precipitation, many of the possible outcomes carry large risk, as the majority of the projections suggest a dryer future. Because non-linear processes dominate the hydrologic response in desert environment, the prospective drying will likely to intensify when rainfall is transformed to runoff and eventually recharged into the aquifers. Thus, even small changes in the precipitation regime can introduce severe ramifications to the region's water resources. In addition to the projected climatic and hydrologic uncertainties an added complication for sustainable water resources management is the inter-dependency of the water systems on both sides of the international border, in which an action on one side of the border influences water availability on the other side. Clearly, sustainable water management in a border setting requires collaboration. A first step for a collaborative transboundary effort is an advancement of the knowledge base, which is the aim of this manuscript. As a final point, given the projected uncertain future and the worrisome observed historical trends, we stress the urgency and the severe risk of water shortages that the region may potentially undergo. This urgent risk for water shortages calls for proactive and collaborative binational planning to achieve a sustainable transboundary aquifer system.

Such a binational effort can be supported by the International Boundary and Water Commission (IBWC), a USA–Mexico joint commission that is responsible for applying the boundary and water treaties between the two countries and settling differences that may arise in their application. Although in its mission statement, the IBWC does not explicitly mention sustainable management of water resources as an objective, it issued in the past several binding interpretations (hereinafter referred to as Minutes) to the 1944 Water Treaty regarding the utilization of waters of the Colorado River, Tijuana River, and Rio Grande that addressed binational collaborative management schemes. Some examples of Minutes that are relevant to the USCRB are the groundwater pumping limits within eight kilometers of the Arizona–Sonora border near San Luis (Minute 242) and the conveyance, treatment, and disposal of the Ambos Nogales' sewage (Minutes 206, 227, 276, and 294). Minute 323, signed in 2017, is the first IBWC Minute that addresses binational adaptation strategies for allocating water resources during water scarcity periods. The Minute specifies cooperative measures and a contingency plan during years of water scarcity in the Colorado River basin. These Minutes can provide a framework for collaborative water resources management in the USCRB to ensure sustainable conditions and provide mitigation measures that address the projected climatic changes.

Author Contributions: Conceptualization, E.S., E.M.T.-V., M.-B.C.-A. and S.B.M.; supervision, S.B.M. and E.M.T.-V.; methodology E.S., E.M.T.-V. and M.-B.C.-A.; validation, E.S.; formal analysis, E.S.; investigation, E.S. and M-.B.C-.A.; resources, E.S. and M-.B.C-.A.; data curation, E.S. and M.-B.C.-A.; writing—original draft preparation, E.S.; writing—review and editing, E.M.T.-V., M.-B.C.-A. and S.B.M.; visualization, E.S. and E.M.T.-V.; project administration, S.B.M.; funding acquisition, S.B.M. All authors have read and agreed to the published version of the manuscript.

Funding: This work was funded by the U.S. Geological Survey (funding authorized by P.L. 109–448) Award Number G17AC00439 for the Transboundary Aquifer Assessment Program.

Institutional Review Board Statement: Not applicable.

Informed Consent Statement: Not applicable.

Data Availability Statement: Not applicable.

Acknowledgments: The authors would like to acknowledge the Transboundary Aquifer Assessment Program (TAAP) and the administrative support of the Water Resources Research Center, University of Arizona.

Conflicts of Interest: The authors declare no conflict of interest. The funders had no role in the design of the study; in the collection, analyses, or interpretation of data; in the writing of the manuscript, or in the decision to publish the results.

Abbreviations

The following abbreviations are used in this article:

AET	Actual Evapotranspiration
AR	Atmospheric River
CMIP3 or 5	Coupled Model Intercomparison Project (CMIP) phase 3 or 5
ENSO	El Niño and the Southern Oscillation
ET	Evapotranspiration
GCM	Global Climate Model
GHG	Green House Gas
IBWC	International Boundary and Water Commission
MCM	Million Cubic Meter
NAM	North American Monsoon
NMSA	Northern Mexico and Southern Arizona
PET	Potential Evapotranspiration

RCP	Representative Concentration Pathway
RCM	Regional Climate Model
SST	Sea Surface Temperature
TC	Tropical Cyclone
USCRB	Upper Santa Cruz River Basin

References

1. IBWC. Joint Report of the Principal Engineers Regarding Joint Cooperative Process. United States-Mexico for the Transboundary Aquifer Assessment Program. 2009. Available online: Wrrc.arizona.edu/sites/wrrc.arizona.edu/files/images/Joint-Report-TAAP-8-19-09.pdf (accessed on 13 May 2021).
2. Megdal, S.B.; Scott, C.A. The Importance of Institutional Asymmetries to the Development of Binational Aquifer Assessment Programs: The Arizona-Sonora Experience. *Water* **2011**, *3*, 949–963. [CrossRef]
3. Callegary, J.; Megdal, S.; Villaseñor, E.T.; Petersen-Perlman, J.; Sosa, I.M.; Monreal, R.; Gray, F.; Noriega, F.G. Findings and lessons learned from the assessment of the Mexico-United States transboundary San Pedro and Santa Cruz aquifers: The utility of social science in applied hydrologic research. *J. Hydrol. Reg. Stud.* **2018**, *20*, 60–73. [CrossRef]
4. Tapia-Villaseñor, E.; Megdal, S. The U.S.-Mexico Transboundary Aquifer Assessment Program as a Model for Transborder Groundwater Collaboration. *Water* **2021**, *13*, 530. [CrossRef]
5. Garfin, G.M.; Gonzalez, P.; Breshears, D.; Brooks, K.; Brown, H.E.; Elias, E.; Gunasekara, A.; Huntly, N.; Maldonado, J.K.; Mantua, N.J.; et al. Chapter 25: Southwest. Impacts, Risks, and Adaptation in the United States. In *The Fourth National Climate Assessment*; U.S. Global Change Research Program: Washington, DC, USA, 2018; Volume 2, pp. 1101–1184.
6. Abatzoglou, J.T.; Brown, T.J. A comparison of statistical downscaling methods suited for wildfire applications. *Int. J. Clim.* **2011**, *32*, 772–780. [CrossRef]
7. Shamir, E.; Halper, E. *The Role of Downscaling Methodology in Projected Climate Change Impacts on Water Resources in Arid Environments*; U.S. Department of Interior, Bureau of Reclamation Research and Development Office: Denver, CO, USA, 2019.
8. Shamir, E.; Halper, E.; Modrick, T.; Georgakakos, K.P.; Chang, H.-I.; Lahmers, T.M.; Castro, C. Statistical and dynamical downscaling impact on projected hydrologic assessment in arid environment: A case study from Bill Williams River basin and Alamo Lake, Arizona. *J. Hydrol. X* **2019**, *2*, 2589–9155. [CrossRef]
9. Shamir, E.; Megdal, S.B.; Carrillo, C.; Castro, C.L.; Chang, H.-I.; Chief, K.; Corkhill, F.E.; Eden, S.; Georgakakos, K.P.; Nelson, K.M.; et al. Climate change and water resources management in the Upper Santa Cruz River, Arizona. *J. Hydrol.* **2015**, *521*, 18–33. [CrossRef]
10. Meixner, T.; Manning, A.H.; Stonestrom, D.A.; Allen, D.M.; Ajami, H.; Blasch, K.W.; Brookfield, A.E.; Castro, C.L.; Clark, J.F.; Gochis, D.J.; et al. Implications of projected climate change for groundwater recharge in the western United States. *J. Hydrol.* **2016**, *534*, 124–138. [CrossRef]
11. Abatzoglou, J.T. Development of gridded surface meteorological data for ecological applications and modelling. *Int. J. Clim.* **2013**, *33*, 121–131. [CrossRef]
12. Scott, C.; Megdal, S.; Oroz, L.; Callegary, J.; Vandervoet, P. Effects of climate change and population growth on the transboundary Santa Cruz aquifer. *Clim. Res.* **2012**, *51*, 159–170. [CrossRef]
13. CONAGUA. Actualización de la Disponibilidad Media Anual de Agua en el Acuífero Río Santa Cruz (2615) Estado de Sonora. *Diario Oficial de la Federación (DOF)*. 2020. Available online: https://sigagis.conagua.gob.mx/gas1/sections/Edos/sonora/sonora.html (accessed on 17 May 2021).
14. Arizona Department of Water Resources. *Groundwater Flow Model of the Santa Cruz Active Management Area Along the Santa Cruz River and Nogales Wash from the International Border to Tubac*; Arizona Department of Water Resources: Phoenix, AZ, USA, 2020.
15. Tapia-Villaseñor, E.M.; Shamir, E.; Megdal, S.B.; Petersen-Perlman, J.D. Impacts of Variable Climate and Effluent Flows on the Transboundary Santa Cruz Aquifer. *JAWRA J. Am. Water Resour. Assoc.* **2020**, *56*, 409–430. [CrossRef]
16. Shamir, E. The value and skill of seasonal forecasts for water resources management in the Upper Santa Cruz River basin, southern Arizona. *J. Arid. Environ.* **2017**, *137*, 35–45. [CrossRef]
17. Nelson, K. *Risk Assessment of Pumping Impacts on Simulated Groundwater Flow in the Santa Cruz Active Management Area*; ADWR Modeling Report No. 21; Arizona Department of Water Resources: Phoenix, AZ, USA, 2010.
18. Garfin, G.; Jardine, A.; Merideth, R.; Black, M.; LeRoy, S. *Assessment of Climate Change in the Southwest United States: A Report Prepared for the National Climate Assessment*; A report by the Southwest Climate Alliance; Island Press: Washington, DC, USA, 2013.
19. Minjárez Sosa, I.; Monreal Saavedra, R.; Grijalva Noriega, F.J.; Ochoa Granillo, J.A.; Tapia Villaseñor, E.M.; Rangel Medina, M.; Montijo González, A. *Actividades Hidrogeológicas en el Acuífero Río Santa Cruz, Estado de Sonora, Primera Etapa*; Conagua: Hermosillo, Sonora, 2011.
20. Serrat-Capdevila, A.; Scott, R.L.; Shuttleworth, W.J.; Valdés, J.B. Estimating evapotranspiration under warmer climates: Insights from a semi-arid riparian system. *J. Hydrol.* **2011**, *399*, 1–11. [CrossRef]
21. Villarreal, M.L.; Norman, L.M.; Webb, R.H.; Turner, R.M. Historical and Contemporary Geographic Data Reveal Complex Spatial and Temporal Responses of Vegetation to Climate and Land Stewardship. *Land* **2013**, *2*, 194–224. [CrossRef]

22. Shilo, E.; Ziv, B.; Shamir, E.; Rimmer, A. Evaporation from Lake Kinneret, Israel, during hot summer days. *J. Hydrol.* **2015**, *528* 264–275. [CrossRef]
23. Castro, C.L.; Chang, H.-I.; Dominguez, F.; Carrillo, C.; E Schemm, J.; Juang, H.-M.H. Can a Regional Climate Model Improve the Ability to Forecast the North American Monsoon? *J. Clim.* **2012**, *25*, 8212–8237. [CrossRef]
24. Higgins, R.W.; Shi, W. Relationships between Gulf of California Moisture Surges and Tropical Cyclones in the Eastern Pacific Basin. *J. Clim.* **2005**, *18*, 4601–4620. [CrossRef]
25. Adam, D.K.; Comrie, A.C. The North American Monsoon. *Bull. Am. Meteorol. Soc.* **1997**, *78*, 2197–2214. [CrossRef]
26. Wang, B.; Biasutti, M.; Byrne, M.P.; Castro, C.; Chang, C.; Cook, K.; Fu, R.; Grimm, A.M.; Ha, K.; Hendon, H. Climate Change Assessment. *Bull. Am. Meteorol. Soc.* **2021**, *102*, E1–E19. [CrossRef]
27. Arriaga-Ramírez, S.; Cavazos, T. Regional trends of daily precipitation indices in northwest Mexico and southwest United States. *J. Geophys. Res. Space Phys.* **2010**, *115*, 14111. [CrossRef]
28. Abatzoglou, J.T.; Dobrowski, S.Z.; Parks, S.A.; Hegewisch, K.C. TerraClimate, a high-resolution global dataset of monthly climate and climatic water balance from 1958–2015. *Sci. Data* **2018**, *5*, 170191. [CrossRef]
29. Castro, C.L.; McKee, T.B.; Pielke, R.A. The Relationship of the North American Monsoon to Tropical and North Pacific Sea Surface Temperatures as Revealed by Observational Analyses. *J. Clim.* **2001**, *14*, 4449–4473. [CrossRef]
30. Shamir, E.; Meko, D.M.; Graham, N.E.; Georgakakos, K.P. Hydrologic Model Framework for Water Resources Planning in the Santa Cruz River, Southern Arizona. *JAWRA J. Am. Water Resour. Assoc.* **2007**, *43*, 1155–1170. [CrossRef]
31. Chang, H.-I.; Castro, C.L.; Carrillo, C.M.; Dominguez, F. The more extreme nature of U.S. warm season climate in the recent observational record and two "well-performing" dynamically downscaled CMIP3 models. *J. Geophys. Res. Atmos.* **2015**, *120*, 8244–8263. [CrossRef]
32. Pascale, S.; Carvalho, L.M.V.; Adams, D.K.; Castro, C.L.; Cavalcanti, I.F.A. Current and Future Variations of the Monsoons of the Americas in a Warming Climate. *Curr. Clim. Chang. Rep.* **2019**, *5*, 125–144. [CrossRef]
33. Luong, T.M.; Castro, C.L.; Chang, H.-I.; Lahmers, T.; Adams, D.K.; Ochoa-Moya, C.A. The More Extreme Nature of North American Monsoon Precipitation in the Southwestern United States as Revealed by a Historical Climatology of Simulated Severe Weather Events. *J. Appl. Meteorol. Clim.* **2017**, *56*, 2509–2529. [CrossRef]
34. Castro, C.L.; Pielke, R.A.; Adegoke, J.O.; Schubert, S.D.; Pegion, P.J. Investigation of the Summer Climate of the Contiguous United States and Mexico Using the Regional Atmospheric Modeling System (RAMS). Part II: Model Climate Variability. *J. Clim.* **2007**, *20*, 3866–3887. [CrossRef]
35. DeMaria, E.M.C.; Hazenberg, P.; Scott, R.L.; Meles, M.B.; Nichols, M.; Goodrich, D. Intensification of the North American Monsoon Rainfall as Observed From a Long-Term High-Density Gauge Network. *Geophys. Res. Lett.* **2019**, *46*, 6839–6847. [CrossRef]
36. Thomas, B.E.; Pool, D.R. Trends in streamflow of the San Pedro River, southeastern Arizona, and regional trends in precipitation and streamflow in southeastern Arizona and southwestern New Mexico. *Prof. Paper US Geol. Surv.* **2006**, *1712*, 79.
37. Shamir, E.; Wang, J.; Georgakakos, K.P. Probabilistic Streamflow Generation Model for Data Sparse Arid Watersheds. *JAWRA J. Am. Water Resour. Assoc.* **2007**, *43*, 1142–1154. [CrossRef]
38. Castro, L.C.; Chang, H.; Leuthold, M.; Luong, T.; Carrillo, C.; Mazon, J.; Stutler, J.; Jares, M.; Lahmers, T.; Cassell, W. *Assessing Climate Change Impacts for DoD Installations in the Southwest United States during the Warm Season*; University of Arizona Tucson United States: Tucson, AZ, USA, 2017; p. 128.
39. Bukovsky, M.S. Temperature Trends in the NARCCAP Regional Climate Models. *J. Clim.* **2012**, *25*, 3985–3991. [CrossRef]
40. Bukovsky, M.S.; Carrillo, C.M.; Gochis, D.J.; Hammerling, D.M.; McCrary, R.R.; Mearns, L.O. Toward Assessing NARCCAP Regional Climate Model Credibility for the North American Monsoon: Future Climate Simulations. *J. Clim.* **2015**, *28*, 6707–6728. [CrossRef]
41. Geil, K.L.; Serra, Y.L.; Zeng, X. Assessment of CMIP5 Model Simulations of the North American Monsoon System. *J. Clim.* **2013**, *26*, 8787–8801. [CrossRef]
42. Pascale, S.; Boos, W.R.; Bordoni, S.; Delworth, T.L.; Kapnick, S.B.; Murakami, H.; Vecchi, G.A.; Zhang, W. Weakening of the North American monsoon with global warming. *Nat. Clim. Chang.* **2017**, *7*, 806–812. [CrossRef]
43. Maloney, E.D.; Camargo, S.J.; Chang, E.; A Colle, B.; Fu, R.; Geil, K.L.; Hu, Q.; Jiang, X.; Johnson, N.C.; Karnauskas, K.B.; et al. North American Climate in CMIP5 Experiments: Part III: Assessment of Twenty-First-Century Projections. *J. Clim.* **2014**, *27*, 2230–2270. [CrossRef]
44. Torres-Alavez, A.; Cavazos, T.; Turrent, C. Land–Sea Thermal Contrast and Intensity of the North American Monsoon under Climate Change Conditions. *J. Clim.* **2014**, *27*, 4566–4580. [CrossRef]
45. Cook, B.; Seager, R.D. The response of the North American Monsoon to increased greenhouse gas forcing. *J. Geophys. Res. Atmos.* **2013**, *118*, 1690–1699. [CrossRef]
46. Colorado-Ruiz, G.; Cavazos, T.; Salinas, J.A.; De Grau, P.; Ayala, R. Climate change projections from Coupled Model Intercomparison Project phase 5 multi-model weighted ensembles for Mexico, the North American monsoon, and the mid-summer drought region. *Int. J. Clim.* **2018**, *38*, 5699–5716. [CrossRef]
47. Maddox, R.A.; Mccollum, D.M.; Howard, K.W. Large-Scale Patterns Associated with Severe Summertime Thunderstorms over Central Arizona. *Weather. Forecast.* **1995**, *10*, 763–778. [CrossRef]

48. Magirl, C.S.; Webb, R.H.; Griffiths, P.G.; Schaffner, M.; Shoemaker, C.; Pytlak, E.; Yatheendradas, S.; Lyon, S.W.; Troch, P.A.; Desilets, S.L.E.; et al. Impact of recent extreme Arizona storms. *Eos Trans. Am. Geophys. Union* **2007**, *88*, 191–193. [CrossRef]
49. Redmond, K.T.; Koch, R.W. Surface Climate and Streamflow Variability in the Western United States and Their Relationship to Large-Scale Circulation Indices. *Water Resour. Res.* **1991**, *27*, 2381–2399. [CrossRef]
50. Sheppard, P.R.; Comrie, A.C.; Packin, G.D.; Angersbach, K.; Hughes, M.K. The climate of the US Southwest. *Clim. Res.* **2002**, *21*, 219–238. [CrossRef]
51. Kim, S. An Assessment of Atmospheric Rivers as Flood Produces in Arizona. Master's Thesis, Hydrology and Water Resources, University of Arizona, Tucson, AZ, USA, 2015; 134p.
52. Neiman, P.J.; Ralph, F.M.; Moore, B.J.; Hughes, M.; Mahoney, K.M.; Cordeira, J.M.; Dettinger, M.D. The Landfall and Inland Penetration of a Flood-Producing Atmospheric River in Arizona. Part I: Observed Synoptic-Scale, Orographic, and Hydrometeorological Characteristics. *J. Hydrometeorol.* **2013**, *14*, 460–484. [CrossRef]
53. Rutz, J.J.; Steenburgh, W.J. Quantifying the role of atmospheric rivers in the interior western United States. *Atmos. Sci. Lett.* **2012**, *13*, 257–261. [CrossRef]
54. Rutz, J.J.; Steenburgh, W.J.; Ralph, F.M. The Inland Penetration of Atmospheric Rivers over Western North America: A Lagrangian Analysis. *Mon. Wea. Rev.* **2015**, *143*, 1924–1944. [CrossRef]
55. Dettinger, M.D. *Fifty-Two Years of "Pineapple-Express" Storms Across the West Coast of North America US Geological Survey, Scripps Institution of Oceanography for the California Energy Commission*; California Energy Commission: Sacramento, CA, USA, 2004.
56. Neelin, J.D.; Langenbrunner, B.; Meyerson, J.E.; Hall, A.; Berg, N. California Winter Precipitation Change under Global Warming in the Coupled Model Intercomparison Project Phase 5 Ensemble. *J. Clim.* **2013**, *26*, 6238–6256. [CrossRef]
57. Cook, B.I.; Ault, T.R.; Smerdon, J.E. Unprecedented 21st century drought risk in the American Southwest and Central Plains. *Sci. Adv.* **2015**, *1*, e1400082. [CrossRef] [PubMed]
58. Dettinger, M.D. Climate Change, Atmospheric Rivers, and Floods in California - A Multimodel Analysis of Storm Frequency and Magnitude Changes1. *JAWRA J. Am. Water Resour. Assoc.* **2011**, *47*, 514–523. [CrossRef]
59. Gao, Y.; Lu, J.; Leung, L.R.; Yang, Q.; Hagos, S.; Qian, Y. Dynamical and thermodynamical modulations on future changes of landfalling atmospheric rivers over western North America. Geophys. *Res. Lett.* **2015**, *42*, 7179–7186. [CrossRef]
60. Warner, M.D.; Mass, C.F.; Salathé, E.P. Changes in Winter Atmospheric Rivers along the North American West Coast in CMIP5 Climate Models. *J. Hydrometeorol.* **2015**, *16*, 118–128. [CrossRef]
61. Payne, A.E.; Magnusdottir, G. An evaluation of atmospheric rivers over the North Pacific in CMIP5 and their response to warming under RCP 8.5. *J. Geophys. Res. Atmos.* **2015**, *120*, 11–173. [CrossRef]
62. Gershunov, A.; Shulgina, T.; Clemesha, R.E.S.; Guirguis, K.; Pierce, D.W.; Dettinger, M.D.; Lavers, D.A.; Cayan, D.R.; Polade, S.D.; Kalansky, J.; et al. Precipitation regime change in Western North America: The role of Atmospheric Rivers. *Sci. Rep.* **2019**, *9*, 1–11. [CrossRef]
63. Shields, C.A.; Kiehl, J.T. Atmospheric river landfall-latitude changes in future climate simulations. *Geophys. Res. Lett.* **2016**, *43*, 8775–8782. [CrossRef]
64. Ritchie, E.A.; Wood, K.M.; Gutzler, D.S.; White, S.R. The Influence of Eastern Pacific Tropical Cyclone Remnants on the Southwestern United States. *Mon. Weather. Rev.* **2011**, *139*, 192–210. [CrossRef]
65. Knutson, T.; Camargo, S.J.; Chan, J.C.L.; Emanuel, K.; Ho, C.-H.; Kossin, J.; Mohapatra, M.; Satoh, M.; Sugi, M.; Walsh, K.; et al. Tropical Cyclones and Climate Change Assessment: Part I: Detection and Attribution. *Bull. Am. Meteorol. Soc.* **2019**, *100*, 1987–2007. [CrossRef]
66. Knutson, T.; Camargo, S.J.; Chan, J.C.L.; Emanuel, K.; Ho, C.-H.; Kossin, J.; Mohapatra, M.; Satoh, M.; Sugi, M.; Walsh, K.; et al. Tropical Cyclones and Climate Change Assessment: Part II: Projected Response to Anthropogenic Warming. *Bull. Am. Meteorol. Soc.* **2020**, *101*, E303–E322. [CrossRef]
67. Magaña, V.; Zermeño, D.; Neri, C. Climate change scenarios and potential impacts on water availability in northern Mexico. *Clim. Res.* **2012**, *51*, 171–184. [CrossRef]
68. Dominguez, F.; Cañon, J.; Valdes, J. IPCC-AR4 climate simulations for the Southwestern US: The importance of future ENSO projections. *Clim. Chang.* **2010**, *99*, 499–514. [CrossRef]
69. Ajami, H.; Meixner, T.; Dominguez, F.; Hogan, J.; Maddock, T. Seasonalizing mountain system recharge in semi-arid basins—climate change impacts. *Groundwater* **2012**, *50*, 585–597. [CrossRef]
70. Serrat-Capdevila, A.; Valdés, J.B.; Pérez, J.G.; Baird, K.; Mata, L.J.; Maddock III, T. Modeling climate change impacts–and uncertainty–on the hydrology of a riparian system: The San Pedro Basin (Arizona/Sonora). *J. Hydrol.* **2007**, *347*, 48–66. [CrossRef]

water

MDPI

Article

Assessing Groundwater Withdrawal Sustainability in the Mexican Portion of the Transboundary Santa Cruz River Aquifer

Elia M. Tapia-Villaseñor [1,*], Eylon Shamir [2], Mary-Belle Cruz-Ayala [3] and Sharon B. Megdal [3]

1 Departamento de Geología, Universidad de Sonora, Hermosillo 83000, Mexico
2 Hydrologic Research Center, San Diego, CA 92127, USA; eshamir@hrcwater.org
3 Water Resources Research Center, The University of Arizona, Tucson, AZ 85719, USA; marybelca@email.arizona.edu (M.-B.C.-A.); smegdal@arizona.edu (S.B.M.)
* Correspondence: elia.tapia@unison.mx

Abstract: The impact of climate uncertainties is already evident in the border communities of the United States and Mexico. This semi-arid to arid border region has faced increased vulnerability to water scarcity, propelled by droughts, warming atmosphere, population growth, ecosystem sensitivity, and institutional asymmetries between the two countries. In this study, we assessed the annual water withdrawal, which is essential for maintaining long-term sustainable conditions in the Santa Cruz River Aquifer in Mexico, which is part of the U.S.–Mexico Transboundary Santa Cruz Aquifer. For this assessment, we developed a water balance model that accounts for the water fluxes into and out of the aquifer's basin. A central component of this model is a hydrologic model that uses precipitation and evapotranspiration demand as input to simulate the streamflow into and out of the basin, natural recharge, soil moisture, and actual evapotranspiration. Based on the precipitation record for the period 1954–2020, we found that the amount of groundwater withdrawal that maintains sustainable conditions is 23.3 MCM/year. However, the record is clearly divided into two periods: a wet period, 1965–1993, in which the cumulative surplus in the basin reached ~380 MCM by 1993, and a dry period, 1994–2020, in which the cumulative surplus had been completely depleted. Looking at a balanced annual groundwater withdrawal for a moving average of 20-year intervals, we found the sustainable groundwater withdrawal to decline from a maximum of 36.4 MCM/year in 1993 to less than 8 MCM/year in 2020. This study underscores the urgency for adjusted water resources management that considers the large inter-annual climate variability in the region.

Keywords: Santa Cruz River Aquifer; Mexico; water balance model; climate uncertainty; transboundary aquifer; transboundary aquifer assessment; Arizona; Sonora

Citation: Tapia-Villaseñor, E.M.; Shamir, E.; Cruz-Ayala, M.-B.; Megdal, S.B. Assessing Groundwater Withdrawal Sustainability in the Mexican Portion of the Transboundary Santa Cruz River Aquifer. *Water* **2022**, *14*, 233. https://doi.org/10.3390/w14020233

Academic Editor: Pankaj Kumar

Received: 30 September 2021
Accepted: 3 January 2022
Published: 13 January 2022

1. Introduction

According to the International Groundwater Resources Assessment Centre (IGRAC), a total of 468 transboundary aquifers have been identified worldwide [1], a figure that has steadily increased over the last decade due to advances in transboundary aquifer assessment. Groundwater from transboundary aquifers constitutes a significant source of fresh water for the environment and numerous communities in almost every nation [2,3], representing a valuable, invisible, and finite resource that needs to be managed sustainably.

Historically, the United States and Mexico have engaged in insightful binational cooperation and dialogue regarding water resources. A vivid example of such cooperation, the 1994 Treaty for the Utilization of Waters of the Colorado and Tijuana Rivers and of the Rio Grande, along with its interpretations (Minutes), addresses specific border, environmental, and water-related issues. Yet, U.S.–Mexico relations surrounding water resources have not been exempted from conflict, such as the diplomatic dispute regarding the United States unilateral decision to build the All-American Canal in California that affected groundwater recharge in Mexican territory. In addition, the institutional asymmetries between the two countries, which are detailed in [2,4,5], could also jeopardize possible cooperation on water resources management, as described by [6]. Fortunately, among other outcomes, cooperation between the United States and Mexico has resulted in transboundary-aquifer assessment efforts to improve the understanding of their shared water resources.

A solid scientific foundation on groundwater resources is a needed first step in developing groundwater management strategies in transboundary settings [2]. It is also essential in places that rely on groundwater resources for their basic activities or are currently affected by climate uncertainties, such as the Transboundary Santa Cruz Aquifer (TSCA) shared between the United States and Mexico [3] (Figure 1). Water supply in the TSCA, the binational aquifer recharged by the Santa Cruz River, is highly sensitive to climate variability and largely depends on compliance of local and international water and wastewater transfer agreements (e.g., [3,7–9]). The TSCA recharge results from riverbed infiltration and mountain front recharge in Mexico and the United States. Thus, the TSCA is a binational aquifer in which the water-resources management and natural processes on one side of the border directly impact the neighboring country.

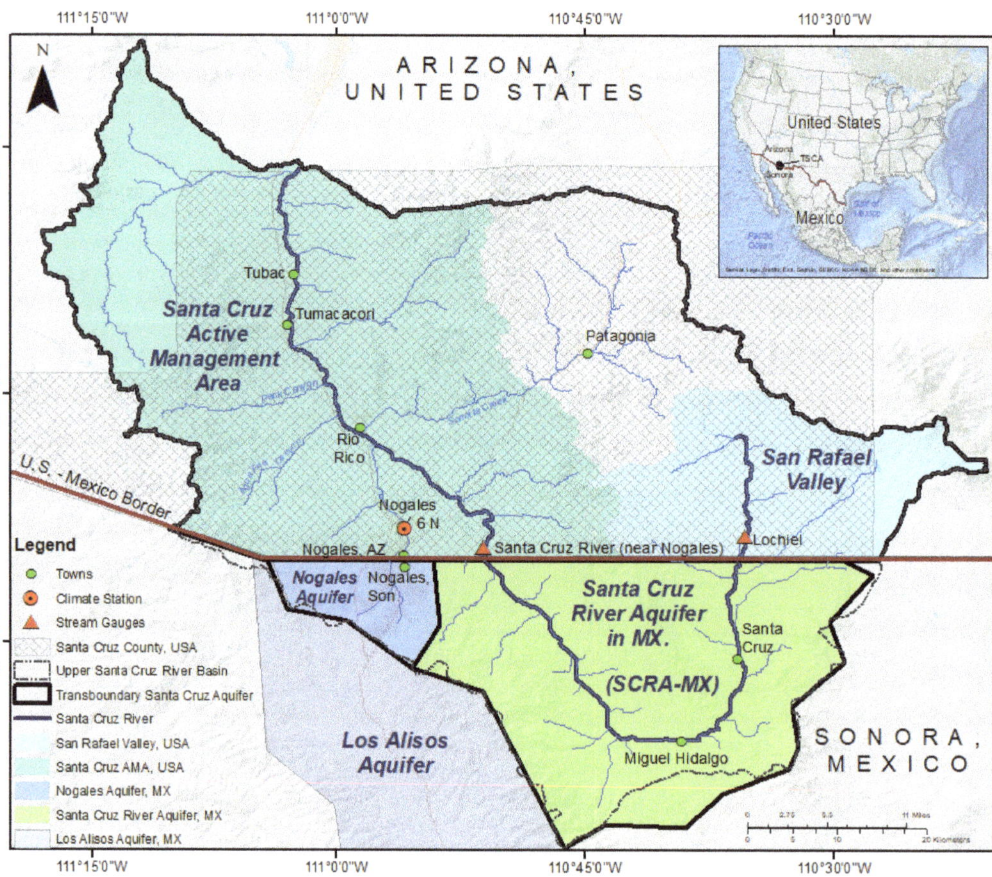

Figure 1. Political-administrative domains in the Transboundary Santa Cruz Aquifer.

Because of the region's scarce water resources, population increase, and growing groundwater demands on both sides of the border, the TSCA was selected for the U.S.–

Mexico Transboundary Aquifer Assessment Program (TAAP). The TAAP was signed in 2009 by the principal engineers of the International Boundary and Water Commission (IBWC) and aimed to improve the knowledge of U.S.–Mexico transboundary aquifers [10]. The principles of the TAAP Cooperative Framework include elements that promote trust between the United States and Mexico (e.g., data sharing, development of binational aquifer assessment activities, the establishment of technical advisory committees, and the establishment of technical groups). These elements are crucial to maintaining the binational cooperation necessary when researching shared aquifers. Transboundary aquifer assessments worldwide have effectively employed these elements, including the Guarani, Nubian Sandstone, Saharan Aquifer, and Genevese Aquifer [2]. This study is part of the TAAP's effort to better understand the TSCA, particularly in the Mexican portion of the aquifer.

The TSCA comprises four political-administrative domains: the Santa Cruz Active Management Area (SCAMA) in Arizona, with an areal extent of 1,854.43 square kilometers (km^2); the San Rafael Valley, with an areal extent of approximately 465 km^2; the Nogales Aquifer in Mexico, with an areal extent of 120 km^2; and the Santa Cruz River Aquifer in Mexico (SCRA-MX), with an areal extent of 952 km^2 (Figure 1). The region's water supply relies on a relatively limited-storage, alluvial aquifer system underneath the Santa Cruz River Valley. The dominant source of recharge for the aquifer is the episodic streamflow events in the intermittent Santa Cruz River and its ephemeral desert tributaries. These episodic streamflow events are triggered by highly variable, seasonal (winter and summer) precipitation events (e.g., [7]). Thus, due to this region's limited groundwater storage and its reliance on episodic streamflow events, even small changes in groundwater recharge patterns coupled with increased water demand from border communities can adversely affect the water-supply reliability. Additionally, precipitation projections for the Upper Santa Cruz River Basin point to significant uncertainty and increased interannual variability, which will likely challenge water providers in meeting the water demands of the border communities [3,7,9,11].

Though previous studies have analyzed water resources in different portions of the TSCA, only a few have addressed the Santa Cruz River Aquifer in Mexico (SCRA-MX). For instance, studies have assessed the impact of urban growth on water resources, focusing on the "Ambos Nogales" region, which is located within the Nogales Aquifer and the SCAMA regions in Mexico and the United States [12,13]. Other studies developed ecosystem-services tools to assess the impacts of climate change and urban growth in the U.S. portion of the Santa Cruz Watershed [14] and to evaluate flood risk in the Ambos Nogales region, considering various scenarios of land-use changes [15]. In addition, climate change and water-resources assessments through hydrologic frameworks have also been developed for the SCAMA, attempting to bridge the gap between scientific findings and stakeholders [3,7,8,16,17].

Studies focusing on the SCRA-MX include hydrogeological characterizations of the aquifer [18], regional studies that assessed the impacts of climate change on local water resources [11,19], and the water availability reports published by the National Water Commission in Mexico (CONAGUA) [20–23]. These studies have improved the knowledge of the TSCA and have helped to develop tools that assist with water-resources-management decisions. However, a deeper understanding of the TSCA system, particularly the SCRA-MX, is needed to develop management strategies focused on groundwater sustainability.

Sustainable groundwater withdrawal can be generally defined as the amount of water that can be withdrawn from an aquifer without causing undesirable environmental, economic, or social consequences [24,25]. Undesirable outcomes of unsustainable groundwater withdrawal may include a decrease in water availability for populations and the environment, a deterioration of the groundwater quality, riparian vegetation die-off, an intrusion of contaminated water or seawater, and land subsidence. This study aims to identify, through a water-balance model, the annual groundwater-withdrawal rate from the SCRA-MX that maintains sustainable conditions. Although sustainable groundwater withdrawal can have various definitions and nuances, we define groundwater-withdrawal sustainability as the withdrawal rate that maintains a multi-year balance between the water fluxes into and out of the basin.

2. Study Area

From its headwaters in the San Rafael Valley in Arizona, the Santa Cruz River flows southward to cross the U.S.–Mexico border into Sonora, Mexico. The river then curves northward and returns to the United States, just east of Nogales, Arizona; from there, it flows north to merge with the Gila River, a tributary of the Colorado River (Figure 1).

In the Mexican territory, water from the TSCA is primarily used by the city of Nogales and the town of Santa Cruz. According to Mexico's 2020 census, the number of registered residents was 264,782 and 1,835 in Nogales and the town of Santa Cruz, respectively. These numbers mark a 20.2% population increase for Nogales and an 8.16% decrease for the town of Santa Cruz compared with the 2010 census. On the other side of the border, in the 2020 census for Nogales, Arizona, the population declined from 20,837 (2010) to 19,770 (2020). During the same period, the total population in Santa Cruz County, Arizona, was almost unchanged (47,420 in 2010 and 47,669 in 2020).

In Mexico, the national Law of the Nation's Waters (in Spanish, *Ley de Aguas Nacionales*, or LAN), signed in 1992, defines the role of the National Water Commission (CONAGUA) as the federal agency responsible for water management. Grounded in the Constitution, the LAN ordains in Article 20 that "the exploitation, use, or non-consumptive use [e.g., energy production] of the nation's water resources should be carried out through a concession or *asignación* (in Spanish) granted by the Federal Executive Branch or Basin Councils" [26,27]. *Asignación* is the legal term that the legislation utilizes to describe water appropriation for urban or domestic purposes. This appropriation cannot be transferred to other users. A concession defines the amount of water that can be extracted from a specific well/aquifer. The duration of concessions ranges from five to thirty years, and users can apply for an extension [28]. The concessions and *asignaciones* are registered in the Public Registry of Water Rights (in Spanish, *Registro Público de Derechos de Agua*, or REPDA).

CONAGUA is also responsible for publishing groundwater availability reports for each aquifer in the Official Federal Gazette (in Spanish, *Diario Oficial de la Federación*, or DOF). These reports, which are published every three years, guide the appropriations of water concession and allocation volumes. In CONAGUA reports, water balance models are used to assess groundwater availability. The premise of these water balance models is that the *Mean Annual Groundwater Availability* for a given aquifer is equal to the difference between *Mean Annual Recharge* and the *Mean Annual Groundwater Extractions* and the *Natural Discharge* for environmental needs. For example, in 2020, CONAGUA published groundwater availability reports for 653 aquifers and reported the available volume for appropriation in the SCRA-MX to be 33.85 MCM/year [29]. It should be noted that the actual volume of groundwater withdrawal is often not monitored by CONAGUA and may therefore deviate from REPDA's authorized volumes.

In the SCRA-MX groundwater concessions and *asignaciones* have increased from 19.2 MCM/year in 1995 to 33.85 MCM in 2020 (Figure 2) [30]. This increase is primarily attributed to a gradual increase in appropriated concessions for agriculture, from 0 in 1995 to approximately 9 MCM/year in 2020. Additional appropriation of approximately 2 MCM/year was allocated since 2011 to the industrial sector for supporting copper mining operations.

In the Nogales Aquifer in Mexico (Figure 1), groundwater allocations (concessions and *asignaciones*) have ranged from 0.003 MCM/year to 1.37 MCM, since 1997. It is important to note that additional water has been transferred for decades from both the SCRA-MX and Los Alisos aquifers to supply the water needs of the city of Nogales [3,23,31]. According to CONAGUA, since 1997, most concessions authorized in the Nogales Aquifer have been industrial, consistent with the main economic activity reported by the Ministry of Economy (Figure 2). In comparison, most of the groundwater volume allocated for the SCRA-MX is dedicated to urban-public services. Concessions in the SCRA-MX for livestock and industrial activities have increased since 2011 (Figure 2). Data published by the Ministry of Economy in 2019 show that the agricultural and mining sectors of the Nogales and Santa Cruz municipalities have registered minimal increases in their economic activity during this period [32].

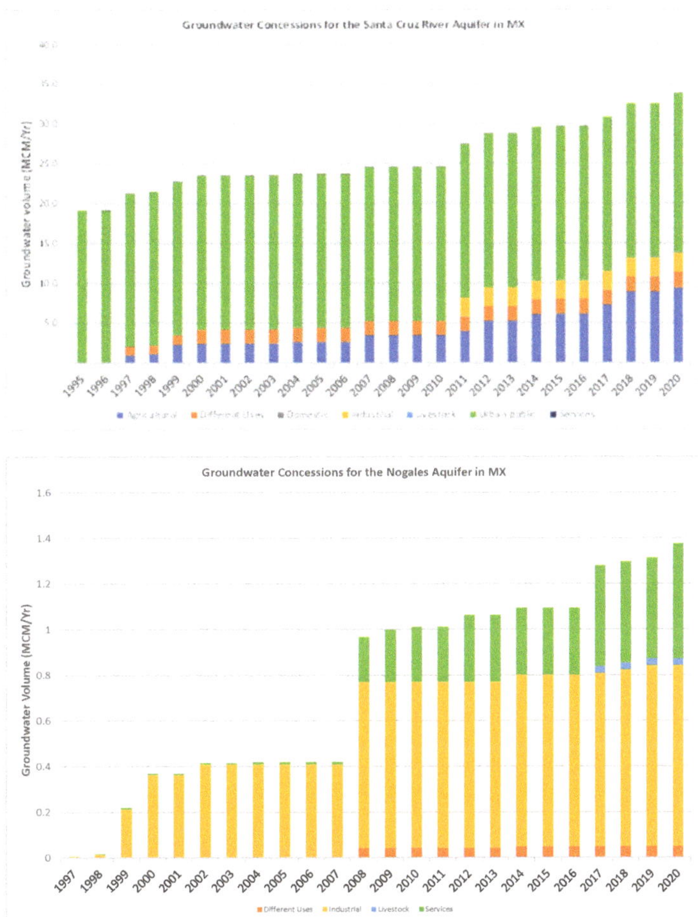

Figure 2. Groundwater Concessions (*Asignaciones)* in the Santa Cruz River Aquifer in Mexico and the Nogales Aquifer. Source: REPDA (2020) [30].

3. Materials and Methods

Our study assessed the amount of annual groundwater withdrawal that maintains long-term sustainable conditions in the SCRA-MX. Sustainable groundwater withdrawal can be generally defined as the amount of water that can be withdrawn from the aquifer without causing undesirable environmental, economic, or social consequences [24,25]. Undesirable implications due to unsustainable groundwater withdrawal may include the decrease in water availability for populations and the environment, deterioration of the groundwater quality, riparian vegetation die-off, intrusion of contaminated water, intrusion of seawater, and land subsidence.

Within the U.S. side of the border, the term safe yield is often used to describe a management goal that maintains sustainable conditions. Safe yield is defined by ADWR as a groundwater management goal that attempts to achieve and maintain a long-term balance between the annual amount of groundwater withdrawn and the annual amount of natural and artificial recharge (A.R.S. § 45–561(12)). The terms "safe yield" and "sustainability", with respect to groundwater management, are often interchangeably used. Safe yield was historically defined as the attainment and maintenance of a long-term balance between the amount of groundwater withdrawn and the amount of recharge (e.g., [33]). Adhering to this definition, in order to reach safe yield conditions, groundwater withdrawal should not exceed natural recharge. This practice, however, ignores other long-term water fluxes out of the basin such as discharge, evapotranspiration, or springs that extract unaccounted for groundwater, which eventually may deplete the aquifer. Regardless of the term selection, the selected term should be clearly defined for each specific aquifer considering its management goals and the potential hydrologic, economic, or ecologic harms inflicted by unsustainable management [34].

To estimate the amount of annual withdrawal that maintains sustainable conditions, we used a modeling framework that consisted of a water balance model (WBM) and a hydrologic model. The WBM was developed to account for all annual water fluxes into and out of the basin of the SCRA-MX and to calculate the long-term cumulative water deficits or surpluses. In an arid environment that relies on highly inter-annual climatic variability and therefore highly variable year-to-year natural recharge, the deficits and surpluses should be assessed over multiple years. For instance, the current ADWR recommendation for a quantitative assessment of safe yield is to consider a 20-year moving average interval for the natural components of the water budget (e.g., natural recharge) and a three-year running average for the artificial components (e.g., groundwater withdrawals and incidental recharge). In our study, we assessed the sustainable withdrawal by first considering the entire period of the historical record (1954–2020) and second, by considering 20 year moving averages, as recommended by ADWR.

3.1. Water Balance Model

Adapted from CONAGUA (2020) [23], the annual mass balance in the SCRA-MX basin is calculated using the following equation:

$$\Delta S = Q_{in} + GW_{in} + Re + Ag - Q_{out} - GW_{out} - ET - Pu \quad (1)$$

where ΔS represents the annual positive or negative water storage changes in the aquifer and vadose zone, Q_{in} and Q_{out} are the Santa Cruz River streamflow in and out of the basin. GW_{in} and GW_{out} are the groundwater fluxes into and out of the basin; Re is the natural groundwater recharge component; Ag is the return flow from irrigated agriculture; ET is the actual evapotranspiration losses; and Pu is the groundwater withdrawal. The units for all the terms in Equation (1) are million cubic meters per year (MCM/year). CONAGUA's water balance model results for the SCRA-MX basin are in Table 1.

Table 1. CONAGUA's WBM components for 2020 (MCM/year). Source: CONAGUA (2020) [23].

Inflows			Outflows			ΔS
GW_{in}	Ag	Re	Pu	GW_{out}	ET	
10.2	4.1	23.8	26.4	2.0	8.8	0.9

In this study, we solved the WBM equation to determine the groundwater withdrawal (Pu) that maintains the long-term changes of ΔS in sustainable conditions. This simulation was implemented at an annual time step to assess the overall long-term balance. In the following section, we describe the WBM components considered in this study.

3.2. Precipitation

Hourly precipitation time series are needed as input to the hydrologic model. Hourly precipitation records since 1949 are available from the Nogales 6N station (USC00025924;

W110.968, N31.4554, 1055 m) from the US National Weather Service (Figure 1). However, we found many disagreements when this record was compared to the 1954–2020 daily quality-controlled dataset from the same station. We decided, therefore, to use the daily time series and disaggregate it to hourly (see Figures S1–S6). The disaggregation was carried out by using the hourly time series to identify the hourly diurnal distribution with reported daily precipitation days. If hourly events were unavailable for the target date, we selected from the hourly time series a rainy day within a short duration from the target date.

The hourly precipitation was then spatially interpolated over the study area using the 1958–2020, ~4 km^2 gridded monthly rainfall, available from the TerraClimate dataset [35]. The interpolation was carried out by using the ratios of the station's grid cell with the other TerraClimate grid cells for the matching months. These ratios were used as multipliers for the interpolation to derive 4 km^2 hourly time series. Prior to 1958, a randomly selected month from the same wetness tercile as the station's record was used for the interpolation. The interpolated 4 km^2 grid was then averaged over the area of the modeling units to derive the hourly Mean Areal Precipitation (MAP) time series, which were used as input to the hydrologic model. This spatial interpolation method assumes that the Nogales gauge well represents the occurrence of hourly events over the study area, and that the hourly rainfall distribution throughout the month is uniformly distributed in space. These assumptions are particularly challenged during the North American Monsoon summer rainfall characterized by small-scale local convective thunderstorms.

3.3. Streamflow (Q_{in} and Q_{out})

Observations of surface inflow and outflow to and from the Mexican portion of the Santa Cruz River are available from the USGS hydrometric stations at Lochiel (USGS 09480000) and near Nogales (USGS 09480500). The Lochiel hydrometric station, approximately 2.5 km north of the international border (Latitude 31°21′19″, Longitude 110°35′20″, 1400 m above sea level), drains 209 km^2 of the Santa Cruz River headwater at the San Rafael Valley and parts of the Patagonia and Huachuca Mountains. It has a daily streamflow record for the period January 1949–August 2014 and from May 2019 to present. Approximately one kilometer north of the international border, the Nogales hydrometric station (Latitude 31°20′40″ N 110°51′03″ W, 1120 m above sea level) drains an area of 1364 km^2. It has a daily streamflow record from 1913 to the present, with some missing years during the 1920s. We note that although the 1954–2020 observed average streamflow out of the basin was 22.1 MCM/year (range 0–181 MCM/year), the streamflow out of the basin was likely generated from rainfall over the basin and therefore was not considered as a negative flux in Equation (1).

For this study, a hydrologic model was used to simulate the inflow and outflow (i.e., Q_{in} and Q_{out}) as a function of precipitation. The hydrologic model we used is the Sacramento Soil Moisture Accounting (SAC-SMA) model [36], as it was configured for this basin by the Colorado Basin River Forecast Center (CBRFC), U.S. National Weather Service (see Figures S7 and S8).

The SAC-SMA model is a continuous hydrologic model that keeps track of the water content at the basin's top and subsoil layers. It uses precipitation and evapotranspiration (ET) demand as input to simulate runoff, recharge, actual evapotranspiration, and soil moisture. The CBRFC's primary purpose is to warn for high-flow events. Therefore, they focused their SAC-SMA model calibration on capturing episodic flow events. In our study, the model was used to account for the overall streamflow influx into the area of interest. Therefore, the model required additional calibration to capture the range of flow regimes. The calibration was carried out by comparing the simulated streamflow on the Santa Cruz River in Lochiel and near Nogales to observed flow from the USGS gauges. The assessment was carried out for ranging time scales of daily, seasonal, and annual flows. The CBRFC SAC-SMA model configuration for the SCRA-MX basin is based on three hydrologic units. The first hydrologic unit (210 km^2) drains the headwater of the Santa Cruz River to the US–Mexico border crossing. The second and third hydrologic units are the upper and lower parts of the SCRA-MX basin, respectively. The upper part of the basin (617 km^2) drains areas higher than 1515 meters, while the lower part of the basin (537 km^2) drains areas lower than 1515 meters. In our implementation, the surface runoff generated by the lower part of the basin was considered for the flow simulation at the outlet.

The runoff from the upper part, below 50 m^3/sec was considered as the groundwater recharge component. This assumption is warranted, as it is seen that most of the flow at the Nogales gauge is attributed to local rainfall events. During large storms in the upper basin, the flow contribution to the basin's outlet is delayed and later appears as baseflow [7,16].

In Table 2, summary statistics for the 1954–2020 estimated annual recharge are provided. Notice that CONAGUA (2020) [23] estimated the vertical recharge at 23.8 MCM/year, comparable to our estimated annual average. However, as it is apparent from the values presented in Table 2, the large inter-annual variability of the groundwater recharge may not be well represented by the sample's first-moment indicator.

Table 2. The 1954–2020 estimated recharge in the SCRA-MX.

Estimated Recharge (MCM/Year)	
Average	25.8
Median	20.1
Standard Deviation	22.8
Minimum	0.7
Maximum	104.5

3.4. Groundwater (GW_{in} and GW_{out})

The border crossing groundwater inflow and outflow mainly occur at the alluvial aquifer underneath the river's channel bed. These fluxes are not measured and are estimated from previous studies. Although these fluxes are likely dependent on the aquifer pressure gradients near the international border, we assume constant groundwater fluxes. In our analysis, we adopted CONAGUA (2020) [23] estimate of +10.2 and −2.0 MCM/year for the GW_{in} and GW_{out}, respectively ('+' indicates a flow from the United States to Mexico and '-' indicates a flow from Mexico to the United States). Other studies estimated GW_{out} to be 3.5 MCM/year [37], 1.54 MCM/year [38], and 1.66 MCM/year [39].

3.5. Evapotranspiration (ET)

Evapotranspiration (ET) from the basin can be divided into ET from the soil, ET from irrigated agriculture fields, and ET from the shallow groundwater aquifer through riparian vegetation and exposed surface water sections of the stream. In Equation (1), the ET variable refers to the latter component. The hydrologic model calculates the ET from the soil, and it is implicitly accounted for in the recharge and streamflow terms. The ET from the agricultural field is considered in the calculation of the agricultural return term. In CONAGUA (2020) [23], the total ET losses from the aquifer were estimated as 8.8 MCM/year. This estimate assumes that ET from the groundwater is linearly reduced with depth-to-water up to an extinction depth of 10 m. In CONAGUA (2020) [23], the surface area estimate of the aquifer's water levels was provided as a base for the ET estimate. This procedure assumes that the aquifer's water level and the potential evapotranspiration are not changing from year to year.

Using the hydrologic model simulations, we found that the average actual ET from the soil is 314 MCM/year, and the average actual ET is 88% of the annual precipitation. The actual ET is highly correlated with precipitation and ranges from 130 to 530 MCM/year, 62 to 103 percent of the annual precipitation, respectively. These actual ET estimates are comparable to findings by Minjarez et al. (2011) [19].

3.6. Agricultural Return Flow (Ag)

To estimate the agricultural return flow, we used the CONAGUA (2020) [23] procedure. It was based on calculations of crop consumptive use, which is the amount of transpired water during the growth period of the crop. The agricultural return is then calculated as the irrigated water and precipitation that is in excess of the estimated consumptive use. In CONAGUA (2009 to 2020) [20–23], the irrigated agriculture area was estimated as 8.3 km^2 of alfalfa (60%), oat (30%), and sorghum (10%). Using the modified Blaney–Criddle equation [40], the integrated consumptive use of these crops was estimated as 901 mm/year (7.5 MCM/year), and the agricultural return was estimated as ~4.1 MCM/year. In our implementation of the WBM, we used CONAGUA's estimate of consumptive use and the dynamic year-to-year change in precipitation to estimate the groundwater withdrawal that was needed for irrigation. The 1954–2020 average annual precipitation over the agricultural fields was 2.6 MCM/year (range 0.9–5.5 MCM/year), and the average groundwater withdrawal that satisfied the irrigation demand was 4.9 MCM/year, ranging from 1.9 to 6.6 MCM/year. This demand calculation assumes that precipitation occurred during the growing season, and the irrigation was optimized to satisfy the crops' consumptive use. It is important to note that the National Institute of Statistics and Geography (*Instituto Nacional de Estadística y Geografía*) estimated the irrigated agriculture area in the basin to be 15.7 km^2 [41]. Using a 30 m near-infra-red band of Landsat-8 images from May 2018 and May 2019, our team estimated an area of approximately 17 km^2 of agricultural fields. Thus, the water consumption, as well as the areal extent of irrigated agriculture in the basin, is uncertain and requires a comprehensive survey.

4. Results

Using 1954–2020 climate dependent recharge, Qin and Ag (as explained above), we solved Equation (1) for the amount of groundwater withdrawal (Pu) yielding a ΔS annual average of zero. The Pu that maintains a 1954–2020 average ΔS of zero is 23.3 MCM/year. This Pu is in addition to the Pu used for irrigation that satisfies the estimated consumptive use of the cultivated fields, as described in CONAGUA (2020) [23]. Using this estimated Pu, the average fraction of the inflow and outflow fluxes from the basin are presented in Figure 3, and the average quantities of these various fluxes are presented in Figure 4. The largest influx to the basin is the natural recharge, a highly variable flux (see Table 2) that is mainly controlled by the inter-annual variability of precipitation over the SCRA-MX basin.

Figure 3. The 1954–2020 average annual percentages of inflow and outflow fluxes to the SCRA-MX basin.

The cumulative changes of the ΔS using the estimated Pu of 23.33 MCM are shown in Figure 5a. It is seen that out of the 67 water years, approximately 33% have shown a surplus while most years ended with a deficit. The cumulative surplus consistently increased from 1965 to reach a surplus of approximately 385 MCM in 1992. These surplus years can be related to frequent El Niño-Southern Oscillation conditions and positive Pacific Decadal Oscillation [8]. However, since 1992, only two years showed an annual surplus (positive ΔS) and in 2020, the entire surplus that had been gained until 1992 was depleted. These long periods of accrued surplus (1965–1992) and deficit (1995–2020) exemplify the dependence of the sustainable Pu on the period of analysis. The increasing and decreasing trends shown in Figure 5 seem to support ADWR recommendations for examining 20-year intervals, a duration sufficiently long to capture the observed multi decadal trends.

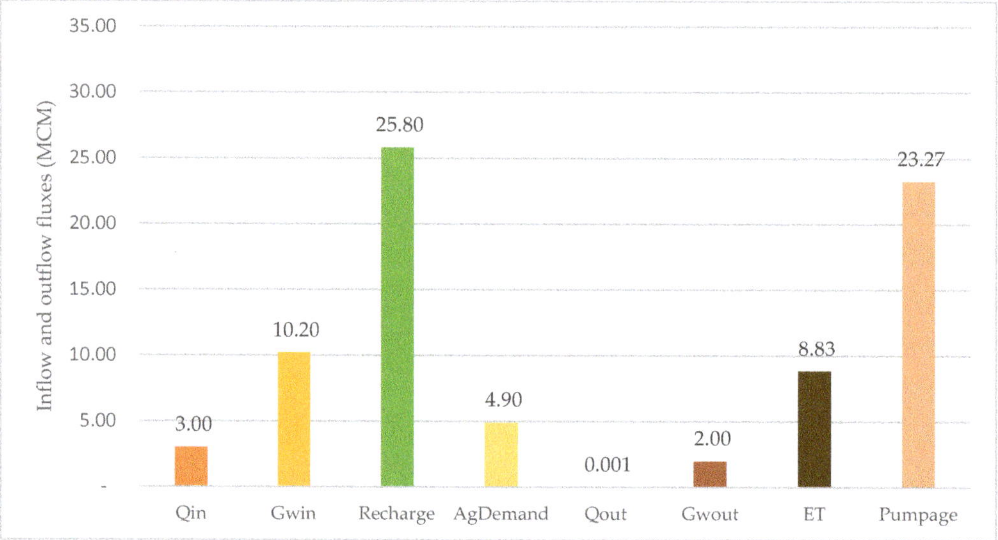

Figure 4. The 1954–2020 average inflow and outflow fluxes for maintaining sustainable conditions.

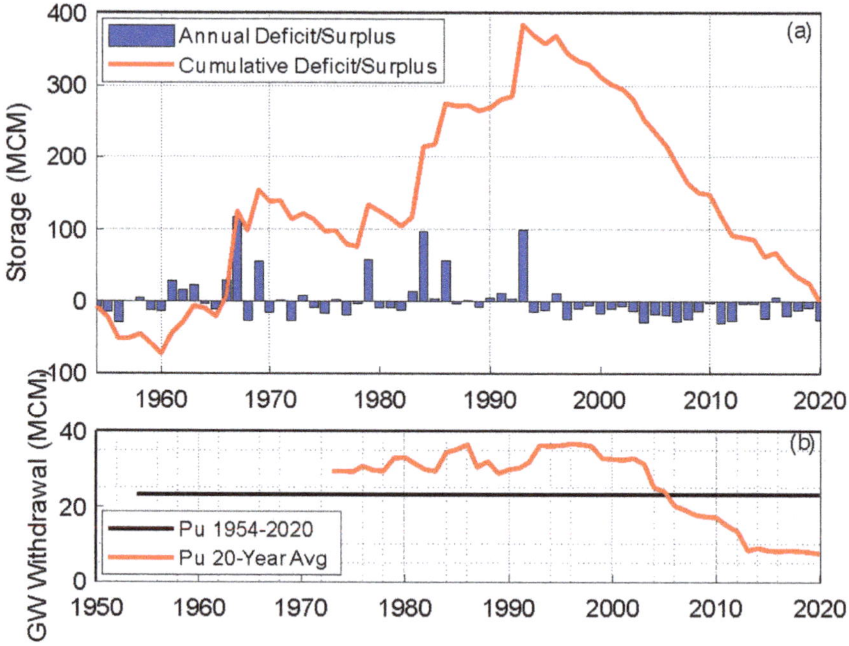

Figure 5. (**a**) Estimated annual and cumulative storage with annual groundwater withdrawal rate that yields an average of zero storage during the period 1954–2020. (**b**) Estimated sustainable groundwater withdrawal (Pu) using a 20-year moving average interval (red) and the 1954–2020 duration.

The fluxes estimated for the WBM can be grouped to three general categories: climate driven annually variable fluxes (Q_{in}, Q_{out}, Re), satisfying water demand fluxes (Ag, Pu), and constant annual fluxes (GWin, GWout, ET). The first category is based on a hydrologic model that uses sub-daily precipitation and evapotranspiration demand time series as input to simulate the fluxes needed for the WBM. While the simulated Q_{in} and Q_{out} were compared to observed streamflow records, the Re estimate cannot be compared to observations. As discussed in the results section, the Re is the largest flux into the basin (Figures 3 and 4) and has large inter-annual variability (Table 2).

Considering moving averages of 20-year intervals, the estimated Pu is shown in Figure 5b (average Pu of 26.3 MCM/year (ranging from 8.1 to 36.8 and an S.D of 9.6 MCM/year). As expected, the estimated 20-year annual Pu has continuously declined since the mid-1990s to approximately 9 MCM/year since 2012.

5. Discussion

Overall, there has been constant cooperation and dialogue over water resources shared between the United States and Mexico. A remarkable example is the 1944 Water Treaty that has allowed sharing surface water among both countries. However, this agreement does not include groundwater management. This absence has not been addressed, although some steps have been taken—for instance, creating the TAAP that allows technical cooperation between both countries and sharing information on groundwater resources.

The Santa Cruz River Aquifer in Mexico (SCRA-MX) is part of the Transboundary Santa Cruz Aquifer (TSCA), an aquifer shared by the United States and Mexico. The TSCA is located in a semi-arid region characterized by limited groundwater storage, dependency on climate variability, and physical water and wastewater transfers within Mexican territory and between the two countries [3,11,19]. Because of this region's limited groundwater storage and the border communities' reliance on groundwater as their sole resource, even small changes in groundwater recharge patterns coupled with increased water demands can detrimentally impact the water-supply reliability.

Previous efforts on the TAAP have focused on understanding the aquifer characteristics of the TSCA, particularly the U.S. portion of the aquifer e.g., [3,12]. Our study improves the understanding of the SCRA-MX, contributes to the overall knowledge of the binational TSCA, and provides information that could serve as a reference for developing a fully binational water budget model.

This analysis, along with previous studies for the TSCA (e.g., [3,11]), reported a substantial decline in regional precipitation since the early 21st century. For example, summer and winter precipitation has declined by 10% and 33%, respectively, according to comparisons of precipitation records from 1955–2000 to 2001–2020. These declines are substantially larger when comparing the same periods of the observed streamflow records out of the SCRA-MX basin (65% and 78% for summer and winter flow, respectively) [8].

Moreover, climate model projections for the mid-21st century for the SCRA-MX basin point to changes in precipitation regime, although these changes are highly uncertain e.g., [8]. These projections will pose additional challenges for water providers in meeting the demands for border communities [3,7,9,11]. To date, most water resources studies in the TSCA have focused on the *Ambos Nogales* region or the SCAMA (e.g., [3,7–9]). Excluding the CONAGUA water availability reports, only a few studies have examined the impact of groundwater extractions in the SCRA-MX (i.e., [18,19,38]). In our study, we used a water balance model approach to estimate the amount of groundwater that could be withdrawn from SCRA-MX, while maintaining a long-term balance between water flowing into and out of the basin. Our study only assessed long-term water resources availability while not examining other potential ecological, economic, water quality degradation, or other harms that water resources management practices may cause. Although our analysis yields deterministic estimates for sustainable annual groundwater withdrawal, based on the best available data and information to derive the input for the WBM equation, it is important to note the analysis' main assumptions and the known sources of uncertainties that may have influenced the results. The main assumption that may require additional examination is that multi-year cumulative surpluses can be indefinitely stored in the basin and used to compensate for shortages in deficit years. It is likely that the aquifer's storage of surpluses is limited by the size of the aquifer and the dynamic of the groundwater interaction with the stream and atmosphere.

Since natural groundwater recharge is highly variable in space and time, an accurate measure of this dominant flux is impractical. However, additional hydrological and hydrogeological measurements could advance understanding of the basin's hydrological process to potentially reduce the uncertainty in the natural recharge estimate. The uncertainty source in the second category stems from a lack of groundwater withdrawal monitoring. Following CONAGUA's procedure we assumed that the groundwater withdrawal was equal to the appropriated concessions and *asignaciones*, as reported by REPDA. Additional information is needed to understand how well the appropriated concessions represent the actual groundwater withdrawal in the basin.

An additional source of uncertainty, as discussed before, is the areal appraisal of the cultivated and irrigated fields. The third category of fluxes, which were assigned as constants following CONAGUA's estimates, is also likely to vary in time. The main reason for assigning them as constants is the lack of information and data to understand their temporal variability. With the available information on the economic activities in the Nogales and Santa Cruz municipalities, it is possible to identify a positive relationship between increased industrial activities and water allocations from 2009 to 2020. While the groundwater surplus has reduced since 1995, allocations for agricultural and industrial activities have increased. Considering this trend, it would be desirable that the national authority assess the potential negative impacts of groundwater over-allocation and its availability to maintain a long-term balance between water flowing into and out of the basin.

It is generally possible to monitor groundwater extraction for *asignaciones* because they are dedicated to public services, for which municipal and state government agencies are responsible for reporting to CONAGUA. However, for groundwater concessions, the monitoring is limited. Additionally, as mentioned above, concessions can be transferred to other users, and although these changes must be reported to CONAGUA, they are often not being promptly reported. Future TAAP efforts on transboundary aquifer assessment include the evaluation of the uncertainty associated with the water balance model that was developed for the TSCA and the identification of specific actions that can substantially reduce uncertainty in WBM simulations. In addition, development of recommendations for a model and data management framework for binational watersheds with similar setting to the TSCA.

6. Conclusions

In this study, we assessed the amount of groundwater withdrawal that maintains sustainable conditions in the SCRA-MX, which is part of the TSCA. In this part of the aquifer, the regulatory allocated groundwater concessions had steadily increased from approximately 18 MCM/year in 1995 to approximately 34 MCM/year in 2020. The increase in groundwater withdrawal concessions was primarily attributed to new allocations for agricultural and industrial usage. In this study, we used a water balance model (WBM) that accounts for all the annual water fluxes into and out of the basin to determine the amount of multi-year groundwater withdrawal that maintains sustainable conditions. In our study, "sustainable conditions" is defined as the amount of annual groundwater withdrawal that maintains a long-term difference of zero between the water fluxes into and out of the basin. We developed a hydrologic model to estimate the year-to-year WBM fluxes of natural recharge and streamflow into and out of the basin (i.e., Sacramento Soil Moisture Accounting). This contribution adds information to current CONAGUA publications. The SAC-SMA model, which was constructed for the region as three sub-basins, uses hourly precipitation and evapotranspiration demand as model input to continuously simulate streamflow, soil moisture, actual evaporation from the soil, and groundwater recharge. The hourly precipitation time series for the SAC-SMA model was developed for 1954–2020 using a gauge located near the border and interpolated using monthly gridded climatology.

The average annual groundwater withdrawal amount that maintained sustainable conditions from 1954–2020 was 23.3 MCM/year. However, by implementing this constant annual withdrawal, there was a period of accrued surplus (1965–1993) followed by an accrued deficit (1994–2020). We also estimated the annual groundwater withdrawals that maintain sustainable conditions in a moving average of 20-year intervals, as recommended by ADWR for safe yield assessment in the SCAMA. For the analysis of the moving average of 20-year intervals, the groundwater withdrawal that maintained sustainable conditions peaked in 1993 at 36.4 MCM/year and had since declined to less than 8 MCM/year in 2020. CONAGUA, in their latest groundwater availability report [23], estimated that groundwater withdrawal of 26.4 MCM/year yields an additional 2.2. MCM/year of available water that could be allocated.

This study demonstrates the sensitivity of water resources management in the Mexican part of the Santa Cruz River basin and its high dependence on natural recharge, which depends on precipitation variability. It points to the challenge of identifying a management scheme that yields sustainable conditions. These challenges are exacerbated by the recent dry period and the projected uncertain precipitation in the region [12]. These mounting challenges call for careful adaptive management and planning of the aquifer to maintain sustainable conditions and long-term reliable water supply into the future.

Supplementary Materials: The following supporting information can be downloaded at: https://www.mdpi.com/article/10.3390/w14020233/s1, Figure S1: Average annual, summer (JJAS), and winter (NDJFM) rainfall over the USCRB (outlined in red). Data are available from the TerraClimate monthly 4 km^2 1958–2020 dataset, Figure S2: A scattergram of 1958–2020 monthly rainfall of the gauge record from Nogales 6N and the matching grid cell available from the TerraClimate dataset, Figure S3: Locations of the of daily rain gauges, Figure S4: A comparison between the total summer rainfall cumulative distributions of the spatially disaggregated Nogales gauge (black) and seven rain gauge records (red). Note that the seven gauges have different durations, as indicated at the subplots' titles, Figure S5: A comparison between the total winter rainfall cumulative distributions of the spatially disaggregated Nogales gauge (black) and seven rain gauge records (red). Note that the seven gauges have different durations, as indicated at the subplots' titles, Figure S6: A scattergram of the summer (red) and winter (blue) total precipitation of the spatially disaggregated observed Nogales record compared with matched observed record from the seven gauges, Figure S7: The 1949–2020 cumulative distributions of simulated (red) and observed (black) streamflow on the Santa Cruz River at Lochiel for the summer, winter, and annual, Figure S8: The 1949–2020 cumulative distributions of simulated (red) and observed (black) streamflow on the Santa Cruz River at the Nogales gauge for the summer, winter, and annual.

Author Contributions: Conceptualization, E.M.T.-V., E.S., M.-B.C.-A. and S.B.M.; supervision, S.B.M. and E.M.T.-V.; methodology E.S., E.M.T.-V., M.-B.C.-A.; validation, E.S.; formal analysis, E.S.; investigation, E.S. and M.-B.C.-A.; resources, E.S. and M.-B.C.-A.; data curation, E.M.T.-V., E.S. and M.-B.C.-A.; writing—original draft preparation, E.M.T.-V., E.S.; writing—review and editing, E.S., M.-B.C.-A. and S.B.M.; visualization, E.S., E.M.T.-V.; project administration, S.B.M.; funding acquisition, S.B.M. All authors have read and agreed to the published version of the manuscript.

Funding: This work was funded by the U.S. Geological Survey (funding authorized by P.L. 109–448) Award Number G17AC00439 for the Transboundary Aquifer Assessment Program.

Institutional Review Board Statement: Not applicable.

Informed Consent Statement: Not applicable.

Data Availability Statement: Not applicable.

Acknowledgments: The authors would like to acknowledge the Transboundary Aquifer Assessment Program (TAAP) and the administrative support of the Water Resources Research Center, University of Arizona.

Conflicts of Interest: The authors declare no conflict of interest. The funders had no role in the design of the study; in the collection, analyses, or interpretation of data; in the writing of the manuscript, or in the decision to publish the results.

References

1. International Groundwater Resources Assessment Centre (IGRAC); UNESCO International Hydrological Programme (UNESCO-IHP). Transboundary Aquifer of the World [Map]. Update 2021. 2020. Available online: https://www.un-igrac.org/news/igrac-launches-new-transboundary-aquifers-world-map-2021 (accessed on 6 November 2017).
2. Tapia-Villaseñor, E.M.; Megdal, S.B. The U.S.-Mexico Transboundary Aquifer Assessment Program as a Model for Transborder Groundwater Collaboration. *Water* **2021**, *13*, 530. [CrossRef]
3. Tapia-Villaseñor, E.M.; Shamir, E.; Megdal, S.B.; Petersen-Perlman, J.D. Impacts of Variable Climate and Effluent Flows on the Transboundary Santa Cruz Aquifer. *JAWRA J. Am. Water Resour. Assoc.* **2020**, *56*, 409–430. [CrossRef]
4. Megdal, S.B.; Scott, C.A. The Importance of Institutional Asymmetries to the Development of Binational Aquifer Assessment Programs: The Arizona-Sonora Experience. *Water* **2011**, *3*, 949–963. [CrossRef]
5. Callegary, J.B.; Megdal, S.B.; Tapia Villaseñor, E.M.; Petersen-Perlman, J.D.; Minjárez Sosa, I.; Monreal, R.; Gray, F.; Grijalva Noriega, F. Findings and lessons learned from the assessment of the Mexico-United States transboundary San Pedro and Santa Cruz aquifers: The utility of social science in applied hydrologic research. *J. Hydrol. Reg. Stud.* **2018**, *20*, 60–73. [CrossRef]
6. Skoulikaris, C. Transboundary Cooperation Through Water Related EU Directives Implementation Process. The Case of Shared Waters Between Bulgaria and Greece. *Water Resour. Manag.* **2021**, *35*, 4977–4993. [CrossRef]
7. Shamir, E.; Megdal, S.B.; Carrillo, C.; Castro, C.L.; Chang, H.-I.; Chief, K.; Corkhill, F.E.; Eden, S.; Georgakakos, K.P.; Nelson, K.M.; et al. Climate change and water resources management in the Upper Santa Cruz River, Arizona. *J. Hydrol.* **2015**, *521*, 18–33. [CrossRef]
8. Shamir, E. The value and skill of seasonal forecasts for water resources management in the Upper Santa Cruz River basin, southern Arizona. *J. Arid Environ.* **2017**, *137*, 35–45. [CrossRef]
9. Shamir, E.; Halper, E. *The Role of Downscaling Methodology in Projected Climate Change Impacts on Water Resources in Arid Environments*; Project ST-2016-9039; Research and Development Office, U.S. Department of Interior, Bureau of Reclamation: Denver, CO, USA, 2019.
10. IBWC. *Joint Report of the Principal Engineers Regarding the Joint Cooperative Process United States-Mexico for the Transboundary Aquifer Assessment Program*; International Boundary and Water Comission: El Paso, TX, USA, 2009.
11. Shamir, E.; Tapia-Villaseñor, E.M.; Cruz-Ayala, M.-B.; Megdal, S.B. A Review of Climate Change Impacts on the USA-Mexico Transboundary Santa Cruz River Basin. *Water* **2021**, *13*, 1390. [CrossRef]
12. Norman, L.M.; Feller, M.; Phillip Guertin, D. Forecasting urban growth across the United States–Mexico border. *Comput. Environ. Urban Syst.* **2009**, *33*, 150–159. [CrossRef]
13. Wilder, M.; Scott, C.A.; Pineda-Pablos, N.; Varady, R.G.; Garfin, G.M. *Moving Forward from Vulnerability to Adaptation: Climate Change, Drought, and Water Demand in the Urbanizing Southwestern United States and Northern Mexico. (Avanzando desde la Vulnerabilidad hacia la Adaptación: El Cambio Climático, la Sequía, y la Demanda del Agua en Áreas Urbanas del Suroeste de los EEUU y el Norte de México*; Udall Center for Studies in Public Policy, The University of Arizona: Tucson, AZ, USA, 2012.
14. Norman, L.; Tallent-Halsell, N.; Labiosa, W.; Weber, M.; McCoy, A.; Hirschboeck, K.; Callegary, J.; van Riper, C.; Gray, F. Developing an Ecosystem Services Online Decision Support Tool to Assess the Impacts of Climate Change and Urban Growth in the Santa Cruz Watershed; Where We Live, Work, and Play. *Sustainability* **2010**, *2*, 2044–2069. [CrossRef]
15. Norman, L.M.; Huth, H.; Levick, L.; Shea Burns, I.; Phillip Guertin, D.; Lara-Valencia, F.; Semmens, D. Flood hazard awareness and hydrologic modelling at Ambos Nogales, United States-Mexico border: Flood hazard awareness and hydrologic modelling at Ambos Nogales. *J. Flood Risk Manag.* **2010**, *3*, 151–165. [CrossRef]
16. Shamir, E.; Meko, D.M.; Graham, N.E.; Georgakakos, K.P. Hydrologic Model Framework for Water Resources Planning in the Santa Cruz River, Southern Arizona. *J. Am. Water Resour. Assoc.* **2007**, *43*, 1155–1170. [CrossRef]
17. Eden, S.; Megdal, S.B.; Shamir, E.; Chief, K.; Mott Lacroix, K.E. Opening the Black Box: Using a Hydrological Model to Link Stakeholder Engagement with Groundwater Management. *Water* **2016**, *8*, 216. [CrossRef]
18. Minjarez Sosa, I.; Monreal Saavedra, R.; Rangel Medina, M.; Grijalva Noriega, F.; Ochoa Granillo, J.A.; Tapia Villaseñor, E.M.; Ramírez Díaz, L.L.; Cirett Galán, S.; Moncada García, G.; Monijo González, A. *Actividades Hidrogeológicas en el Acuífero Río Santa Cruz, Estado de Sonora, Primera Etapa*; Universidad de Sonora: Hermosillo, Mexico, 2011.
19. Christopher, A. Scott, S.M.E.A. Effects of climate change and population growth on the transboundary Santa Cruz aquifer. *Inter-Res. Sci. Cent.* **2012**, *51*, 159–170.
20. CONAGUA. *Determinación de la Disponibilidad de Agua en el Acuífero 2615 Río Santa Cruz, Estado de Sonora*; Diario Oficial de la Federación (DOF): Ciudad de México, Mexico, 2009.
21. CONAGUA. *Actualización de la Disponibilidad Media Anual de Agua en el Acuífero Río Santa Cruz (2615) Río Santa Cruz, Estado de Sonora*; Diario Oficial de la Federación (DOF): Ciudad de México, Mexico, 2015; Available online: http://www.dof.gob.mx/nota_detalle.php?codigo=5389380&fecha=20/04/2015 (accessed on 2 January 2021).

22. CONAGUA. *Actualización de la Disponibilidad Media Anual de Agua en el Acuífero Río Santa Cruz (2615) Río Santa Cruz, Estado de Sonora*; Diario Oficial de la Federación (DOF): Ciudad de México, Mexico, 2018; Available online: https://dof.gob.mx/nota_detalle.php?codigo=5510042&fecha=04/01/2018 (accessed on 2 January 2021).
23. CONAGUA. *Actualización de la Disponibilidad Media Anual de Agua en el Acuífero Río Santa Cruz (2615) Río Santa Cruz, Estado de Sonora*; Diario Oficial de la Federación (DOF): Ciudad de México, Mexico, 2020; Available online: https://sigagis.conagua.gob.mx/gas1/Edos_Acuiferos_18/sonora/DR_2615.pdf (accessed on 2 January 2021).
24. Alley, W.M.; Reilly, T.E.; Franke, O.L. *Sustainability of Ground-Water Resources*; U.S. Dept. of the Interior, U.S. Geological Survey: Denver, CO, USA, 1999.
25. Alley, W.M.; Leake, S.A. The Journey from Safe Yield to Sustainability. *Ground Water* **2004**, *42*, 12–16. [CrossRef] [PubMed]
26. de Aguas Nacionales, L. *Decreto por el que se Reforman, Adicionan y Derogan Diversas Disposiciones de la Ley de Aguas Nacionales*; DOF: Ciudad de México, Mexico, 2004.
27. CONAGUA. *Water Banks in Mexico*; Ministry of the Environment and Natural Resources: Ciudad de México, Mexico, 2012.
28. OECD. *Making Water Reform Happen in Mexico*; OECD Studies on Water; OECD Publishing: Paris, France, 2013.
29. DOF. *ACUERDO por el que se Actualiza la Disponibilidad Media Anual de Agua Subterránea de Los 653 Acuíferos de los Estados Unidos Mexicanos*; DOF: Ciudad de México, México, 2020.
30. CONAGUA. Consulta a la Base de Datos del Registro Público de Derechos de Agua (REPDA). 2020. Available online: https://app.conagua.gob.mx/consultarepda.aspx (accessed on 15 July 2021).
31. Prichard, A.H.; Scott, C.A. Interbasin water transfers at the US–Mexico border city of Nogales, Sonora: Implications for aquifers and water security. *Int. J. Water Resour. Dev.* **2014**, *30*, 135–151. [CrossRef]
32. Data Mexico. Nogales: Economy, Employment, Equity, Quality of Life, Education, Health and Public Safety. Available online: https://datamexico.org/en/profile/geo/nogales-30115 (accessed on 15 July 2021).
33. Sophocleous, M. From safe yield to sustainable development of water resources—The Kansas experience. *J. Hydrol.* **2000**, *235*, 27–43. [CrossRef]
34. Megdal, S.B. Arizona's 1980 Groundwater Management Act. *Water Rep.* **2012**, *2012*, 7.
35. Abatzoglou, J.; Dobrowski, S.; Parks, S.; Hegewisch, K. TerraClimate, a high-resolution global dataset of monthly climate and climatic water balance from 1958–2015. *Sci. Data* **2018**, *5*, 170191. [CrossRef] [PubMed]
36. Burnash, R.J.; Ferral, R.L.; McQuire, R.A. *A Generalized Stream Flow Simulation System in Conceptual Modeling for Digital Computers*; US National Weather Services: Sacramento, CA, USA, 1973.
37. ADWR. *Groundwater Flow Model of the Santa Cruz Active Management Area along the Santa Cruz River and Nogales Wash from the International Border to Tubac*; ADWR: Phoenix, AZ, USA, 2020.
38. Tapia Padilla, G. *Aplicación de Modflow para la Simulación Hidrogeológica Estacionaria de una Porción de la Cuenca del río Santa Cruz, Sonora, México*; Universidad de Sonora: Hermosillo, Mexico, 2005.
39. Técnicas Geológicas y Mineras. *Estudio de Evaluación de la Disponibilidad De agua y Definición de Explotación en el Valle del Río Santa Cruz con Fines de Abastecimiento de Agua a la Ciudad de Nogales, Son*; CONAGUA: Ciudad de México, México, 1989.
40. Aguilera, M.; Martínez, R. *Relaciones Agua Suelo Planta Atmósfera*; Universidad Autónoma de Chapingo: Estado de México, Mexico, 1996.
41. INEGI. Conjunto de datos vectoriales de la carta de Uso del suelo y vegetación. Escala 1:250,000. Serie VI. Conjunto Nacional. 2017. Available online: https://www.inegi.org.mx/app/biblioteca/ficha.html?upc=889463598459 (accessed on 2 January 2021).

water

MDPI

Article

Gradient Self-Potential Logging in the Rio Grande to Identify Gaining and Losing Reaches across the Mesilla Valley

Scott Ikard [1,*], Andrew Teeple [1] and Delbert Humberson [2]

1 U.S. Geological Survey Oklahoma-Texas Water Science Center, 1505 Ferguson Lane, Austin, TX 78754, USA; apteeple@usgs.gov
2 International Boundary and Water Commission—United States Section, 4191 North Mesa St., El Paso, TX 79902, USA; Delbert.humberson@ibwc.gov
* Correspondence: sikard@usgs.gov; Tel.: +1-(512)-927-3500

Abstract: The Rio Grande/Río Bravo del Norte (hereinafter referred to as the "Rio Grande") is the primary source of recharge to the Mesilla Basin/Conejos-Médanos aquifer system in the Mesilla Valley of New Mexico and Texas. The Mesilla Basin aquifer system is the U.S. part of the Mesilla Basin/Conejos-Médanos aquifer system and is the primary source of water supply to several communities along the United States–Mexico border in and near the Mesilla Valley. Identifying the gaining and losing reaches of the Rio Grande in the Mesilla Valley is therefore critical for managing the quality and quantity of surface and groundwater resources available to stakeholders in the Mesilla Valley and downstream. A gradient self-potential (SP) logging survey was completed in the Rio Grande across the Mesilla Valley between 26 June and 2 July 2020, to identify reaches where surface-water gains and losses were occurring by interpreting an estimate of the streaming-potential component of the electrostatic field in the river, measured during bankfull flow. The survey, completed as part of the Transboundary Aquifer Assessment Program, began at Leasburg Dam in New Mexico near the northern terminus of the Mesilla Valley and ended ~72 kilometers (km) downstream at Canutillo, Texas. Electric potential data indicated a net losing condition for ~32 km between the Leasburg Dam and Mesilla Diversion Dam in New Mexico, with one ~200-m long reach showing an isolated saline-groundwater gaining condition. Downstream from the Mesilla Diversion Dam, electric-potential data indicated a neutral-to-mild gaining condition for 12 km that transitioned to a mild-to-moderate gaining condition between 12 and ~22 km downstream from the dam, before transitioning back to a losing condition along the remaining 18 km of the survey reach. The interpreted gaining and losing reaches are substantiated by potentiometric surface mapping completed in hydrostratigraphic units of the Mesilla Basin aquifer system between 2010 and 2011, and corroborated by surface-water temperature and conductivity logging and relative median streamflow gains and losses, quantified from streamflow measurements made annually at 16 seepage-measurement stations along the survey reach between 1988 and 1998 and between 2004 and 2013. The gaining and losing reaches of the Rio Grande in the Mesilla Valley, interpreted from electric potential data, compare well with relative median streamflow gains and losses along the 72-km long survey reach.

Keywords: self-potential; temperature; conductivity; surface water; groundwater; groundwater and surface water interactions; rivers; resistivity; streamflow

Citation: Ikard, S.; Teeple, A.; Humberson, D. Gradient Self-Potential Logging in the Rio Grande to Identify Gaining and Losing Reaches across the Mesilla Valley. *Water* **2021**, *13*, 1331. https://doi.org/10.3390/w13101331

Academic Editors: Sharon B. Megdal and Anne-Marie Matherne

Received: 9 March 2021
Accepted: 8 May 2021
Published: 11 May 2021

1. Introduction

In 2006, the United States (U.S.)–Mexico Transboundary Aquifer Assessment Act (Public Law 109–448, herein referred to as the "Act") authorized collaboration between the U.S. and Mexico in conducting hydrogeologic characterization, mapping, and groundwater-flow modeling for priority transboundary aquifers that are internationally shared [1,2]. The following criteria were used to identify priority transboundary aquifers along the U.S.–Mexico border region: (1) the proximity of a transboundary aquifer to metropolitan areas with high population density, (2) the extent to which an aquifer would be utilized as a source of water supply, and (3) the vulnerability of an aquifer to anthropogenic or environmental contamination. Based on these criteria, the Mesilla Basin/Conejos-Médanos aquifer system (Figure 1) was designated a priority transboundary aquifer. The Mesilla Basin aquifer system is the U.S. part of the Mesilla Basin/Conejos-Médanos aquifer system (Figure 1). The Mesilla Basin aquifer system (hereinafter referred to as the "Mesilla Basin aquifer") is hydraulically connected to the Conejos-Médanos aquifer system in Chihuahua, Mexico, and there are no natural barriers to inhibit groundwater flow across the border [2]. Many communities along the U.S.–Mexico border in and near the Mesilla Valley rely partially or completely on groundwater in the Mesilla Basin aquifer for industry, agriculture, and drinking water.

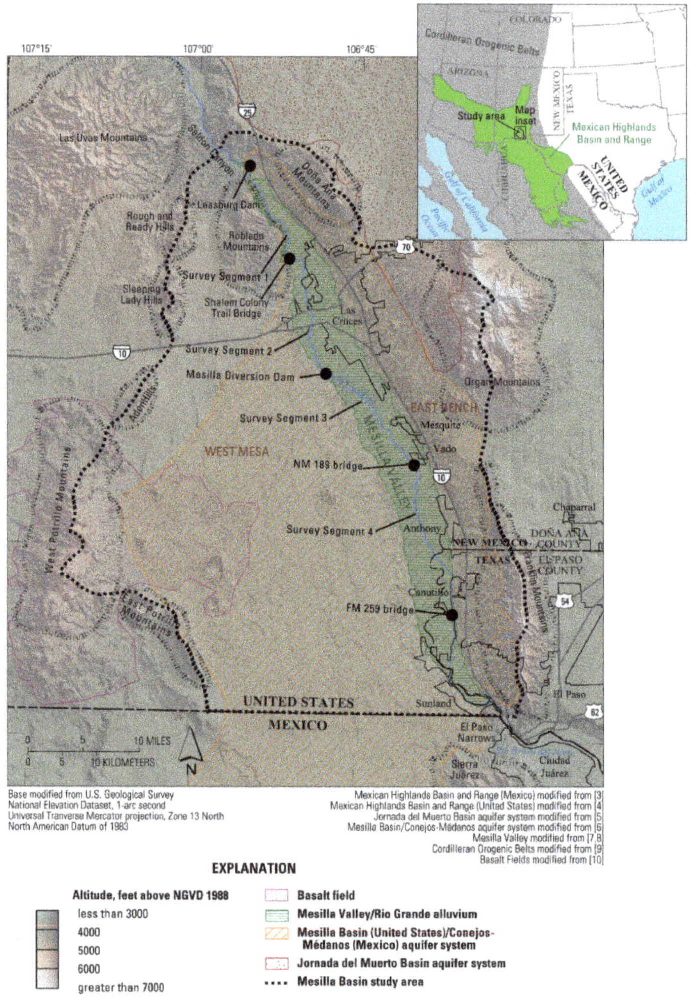

Figure 1. Location map of the entire surveyed reach of the Rio Grande across the Mesilla Valley in Doña Ana County, New Mexico (N. Mex.) and El Paso County, Texas (Tex.). The surveyed reach began at Leasburg Dam, N. Mex., and ended downstream at the Farm-to-Market (FM) 259 bridge in Canutillo, Tex., and was divided into survey segments 1–4. Modified from Figure 3 in [2–10].

There is active litigation and adjudication of water rights associated with transboundary aquifers along the U.S.–Mexico border [1]. Adjudication is often complicated by disparate policy frameworks for groundwater and surface-water resources, even though they are often interdependent and function as a single resource [11]. Among the unique policy and management-related challenges for these resources are the need to develop a shared definition of aquifer boundaries and to develop methods to assess whether the aquifers on both sides of the border are indeed hydraulically connected and internationally shared. These challenges are exacerbated by uncertainty about the interdependence between transboundary aquifers and regional surface-water resources, which is also necessary to understand how to adequately manage and sustain water resources in the border region.

Addressing the complex challenges associated with transboundary aquifers depends upon a scientific approach to inform the management practices and policies that are enacted. The Act outlines specific scientific objectives for transboundary aquifers including: (1) establishing relevant hydrogeological, geochemical, and geophysical field studies that integrate ongoing monitoring and metering, (2) developing and enhancing geographic information systems databases pertaining to priority transboundary aquifers, and (3) developing groundwater-flow models of priority transboundary aquifers [2,12].

This paper describes a contribution to scientific objectives (1) and (2), and is an extension of the scientific investigation of [2]. The self-potential (SP) method of geophysical prospecting was applied in this investigation to study regional-scale groundwater and surface-water exchanges between the Rio Grande and the Mesilla Basin aquifer. SP measurements consist of naturally occurring electrical voltages at the land-surface, in surface-water bodies, and in boreholes, that contain information about coupled thermodynamic flows in near-surface aquifers [13–17] that are driven by hydraulic gradients (associated with streaming potentials), temperature gradients (thermo-electric potentials), concentration gradients (diffusion potentials), redox gradients (mineralization potentials), and certain biogeochemical reactions [18–22].

The physics of streaming potentials is well understood [15,23–32]. Streaming potentials are generated by streaming currents, flowing in the pore spaces of an aquifer in opposition to the groundwater-flow direction. Streaming currents originate by the advection of accumulated electrical charges in the electrical double layer (EDL) along the solid–fluid interface [33]. As a condition of electroneutrality, the streaming current is counterbalanced by a conduction current that flows through the heterogeneous electrical conductivity structure of the aquifer and ensures that the physical divergence of the total current in the aquifer (the sum of streaming currents and conduction currents) is zero. In the case of an unconfined saturated aquifer, the streaming and conduction currents are nearly balanced and the electric equipotentials tend to mimic the hydraulic equipotentials [32]. The transport of accumulated ions in the EDL by advection in the direction of groundwater flow (the direction of decreasing hydraulic gradient) results in a dipolar electrical-potential field, with positive regions that correspond to locations where groundwater exits the aquifer pore space (i.e., surface-water gains), and negative regions that correspond to locations where surface water enters the aquifer pore space (surface-water loss) and flows along preferential groundwater-flow paths [34,35].

Numerical simulations by [36,37] showed that fixed-reference land-based SP profiles, measured perpendicular to a river reach, can likely identify whether the river reach is losing, gaining, or flow-through, based on the polarity of the streaming-potential field in the surface-water relative to the polarity of the field on either side of the floodplain. However, fixed-reference SP surveys of surface-water gains and losses are generally impractical along reaches longer than a few river-kilometers because the fixed reference SP electrode must be continuously revisited and relocated, and because the SP electrodes must contact porous earth at every measurement location to complete an electrical circuit.

Waterborne gradient SP logging is an alternative approach to fixed-reference land-based SP mapping. The waterborne gradient SP logging approach differs from most land-based SP surveys in that both SP electrodes are mobile (no fixed reference electrode) along a profile or grid in a river or lake, and the electrical circuit is completed at each measurement location by the contact between the SP electrodes and the surface water instead of the aquifer [37]. Immersion in surface water reduces contact resistance between the electrodes and the surface water, which enhances the signal-to-noise ratio, and meaningful anomalies of less than a few tens of microvolts have been measured and published [38–43]. Recently, waterborne gradient SP logging enabled the characterization of reach-scale heterogeneous hyporheic-driven groundwater and surface-water exchange between the lower Guadalupe River and the Carrizo–Wilcox aquifer in central Texas [37] (in their Figure 1), identified meter-scale groundwater discharge locations in the Quashnet River of Cape Cod, Massachusetts [44], and identified gaining and losing reaches of the Colorado River where it crosses the Bee Creek Fault, and is incised into the surficial exposures of the rocks that comprise the lower and middle zones of the Trinity Aquifer in central Texas [43,45].

The waterborne gradient SP logging approach is extended herein to identify gaining and losing reaches of the Rio Grande at the basin-scale in the Mesilla Valley. The waterborne gradient SP logging data presented herein are processed into electric potential and interpreted in the context of streaming potential by assuming that streaming potential attributed to surface-water gain and loss through the riverbed and floodplain was the predominant contribution to the electric-potential field in the river. Interpretations of surface-water gain and loss are supported by surface-water temperature and conductivity logging, and geophysical and hydraulic datasets presented by [2] that consist of: (1) profiles of resistivity beneath the Rio Grande channel to depths of 50 m [2], (2) relative median gains and losses in streamflow, quantified at 16 seepage-measurement stations along the survey reach by annual streamflow measurements between 1988 and 1998 and between 2004 and 2013 [2,46], and (3) water-level differences and inferred vertical hydraulic gradients beneath the Rio Grande, quantified by potentiometric surface mapping in wells completed in the Rio Grande alluvium and upper part of the Santa Fe Group between November 2010 and April 2011 [2].

2. Description of the Study Area

The geographic, geologic, and hydrogeologic settings and geochemistry of the Mesilla Basin are described comprehensively by [2] and in references cited therein and are summarized here from those sources. The Mesilla Valley (Figures 1 and 2) is in the region of the Mesilla Basin that is incised by the Rio Grande, between Selden Canyon at the northern end and the El Paso Narrows at the southeastern end. The Mesilla Basin aquifer is heavily relied upon in the Mesilla Valley and in the greater Mesilla Basin for irrigation water and as a primary source of municipal and domestic water supply for numerous communities along the United States–Mexico border including Las Cruces, New Mexico (N. Mex.), El Paso, Texas (Tex.), and Ciudad Juárez, Mexico.

2.1. Hydrogeology of the Mesilla Basin

The Mesilla Basin aquifer is divided into four distinct hydrogeologic units, each of which is recharged primarily by the Rio Grande. The hydrogeologic units are (from youngest to oldest) the middle-to-late Quaternary (Holocene) channel and floodplain deposits of the Rio Grande (referred to as the Rio Grande alluvium), and the poorly consolidated middle-Miocene to late-Pleistocene basin-fill deposits of the Santa Fe Group, which are divided into upper, middle, and lower lithofacies assemblages based on differing granulometric, hydraulic, and geochemical properties. The base of the Mesilla Basin is underlain primarily by lower-to-middle Tertiary volcanic and volcaniclastic bedrock that is block-faulted and influences the groundwater-flow system at depth within the Mesilla Basin aquifer [2].

The Rio Grande is the primary depositional feature in the Mesilla Basin. The river has deposited the Rio Grande alluvium on the Mesilla Valley floor by continued channel avulsion and overbank deposition [2,46–48]. The alluvium is a relatively thin surface layer (generally about 24-m thick, with a maximum thickness of about 46 m) of fluvial sediments derived from outwash fan deposits, eolian sands, and re-worked basin-fill eroded from nearby mountains [47,49]. Recharge to the Mesilla Basin aquifer occurs primarily by vertical flow through the riverbed into the Rio Grande alluvium, and from associated canals, laterals, and drains.

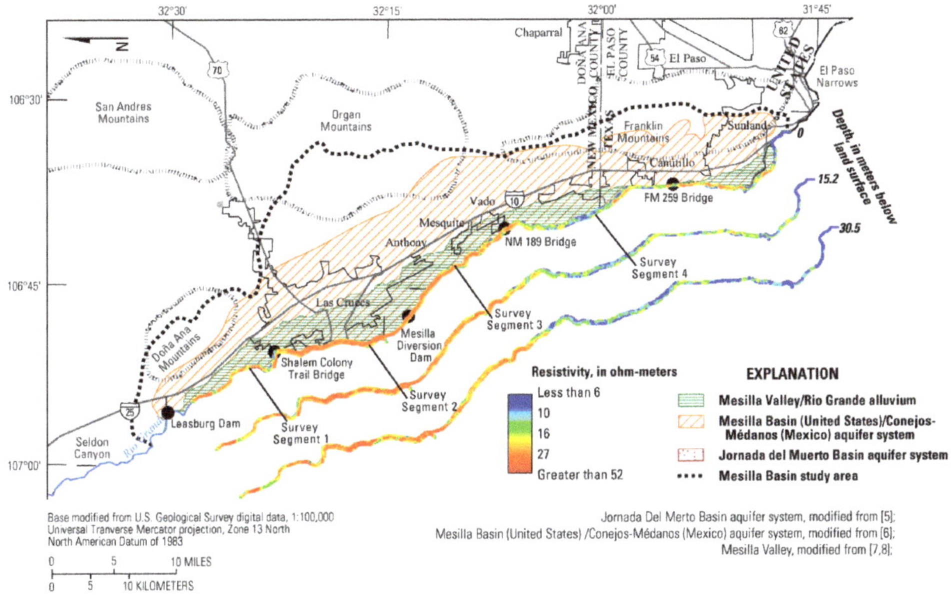

Figure 2. Electric resistivity of the Rio Grande channel and subsurface beneath the channel, determined from helicopter frequency-domain electromagnetic surveys of the levee system (modified from Figure 13 in [2]). Resistivity profiles of the channel are shown at or about the land surface (0 m below land surface) and at depths of 15.2 m and 30.5 m.

The Rio Grande alluvium is hydraulically connected to the thick unconsolidated to semi-consolidated basin-fill deposits of the upper, middle, and lower parts of the Santa Fe Group [2]. Santa Fe Group deposits are composed of alluvium from adjacent uplifts, eolian sediments, and some fluvial sediments from the ancestral (pre-Pleistocene) Rio Grande. In general, the Santa Fe Group consists of sand lenses interbedded with clays and silty clays that exhibit internal discontinuities attributed to basin-and-range extensional faulting [47,49]. The Santa Fe Group is relatively thin in the Mesilla Basin compared to adjacent basins; the saturated thickness is between 610 m and 914 m [2,49,50]. The upper part of the Santa Fe Group is the most productive zone of the Mesilla Basin aquifer but is only partially saturated throughout most of the Mesilla Basin [2]. The upper part of the Santa Fe Group is a thick sequence of fine- to coarse-grained fluvial deposits of gravel and sand, interbedded with fine-grained basin fill (silt and clay over-bank muds) deposited by the ancestral Rio Grande [2]. The middle part of the Santa Fe Group is the primary water-bearing zone of the Mesilla Basin aquifer and is generally fully saturated [2]. The fine-grained lacustrine-playa sediments of the middle part of the Santa Fe Group consist of alternating beds of sand, silty sand, and silty clay, and represent a terminal depocenter environment of the ancestral Rio Grande. The lower part of the Santa Fe Group constitutes the least productive zone of the aquifer; sediments consist of fine-grained basin-floor playa and fluvial-lacustrine facies deposits interbedded with layers of bentonitic claystone and siltstone, with some discontinuous sand lenses [7].

2.2. Electric Resistivity of the Mesilla Basin Aquifer

A three-dimensional resistivity model of the Mesilla Basin aquifer was published by [2] (Figure 15, p. 25), showing horizontal resistivity depth-slices in the southern Mesilla Basin between depths of 0 m and 530 m beneath the land surface. The resistivity model was developed by kriging inverted resistivity from a compilation of historical resistivity datasets, using a horizontal grid spacing of 100 m and a vertical grid spacing of 3 m. The historical resistivity datasets were obtained from helicopter frequency-domain electromagnetic (HFEM) induction surveying of the Rio Grande levee system [51], 12 ground-based time-domain electromagnetic (TDEM) induction soundings [2], and 65 vertical electrical soundings [52] completed within the Mesilla Basin. The locations of HFEM flight paths and TDEM and vertical electrical soundings were mapped by [2] (Figure 9, p. 19).

The HFEM-derived resistivity data were incorporated into this work to assist in the interpretation of the electric potential processed from waterborne gradient SP data, described in Section 3. HFEM data were acquired along three flight paths over the levees along the Rio Grande in the Mesilla Valley. The flight paths consisted of one path along each levee and two additional paths offset 50 m on each side of the levees at the toe of each levee (the flight paths are depicted in Figure 9 of [2]). The rate of data collection along each flight path was 10 samples per second, such that the horizontal resolution of the HFEM resistivity data was one sounding every 3 m along the flight paths. Technical information pertinent to the HFEM survey and quality assurance, and ground-truthing results, are provided by [51].

Figure 2 shows three subsurface profiles of HFEM resistivity data beneath the riverbed of the Rio Grande at depths of 0, 15.2, and 30.5 m. Near the land surface (at or about 0 m below the land surface), the HFEM resistivity profiles show that the resistivity beneath the Rio Grande is generally greater than 20 ohm-m north of Anthony, N. Mex., and resistivity values of less than 10 ohm-m begin to appear south of Anthony, N. Mex. Resistivity values of less than 10 ohm-m are also increasingly prevalent with increasing depth beneath the channel; about half of the resistivity values were less than 10 ohm-m at depths of 15.2 m and 30.5 m (Figure 2), and there were transitions from relatively high resistivity (greater than 20 ohm-m) to relatively low resistivity (less than 10 ohm-m) at depths of 15.2 and 30.5 m near Vado, N. Mex. These low-resistivity areas were interpreted by [2] in combination with water-quality data from 239 wells [2] (Figure 16, p. 30) as sand and gravel deposits saturated with dense saline water upwelling through fractures within the deeper bedrock of the Mesilla Basin.

2.3. Groundwater-Surface Water Connectivity in the Mesilla Valley

The hydraulics of groundwater and surface-water connectivity in the Mesilla Valley are complex (Figure 3). Horizontal hydraulic gradient maps published by [2] (Figure 48, p. 80) indicate that groundwater within the Mesilla Basin aquifer is generally unconfined and flows southward, longitudinal to the Rio Grande, along an average gradient of 0.75–1.1 m per kilometer [2,7]. The Leasburg and Mesilla Diversion Dams in the Mesilla Valley both steepen the local hydraulic gradient from upstream to downstream, and potentially alter regional horizontal and vertical hydraulic gradients and groundwater-flow patterns. Generalized numerical modeling performed by [53–56] indicates that the dams may produce a localized losing condition on the riverbed upstream, and a localized gaining condition on the riverbed downstream, because of steep reductions of the hydraulic gradient across the dams from upstream to downstream.

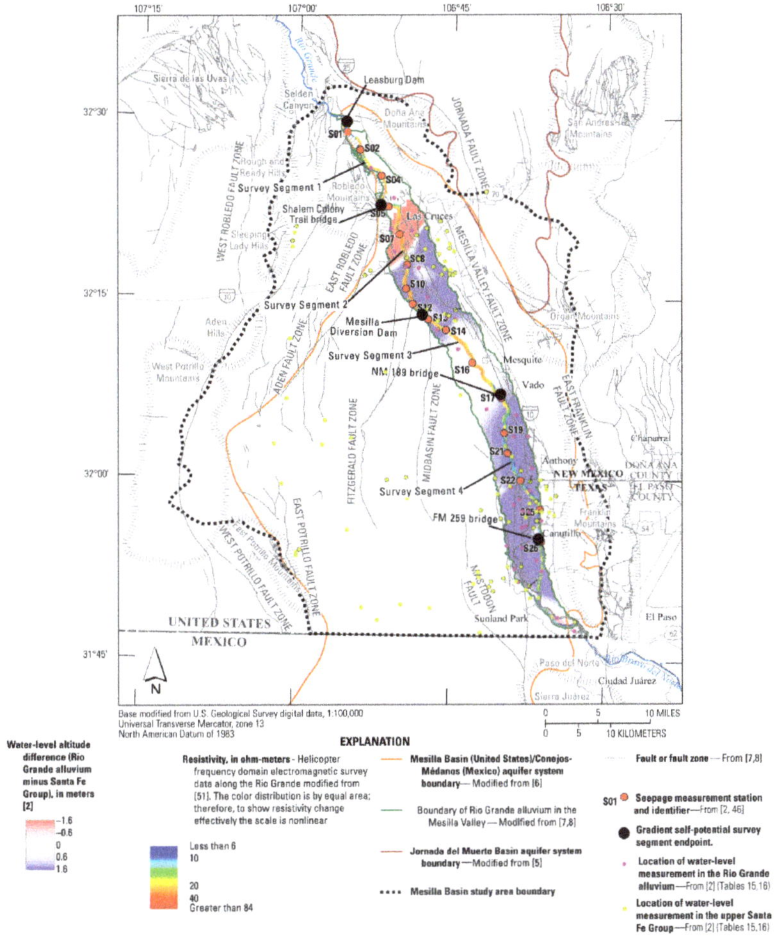

Figure 3. Water-level altitude differences between the Rio Grande alluvium and the Santa Fe Group, with electric resistivity at a depth of 30.5 m beneath the Rio Grande. Water-level differences greater than 0 m indicate that the hydraulic head in the Rio Grande alluvium is greater than the hydraulic head in the upper part of the Santa Fe Group, whereas differences less than 0 m indicate the opposite. Other hydrogeologic features are shown, such as fault zones that intersect the river channel, and the locations of seepage-measurement stations and gradient self-potential survey segment endpoints.

Vertical hydraulic gradients beneath the Rio Grande vary substantially across the Mesilla Valley from upstream to downstream (Figure 3). Figure 3 shows a comparison of differences in 2010–2011 water-levels in the Rio Grande alluvium (measurement locations shown by pink dots) and upper part of the Santa Fe Group (measurement locations shown by yellow dots) with the sub-channel resistivity at a depth of 30.5 m below the land surface. The locations of seepage-measurement stations (red dots, S01–S26) corresponding to seepage investigations of the Rio Grande by [46] are also shown. The 2010–2011 water-level altitudes [2] (Table 15; p. 155) in the Rio Grande alluvium were higher than the water-level altitudes in the upper part of the Santa Fe Group throughout most of the Mesilla Valley (Figure 3, blue sections corresponding to water-level differences greater than 0 m), indicating that the vertical hydraulic gradient was oriented downward and surface-water losses were likely to occur in locations where the differences were greater than 0 (e.g., S08–S14 and downstream from S17). The reductions (white sections, e.g.,

S14–S17) and reversals (red sections, e.g., S05–S08) of the vertical hydraulic gradient were also mapped by [2]. In these locations, the vertical hydraulic gradient either decreased substantially (Figure 3, white sections corresponding to water-level differences of about 0 m) or flattened across faults that intersected the channel, or reversed direction, indicating a likelihood for surface-water gains (Figure 3, red sections corresponding to water-level differences less than 0 m).

Seepage investigations were conducted in the Rio Grande by the U.S. Geological Survey annually between 1988 and 1998, and between 2004 and 2013. During each annual seepage investigation, streamflow was measured over a period of 1–2 days during low-flow conditions in the non-irrigation season (February) at the seepage-measurement stations shown in Figure 3. Net seepage gains or losses were quantified at each station by subtracting the streamflow measured at each station from the streamflow measured at the closest upstream station, and then subtracting inflows to the Rio Grande within the survey segment bounded by the two stations. As indicated by [46], outflows from the river did not occur during the seepage investigations, and inflows were gaged. The annual net streamflow gain/loss data were published by [46] and [2] (App. 1, p. 176), and processed into relative median streamflow gain/loss between seepage-measurement stations in Figure 3 by [2]. These data are plotted in Section 4.

3. Materials and Methods

The waterborne gradient self-potential (SP) survey was completed between 26 June and 2 July 2020 during peak releases of surface water (54 to 65 cubic meters per second) from the Elephant Butte and Caballo Dams upstream from the survey reach. The purpose of the survey was to produce profiles of electric potential, surface-water temperature, and specific conductance measurements in the Rio Grande that could be interpreted in the context of surface-water gains and losses and compared with resistivity and water-level data shown in Figures 2 and 3 and relative median gains and losses in streamflow, determined from seepage investigations of [46] (Section 4). The survey was completed along the left (east) bank of the Rio Grande during bankfull flow conditions (see photographs in Figures 3 and 4). The time period corresponding to bankfull flow was chosen for surveying based on the assumption that vertical hydraulic gradients, and therefore rates of loss in losing reaches, would be optimized during bankfull flow conditions, better enabling their identification. The survey was completed along the left bank instead of the center of the channel because of shallow submerged gravel bars (a few centimeters deep) in the center that were hidden by high suspended bed load during bankfull flow conditions. The average and maximum flow depths during the survey were about 0.5 m and 1 m, respectively.

The survey began at Leasburg Dam, N. Mex. and ended near the Farm-to-Market (FM) 259 bridge in Canutillo, Tex. approximately 72 km downstream (Figure 1). The overall survey reach was subdivided into four 15- to 25-km long segments that were surveyed individually and combined during data processing into two longer reaches for interpretation; one reach between Leasburg Dam and Mesilla Diversion Dam, and a second reach downstream from Mesilla Diversion Dam to Canutillo, Tex. The first survey segment began at the Leasburg Dam and ended a few meters downstream from the Shalem Colony Trail bridge in Las Cruces, N. Mex. (Figures 1 and 4a). The second began a few meters downstream from the Shalem Colony Trail bridge and ended about 160 m upstream from the Mesilla Diversion Dam, between Mesilla and Mesquite, N. Mex. (Figure 4b). The third began a few meters downstream from the Mesilla Diversion Dam and ended a few meters downstream from the New Mexico State Road 189 bridge in Vado, N. Mex., and the fourth began a few meters downstream from the New Mexico State Road 189 bridge and ended a few meters downstream from the FM 259 bridge in Canutillo, Tex.

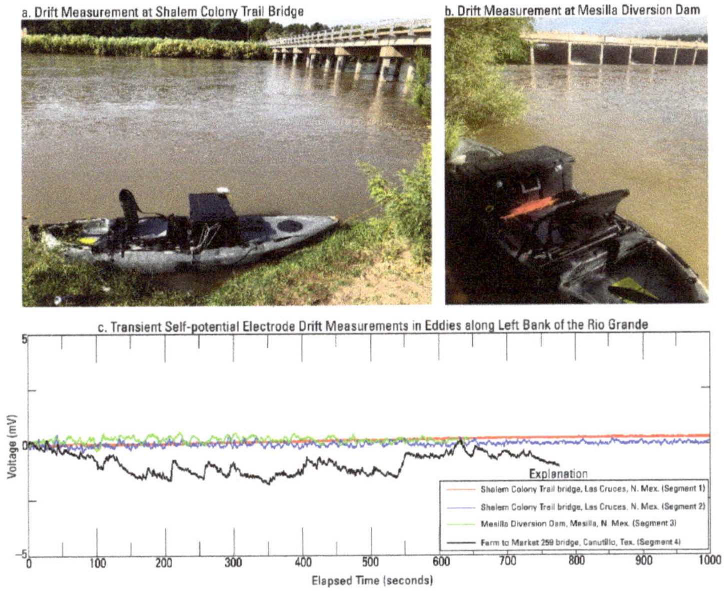

Figure 4. Electrode-drift measurements performed in eddies along the banks of the Rio Grande. (**a**) Photograph of the kayak equipped with GPS, power, and logging equipment used to make drift measurements (red and blue, panel c) at Shalem Colony Trail bridge in Las Cruces, New Mexico (N. Mex.) near the left bank (looking downstream toward the southwest). (**b**) Photograph of the kayak equipped with GPS, power, and logging equipment used to make drift measurements (green, panel c) at Mesilla Diversion Dam in Mesilla, N. Mex., near the right bank (looking upstream toward the northeast). (**c**) Electrode drift data measured for each survey segment used to determine electrode-drift corrections for gradient SP data. The start and end point of survey segments 1–4 are depicted in Figures 1–3.

All measurements were made from a kayak at 0.15-m depth in the water, by positioning sensors through the drain ports in the kayak hull. Two freshwater-submersible, non-polarizing copper-sulfate electrodes were used to create a 0.5-m long electric dipole, which was oriented with the reference electrode upstream from the potential electrode. An Onset HOBO temperature and conductivity logger was placed into the Rio Grande surface water through a drain port adjacent to the reference SP electrode. GPS, power, and logging equipment were transported on board the kayak, and the geospatial coordinates of the dipole midpoint were logged with a Trimble DSM232 differential GPS with horizontal accuracy between 5 cm and 10 cm. The geophysical datasets and processing codes are available online by [57]. The resistivity, water-level altitude, relative median streamflow gain, and loss data, and other related hydraulic, geophysical, and geochemical datasets are provided [2,57,58].

3.1. Gradient Self-Potential Logging

The raw gradient SP data measured in the Rio Grande consisted of voltages between the reference and potential electrodes of the dipole, which were logged at a period of 1 s per measurement by an Agilent U1252B multimeter as the dipole floated downstream in the river. The raw measurements were contaminated by transient electrode-drift voltages (Figure 4c), which were removed from the data during processing. During the survey, time-lapse electrode-drift measurements were logged in eddies along each survey segment at either the beginning or end of the segments to estimate the drift characteristics of the electrodes and enable electrode-drift corrections to the raw gradient SP data. Drift measurement locations were re-occupied for several drift measurements to evaluate the repeatability of the electrode drifts during the course of the survey. For example, the survey segment 1 drift measurement was performed at the Shalem Colony Trail bridge at the end of the survey day after acquiring segment 1 data, and the survey segment 2 drift measurement was made at the same location on a different survey date before acquiring segment 2 data. Photographs of two electrode-drift measurements along the overall survey reach are shown in Figure 4a,b (one taken on the downstream side of the Shalem Colony Trail bridge (Figure 4a) and another taken downstream from the Mesilla Diversion Dam (Figure 4b)). The electrode-drift measurements corresponding to each survey segment are shown in Figure 4c. During all electrode-drift measurements, the electrode drifts were approximately linear and characterized by relatively flat slopes and small voltages. Electrode-drift measurements corresponding to survey segment 4 (Figure 4c, black), which are attributed to turbulence in the channel at the location of the survey segment 4 electrode-drift measurement, show more noise and larger total electrode drift compared to the other survey segments.

To correct the raw gradient SP data for transient electrode drift, ordinary least-squares (OLS) regression lines were fitted to the electrode-drift data corresponding to each survey segment and then subtracted from the gradient SP data corresponding to the survey segment. The results of electrode-drift corrections are shown in Figure 5 for gradient SP data measured along each individual survey segment. The slopes, intercepts, and coefficients of determination of the fitted OLS regression lines that define the electrode drift patterns are summarized in Table 1.

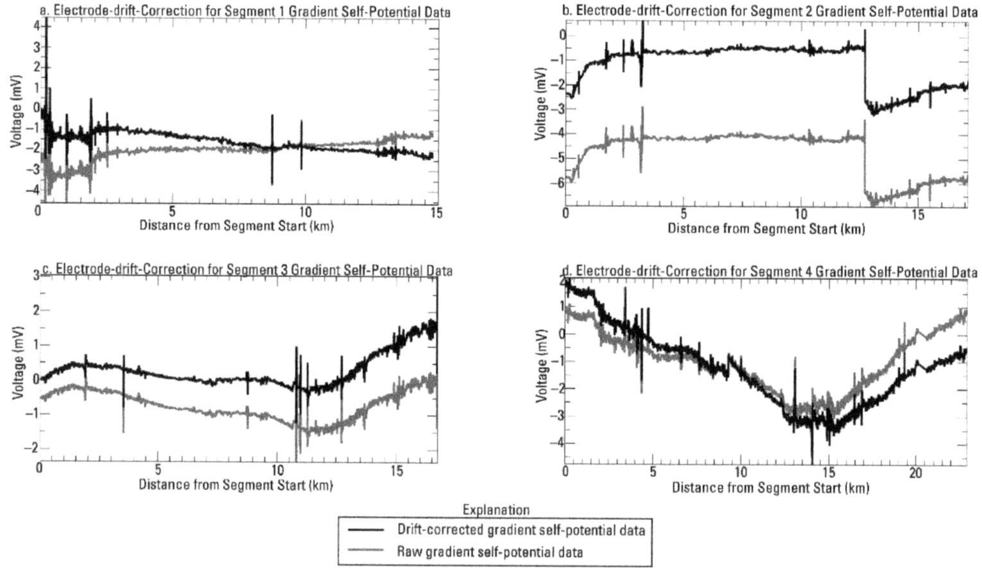

Figure 5. Demonstration of electrode-drift correction applied to raw gradient SP data along individual survey segments. The starting and ending points of survey segments 1–4 are depicted in Figures 1–3.

The drift corrections aligned individual survey segments (Figure 5) into two longer continuous profiles (Figure 6); one upstream from the Mesilla Diversion Dam composed of data from survey segments 1 and 2, and the second downstream from the Mesilla Diversion Dam composed of data from survey segments 3 and 4. An exact alignment was achieved between the endpoints of survey segments 3 and 4; however, a shift (DC offset, Figure 5b) of +3.42 mV was required for gradient SP data along segment 2 after drift-correction to properly align the beginning of survey segment 2 with the endpoint of survey segment

1. The drift-corrected gradient SP data were then scrutinized to identify and manually remove electrical noise in the form of relatively large-amplitude dipolar spikes caused by approximately 10–15 low bridges (e.g., see Figure 4a) and several cast-iron pipelines that spanned the river (geospatial coordinates of approximate noise locations are provided by [57]). The maximum amplitude of manually removed electrical noise spikes in the data exceeded 50 mV in some locations. This sequence of electrode-drift correction and manual noise removal produced the drift-corrected, gradient SP profiles shown in Figure 6b (black).

Table 1. Summary of slopes, intercepts, and coefficients of determination from ordinary least squares (OLS) regression lines fitted to the gradient SP data (electrode drift voltage), surface-water temperature, and specific conductance.

Data Series	Survey Segment [1]	Slope	Intercept	Coefficient of Determination
Electrode Drift Voltage	1	0.00035	−0.00472	0.9556
	2	−0.00004	0.02242	0.0569
	3	−0.00013	0.25666	0.0318
	4	0.0002	−0.96317	0.0097
Surface-water Temperature	1	0.00026	23.71262	0.9292
	2	0.0002	22.60852	0.9717
	3	0.00045	22.33067	0.9758
	4	0.00035	22.71919	0.9939
Specific Conductance	1	0.00304	623.7121	0.7115
	2	0.00532	605.9887	0.7971
	3	0.00614	611.4147	0.9433
	4	0.00567	614.8888	0.9666

[1] Survey segment 1: Leasburg Dam, New Mexico (N. Mex.) to Shalem Colony Trail bridge, Las Cruces, N. Mex. Survey Segment 2: Shalem Colony Trail bridge to Mesilla Diversion Dam, Mesilla, N. Mex. Survey Segment 3: Mesilla Diversion Dam to New Mexico State Road 189 bridge, Vado, N. Mex. Survey Segment 4: New Mexico State Road 189 bridge to Farm-to-Market 259 bridge, Canutillo, Texas (Figures 1–3).

The drift-corrected gradient SP data (Figure 6b, black), denoted as ΔV (in units of mV), were assumed to be a superposition of large and small-scale spatial components; a large-scale (low-frequency spatial variation) "L" component, ΔV_L (Figure 6b, red), a small-scale (high-frequency spatial variation) "H" component ΔV_H (not shown), and some unknown level of noise, ΔV_N, referred to herein as the "N" component (not shown). This assumption was expressed as $\Delta V = \Delta V_L + (\Delta V_H + \Delta V_N)$, where the combination of the H and N components are considered the "HN" component and shown in Figure 6c (an intermediate step in the decomposition). The drift-corrected gradient SP data were decomposed into each of these components by signal processing, following the approach of [37].

The L component of the drift-corrected gradient SP data was estimated by convolution of the drift-corrected data with the Gaussian filter in Equations (1) and (2). In the convolution equation (Equation (1)), g[k] is the Gaussian-shaped impulse response function given in Equation (2), $\sigma = 30$ is the number of gradient SP measurements that define the half-width of g[k], n is an index for the raw gradient SP data, and k is an index for the discrete sequence g[k].

$$\Delta V_L[n] = \sum_{k=-3\sigma}^{3\sigma} \Delta V[n-k] g[k] \quad (1)$$

$$g[k] = \left(2\pi\sigma^2\right)^{-1/2} e^{-k^2/2\sigma^2},\ k = -3\sigma, \ldots, 3\sigma \quad (2)$$

The HN component of the drift-corrected gradient SP data (Figure 6c) was produced by subtracting the L component determined by Equation 1 from the drift-corrected gradient SP data by $\Delta V_{HN} = (\Delta V_H + \Delta V_N) = \Delta V - \Delta V_L$. This component represents variability in the gradient SP data that occurs over a much smaller spatial scale in the Rio Grande compared to the L component data, which is smooth compared to the HN component.

The H component (i.e., a signal of possible interest [37]) was partitioned from the HN component by applying the windowed moving average filter formulated by [37] to the HN component data in Figure 6c. The moving average filter is a technique that is commonly used to partition gravity, magnetic, and SP data into residual and regional components [59]. This filter, when applied to the data in Figure 6c, produced an estimate of the N component of the gradient SP data, which was subtracted from the HN component to produce $\Delta V_H = \Delta V_{HN} - \Delta V_N$.

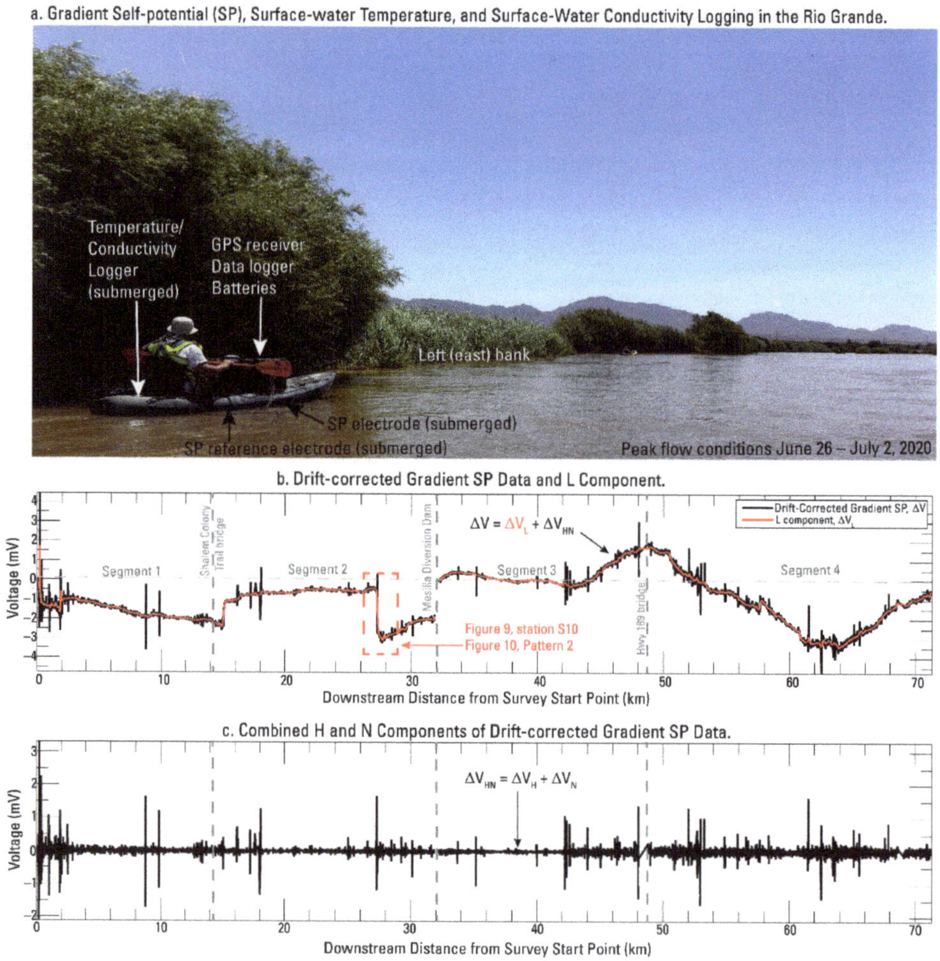

Figure 6. Photograph of data acquisition procedure, and drift-corrected gradient SP data measured in the Rio Grande across the Mesilla Valley with decomposed large- and small-scale spatial components. (**a**) Photograph of data acquisition procedure along the left (east) bank of the Rio Grande (looking downstream to the southeast). (**b**) Drift-corrected gradient SP data (black) and low-frequency (L) component (red) of the data determined by convolution with a Gaussian filter. (**c**) Combination of high-frequency (H) and noise (N) components of the drift-corrected gradient SP data, determined by subtracting the L component from the drift-corrected gradient SP data. The starting and ending points of survey segments 1–4 are depicted in Figures 1–3.

The drift-corrected gradient SP data, and the L, H, and N components, were each converted to electric field strength by $E_j = -\Delta V_j / \Delta L$ (in mV per meter), where ΔV_j (mV) is the gradient SP data of component j and $\Delta L = 0.5$ m is the length of the electric dipole used during data acquisition. The E_j profiles were each numerically integrated into electric-potential profiles, V_j (mV), by Equation (3), where $V_j[n]$ is the integrated electric potential corresponding to component j.

$$V_j[n] = V_j[n-2] + \frac{1}{3}E_j[n] + \frac{4}{3}E_j[n-1] + \frac{1}{3}E_j[n-2] \tag{3}$$

The electric-potential profiles corresponding to each component of the gradient SP data are shown in Figure 7. Figure 7a shows the result of integrating the drift-corrected gradient SP data (black curve, Figure 6b) prior to partitioning the data into the L and HN components. Figure 7b,c show the results of integrating the L and H components of gradient SP data, respectively.

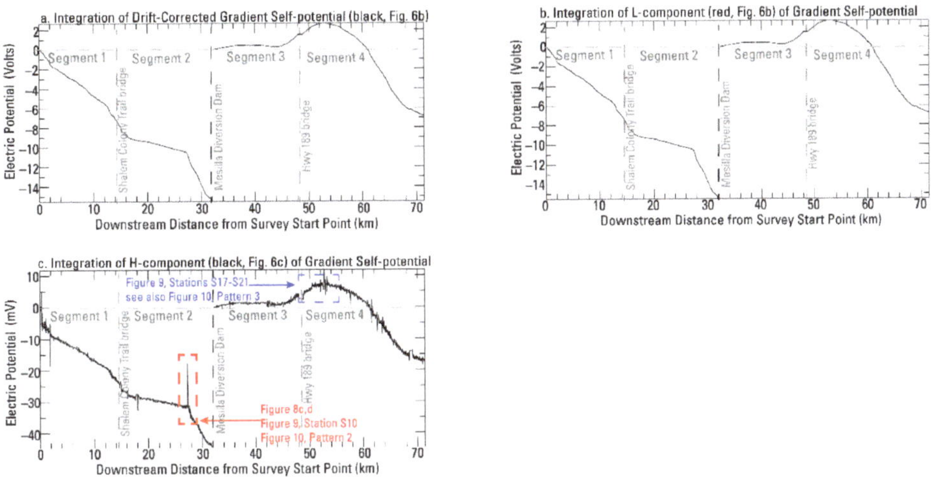

Figure 7. Electric potential profiles in the Rio Grande across the Mesilla Valley, determined by numerical integration of drift-corrected gradient SP data and the corresponding L and H components using Equation (3). The starting and ending points of survey segments 1–4 are depicted in Figures 1–3. (**a**) Integration of drift-corrected gradient SP data prior to decomposition into L and HN components. (**b**) Integration of the L component. (**c**) Integration of the H component after removing an estimate of the N component produced by filtering with a windowed moving average filter described by [37]. The spike observed along survey segment 2 is about 200-m wide and collocated with the drop in gradient SP data along survey segment 2 (Figure 6b) and discrete spikes in measured surface-water temperature and specific conductance (Figure 8). Note the differences in scale of the y-axes based on the integrated component.

3.2. *Surface-Water Temperature and Conductivity Logging*

Surface-water temperature and conductivity data were logged simultaneously with gradient SP data at a period of 2 s per measurement. The surface-water conductivity data were corrected to specific conductance using Equation (4) [60], where σ_s is the specific conductance (conductivity relative to a temperature of 25 degrees Celsius, in microsiemens per centimeter), T is the measured surface-water temperature (degrees Celsius), and σ is the measured surface-water electric conductivity (microsiemens per centimeter).

$$\sigma_s = \frac{\sigma}{1 + 0.02(T - 25)} \tag{4}$$

The temperature and specific conductance data are color-coded by survey segment and plotted in Figure 8a,b relative to the initial measurements at the upstream ends of the segments, to show the total change in temperature and specific conductance along each segment in a comparable manner. Linear increases in temperature and specific conductance data occurred along each of the survey segments, from upstream to downstream, and the temperature and specific conductance data were therefore corrected (before referencing the initial measurements). Data corrections were made by fitting OLS regression lines to the data from each survey segment and subtracting the OLS regression lines corresponding to each survey segment from the corresponding temperature and specific conductance data. Subtracting the OLS regression lines produced profiles that showed the deviations of the respective variables around the linear increases in the data. The deviations are aligned end-to-end by survey segment between the Leasburg Dam and the FM 259 bridge in Canutillo, Tex. (Figure 8c,d). The slopes, intercepts, and coefficients of determination of the fitted OLS regression lines are summarized in Table 1.

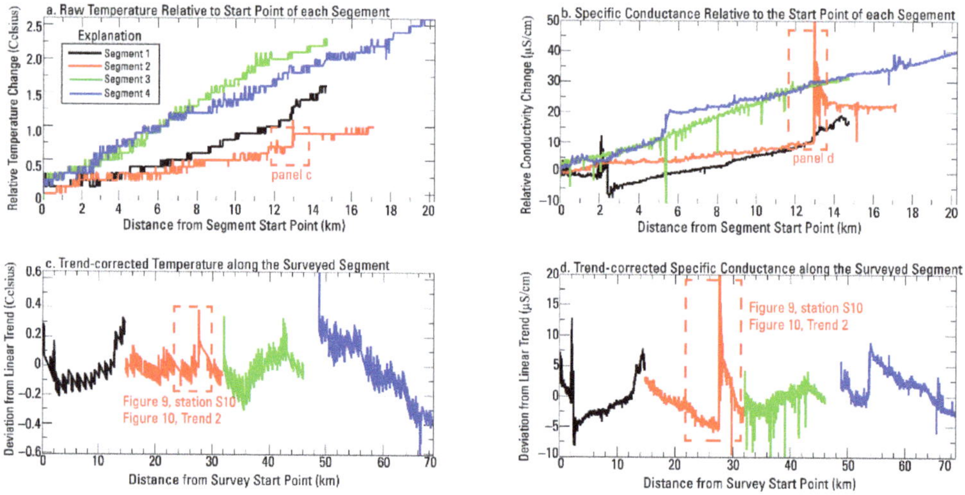

Figure 8. Surface-water temperature and specific conductance data measured in the Rio Grande. (**a**) Raw temperature data measured along individual survey segments, plotted relative to the initial measurements at the upstream ends of the segments. (**b**) Raw specific conductance data measured along individual survey segments, plotted relative to the initial measurements at the upstream ends of the segments. (**c**) Temperature deviations about the linear increases of the data in panel a, versus survey distance. (**d**) Specific conductance deviations about the linear increases of the data in panel b, versus survey distance. The starting and ending points of survey segments 1–4 are depicted in Figures 1–3.

4. Results and Discussion

The locations of surface-water gains and losses are interpreted in this section by a combined analysis of electric-potential data (Figure 7), surface-water temperature and specific conductance data (Figure 8), and relative median gains and losses in streamflow along the survey reach (Figure 9a). Across the Mesilla Valley, hydraulic conditions beneath the Mesilla Valley floor control the vertical hydraulic gradient between the river and the Rio Grande alluvium via the vertical hydraulic gradient between the Rio Grande alluvium and the upper part of the Santa Fe Group. Because the vertical hydraulic gradient varies along the survey reach (Figure 3), streaming potentials appear to make a predominant contribution to the electrostatic field in the surface water, and the electric potentials determined from the gradient SP data correspond notably well to the relative median gain/loss curve determined by annual streamflow measurements.

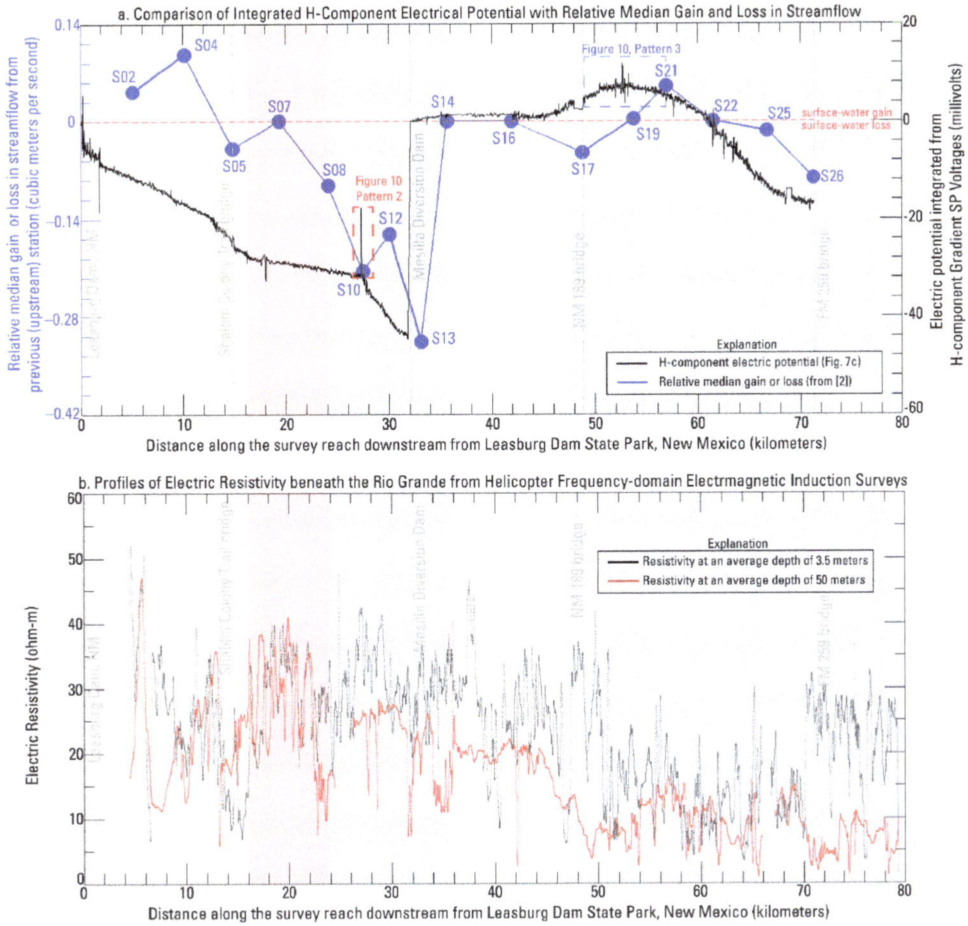

Figure 9. (**a**) Comparison of integrated electric potential (black) integrated from the H component of drift-corrected gradient SP data (see also Figure 2 for locations of seepage-measurement stations) with streamflow data (blue) published by [2,46,61]. Station S02 is relative to station S01 located near 0 km on the horizontal axis (Figure 3). Station S14 is relative to Station S13. Color-shading represents the 2010–2011 water-level differences between the Rio Grande alluvium and the upper part of the Santa Fe Group (Figure 3). Water-level differences greater than 0 m (blue shades) indicate that the hydraulic head in the Rio Grande alluvium is greater than the hydraulic head in the upper part of the Santa Fe Group, whereas differences less than 0 m (red shades) indicate the opposite. (**b**) Comparison of resistivity profiles at average depths of 3.5 m (black) and 50 m (red) beneath the Rio Grande, determined from helicopter frequency-domain electromagnetic surveys.

4.1. Comparison of Electric Potential to Streamflow

Integration of the H component of the gradient SP profile produced a streaming potential profile that contains both low- and high-frequency variations (Figure 7c). This effect was also observed by [37] (see their Figure 5) along 15 km of the lower Guadalupe River across the surficial exposure of the rocks that compose the Carrizo-Wilcox aquifer. The H-component electric potential of the lower Guadalupe River was related by [37] to a superposition of localized bedform-driven hydrodynamic hydraulic gradients and reach-scale hydraulic gradients driving hyporheic flow cells, which were collectively superimposed upon a broader quasi-static regional hydraulic gradient. Superposition of the hydraulic gradients at various spatial scales influenced the exchange processes between the lower Guadalupe River and the Carrizo–Wilcox aquifer at variable spatial scales, from localized gains and losses attributed to channel bedforms, to the regional net gains and losses across the surficial exposure of the rocks that compose the aquifer. Through numerical modeling and waterborne electric resistivity tomography, the low-frequency variation of the data presented by [37] was attributed to net gains and losses influenced by the regional hydraulic gradient in the surficial exposure of the rocks that compose the Carrizo–Wilcox aquifer. Through signal processing and numerical modeling, the high-frequency variation was attributed to localized hydrodynamic gradients, created by riffle and pool sequences along the riverbed that created negative and positive vertical hydraulic gradients, associated with distinct patches of surface-water losses and surface water gains, respectively. The H-component electric potential in the lower Guadalupe River was shown by [37] to be a signal of interest because it reflected gaining and losing conditions at the regional scale, the reach scale, and smaller spatial scales.

If the underlying assumption that the electric potential is predominantly of streaming potential origin is valid, then the electric-potential profiles in Figure 7a–c, in theory, represent net gain or loss in the Rio Grande by changes in the polarity of the steaming-potential component inherent within the data. A comparison of the electric potential integrated from the H component of gradient SP data (Figure 7c) with relative median streamflow gain and loss along the survey reach (Figure 9) indicates that streaming potential was likely the predominant contribution to the electric-potential field in the surface water of the Rio Grande at the time of the geophysical logging survey. Figure 9a depicts the relative median gain or loss (blue curve) at seepage-measurement stations shown in Figure 2, relative to the adjacent (upstream) measurement station along the survey reach of Rio Grande (S02–S04, etc.), in comparison to 2010–2011 water-level differences (color-shading) between the Rio Grande alluvium and upper part of the Santa Fe Group in Figure 3 and the H-component electric-potential in Figure 7c. Profiles of electric resistivity at average depths of 3.5 m and 50 m beneath the Rio Grande are shown in Figure 9b. The relative median net streamflow gain/loss curve reflects long-term conditions, whereas the electric potential is a relatively instantaneous representation by comparison. The double vertical axes in Figure 9a are aligned at 0 cubic meters per second (blue series, left vertical axis) and 0 millivolts (black series, right vertical axis), such that everything above the red line represents a surface-water gain, and everything below the red line represents a surface-water loss, for both data series (assuming the black curve is an adequate representation of the streaming potential). Station S02 is plotted relative to the initial station S01 near x = 0 km on the horizontal axis (see also Figure 3), and station S14 is plotted relative to station S13. Reductions of relative median gain or loss in streamflow between two adjacent stations represent net losses along the survey segments and increases represent net gains, whereas negative electric potential is interpreted as representative of a net losing condition and positive electric potential is interpreted as representative of a net gaining condition [44].

The shape and sign of the relative median streamflow gain/loss curve in Figure 9 resemble quite closely the shape and polarity of the electric-potential profiles upstream and downstream from the Mesilla Diversion Dam (proximal to station S13). The relative median streamflow gain/loss curve indicates not only that the Rio Grande is generally a losing river throughout much of the study area, but also that there are several reaches where a relative gain in streamflow may occur between adjacent stations. Net losses occur between stations S02 and S13 in spite of apparent net gains between S02 and S04, S05 and S07, and S10 and S12. The electric-potential profile has the same general pattern of the streamflow gain/loss curve upstream from the Mesilla Diversion Dam. Electric potential is entirely negative upstream from the Mesilla Diversion Dam and decreases along the entire reach between the Leasburg Dam and Mesilla Diversion Dam. There are several clear slope breaks in the electric potential along this reach (just upstream from S05 and at S10), although it is unclear if or how they may be related to the relative median streamflow gain/loss curve. The intermittent gain between S10 and S12 appears to be a result of the isolated ~200-m long gaining reach, demarcated by the discrete spike in the electric potential at S10. The spike at S10 is coincident with discrete increases in surface-water temperature and specific conductance (Figure 8) and a discrete reduction in gradient SP voltage measured in the river along segment 2 (Figure 6b).

The increase in electric potential at the Mesilla Diversion Dam between S12 and S13 is a result of the discontinuous nature of the electric profile across the dam. Gradient SP data could not be measured over the dam, and so the electric-potential profile downstream from the Mesilla Diversion Dam represents conditions relative to the beginning of the reach at the Mesilla Diversion Dam, which represents a point of zero reference potential. A neutral condition (no apparent gain or loss) is indicated in the relative median streamflow gain/loss curve between S13 and S16. This neutral condition, shown in the streamflow gain/loss curve, corresponds to an approximately constant neutral to mild gaining condition, as shown by the electric-potential profile between the Mesilla Diversion Dam and a point about 6 km downstream from S16. This condition, marked by a positive electric potential of less than 1 mV, begins at the Mesilla Diversion Dam and remains relatively constant for approximately 12 km downstream along most of survey segment 3. The net losing condition between stations S16 and S17 in the relative median streamflow gain/loss curve is not clearly observed in the electric-potential profile and is a possible result of either a true gaining condition at the time of the survey or a strong vertical concentration gradient masking the losing condition in the relative median net streamflow gain/loss curve by a positive diffusion potential. The resistivity profile data in Figure 9b support the latter, where resistivity at an average depth of 50 m shows a notable decrease in resistivity at this location that is not prevalent in the resistivity profile at an average depth of 3.5 m beneath the channel. However, the net gain between S17 and S21 is clearly represented by the electric-potential profile, which shows a steadily increasing potential from negative to positive between 12 km and 22 km downstream from the Mesilla Diversion Dam at the end of segment 3 and into approximately the first half of survey segment 4, before it peaks near station S19 and begins to decrease. The net loss between S21 and S26 is clearly seen in the electric potential profile in the second half of survey segment 4, which decreases over an 18-km segment to the end of segment 4 and indicates a net surface-water loss.

The shaded areas in Figure 9 represent the 2010–2011 water-level differences between the Rio Grande alluvium and the upper part of the Santa Fe Group that are mapped in Figure 3. Water-level differences greater than 0 m (blue shade) indicate that the hydraulic head in the Rio Grande alluvium is greater than the hydraulic head in the upper part of the Santa Fe Group, whereas differences less than 0 m (red shade) indicate the opposite. Under any flow conditions, a losing reach of the river occurs when the hydraulic head in the Mesilla Basin aquifer is less than the hydraulic head supplied by the Rio Grande with the vertical hydraulic gradient oriented downward, and a gaining reach of the river occurs when the hydraulic head in the Mesilla Basin aquifer exceeds the hydraulic head supplied by the river, with the vertical hydraulic gradient oriented upward. With this in mind, it is worth noting that the relative median streamflow gain/loss and electric potential data in Figure 9 represent different seasons (non-irrigation vs. irrigation seasons, respectively) and therefore entirely different flow conditions in the river and vertical hydraulic gradients. Streamflow was measured in February of each survey year (1988–1998, 2004–2013) during a low-flow condition in the non-irrigation season, whereas the electric potential profile was measured in June and July 2020 during a bankfull flow condition at the peak of the irrigation season. During the non-irrigation season, well pumps are off and horizontal hydraulic gradients between the floodplain and the river are reduced relative to the irrigation season, which minimizes surface-water loss into the floodplain in the capture zones of irrigation wells. The river stage is at a minimum, which minimizes the vertical hydraulic gradients and the potential for surface-water losses in losing reaches and enhances the vertical hydraulic gradients in gaining reaches and maximizes the potential for surface-water gains. In contrast, during the irrigation season, groundwater pumping on the floodplain steepens horizontal hydraulic gradients between the floodplain and the river and maximizes surface-water losses into the floodplain in the capture zones of the pumping wells. The river stage is at a maximum (bankfull flow), which maximizes the vertical hydraulic gradients and the potential for surface-water losses in losing reaches and reduces vertical hydraulic gradients in gaining reaches (possibly reversing them) and minimizes the potential for surface-water gains. The implication of the differences in flow conditions represented by the relative median streamflow gain/loss and electric potential profile in Figure 9a is therefore that the relative median streamflow gain/loss curve likely minimizes surface-water losses and enhances surface-water gains, whereas the electric potential profile likely enhances surface-water losses and minimizes surface-water gains. One possible example of this effect in Figure 9a is the noticeable gains in streamflow between S02 and S04, and S05 and S07, which are not apparent in the electric potential profile data.

4.2. Surface-Water Temperature and Specific Conductance Data

Surface-water temperature and specific conductance data show subtle, indirect indicators of surface-water gains at several locations along the survey reach, but do not appear to show clear anomalies attributed to losing reaches. The data (Figure 8a,b) show that differences exist between the survey segments upstream and downstream from Mesilla Diversion Dam. Data corresponding to survey segments upstream from Mesilla Diversion Dam (survey segments 1 and 2) have roughly the same slope, as do those of survey segments downstream from the dam (survey segments 3 and 4); however, the slopes of profile data downstream from the dam are comparatively greater than for those upstream. When plotted against one another (Figure 10), the temperature and specific conductance deviations indicate that survey segments 2 and 4 can be further subdivided into sub-segments on the basis of different relations between surface-water temperature and specific conductance deviations about the linear increases that were removed from the data (Figure 8a,b).

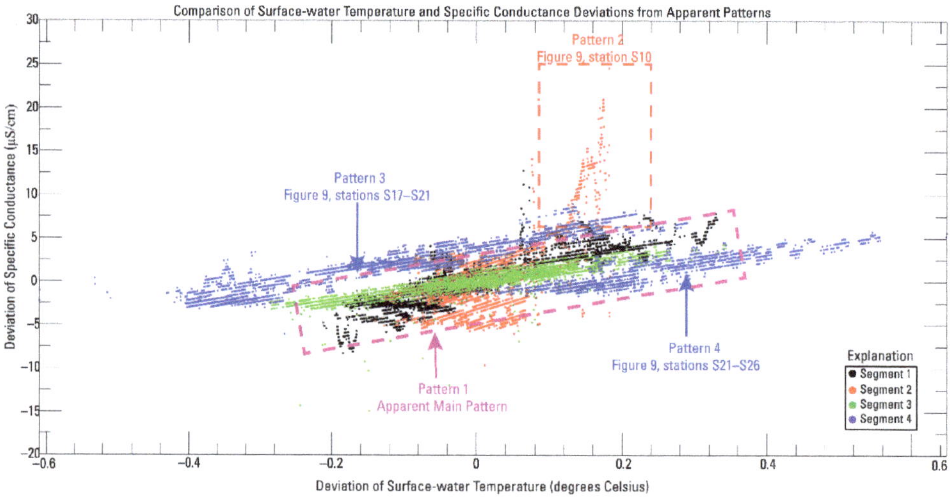

Figure 10. Scatter plot of specific conductance deviations versus temperature deviations showing different relations exist between surface-water temperature and specific conductance along survey segments 2 and 4.

The apparent main pattern in Figure 10 (approximated by the pink box) shows a general linear increase to which the main point cloud of surface-water temperature and specific conductance data appear to adhere. The general adherence to this pattern by data from multiple survey segments shows that the surface-water temperature and specific conductance relations are roughly similar throughout the majority of these segments. However, two individual survey segments show that more than one unique temperature-specific conductance relation exists in the surface water along the survey segments. The upward inflection from the pattern observed in survey segment 2 data (Figure 10, Pattern 2) indicates that temperature and specific conductance vary differently in some parts of survey segment 2 than in others, with specific conductance showing increases (positive deviations) relative to the linear increase that was removed from the data. Data that comprise Pattern 2 coincide with temperature and specific conductance spikes observed along the survey segment 2 (Figure 8c,d), the discrete decrease in gradient SP voltage observed on survey segment 2 (Figures 5b and 6b), and the isolated ~200-m wide positive electric-potential spike on survey segment 2 near seepage-measurement station S10 (Figure 9).

Two distinct surface-water temperature and specific conductance relations exist along survey segment 4 (Figure 10, Patterns 3 and 4). Like survey segment 2, the different patterns indicate that temperature and specific conductance vary differently in some parts of survey segment 4 than in others, and also differently relative to the main apparent pattern and Pattern 2. Both individual increases observed in the data from survey segment 4 have approximately the same apparent slope, which also appears to be consistent with the slope of the main apparent pattern of increases (pink box). Pattern 3 is characterized by larger specific conductance deviations than Pattern 4 and shows both positive and negative temperature deviations relative to the linear increase that was removed (Figure 8a,b). The Pattern 3 data coincide with the positive electric-potential observed between seepage-measurement stations S17 and S21 (Figure 9), which is shown to be a gaining reach in both the electric potential and the relative median streamflow gain/loss curve. This reach of the river is characterized by decreased electrical resistivity (Figures 2, 3 and 9b) beneath the riverbed, attributed to saline groundwater upwelling from the Santa Fe Group into the Rio Grande alluvium and ultimately into the Rio Grande [2,62]. Pattern 4 is characterized generally by smaller positive specific conductance deviations compared to those of Pattern 3, and the temperature deviations that comprise the increase appear to be predominantly, if not entirely, positive. The surface-water temperature and specific conductance data that comprise Pattern 4 coincide with the reach between seepage-measurement stations S21 and S26, which is a losing reach defined by both the streamflow gain/loss curve and the electric potential profile (Figure 9).

5. Conclusions

Gradient SP, surface-water temperature, and surface-water conductivity data were continuously logged along the left bank of the Rio Grande between Leasburg Dam, New Mexico, and Canutillo, Texas, during bankfull flow conditions between 26 June and 2 July 2020. Four survey segments, each 15 km to 25 km in length, were individually surveyed and processed into two longer reaches (one reach upstream and one reach downstream from the Mesilla Diversion Dam) for the interpretation of surface-water gain or loss. Gradient SP profiles were corrected for transient electrode-drift, decomposed into scale-representative components, and numerically integrated separately into electric-potential profiles that were interpreted in the context of surface-water gains and losses by comparison with water-level differences mapped in wells completed in the Rio Grande alluvium and the upper part of the Santa Fe Group, and relative median streamflow gain/loss quantified by streamflow measured at 16 stations along the survey reach.

The electric-potential profiles integrated from gradient SP data each displayed a similar appearance along the survey reaches upstream and downstream from the Mesilla Diversion Dam, but with larger amplitudes corresponding to the larger spatial scale. Integration of the L component produced an electric potential amplitude of ~14 V along the 32-km reach between the Leasburg Dam and Mesilla Diversion Dam, and an amplitude of about 8 V downstream from the Mesilla Diversion Dam, whereas integration of the H component produced an electric potential amplitude of ~40 mV and 25 mV, respectively, along the same reaches. At the time of the survey, the 32-km long reach between the Leasburg Dam and Mesilla Diversion Dam showed a strong propensity for net surface-water losses along the entire reach, with only one location showing indicators of small-scale isolated gain of saline groundwater over a 200-m long reach that coincided with increased surface-water specific conductance (positive deviations relative to the increase observed in the specific conductance data). Downstream from the Mesilla Diversion Dam, electric-potential data indicated a neutral to a mild propensity for surface-water gain for approximately 12 km, that increased between 12 km and 22 km from the Mesilla Diversion Dam, where the gaining condition peaked and began the final transition to a losing condition along the remaining 18 km of the survey reach. The electric potential in the Rio Grande compared notably well with relative median streamflow gain/loss along the reach, and the combination of geophysical and hydraulic data interpreted herein shows the value and usefulness of gradient self-potential logging in rivers for identifying gaining and losing reaches at the regional or basin scale.

The gradient self-potential survey and data processing described herein support the fundamental science objectives of the Transboundary Aquifer Assessment Act (Public Law 109-448) by expanding the available geophysical tools and developing new datasets to assess groundwater and surface-water connectivity between transboundary aquifers and surface-water resources in the United States–Mexico border region. The approach used in this work has great potential for enabling a better understanding of the extent to which transboundary aquifers may be used as sources of water supply, and a means of quickly assessing the vulnerability of transboundary aquifers to anthropogenic or environmental contamination through surface-water connectivity. Gradient self-potential logging is applicable to other transboundary aquifers and is easily adapted to time-lapse monitoring for studying seasonal and annual changes in groundwater and surface-water connectivity in the border region.

Author Contributions: Conceptualization, S.I. and A.T.; methodology, S.I. and A.T.; software, S.I.; validation, S.I., A.T. and D.H.; formal analysis, S.I.; investigation, S.I., A.T.; resources, S.I., A.T., D.H.; data curation, S.I.; writing—original draft preparation, S.I., D.H.; writing—review and editing, S.I.; visualization, S.I.; supervision, S.I.; project administration, A.T.; funding acquisition, A.T. All authors have read and agreed to the published version of the manuscript.

Funding: This work was funded by the U.S. Geological Survey (funding authorized by P.L. 109–448) for the Transboundary Aquifer Assessment Program.

Informed Consent Statement: Not applicable.

Data Availability Statement: The gradient self-potential, surface-water temperature and conductivity logging data are available online at https://doi.org/10.5066/P9GTF1QB (accessed on 10 May 2021). The datasets contributed by [2] are available online at https://doi.org/10.5066/F7PV6HJ3 (accessed on 10 May 2021) and https://doi.org/10.3133/sir20175028 (accessed on 10 May 2021).

Acknowledgments: This work was done in collaboration with the International Boundary and Water Commission (IBWC)–United States Section under the Transboundary Aquifer Assessment Act (Public Law 109–448). We sincerely thank the Hydrologic Technicians at the U.S. Geological Survey New Mexico Water Science Center, Park Rangers at Leasburg Dam State Park, New Mexico, and IBWC staff for logistical assistance in the field. We thank Shawn Carr and William Seelig for their assistance with data acquisition in the field. We thank the reviewers for their assistance with improving this manuscript. Any use of trade, firm, or product names is for descriptive purposes only and does not imply endorsement by the U.S. Government. Writings prepared by U.S. Government employees as part of their official duties, including this paper, cannot be copyrighted and are in the public domain.

Conflicts of Interest: The authors declare no conflict of interest.

References

1. Alley, W.M. *Five-Year Interim Report of the United States-Mexico Transboundary Aquifer Assessment Program: 2007–2012*; U.S. Geological Survey Open-File Report 2013–1059; U.S. Geological Survey: Reston, VA, USA, 2013; pp. 1–31. [CrossRef]
2. Teeple, A.P. *Geophysics- and Geochemistry-Based Assessment of the Geochemical Characteristics and Groundwater-Flow System of the U.S. Part of the Mesilla Basin/Conejos-Médanos Aquifer System in Doña Ana County, New Mexico, and El Paso County, Texas, 2010–2012*; U.S. Geological Survey Scientific Investigations Report 2017-5028; U.S. Geological Survey: Reston, VA, USA, 2017; 183p. [CrossRef]

3. Instituto Nacional de Estadística y Geografía. Subprovincias Fisiográficas. 2001. Available online: https://www.inegi.org.mx/temas/fisiografia/#Descargas (accessed on 18 February 2021).
4. Fenneman, N.M.; Johnson, D.W. Physiographic Divisions of the Conterminous U.S. U.S. Geological Survey Digital Data. 1946. Available online: https://water.usgs.gov/GIS/metadata/usgswrd/XML/physio.xml (accessed on 18 February 2021).
5. Daniel, B. Stephens and Associates, Inc. Evaluation of Rio Grande Salinity, San Marcial, New Mexico to El Paso, Texas. 2010. Available online: http://www.nmenv.state.nm.us/swqb/documents/swqbdocs/LRG/Program/LRG_Salinity_Rpt_6-30-10.pdf (accessed on 18 October 2015).
6. El Paso Water Utilities. Water—Water Resources. 2007. Available online: https://www.epwater.org/our_water/water_resources (accessed on 15 August 2014).
7. Hawley, J.W.; Kennedy, J.F. Creation of a Digital Hydrogeologic Framework Model of the Mesilla Basin and Southern Jornada del Muerto Basin. New Mexico Water Resources Research Institute Technical Completion Report 332; 2004; pp. 1–105. Available online: https://nmwrri.nmsu.edu/wp-content/uploads/TR/tr332.pdf (accessed on 18 February 2021).
8. Hawley, J.W.; Kennedy, J.F.; Ortiz Marquita Carrasco, S. Digital hydrogeologic framework model of the Rincon Valley and adjacent areas of Doña Ana, Sierra and Luna Counties, NM. New Mexico Water Resources Research Institute of New Mexico State University, Addendum to Technical Completion Report 332. 2005. Available online: http://wrri.nmsu.edu/publish/techrpt/tr332/cdrom/addendum.pdf (accessed on 22 October 2009).
9. DeCelles, P.G.; Ducea, M.N.; Kapp, P.; Zandt, G. Cyclicity in Cordilleran orogenic systems. *Nat. Geosci.* **2009**, *2*, 251–257. [CrossRef]
10. Hoffer, J.M. Geology of Potrillo Basalt Field, south-central New Mexico. New Mexico Bureau of Geology and Mineral Resources Circular 149; 1976; pp. 1–30. Available online: https://geoinfo.nmt.edu/publications/monographs/circulars/downloads/149/Circular-149.pdf (accessed on 18 February 2021).
11. Winter, T.C.; Harvey, J.W.; Franke, O.L.; Alley, W.M. Ground water and surface water: A single resource. *Circular* **1998**, *1139*, 88. [CrossRef]
12. Ging, P.B.; Humberson, D.G.; Ikard, S.J. *Geochemical assessment of the Hueco Bolson, New Mexico and Texas, 2016–2017*; U.S. Geological Survey Scientific Investigations Report 2020–5056; U.S. Geological Survey: Reston, VA, USA, 2020; pp. 1–30. [CrossRef]
13. Onsager, L. Reciprocal Relations in Irreversible Processes. I. *Phys. Rev.* **1931**, *37*, 405–426. [CrossRef]
14. Onsager, L. Reciprocal Relations in Irreversible Processes. II. *Phys. Rev.* **1931**, *38*, 2265–2279. [CrossRef]
15. Sill, W.R. Self-potential modeling from primary flows. *Geophysics* **1983**, *48*, 76–86. [CrossRef]
16. Mitchell, J.K. Conduction phenomena: From theory to geotechnical practice. *Géotechnique* **1991**, *41*, 299–340. [CrossRef]
17. Nyquist, J.E.; Corry, C.E. Self-potential: The ugly duckling of environmental geophysics. *Lead. Edge* **2002**, *21*, 446–451. [CrossRef]
18. Naudet, V. A sandbox experiment to investigate bacteria-mediated redox processes on self-potential signals. *Geophys. Res. Lett.* **2005**, *32*, 1–4. [CrossRef]
19. Atekwana, E.A.; Slater, L.D. Biogeophysics: A new frontier in Earth science research. *Rev. Geophys.* **2009**, *47*. [CrossRef]
20. Revil, A.; Mendonca, C.A.; Atekwana, E.A.; Kulessa, B.; Hubbard, S.S.; Bohlen, K.J. Understanding biogeobatteries: Where geophysics meets microbiology. *J. Geophys. Res. Space Phys.* **2010**, *115*. [CrossRef]
21. Revil, A.; Fernandez, P.; Mao, D.; French, H.K.; Bloem, E.; Binley, A. Self-potential monitoring of the enhanced biodegradation of an organic contaminant using a bioelectrochemical cell. *Geophysics* **2015**, *34*, 198–202. [CrossRef]
22. Heenan, J.W.; Ntarlagiannis, D.; Slater, L.D.; Beaver, C.L.; Rossbach, S.; Revil, A.; Atekwana, E.A.; Bekins, B. Field-scale observations of a transient geobattery resulting from natural attenuation of a crude oil spill. *J. Geophys. Res. Biogeosci.* **2017**, *122*, 918–929. [CrossRef]
23. Ishido, T.; Mizutani, H. Experimental and theoretical basis of electrokinetic phenomena in rock-water systems and its applications to geophysics. *J. Geophys. Res. Space Phys.* **1981**, *86*, 1763–1775. [CrossRef]
24. Revil, A. Ionic Diffusivity, Electrical Conductivity, Membrane and Thermoelectric Potentials in Colloids and Granular Porous Media: A Unified Model. *J. Colloid Interface Sci.* **1999**, *212*, 503–522. [CrossRef]
25. Revil, A.; Schwaeger, H.; Cathles, L.M.; Manhardt, P.D. Streaming potential in porous media: 2. Theory and application to geothermal systems. *J. Geophys. Res. Space Phys.* **1999**, *104*, 20033–20048. [CrossRef]
26. Revil, A.; Leroy, P. Hydroelectric coupling in a clayey material. *Geophys. Res. Lett.* **2001**, *28*, 1643–1646. [CrossRef]
27. Revil, A.; Naudet, V.; Nouzaret, J.; Pessel, M. Principles of electrography applied to self-potential electrokinetic sources and hydrogeological applications. *Water Resour. Res.* **2003**, *39*, 1–15. [CrossRef]
28. Revil, A.; Linde, N.; Cerepi, A.; Jougnot, D.; Matthäi, S.; Finsterle, S. Electrokinetic coupling in unsaturated porous media. *J. Colloid Interface Sci.* **2007**, *313*, 315–327. [CrossRef]
29. Revil, A.; Linde, N. Chemico-electromechanical coupling in microporous media. *J. Colloid Interface Sci.* **2006**, *302*, 682–694. [CrossRef] [PubMed]
30. Malama, B.; Kuhlman, K.L.; Revil, A. Theory of transient streaming potentials associated with axial-symmetric flow in unconfined aquifers. *Geophys. J. Int.* **2009**, *179*, 990–1003. [CrossRef]
31. Revil, A.; Woodruff, W.F.; Lu, N. Constitutive equations for coupled flows in clay materials. *Water Resour. Res.* **2011**, *47*, 1–21. [CrossRef]
32. Revil, A.; Ahmed, A.S.; Jardani, A. Self-potential: A Non-intrusive Ground Water Flow Sensor. *J. Environ. Eng. Geophys.* **2017**, *22*, 235–247. [CrossRef]

33. Knight, R.; Pyrak-Nolte, L.J.; Slater, L.; Atekwana, E.; Endres, A.; Geller, J.; Lesmes, D.; Nakagawa, S.; Revil, A.; Sharma, M.M.; et al. Geophysics at the interface: Response of geophysical properties to solid-fluid, fluid-fluid, and solid-solid interfaces. *Rev. Geophys.* **2010**, *48*, 1–30. [CrossRef]
34. Ikard, S.J.; Revil, A.; Jardani, A.; Woodruff, W.F.; Parekh, M.; Mooney, M. Saline pulse test monitoring with the self-potential method to nonintrusively determine the velocity of the pore water in leaking areas of earth dams and embankments. *Water Resour. Res.* **2012**, *48*, 1–17. [CrossRef]
35. Ikard, S.; Revil, A. Self-potential monitoring of a thermal pulse advecting through a preferential flow path. *J. Hydrol.* **2014**, *519*, 34–49. [CrossRef]
36. Valois, R.; Cousquer, Y.; Schmutz, M.; Pryet, A.; Delbart, C.; Dupuy, A. Characterizing Stream-Aquifer Exchanges with Self-Potential Measurements. *Ground Water* **2017**, *56*, 437–450. [CrossRef] [PubMed]
37. Ikard, S.J.; Teeple, A.P.; Payne, J.D.; Stanton, G.P.; Banta, J.R. New Insights on Scale-dependent Surface-Groundwater Exchange from a Floating Self-potential Dipole. *J. Environ. Eng. Geophys.* **2018**, *23*, 261–287. [CrossRef]
38. Heinson, G.; White, A.; Constable, S.; Key, K. Marine self-potential exploration. *Explor. Geophys.* **1999**, *30*, 1–4. [CrossRef]
39. Heinson, G.; White, A.; Robinson, D.; Fathianpour, N. Marine self-potential gradient exploration of the continental margin. *Geophysics* **2005**, *70*, G109–G118. [CrossRef]
40. Kawada, Y.; Kasaya, T. Marine self-potential survey for exploring seafloor hydrothermal ore deposits. *Sci. Rep.* **2017**, *7*, 1–12. [CrossRef]
41. Safipour, R.; Hölz, S.; Halbach, J.; Jegen, M.; Petersen, S.; Swidinsky, A. A self-potential investigation of submarine massive sulfides: Palinuro Seamount, Tyrrhenian Sea. *Geophysics* **2017**, *82*, A51–A56. [CrossRef]
42. Kawada, Y.; Kasaya, T. Self-potential mapping using an autonomous underwater vehicle for the Sunrise deposit, Izu-Ogasawara arc, southern Japan. *Earth Planets Space* **2018**, *70*, 142. [CrossRef]
43. Ikard, S.J.; Sparks, D.D. *Waterborne Gradient Self-Potential, Temperature, and Conductivity Logging of Lake Travis, Texas, Near the Bee Creek Fault, March–April 2020*; U.S. Geological Survey Data Release. 2020; Available online: https://www.sciencebase.gov/catalog/item/5e860cdfe4b01d50927fb71a (accessed on 10 May 2021). [CrossRef]
44. Ikard, S.J.; Briggs, M.A.; Minsley, B.J.; Lane, J.W. *Investigation of Scale-Dependent Groundwater/Surface-Water Exchange in Rivers by Gradient Self-Potential Logging: Numerical Models and Field Experiment Data, Quashnet River, Massachusetts, October 2017 (ver. 2.0, November 2020)*; U.S. Geological Survey Data Release. 2020; Available online: https://www.sciencebase.gov/catalog/item/5e275734e4b014c85308fc30 (accessed on 10 May 2021). [CrossRef]
45. Hunt, B.B.; Cockrell, L.P.; Gary, R.H.; Vay, J.M.; Kennedy, V.; Smith, B.A.; Camp, J.P. Hydrogeologic Atlas of Southwest Travis County, Central Texas: Barton Springs/Edwards Aquifer Conservation District Report of Investigations 2020–0429, April 2020, 80p. With Digital Datasets. Available online: https://repositories.lib.utexas.edu/handle/2152/81562 (accessed on 18 February 2021).
46. Crilley, D.M.; Matherne, A.M.; Thomas, N.; Falk, S.E. *Seepage Investigations of the Rio Grande from below Leasburg Dam, Leasburg, New Mexico, to above American Dam, El Paso, Texas, 2006–2013*; U.S. Geological Survey Open-File Report 2013–1233; U.S. Geological Survey: Reston, VA, USA, 2013; pp. 1–34. [CrossRef]
47. George, P.G.; Mace, R.E.; Mullican, W.F. The hydrology of Hudspeth County, Texas. Texas Water Development Board Report 364; 2005; pp. 1–106. Available online: www.twdb.texas.gov/publications/reports/numbered_reports/doc/R364/R364.pdf (accessed on 18 February 2021).
48. George, P.G.; Mace, R.E.; Petrossian, R. Aquifers of Texas. Texas Water Development Board Report 380; 2011; pp. 1–180. Available online: www.twdb.texas.gov/publications/reports/numbered_reports/doc/R380_AquifersofTexas.pdf (accessed on 18 February 2021).
49. Leggat, E.; Lowry, M.; Hood, J. Ground-water resources of the lower Mesilla Valley, Texas and New Mexico. In *Ground-Water Resources of the Lower Mesilla Valley, Texas and New Mexico*; US Geological Survey: Liston, VA, USA, 1964; pp. 1–53. [CrossRef]
50. Hawley, J.W.; Lozinski, R.P. Hydrogeologic Framework of the Mesilla Basin in New Mexico and Western Texas. New Mexico Bureau of Mines and Mineral Resources Open File Report 323; 1992; pp. 1–55. Available online: https://geoinfo.nmt.edu/publications/openfile/downloads/300-399/323/ofr_323.pdf (accessed on 18 February 2021).
51. Dunbar, J.B.; Murphy, W.L.; Ballard, R.F.; McGill, T.E.; Peyman-Dove, L.D.; Bishop, M.J. *Condition assessment of U.S. International Boundary and Water Commission, Texas and New Mexico Levees—Report 2. U.S.*; Army Corps of Engineers Engineer Research and Development Center: Vicksburg, MS, USA, 2004; p. 121.
52. Zohdy, A.A.; Bisdorf, R.J.; Gates, J.S. *Schlumberger Soundings in the Lower Mesilla Valley of the Rio Grande, Texas and New Mexico*; U.S. Geological Survey Open-File Report 76–324; U.S. Geological Survey: Reston, VA, USA, 1976; pp. 1–77. [CrossRef]
53. Minsley, B.J.; Burton, B.L.; Ikard, S.; Powers, M.H. Hydrogeophysical Investigations at Hidden Dam, Raymond, California. *J. Environ. Eng. Geophys.* **2011**, *16*, 145–164. [CrossRef]
54. Ikard, S.; Revil, A.; Schmutz, M.; Jardani, A.; Karaoulis, M.; Mooney, M. Characterization of focused seepage through an earth-fill dam using geoelectrical methods. *Ground Water* **2014**, *52*, 952–964. [CrossRef] [PubMed]
55. Ikard, S.; Rittgers, J.; Revil, A.; Mooney, M. Geophysical Investigation of Seepage beneath an Earthen Dam. *Ground Water* **2014**, *53*, 238–250. [CrossRef]
56. Ikard, S.; Pease, E. Preferential groundwater seepage in karst terrane inferred from geoelectric measurements. *Near Surf. Geophys.* **2018**, *17*, 43–53. [CrossRef]

57. Ikard, S.J.; Carr, S.M.; Seelig, W.G.; Teeple, A.P. *Waterborne Gradient Self-Potential, Temperature, and Conductivity Logging of the Rio Grande from Leasburg Dam State Park, New Mexico to Canutillo, Texas, during a Bankfull Condition, June–July 2020*; U.S. Geological Survey Data Release; 2021; Available online: https://doi.org/10.5066/P9GTF1QB (accessed on 10 May 2021). [CrossRef]
58. Teeple, A.P. *Time-Domain Electromagnetic Data Collected in the U.S. Part of the Mesilla Basin/Conejos-Médanos Aquifer System in Doña Ana County, New Mexico, and El Paso County, Texas, November 2012*; U.S. Geological Survey Data Release; 2017. Available online: https://www.sciencebase.gov/catalog/item/584adb27e4b07e29c706de08 (accessed on 10 May 2021). [CrossRef]
59. Abdelrahman, E.M.; Gobashy, M.M.; Essa, K.; Abo-Ezz, E.R.; El-Araby, T.M. A least-squares minimization approach to interpret gravity data due to 2D horizontal thin sheet of finite width. *Arab. J. Geosci.* **2016**, *9*, 1–7. [CrossRef]
60. U.S. Geological Survey. *Specific Conductance: U.S. Geological Survey Techniques and Methods*; Book 9; U.S. Geological Survey: Reston, VA, USA, 2019; Chapter A6.3; p. 15. [CrossRef]
61. U.S. Geological Survey [USGS]. USGS Water Data for the Nation: U.S. Geological Survey National Water Information System Database. 2021. Available online: http://waterdata.usgs.gov/nwis/ (accessed on 7 April 2021). [CrossRef]
62. Briggs, M.A.; Lautz, L.K.; Buckley, S.F.; Lane, J.W. Practical limitations on the use of diurnal temperature signals to quantify groundwater upwelling. *J. Hydrol.* **2014**, *519*, 1739–1751. [CrossRef]

water

MDPI

Article

Salinity Contributions from Geothermal Waters to the Rio Grande and Shallow Aquifer System in the Transboundary Mesilla (United States)/Conejos-Médanos (Mexico) Basin

Jeff D. Pepin [1,*], Andrew J. Robertson [1] and Shari A. Kelley [2]

1 U.S. Geological Survey, 6700 Edith Blvd. NE, Building B, Albuquerque, NM 87113, USA; ajrobert@usgs.gov
2 New Mexico Bureau of Geology & Mineral Resources, New Mexico Institute of Mining & Technology, 801 Leroy Place, Socorro, NM 87801, USA; shari.kelley@nmt.edu
* Correspondence: jpepin@usgs.gov; Tel.: +1-505-830-7952

Abstract: Freshwater scarcity has raised concerns about the long-term availability of the water supplies within the transboundary Mesilla (United States)/Conejos-Médanos (Mexico) Basin in Texas, New Mexico, and Chihuahua. Analysis of legacy temperature data and groundwater flux estimates indicates that the region's known geothermal systems may contribute more than 45,000 tons of dissolved solids per year to the shallow aquifer system, with around 8500 tons of dissolved solids being delivered from localized groundwater upflow zones within those geothermal systems. If this salinity flux is steady and eventually flows into the Rio Grande, it could account for 22% of the typical average annual cumulative Rio Grande salinity that leaves the basin each year—this salinity proportion could be much greater in times of low streamflow. Regional water level mapping indicates upwelling brackish waters flow towards the Rio Grande and the southern part of the Mesilla portion of the basin with some water intercepted by wells in Las Cruces and northern Chihuahua. Upwelling waters ascend from depths greater than 1 km with focused flow along fault zones, uplifted bedrock, and/or fractured igneous intrusions. Overall, this work demonstrates the utility of using heat as a groundwater tracer to identify salinity sources and further informs stakeholders on the presence of several brackish upflow zones that could notably degrade the quality of international water supplies in this developed drought-stricken region.

Keywords: salinization; transboundary aquifers; geothermal; international water supplies; water quality; upflow; vertical groundwater flow; heat transport; thermal modeling

Citation: Pepin, J.D.; Robertson, A.J.; Kelley, S.A. Salinity Contributions from Geothermal Waters to the Rio Grande and Shallow Aquifer System in the Transboundary Mesilla (United States)/Conejos-Médanos (Mexico) Basin. *Water* **2022**, *14*, 33. https://doi.org/10.3390/w14010033

Academic Editors: Sharon B. Megdal and Anne-Marie Matherne

Received: 25 June 2021
Accepted: 11 December 2021
Published: 23 December 2021

1. Introduction

Natural and anthropogenic salinization of water supplies challenges sustainable water resource management, particularly in drought-stricken regions such as the southwestern United States and northern Mexico [1]. The transboundary Mesilla (United States)/Conejos-Médanos (Mexico) Basin (referred to herein as the Basin) of New Mexico and Texas (United States) and Chihuahua (Mexico) is one populated region facing these challenges in light of declining water levels, deteriorating water quality, and increased water use on both sides of the international border (Figure 1) [2]. Both groundwater and the Rio Grande (United States)/Rio Bravo (Mexico) are heavily relied upon to meet water demand. Hogan et al. (2007) have shown that Rio Grande chloride concentration more than doubles from around 120 milligrams per liter (mg/L) to 280 mg/L between the inlet and outlet of the Mesilla portion of the Basin (Mesilla) [3]. Driscoll and Sherson (2016) later demonstrated that during periods of minimal upstream reservoir releases (i.e., non-release seasons) within the 2009 to 2013 time period, Rio Grande salinity (as approximated by total dissolved solids (TDS)) averaged about 1500 mg/L at the basin inlet (RG-LB, Figure 1) and increased to approximately 2200 mg/L at the outlet (RG-EP, Figure 1) [4]. This salinity increase, along with similar spatial trends in groundwater salinity, are likely affected by several factors, including (1) runoff and recharge from agricultural activity, (2) wastewater discharge, (3) evapoconcentration, (4) topographically and/or buoyancy driven upwelling (vertical flow/upflow) of geothermal and non-thermal groundwater, and (5) intra-basin groundwater flow from surrounding basins [3,5–11].

Salinity contributions from geothermal waters, meaning salinity from waters of naturally elevated temperature, have not been studied as extensively in this region as some of the other salinity mechanisms and are the focus of this research. Three prominent geothermal systems have been identified in the study area, where previous research has largely focused on geothermal energy production and development rather than salinity contributions [12–16]. Upwelling waters associated with these geothermal systems have naturally elevated salinities ranging from about 1800 to 4800 mg/L and therefore have the potential to degrade surrounding freshwater supplies [15–17]. This work combines previously published geothermal discharge estimates and historical (1972–2018) temperature measurements to identify prominent geothermal groundwater upflow zones and estimate their salinity contribution to the region's primary aquifer system and to the Rio Grande.

Analyzed temperature data includes temperature measured as a function of depth (temperature profile) collected within 379 wells dispersed throughout the Mesilla [18]. Temperature profiles typically portray a linear increase in temperature with depth when groundwater flow rates (i.e., advection) are slow. Systematic curvature is evident in temperature profiles when vertical and/or horizontal advection rates dominate over thermal conduction; the degree of curvature increases with higher rates of advection [19]. This systematic relation between profile curvature and flow rates enables the quantitative estimation of discharge based on temperature data, thereby permitting heat to be used as a groundwater tracer [19]. This research entails the following: (1) classifying temperature profile curvature, (2) calculating 1D vertical flow rates for the temperature profiles that have upflow curvature, (3) estimating corresponding spatial areas of upflow, and (4) coupling estimated vertical flow rates, areas, and groundwater salinity data to estimate potential volumetric salinity contributions to the primary aquifer system and the Rio Grande. This approach identifies prevalent geothermal upflow zones that are localized within more broadly defined and diffuse upwelling geothermal systems. Estimates of salinity flux were also computed for the broad geothermal systems if previously published geothermal discharge estimates were available.

Overall, this work confirms the notable salinity flux associated with geothermal waters upwelling in the Mesilla. Identified localized upflow zones likely contribute upwards of 8500 tons of dissolved solids to the shallow aquifer system annually, or approximately 4% of average annual cumulative Rio Grande dissolved solids leaving the basin from 2009 through 2013 [4]. Coupling previously published estimates of geothermal discharge for the broad geothermal systems with corresponding geothermal groundwater salinity indicates that the total geothermal salinity contributions may be much higher, exceeding 45,000 tons of dissolved solids per year, or 22% of 2009–2013 average annual cumulative Rio Grande mass flux at the basin outlet. Groundwater elevation mapping of water levels measured in 2010 indicates that brackish groundwater from identified upflow zones likely flows towards northern Chihuahua in Mexico, and Las Cruces, the southern Mesilla, and the Rio Grande within the United States [20]. Generally, this work indicates geothermal waters may appreciably affect the salinity budget for this region, both in the United States and Mexico, identifies localized upflow zones that could inform future mitigation efforts, and demonstrates the utility of using heat as a groundwater tracer to evaluate geothermal salinity fluxes.

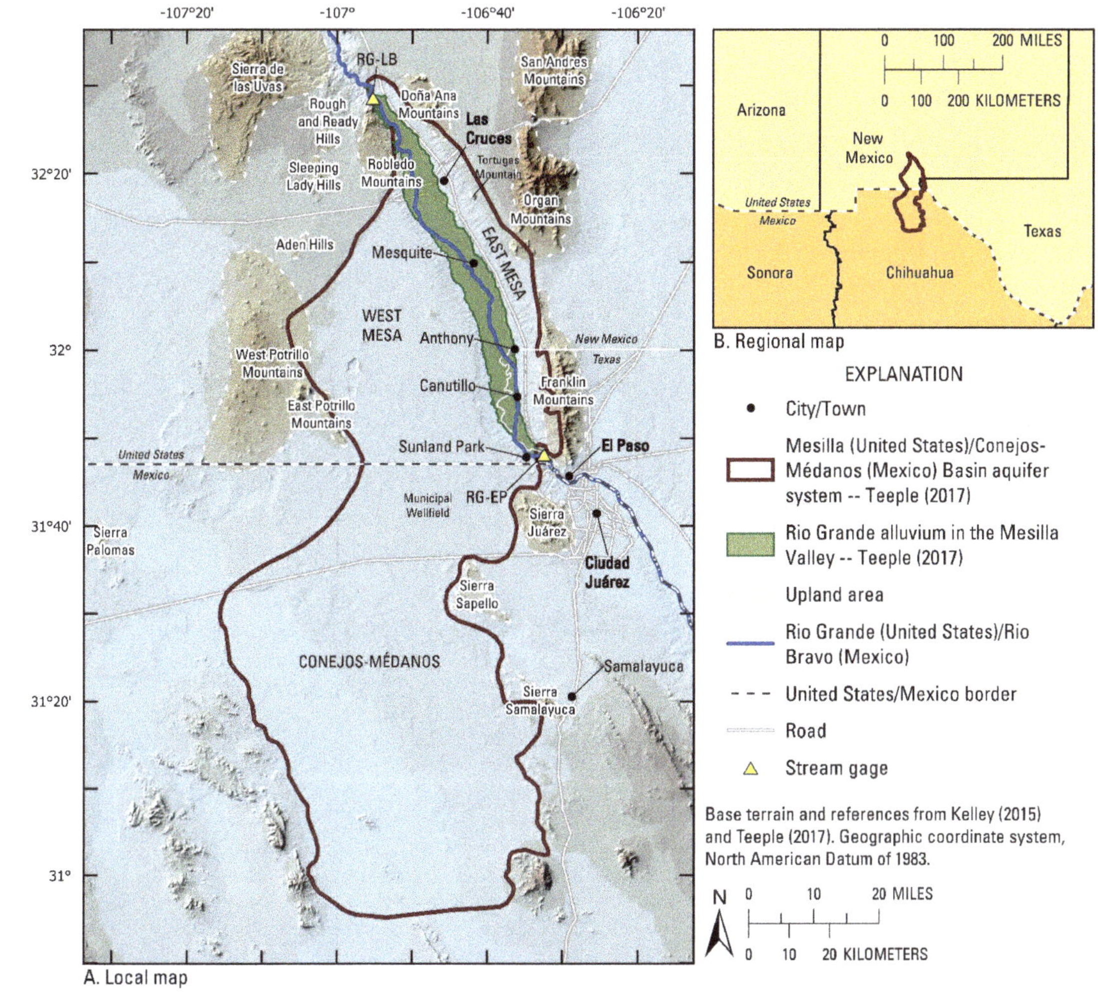

Figure 1. Local (**A**) and regional (**B**) maps showing the location of the study area in the United States (New Mexico and Texas) and Mexico (Chihuahua) [21,22]. All analyzed temperature data in this work were collected in the United States, while interpreted groundwater elevation mapping covers substantial portions of the aquifer system in both countries. Stream gage abbreviations are as follows: RG-LB = USGS 08363510 Rio Grande below Leasburg Dam at Fort Selden, New Mexico; RG-EP = USGS 08364000 Rio Grande at El Paso, Texas [23].

2. Background

2.1. Description of the Study Area

The study area is the Mesilla (United States)/Conejos-Médanos (Mexico) Basin and the corresponding aquifer system, which covers a total area of about 7200 square kilometers (km^2), with around 2700 km^2 (37%) in the United States (U.S.) and 4500 km^2 (63%) in Mexico (Figure 1) [21]. The Mesilla portion of the Basin (Mesilla) is further divided into the West Mesa and the East Mesa, which are separated by the Mesilla Valley. This study area is in the southern portion of the Rio Grande rift, a tectonically active extensional province that stretches from southern Colorado into Mexico and is bound by numerous fault zones and upland areas [24–27]. An extensive basalt field, including Kilbourne Hole and Hunt's Hole (volcanic maars) and several igneous intrusions, is in the southwest Mesilla within and near the West Potrillo Mountains and the East Potrillo Mountains [21,27,28]. The landscape of the Conejos-Médanos portion of the Basin (Conejos-Médanos) is dominated by dune fields [29].

Regional climate has been described as arid and dry with low humidity and precipitation, high evaporation, and a wide range of temperature and vegetation types [21,29]; these characteristics are often strongly elevation dependent. Most precipitation falls during July through September as monsoonal rainfall [29,30].

The main surface water features in the area are the Rio Grande (U.S.)/Rio Bravo (Mexico) and an intricate network of irrigation canals that are primarily fed by diverted Rio Grande water [4,29,31]. The Rio Grande enters the Mesilla through Selden Canyon north of Las Cruces near the RG-LB stream gage and the adjacent Leasburg Diversion Dam (Figure 1). The river flows south-southeast through the Mesilla Valley before exiting the Basin at the Paso del Norte near the RG-EP stream gage at El Paso, where it forms the U.S./Mexico international border. Surface water flow is strongly dependent on releases from upstream reservoirs, with the highest flows typically occurring during the summer growing season [4]. Flows dramatically decline or cease when reservoir releases are halted, typically during the winter months. The Rio Grande alternates between being a losing and gaining stream and water quality generally degrades as the river traverses the Mesilla Valley [4,6,11].

The Quaternary/late Tertiary Santa Fe Group sediments and Quaternary Rio Grande Valley alluvium constitute the Basin's aquifer system and are the primary regional aquifers [4,21,27]. More than 120 million acre-feet of groundwater is estimated to be potentially recoverable from this mixture of unconsolidated sand, gravel, silt, and clay [20,27]. Groundwater salinity varies widely from less than 500 mg/L to about 30,000 mg/L, with a prominent zone of brackish to highly saline (5000 to 30,000 mg/L) groundwater located near the Basin outlet (Figure 1) [3,6,11,21]. This groundwater zone is likely associated, in part, with regional-scale non-thermal groundwater upwelling [3,5,11]. The base of the aquifer system is defined by a variety of consolidated rocks including Precambrian crystalline rocks; Paleozoic and Mesozoic dolomite, limestone, and sandstone; intrusive rocks; and Paleogene sedimentary and volcanic rocks—all of which are referred to as basement rocks herein [27,29]. Exposures of these rocks are largely in upland areas and where horst blocks crop out. Depths to these bedrock units also become shallower at the Basin margin, notably near the Basin outlet at Paso del Norte.

Recharge to the aquifer system is mainly within the Mesilla Valley along losing reaches of the Rio Grande and irrigation canals, with smaller amounts of mountain front recharge near upland areas [6,11,17,21]. Groundwater salinity is often less than 250 mg/L in local mountain front recharge areas, whereas surface water recharge commonly ranges from around 400 to 2200 mg/L [4,11].

International groundwater elevation mapping indicates that groundwater generally flows towards the Rio Grande and that some groundwater flows from the Conejos-Médanos in Mexico into the Mesilla in the United States [20,29]. Groundwater generally flows east-southeast in the West Mesa, south-southeast in the Mesilla Valley, and south-southwest in the East Mesa. In contrast, groundwater flows north-northwest from the southern Conejos-Médanos towards lowlands in the western part of the basin. From the lowlands, groundwater slowly moves north-northeast towards the Mesilla and Rio Grande.

The area has largely been developed for agriculture since the 1900s, but also contains relatively large population centers in Las Cruces, El Paso, and Ciudad Juárez (Figure 1) [29,31]. Groundwater is the primary drinking water supply and supplements surface water for irrigation [29,31]. Some of the groundwater that flows towards the United States from Mexico is intercepted by a municipal wellfield that supplements water supply to Ciudad Juárez [20,29]. Generally, dependence on groundwater in the region results in notable water level fluctuations, particularly in the Mesilla Valley where agricultural development is most prominent [31].

2.2. Known Geothermal Systems within the Study Area

There are at least three known geothermal systems in the study area, all of which are in the Mesilla. The East Mesa geothermal system is thought to be one of the largest low-temperature (less than 90 °C) systems in the United States, spanning from east of Las Cruces southward to nearly the Texas Stateline (Figure 1) [13]. Geothermal upwelling associated with this system is fault-controlled and focused along a largely buried horst block. Estimated natural groundwater discharge from heat flow analyses for the broad footprint of this system is upwards of 15,000 acre-feet per year [13]. A portion of this system was developed east of Las Cruces near Tortugas Mountain for college campus heating, greenhouse heating, and aquaculture [15]. Produced waters are typically around 64 °C with an average TDS of about 1800 mg/L [15]. Groundwater volumes between 1225 and 1780 acre-feet per year may naturally discharge from this portion of the system alone [6,32].

The Radium Springs geothermal system, located near the basin inlet adjacent to the RG-LB stream gage (Figure 1), is another developed geothermal system in the Mesilla [13,14,16]. This system serves one of the largest geothermal greenhouses in the United States at Masson Farms of New Mexico [13,16]. The geothermal anomaly is thought to cover an area of about 78 km^2 [13]. Geothermal upwelling is associated with Quaternary faulting and outflow within a highly fractured rhyolitic intrusion [13,16]. No previously published estimates of natural discharge are known to the authors, possibly due to data scarcity because the system is largely developed on private land. Temperatures of produced waters are about 99 °C with TDS around 3650 mg/L [16].

Lastly, the low-temperature East Potrillo geothermal system is an undeveloped resource in the southern foothills of the East Potrillo Mountains (Figure 1) [12]. Geothermal upwelling is controlled by the East Potrillo fault zone and corresponding highly fractured East Potrillo Mountain horst block [12]. Heat flow analysis indicates a groundwater discharge rate of approximately 970 acre-feet per year over an area of about 2.4 km^2 [12]. Groundwater chemistry of produced water is unknown because this system is undeveloped; however, historical data collected nearby along the same fault zone indicates specific conductance (SC) values of 7400 microsiemens per centimeter (μS/cm) [23]. Using the SC to TDS conversion factor from Driscoll and Sherson (2016) of 0.6518 yields a TDS estimate of 4823 mg/L for the East Potrillo geothermal waters [4].

Provided the elevated salinity of the geothermal waters in the study area (1800 to over 4800 mg/L) and notable corresponding discharge rates, these three systems alone have the potential to adversely affect the groundwater chemistry of the shallow aquifer system and of the Rio Grande in this region.

2.3. Using Heat as a Groundwater Tracer

Water carries heat with it as it flows, which enables temperature measurements to be used to trace groundwater flow. When advection rates are high enough, the water possesses a temperature signature that is characteristic of its flow history. For example, hot water upwelling from deep within the earth may remain hot when it reaches the shallow aquifer system or land surface (as hot springs), given appropriate advection, conduction, and mixing conditions. Figure 2 illustrates how the advective transport of heat in vertical (upflow and downflow) and horizontal (lateral) groundwater flow can perturb a typical linear conductive geothermal gradient in measured temperature profiles. This concept gives rise to the idea of using heat as a groundwater tracer. Anderson (2005) provides a detailed review of the extensive work that has been done using heat as a tracer dating back to the 1950s [19]. This technique has been successfully used in many ways, including the estimation of recharge and discharge rates, hydraulic conductivities of streambeds, basin-scale permeabilities, and hyporheic zone flow patterns [19]. Herein, heat is used as a tracer to estimate vertical salinity fluxes at locations that have a temperature profile curvature that is indicative of groundwater upflow.

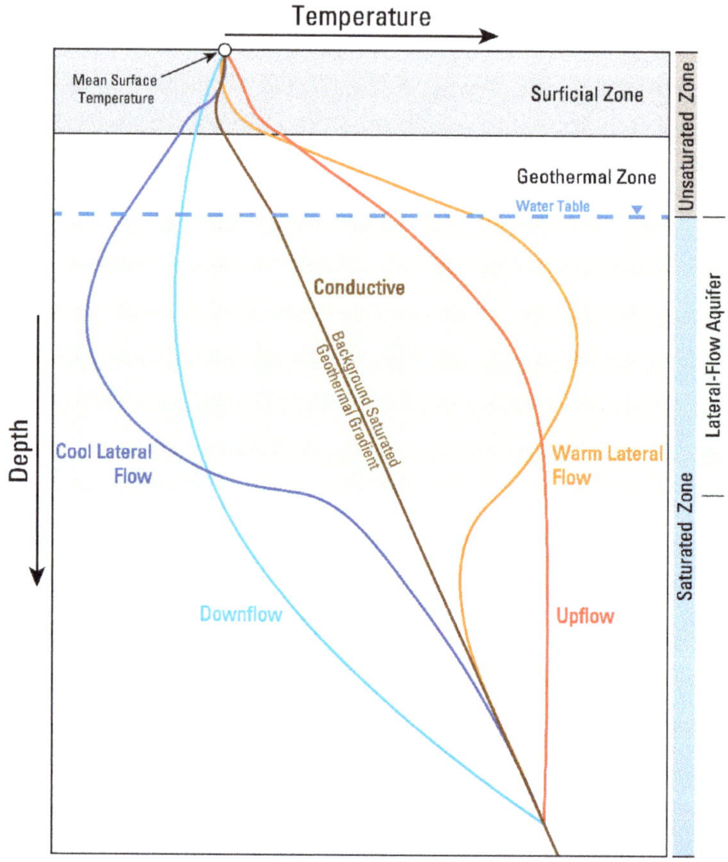

Figure 2. Schematic showing typical temperature profile deviations from the conductive case because of horizontal or vertical groundwater flow. The surficial zone refers to depths in which temperatures are influenced by relatively short-term (e.g., daily, seasonal) temperature variations.

3. Materials and Methods

The analyses presented herein include using legacy temperature data to estimate the salinity flux associated with groundwater upflow zones within the study area. This methodology included data preparation, temperature profile curvature classification to identify upflow zones, and estimation of salinity fluxes from identified upflow zones and their host geothermal systems.

3.1. Description of Data

Data used in this study were historical industry, academic, and researcher data that were collected and/or compiled by the New Mexico Bureau of Geology and Mineral Resources. The analyzed dataset included 379 temperature profiles made up of 11,161 individual temperature measurements. Corresponding lithology records, which included thermal conductivity and porosity estimates in some cases, were identified for 199 (52.5%) profiles. Much of these data (98%) were collected during a period of extensive geothermal exploration within the study area in the late 1970s and early 1980s. Of all the profiles, 116 (31%) were measured in the 1970s, 253 (67%) were measured in the 1980s, and 8 (2%) were measured in 2018. Two profiles had unknown collection dates but were most likely measured in the late 1970s or early 1980s. Overall, measurement dates ranged from

21 April 1972 to 23 February 2018, with a median measurement date of 27 March 1980. Data preprocessing included correction of obvious typographical errors by consulting original records; conversion of depth and temperature units to meters and degrees Celsius, respectively; and removal of spaces, commas, periods, slashes, apostrophes, and personally identifiable information from well names. Per standard practice in the geothermal industry, the most recently measured profile was used to favor thermal equilibrium in instances where multiple temperature profiles were available at the same location. The final dataset of temperature profiles and corresponding well records was published as a U.S. Geological Survey (USGS) data release in 2019 [18]; the profile ID numbers used in this research correspond to those from the data release

In the final dataset, measured depths ranged from 1 to 910 m (median = 67 m), while measured temperatures ranged from 10.5 to 86.4 °C (median = 24.8 °C). Measurement intervals within the boreholes ranged from about 0.5 m to 20 m, with a median of about 3 m (10 feet). Reported measurement precision for temperatures were within 1 °C or better, whereas reported measurement precision for depth measurements were 1 m or less. A table with additional relevant details for the profiles is provided in Table S1.

3.2. Classification Analysis

Temperature profiles were classified based on their curvature (Figure 2) to facilitate identification of upflow zones and regions of warm lateral flow, which would be proximal to upflow zones. The analysis began by plotting the profiles for visual inspection (profile plots are provided in Figure S1). Temperatures measured near the land surface are subject to relatively short-term (e.g., daily, seasonal) temperature variations, and were considered data noise for this study's objectives [33–35]. Temperatures measured within the first 20 m of the subsurface (surficial zone) were therefore omitted to avoid this interference; this is a conservative depth threshold that was chosen based on visual inspection of the profiles. Because this research seeks salinity flux estimates of upwelling groundwater, it was necessary to identify portions of profiles that were measured below the water table (i.e., saturated zone). Well records, smoothed profiles and their derivatives, and nearby USGS water level data were conjunctively used to estimate water table depth where feasible and thereby identify profiles warranting further analysis. Smoothing methods were used to aid in the identification of dominant profile and derivative characteristics. These methods included cubic smoothing splines and 2nd degree local regression (LOESS) fits. LOESS fits were computed with a smoothing parameter that was (1) held constant for all profiles (0.75), (2) determined by using a bias-corrected Akaike information criterion (AICC), and (3) determined by using generalized cross-validation (GCV), whereas spline fits all used leave-one-out cross-validation (LOOCV) to determine their degree of smoothness; the range of smoothing approaches was implemented to give an ensemble of reasonable smoothed profiles and derivatives. Smoothing fits were computed using the 'stats' (version 3.5.3) and fANCOVA (version 0.5-1) packages of the open-source R programming language (version 3.5.3) [36,37]; plots of the smoothed results overlain by the raw data for saturated profiles are provided in Figure S2. Profiles with less than four measurements below the estimated water table elevation at any given location were not further analyzed because of the insufficient amount of data to confidently assess profile curvature. Final classifications of profile curvature were plotted spatially and included: upflow, warm lateral flow, downflow/cool lateral flow, conductive, undetermined, and not analyzed.

3.3. Flux Estimation

Computing salinity flux estimates required coupling groundwater salinity data with volumetric upflow rate estimates. The Bredehoeft and Papadopulos (1965) 1D vertical heat transport analytical solution was applied to estimate a vertical specific discharge rate for each upflow profile [38]. This solution uses the thermal Peclet number (Pe), which is defined as the ratio of thermal advection to conduction as follows:

$$Pe = \rho_f c_f q_z L / K_e \quad (1)$$

where ρ_f is fluid density, c_f is fluid specific heat capacity, q_z is vertical specific discharge, L is the saturated thickness over which the temperature data were analyzed, and K_e is the effective thermal conductivity. Peclet numbers of larger magnitude correspond to higher vertical flux rates and more extensive profile curvature. Negative Peclet numbers indicate groundwater upflow, whereas positive Peclet numbers are associated with groundwater downflow. The practical minimum detectable Peclet number is typically considered to be around $^+/_-$ 0.2 given thermal conductivity variations and measurement accuracy limitations [33]. Rearranging Equation (1) to solve for vertical flux yields the following expression:

$$q_z = K_e Pe / \rho_f c_f L \quad (2)$$

Vertical flux can be estimated by specifying the thermal properties listed in Equation (2) and iteratively solving for the Peclet number that best matches measured temperatures when used in the Bredehoeft and Papadopulos (1965) analytical solution [38]. In this study, best-fit Peclet numbers were determined by minimizing root-mean-squared error (RMSE) between the analytical solution and the measured data. Fluid specific heat capacity was specified to be 4180 joules per kilogram per degrees Celsius (J/kg °C) at all locations. Fluid density was estimated by using the median analyzed profile temperature and the temperature-dependent water density relation of Kell (1975), which has been shown to be valid for water ranging in temperature from 0 to 150 °C [39]. Salinity effects on water properties were neglected because detailed water chemistry was not known for all evaluated waters. A sensitivity analysis performed in this study showed temperatures ranging from 25 to 100 °C affected fluid density by about 4%, whereas salinities ranging from 0 to 5000 TDS altered fluid density by 0.5% or less. Therefore, neglecting salinity effects on fluid density is acceptable for the conditions considered in this study. Effective thermal conductivities were estimated from well records and/or computed by using the geometric mean of reported solid and fluid thermal conductivities, as follows:

$$K_e = {k_s}^{(1-n)} {k_f}^{n} \quad (3)$$

where k_s is the thermal conductivity of the solid phase (e.g., sediment grains), k_f is the thermal conductivity of the fluid phase (e.g., air or water), and n is porosity. This relation has been shown to well approximate the effective thermal conductivity in previous studies [40]. Porosities were obtained from well records or previously published literature.

Associated spatial areas of upflow were then estimated by evaluating spatial temperature patterns, thermal cross sections, and the spatial distribution of profile curvature classifications. Thermal cross sections included temperature profiles that were projected onto cross-section profile lines. These data were overlain onto the basement stratigraphy and faults from Sweetkind (2017), along with topography from a USGS 1/3 arc-second (about 10 m) digital elevation model (DEM) [27,41]. Estimated areas were combined with the vertical flux estimates and salinities to determine the salinity flux associated with each local upflow zone, as follows:

$$J_{TDS} = q_z A C_{TDS} = Q_z C_{TDS} \quad (4)$$

where J_{TDS} is mass transfer rate (salinity flux in mass per time), q_z is vertical specific discharge from the application of Bredehoeft and Papadopulos (1965) [38], A is upflow zone area, C_{TDS} is the TDS concentration of upwelling groundwater, and Q_z is volumetric vertical groundwater flux ($Q_z = q_z A$).

This approach for estimating salinity flux has limitations. For example, this approach considers only salinity contributions from localized upflow zones within the broader geothermal systems—salinity flux estimates are therefore conservative and flux from the entire geothermal system is higher. In addition to the localized analyses, estimates of salinity flux were therefore also computed for the broad geothermal systems if previously published geothermal volumetric flux values were available (Equation (3)). Identification of upflow zone locations was limited to portions of the study area where temperature profiles had been measured. It is therefore likely that other unidentified upflow zones contribute salinity to the area. Unaccounted-for fluxes could be evaluated in the future by using more comprehensive 3D heat and solute transport modeling and with additional data collection. Other key assumptions associated with this approach include steady-state thermal equilibrium between the well bore and its subsurface surroundings, constant groundwater and aquifer properties along the analyzed temperature profile interval, and that vertical flow dominates over horizontal flow within the upflow zones. Despite these assumptions, previously published uncertainty research that used synthetic temperature data indicates that reliable flux estimates can be obtained from the Bredehoeft and Papadopulos (1965) solution [38] in heterogenous media and when horizontal fluxes are 10 times greater than vertical fluxes, provided the temperature at the upper boundary is steady through time [35]. Overall, this methodology is thought to be a straightforward way of obtaining reasonable salinity flux estimates associated with geothermal systems and their localized upflow zones.

Previously published geothermal volumetric flux estimates were available for the East Potrillo and East Mesa geothermal systems and were coupled with groundwater salinities to make additional salinity flux estimates in this work. The previously reported estimates were typically for the broad footprint of the geothermal systems, rather than the localized upflow zones of those systems, thereby providing insight into the potential salinity contributions from the host systems. These estimates were derived by using heat-flow modeling techniques [6,12,13]. A typical workflow for obtaining these estimates included estimating heat flow from temperature profiles, constructing contour maps of heat flow from those results, integrating a total heat flux by using the newly constructed map, and subtracting off an assumed background heat flux to compute an amount of excess energy flux (energy flux less background) at the site of interest. That excess energy flux was then assumed to be a result of advection and was used to estimate a corresponding required volumetric groundwater flux to account for the estimated excess energy flux. A detailed mathematical description of this type of modeling is provided in Snyder (1986) [12]. This approach has its limitations, namely that it is tied directly to contoured maps of heat flow that may change appreciably based on contouring techniques and data coverage. Additional uncertainty comes from the common assumption of a reservoir temperature when converting excess energy flux to volumetric groundwater flux; this value is usually conservatively selected as the maximum measured temperature in any given area, which can lead to the underestimation of volumetric groundwater flux. Generally, heat flow modeling techniques are based on fundamental energy balance relations and provide a means to practically estimate geothermal groundwater flux over large areas.

4. Results

4.1. Classifications

Figure 3 shows the spatial distribution of profile classifications. Profiles with upflow curvature (eight profiles) were identified in the eastern portion of the study area just south of Las Cruces and along the East Potrillo Mountains in the southwestern part of the study area. Upflow in the east is associated with the East Mesa geothermal system and the upflow profiles are within the developed portion of this system near Tortugas Mountain. Southwestern upflow is associated with the undeveloped East Potrillo geothermal system, where two upflow zones were identified along the east side of the East Potrillo Mountains. The northernmost of these two upflow zones, located about 12 km (km) north of the main

East Potrillo upflow zone, has not been extensively studied by previous researchers because of its relatively low heat flow [12]. Nevertheless, profile curvature in this area indicated upwelling groundwater, albeit at lower temperatures relative to the southern portion of the geothermal system.

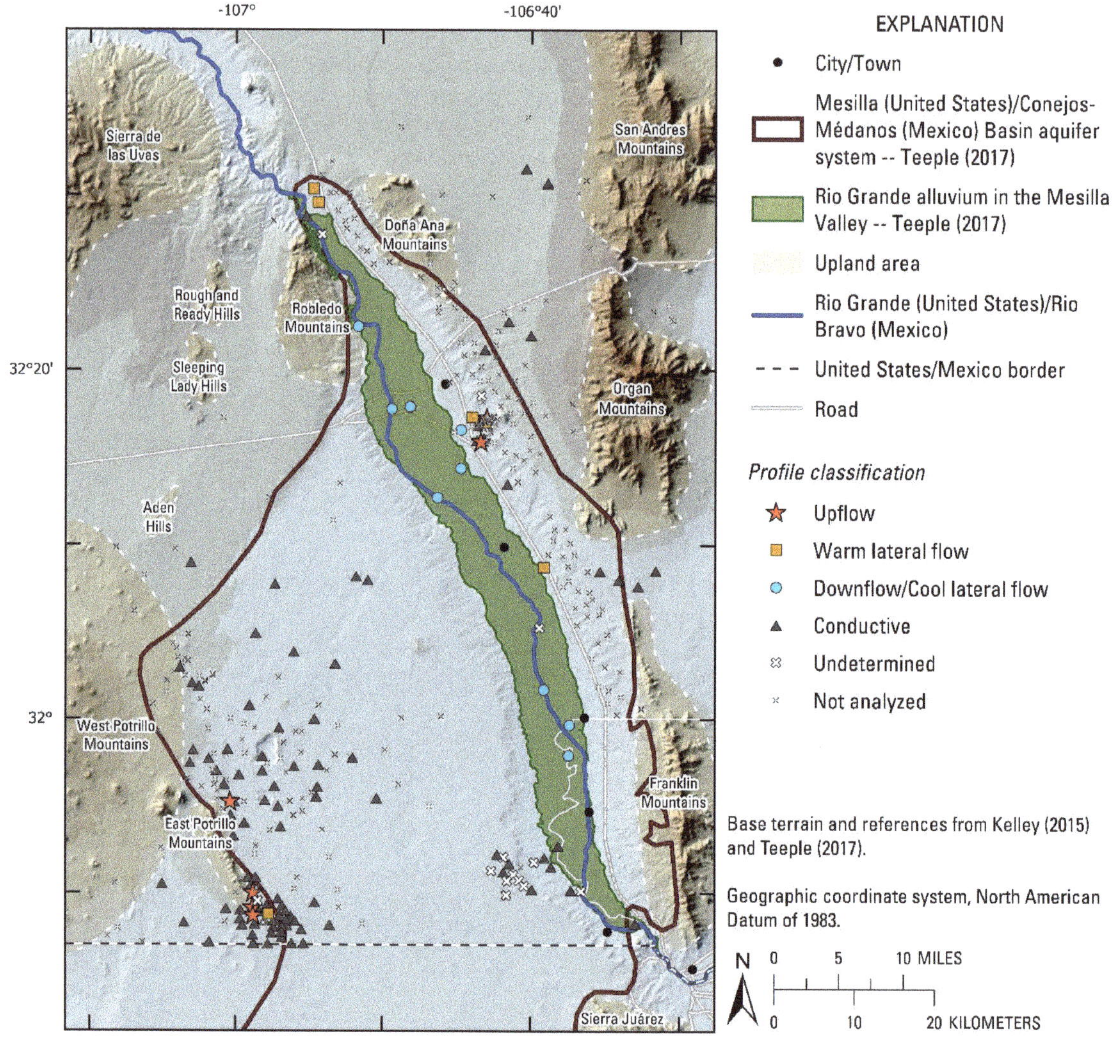

Figure 3. Spatial distribution of profile classifications [21,22].

Profiles with curvature indicating warm lateral flow (seven profiles) were found within the East Potrillo and East Mesa geothermal systems. Interestingly, lateral flow profiles were not identified near the northern East Potrillo upflow zone, thereby indicating relatively slow horizontal groundwater flow rates within the aquifer system. An isolated warm lateral flow profile was identified about 17 km south of the developed East Mesa upflow zone. Measured saturated geothermal gradients (rate of temperature change with depth) were very high (about 125 °C/km) above the horizontal flow horizon at this site relative to typical conductive gradients in the study area (around 35 °C/km), thereby indicating a proximity to warm upwelling groundwater [42]. The precise location of the upflow zone is unknown due to limited data coverage, but it is within the extensive footprint of the East Mesa geothermal system and therefore likely has similar groundwater chemistry (TDS around 1800 mg/L) [15]. Warm lateral flow profiles were also associated with the Radium Springs geothermal system in the northern part of the study area near the RG-LB stream gage (Figure 1). Although no upflow profiles were measured, these warm lateral flow profiles are certainly in the vicinity of upwelling geothermal fluids with appreciable salinity (likely around 3650 mg/L based on produced water TDS at Radium Springs).

Downflow/cool lateral flow profiles (nine profiles) were all within or near the Mesilla Valley. This indicates surface water recharge and cool lateral groundwater flow within the permeable Rio Grande alluvium in the valley. This finding also agrees with previous work that indicated negligible recharge outside of the Mesilla Valley because of the depths to groundwater, effective water consumption by desert vegetation, and the presence of caliche [17,20,29].

The remaining profiles were either linear (conductive, 101 profiles) with little evidence of advective disturbance, too difficult to confidently classify (undetermined, 13 profiles), or simply not further analyzed because of insufficient data below the water table or surficial zone or inadequate data to estimate the depth to the water table (not analyzed, 241 profiles).

Overall, these results indicate the presence of three primary upflow zones and at least two more isolated upflow zones in the study area, all of which are associated with the known geothermal systems in the region.

4.2. Flux Estimates

Several interrelated lines of data were used to estimate salinity fluxes from identified upflow zones. Insets of the regions with upflow profiles are presented in Figure 4, with corresponding thermal cross sections provided in Figure 5, and vertical heat transport analytical solution fits shown in Figure 6. A summary of the flux estimates and input parameters is provided in Table 1.

4.2.1. East Mesa Upflow

The main East Mesa upflow zone was indicated by three upflow profiles in proximity to one another, while an additional localized upflow zone was denoted by an isolated fourth upflow profile about 2 km to the southwest (Figure 4B). Thermal cross-section A-A′ clearly shows a zone of high temperatures that are associated with the upflow profiles at a horizontal distance from A of around 2000 m (Figure 5A). Measured temperatures were cooler to the west where profile classifications indicated lateral flow of upwelling groundwater. The upflow profiles were measured in the Santa Fe Group sediments that overlie the basement, which has been offset locally by the Mesilla Valley fault zone. This fault zone, and the resulting enhanced permeability and irregular basement geometry, are no doubt key hydrogeologic controls on the location of the main upflow zone. Elevated temperatures on the east side of Tortugas Mountain indicated the presence of an additional upflow zone or continuation of the main upflow zone beneath Tortugas Mountain, though this remains uncertain due to the scarcity of deep temperature measurements near Tortugas Mountain (Figure 5A). A similar ambiguity exists in the B-B′ thermal cross section where only shallow temperature profiles separated the main upflow zone from the isolated upflow profile (Figure 5B). The isolated upflow profile at the southern end of B-B′ was surrounded by elevated temperatures, but they were not as high as those at the main upflow zone, thereby making it unknown how far the main upflow zone extends. The main upflow zone area was therefore conservatively estimated to be 53,900 square meters (m^2). This estimate ignored the isolated upflow zone because a continuous connection or corresponding area could not be confidently estimated.

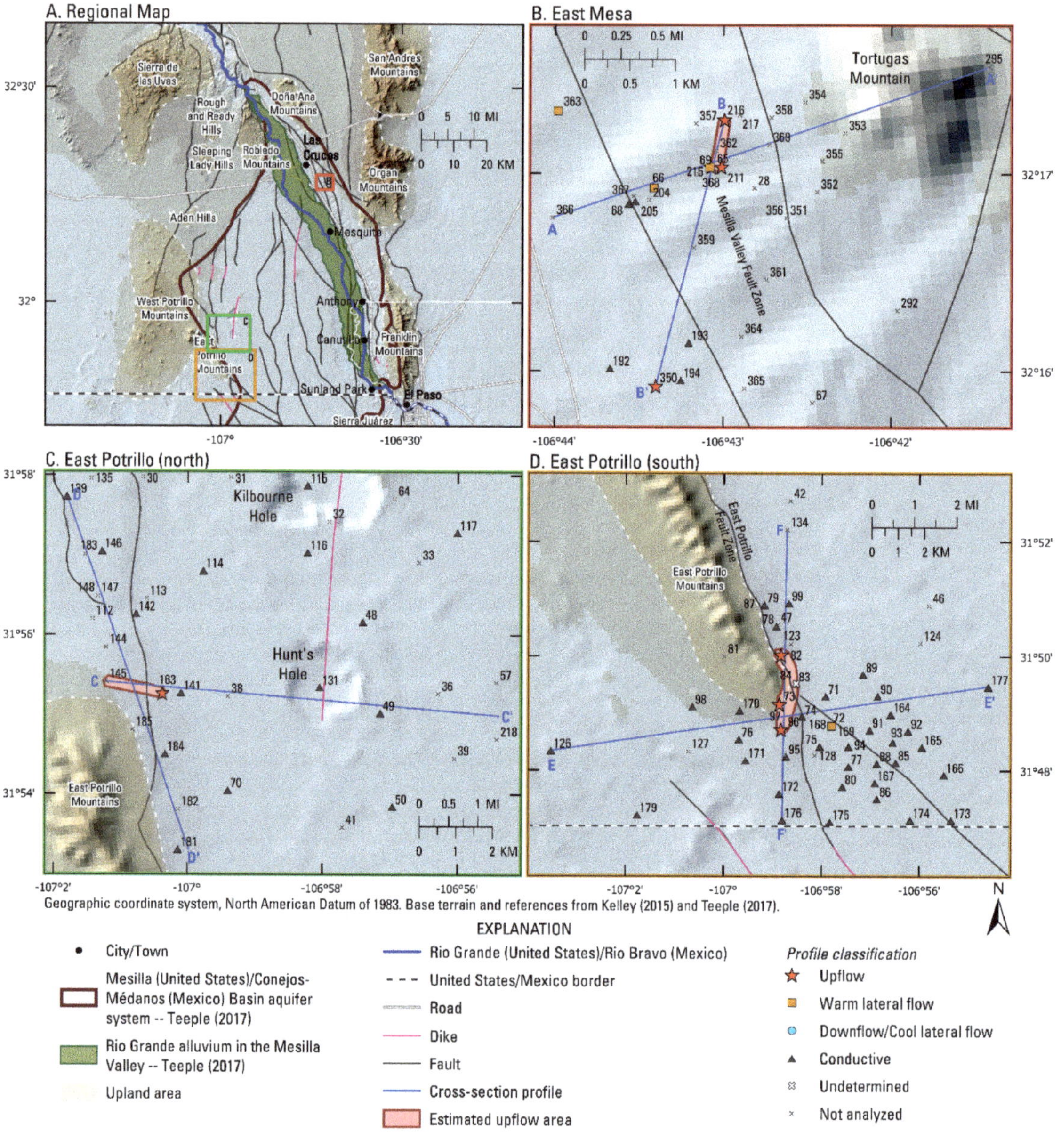

Figure 4. Areal estimates of upflow and relation of faults and dikes to profile classifications for the East Mesa (**B**), East Potrillo (north) (**C**) and East Portrillo (south) (**D**) upflow zones [21,22]. Extents of the insets are given on the regional map inset (**A**). Faults and dikes are from Sweetkind (2017) [27]. Profile locations are labeled with their profile ID number from Pepin et al. (2019) [18].

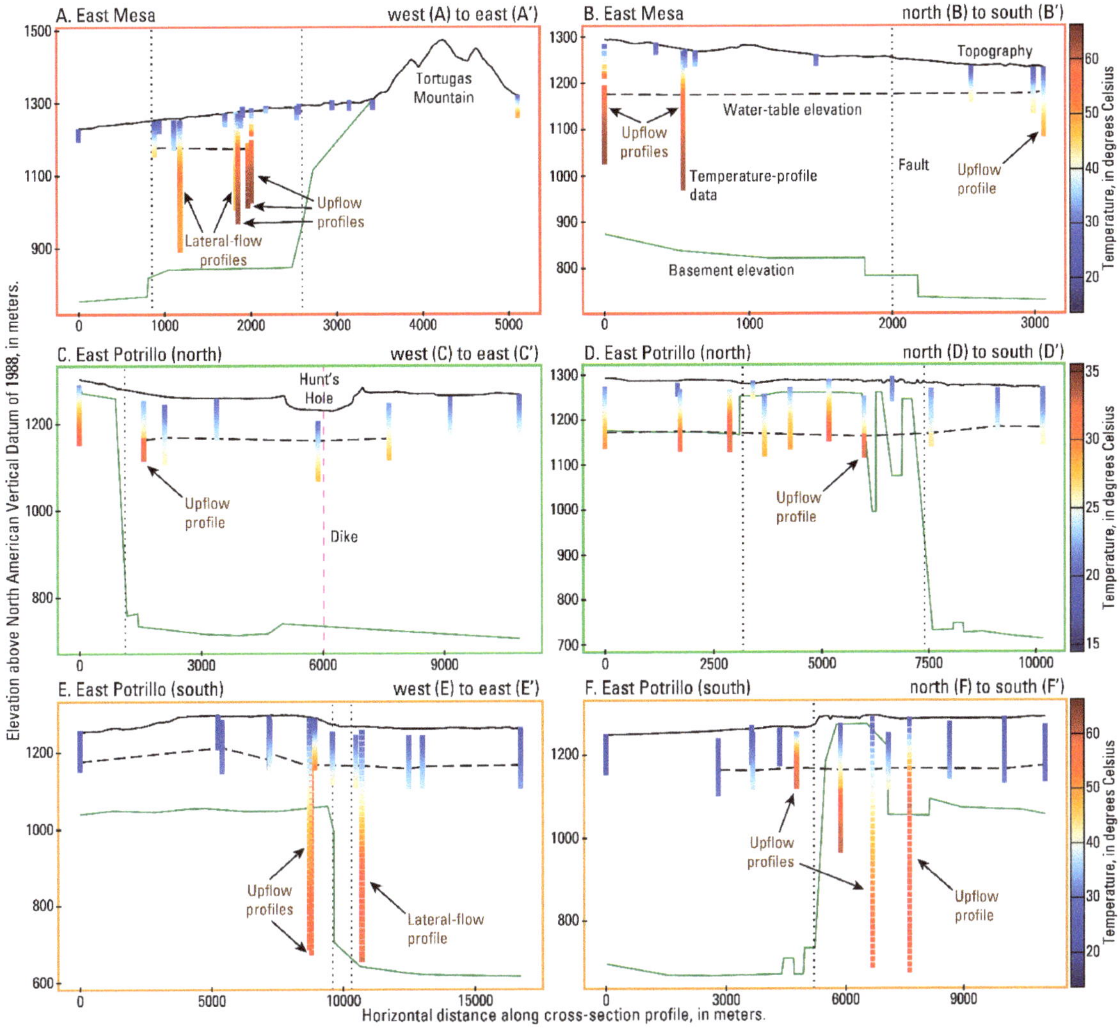

Figure 5. East–west and north–south thermal cross sections for the East Mesa (**A**,**B**), East Potrillo (north) (**C**,**D**) and East Potrillo (south) (**E**,**F**) upflow zones. These plots include temperature observations overlain onto basement stratigraphy, dikes, and faults from Sweetkind (2017) [27]. Additionally, topography from a USGS 1/3 arc-second (about 10 m) digital elevation model [41] is shown along with water table elevations estimated from the well records, smoothed temperature profiles and their derivatives, and nearby USGS water level data [23]. The temperature scale differs for the East Potrillo (north) cross sections (**C**,**D**) relative to the other cross sections. Surface projections are used to plot dikes and faults and their corresponding dips are not depicted.

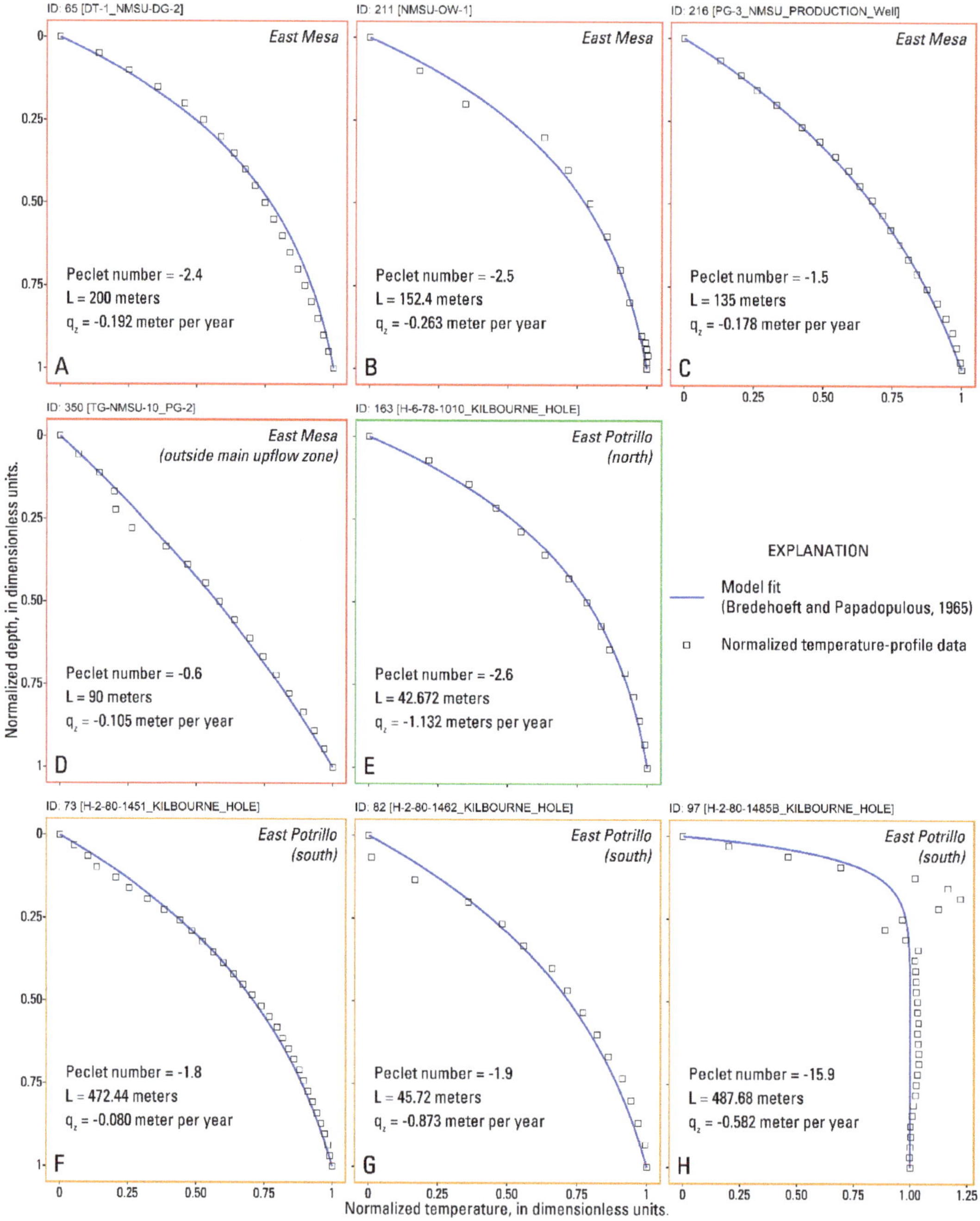

Figure 6. Normalized temperature profiles and their best-fit Bredehoeft and Papadopulos (1965) analytical solution [38] for profiles associated with the East Mesa (**A–D**), East Potrillo (north) (**E**) and East Potrillo (south) (**F–H**) upflow zones. Best-fit Peclet numbers, as determined from root-mean-squared error (RMSE) minimization, along with the length over which temperatures were analyzed (L) and estimated vertical specific discharge (q_Z) are displayed on each plot for reference. The Peclet

number and L are inversely related in equation 2, meaning that high Peclet numbers across small depth intervals will maximize vertical flux rates. Negative Peclet numbers and vertical specific discharge values correspond to upflow curvature. Precision of displayed values is for research reproducibility purposes and does not reflect value uncertainty. Profiles are labeled with their profile ID number from Pepin et al. (2019) [18].

Table 1. Summary of flux estimates and input parameters used in modeling. Fluid specific heat was specified to be 4180 joules per kilogram per degree Celsius at all locations in the modeling. Modeled values for the thickness over which temperature data were analyzed (L) and best-fit Peclet numbers (Pe) are provided in Figure 6. Salinities of 1800 mg/L and 4823 mg/L were used to estimate salinity fluxes for the East Mesa and East Potrillo regions, respectively. Reported precision of tabulated values is for research reproducibility purposes and does not reflect value uncertainty. (Column headings and abbreviations: Region = region of upflow and associated geothermal system; Subregion = localized area within the larger upflow region; ID = temperature profile identification number from Pepin et al. (2019) [18]; K_e = effective thermal conductivity in watts per meter per degree Celsius; n = porosity in dimensionless units; ρ_f = fluid density in kilograms per cubic meter; A = upflow area in square meters; q_z = vertical specific discharge in meters per year; Q_z = volumetric vertical specific discharge in acre-feet per year; J_{TDS} is salinity flux in tons of dissolved solids per year; N/A = not available.)

Region	Subregion	ID	K_e	n	ρ_f	A	q_z	Q_z	J_{TDS}
East Mesa	Main	65	2.083	0.125	983.31	53,900	−0.192	8.4	21
	Main	211	2.084	0.125	981.86	53,900	−0.263	11.5	28
	Main	216	2.084	0.125	982.59	53,900	−0.178	7.8	19
	Isolated	350	2.076	0.125	991.05	N/A [1]	−0.105	N/A [1]	N/A [1]
East Potrillo	North	163	2.45	0.25	995.01	361,700	−1.132	332	2177
	South	73	2.74	0.25	988.59	1,357,700	−0.080	88	575
	South	82	2.74	0.20	983.98	1,357,700	−0.873	961	6302
	South	97	2.33	0.20	984.24	1,357,700	−0.582	641	4203

[1] Not enough proximal temperature data to confidently estimate value.

The data indicate the potential for a much larger area of upflow, so the areal estimate, and the corresponding fluxes, are considered minimum values for the main upflow zone. Application of the Bredehoeft and Papadopulos (1965) 1D vertical heat transport analytical solution [38] to each profile yielded a range in vertical specific discharge rates at the main upflow zone of −0.178 to −0.263 m per year (m/y), where the negative sign denotes upflow (Figure 6A–C). Computed upflow rates for the isolated upflow profile were lower at −0.105 m/y (Figure 6D), which may explain why associated temperatures were cooler at this location, because slower upflow rates typically yield more conductive cooling during groundwater ascent. Multiplying vertical flow rates from the main discharge zone by its estimated area provided a volumetric flux range of 4.8 to 7.1 gallons per minute (gpm), or 7.8 to 11.5 acre-feet per year (afpy). Coupling this vertical flux with the typical groundwater salinity of the East Mesa geothermal system (1800 mg/L) yielded an estimated salinity flux of 19 to 28 tons of dissolved solids per year (t/y).

Previous researchers using heat flow modeling techniques have estimated that total groundwater flux from the entire East Mesa geothermal system, rather than its localized upflow zones estimated here, exceeds 15,000 afpy, with between 1225 and 1780 afpy coming from the region surrounding Tortugas Mountain [6,13,32]. Coupling these previous groundwater flux estimates with the typical salinity of the East Mesa waters (1800 mg/L) yielded an estimated salinity flux of 36,713 t/y for the entire East Mesa geothermal system and 3000 to 4362 t/y for the Tortugas Mountain region. Each of these salinity flux estimates greatly exceed the estimated flux range for the main localized upflow zone. This indicates that the diffuse and more spatially distributed salinity flux from this system is substantially higher than that of its localized upflow zones. The large difference between the estimated salinity flux from the localized upflow zone near Tortugas Mountain (19 to 28 t/y) and the estimates for the more extensive Tortugas Mountain region (3000 to 4362 t/y) further highlights this concept and indicates that additional deep temperature data could be useful in identification of additional upflow profiles in the Tortugas Mountain area.

4.2.2. East Potrillo Upflow

The main East Potrillo upflow zone (East Potrillo (south)) was indicated by three upflow profiles located near each other, whereas the more isolated East Potrillo upflow zone (East Potrillo (north)) had just one upflow profile (Figure 4C,D). Thermal cross sections of the northern upflow zone indicated upflow along the East Potrillo fault zone with likely lateral flow to the east, as evidenced by the temperature distribution even though no lateral flow profiles were identified (Figure 5C,D). The lack of lateral flow curvature to the east indicates that groundwater flow rates may slow once the waters enter the shallow aquifer system. Elevated temperatures to the north of the northern upflow zone, indicated around 2500 m of horizontal distance on Figure 5D (profile ID 148), suggested the probable presence of an additional upflow zone, although no upflow profiles were observed. Thermal cross sections of the southern East Potrillo upflow zone (Figure 5E,F) showed higher temperatures than the northern upflow zone, with upflow along the East Potrillo fault zone (note that the temperature scales differ between Figure 5C–F). Upflow profiles were spatially distributed in a north–south trend with a lateral flow profile indicating eastward groundwater flow of upwelling waters. In addition to the clear association of both upflow zones with the East Potrillo fault zone, upflow profiles were also associated with bedrock highs (Figure 5C–F). Like the main East Mesa upflow zone, faulting and resulting enhanced permeability and bedrock geometry certainly play strong roles in the location of these upflow zones. The northern East Potrillo upflow area was estimated to be 361,700 m^2, whereas the southern upflow zone was estimated at 1,357,700 m^2. Estimated areas were conservatively estimated to avoid overestimation of salinity flux.

Areas associated with both East Potrillo upflow zones were much greater than the area of the main East Mesa upflow zone, which resulted in substantially larger associated fluxes. The 1D vertical groundwater flux estimate for the northern upflow zone was −1.132 m/y (Figure 6E), while estimates for profiles in the southern upflow zone ranged from −0.080 to −0.873 m/y (Figure 6F–H). One profile (ID 97) showed appreciable warm lateral flow effects within the shallowest quarter of the profile that were essentially ignored during flux estimation (Figure 6H); well records showed drillers lost drilling fluid circulation in the vicinity of the lateral flow effects, thereby indicating fracture-controlled lateral flow may be important here. As a result of the lateral flow effects, this profile had greater uncertainty in the flux estimation, but the computed value (−0.582 m/y) was still bracketed by the overall flux range for the upflow zone. Multiplying the 1D flux estimated by the upflow areas yielded volumetric flux estimates of 206 gpm for the northern zone and 54 to 596 gpm for the southern zone. These estimates corresponded to 332 afpy for the northern zone and 87 to 961 afpy for the southern zone. Coupling these groundwater fluxes with the estimated groundwater salinity of the East Potrillo geothermal system (4823 mg/L) yielded a salinity flux of 2177 t/y for the northern zone and 575 to 6302 t/y for the southern zone, or a combined total of 2752 to 8479 t/y.

Snyder (1986) estimated total groundwater flux from the southern portion of the geothermal system to be 970 afpy, which agrees well with the upper estimate from this study of 961 afpy [12]. Snyder's flux estimate corresponded to a salinity flux of 6347 t/y. While this estimate ignored contributions from the northern upflow zone, it indicates that most upwelling salinity at the East Potrillo geothermal system is likely associated with somewhat localized upflow zones rather than broad diffuse upflow. This is in contrast to the East Mesa salinity contributions, which are likely much more distributed throughout the associated geothermal system.

5. Discussion

Salinity fluxes from geothermal systems within the study area could account for a notable amount of Rio Grande salinity if the geothermal waters eventually discharged into the Rio Grande. From 2009 through 2013, the Rio Grande, on average, delivered about 205,000 t/y to the Mesilla outlet near the El Paso stream gage (RG-EP; Figure 1) [4]. Assuming all geothermal salinity contributions are more or less constant through time and eventually make their way to the Rio Grande, the 36,713 t/y from the East Mesa geothermal system as a whole could account for around 18% of average annual Rio Grande salinity, while the 8479 t/y from the East Potrillo geothermal system may contribute about 4% of average annual Rio Grande salinity. Identified local upflow zones associated with these geothermal systems expectedly could contribute less salinity, with the main identified East Mesa upflow zone potentially accounting for only about 0.01%, the northern East Potrillo zone contributing around 1%, and the southern East Potrillo zone adding 0.3 to 3% of Rio Grande salinity. The localized East Mesa upflow zone was located within the more extensive Tortugas Mountain region that had an estimated salinity flux of 3000 to 4362 t/y, which would account for about 1.5 to 2% of Rio Grande salinity. The Basin is a dynamic groundwater region, thereby making it uncertain whether these solids do indeed eventually make their way to the Rio Grande; potential flowpaths are considered later in this discussion section.

These proportions could be exacerbated in periods of low streamflow due to reduced dilution. In these periods, geothermal inputs have the potential to account for a much larger percentage of Rio Grande salinity. For instance, in 2013 the Rio Grande salinity delivery to the RG-EP stream gage at the basin outlet lessened to around 55,000 tons of dissolved solids because of reduced upstream reservoir releases [4]. Geothermal salinity contributions in that particular year could have amounted to 67% from the East Mesa geothermal system, with 5.5% to 8% from the Tortugas Mountain region, and about 15.5% from the East Potrillo geothermal system. Additional salinity could be contributed to the Rio Grande from the Radium Springs geothermal system within the study area. Previously published groundwater flux estimates were unavailable at the time of this study and only warm lateral flow profiles were identified near this geothermal system due to data coverage limitations. This system is known to produce waters with salinities around 3650 mg/L and could be an additional noteworthy natural salinity source that was not accounted for in this work. Overall, this study shows the appreciable potential geothermal salinity contributions to the Rio Grande, especially during periods of low streamflow.

Previously published water level mapping provides further insight into the regions influenced by identified upwelling and laterally flowing geothermal waters. Figure 7 presents upflow and warm lateral flow profile locations with interpolated groundwater elevations from measurements made in 2010 in the shallow aquifer system [20]. Warm lateral flow associated with the Radium Springs geothermal system near the basin inlet is predicted to flow south towards the Rio Grande. Similarly, warm groundwater associated with an isolated lateral flow profile near Mesquite within the footprint of the East Mesa geothermal system is projected to flow to the southwest towards the Rio Grande. Groundwater upwelling near Tortugas Mountain is thought to follow a west-southwest trajectory towards the Rio Grande, with evidence of some upwelling groundwater laterally flowing to the northwest where it is intercepted by wells near Las Cruces. Upflow along the East Potrillo Mountains is predicted to gradually flow eastward toward the Rio Grande and southern Mesilla with a portion of the flow crossing the United States/Mexico international border before being intercepted by municipal wells in the Conejos-Médanos; water quality data was not available for that particular portion of the municipal wellfield that would have allowed further evaluation. Generally, water level mapping underscores the likelihood that upwelling geothermal groundwater affects the Rio Grande and indicates that groundwater supplies in Las Cruces, the southern Mesilla, and municipal production in the northern Conejos-Médanos could be adversely affected by these geothermal systems.

Where does this upwelling brackish groundwater originate? Most geothermal systems in New Mexico, with exception of the active Valles Caldera volcanic system in northern New Mexico, are thought to result from amagmatic (non-magmatic) heating of infiltrating recharge [13,43]. A common multi-step conceptual model is as follows:

1. Upland precipitation infiltrates;
2. Infiltrated groundwater is heated by the Earth's natural geothermal gradient as it flows deeper within the Earth's crust;
3. Salinity of heated waters increases as the groundwater interacts with sediments and rocks along its flowpath;
4. Resulting brackish waters discharge at regional topographic lows, through zones of enhanced permeability (commonly caused by faults), and/or through gaps in overlying lower-permeability stratigraphic layers.

Previous researchers have linked the East Mesa and Radium Springs geothermal systems with geothermal upwelling within fault zones along uplifted bedrock and fractured igneous intrusions, respectively [13–16]. This agrees well with the strong correlation of identified upflow zones with fault zones and uplifted bedrock and further supports the conceptual model presented above. Produced water temperatures from the East Mesa geothermal system are typically around 64 °C, whereas Radium Springs geothermal system temperatures are commonly higher, at approximately 99 °C [15,16]. By assuming an average annual surface temperature of 17 °C and background geothermal gradient of 35 °C/km, measured groundwater temperatures indicate that these upwelling waters ascend from depths of at least 1.3 and 2.3 km, respectively [42]. These are minimum circulation depths because conductive cooling and mixing with shallower cool waters during ascent are likely to occur but are not considered in this study. Maximum measured temperatures in the East Potrillo geothermal system were around 60 °C at depths of less than 215 m, indicating a minimum circulation depth of about 1 km. Geothermal recharge sources are currently unknown but could be evaluated in future work. More specifically, efforts using advanced modeling techniques and geochemical and isotopic tracers could further interrogate geothermal flowpaths, recharge locations, and geothermal groundwater residence times to provide a more complete conceptual model for these systems. Overall, it can confidently be stated that these geothermal waters upwell from depths exceeding 1 km, and in some cases 2 km, along preferential flowpaths caused by fault zones that affect subsurface stratigraphy and permeability.

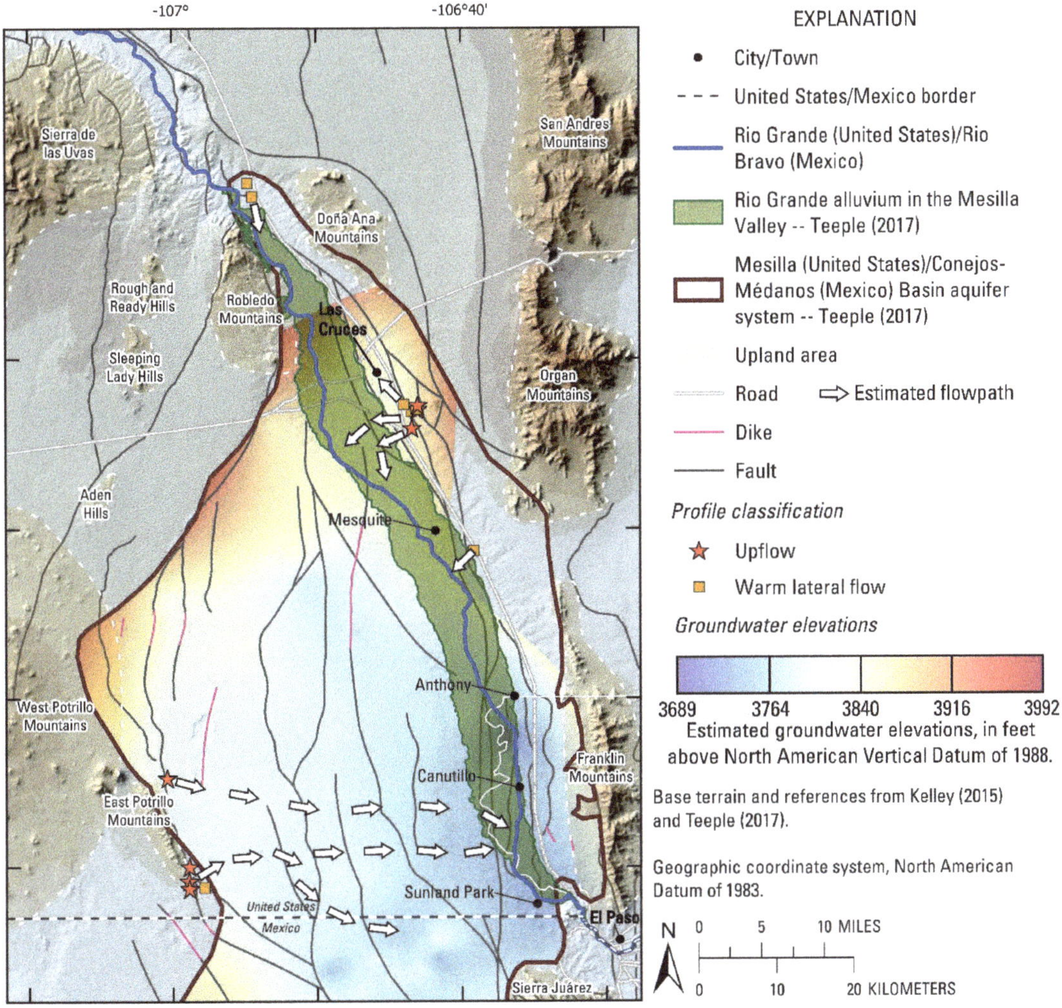

Figure 7. Anticipated flow patterns of upwelling and warm lateral flowing groundwater, as informed by groundwater-elevation mapping from Robertson et al. (2021) [20–22]. Estimated flowpath arrows are oriented perpendicularly to groundwater elevation contours with flow directed down hydraulic gradient. Groundwater elevations depict groundwater flow towards the Rio Grande and southern Mesilla, along with flow intercepted by groundwater wells in Las Cruces and northern Chihuahua. Estimated groundwater elevations vary through time, thereby affecting the estimated flowpaths of upwelling waters through time as well. Generally, upflow zones that are nearest the Mesilla Valley are the most likely to be affected by changing groundwater conditions and their associated flux estimates and flowpaths are therefore more uncertain.

Geothermal characterization and exploration researchers have long established the importance of fault zones, especially fault intersections, on controlling the locations of upwelling geothermal fluids [43,44]. These flow dynamics are common in extensional physiographic provinces, such as the Rio Grande rift and Basin and Range of the western United States [43–45]. Geothermal developers rely on these upwelling fluids being hot, but upflow zones with slightly elevated or background temperatures can still contribute substantial salinity to the shallow aquifer system—similar to the northern East Potrillo upflow zone identified in this study. Relying on this concept and others borrowed from the geothermal research field could prove to be an effective means of locating additional upflow zones within the study area. For example, fault and subsurface stratigraphy could be used to locate areas of potential upflow, particularly in areas with limited thermal data coverage, which could then be further evaluated with targeted data collection and modeling. Identification of upflow zones in this manner could be a cost-effective way to further inform stakeholders on salinity sources and possible mitigation strategies.

This study demonstrates the utility of using heat as a groundwater tracer to further understand sources of salinity and their associated fluxes, along with regional and local groundwater movement. Legacy thermal datasets exist in many areas of the world because of energy exploration and development and new data collection is rather straightforward. These data and the strong and well-understood relation between fluid flow and heat argues for the more common inclusion of heat transport and thermal data calibration techniques in groundwater modeling efforts to improve model accuracy. Researchers could consider evaluating thermal data when assessing groundwater flow patterns and salinity sources on regional and local scales.

Additional future research to further evaluate geothermal salinity contributions and their effects might include mass balance streamgaging; multi-dimensional solute, heat, and mass transport modeling; and additional thermal data collection. Mass balance streamgaging techniques could most readily be used to estimate flux contributions to the Rio Grande from the Radium Springs geothermal system. This would entail measuring salinity loads in conjunction with streamflow upstream and downstream from the Radium Springs geothermal system. Discharge and corresponding salinity contributions from the geothermal system, assuming the system is the dominant salinity source through that river section, could then be back calculated by coupling the values with Radium Springs geothermal groundwater salinity. Multi-dimensional modeling that incorporates temperature and salinity could be used to predict upwelling flux rates and flow patterns more accurately. Lastly, the collection of more thermal data below the surficial zone and water table, particularly within the footprint of the East Mesa geothermal system, could significantly reduce the uncertainty associated with estimated geothermal salinity fluxes. This list of possible study directions is not comprehensive but provides practical avenues that could build upon this study.

Limitations

The approaches used herein have limitations. The most likely complicating factors are related to the following:

1. Sparse data availability affecting the ability to comprehensively identify upflow zones and more accurately estimate upflow areas;
2. Misidentification of profile curvature. For example, upflow profile curvature can look very similar to the upper half of profiles showing warm lateral flow (see Figure 2);
3. Coupling computed 1D vertical fluxes to entire areas of upflow when estimating volumetric fluxes;
4. Potential lateral flow effects in profiles with dominant upflow curvature;
5. Non-steady state conditions of groundwater flow and temperature in the subsurface over the period of investigation (1972–2018).

This study is not intended to be comprehensive but is instructive about the locations of upwelling geothermal waters and their associated salinity fluxes within the study area. The above limitations highlight opportunities to improve upon this research in the future, particularly with additional data collection and more advanced modeling tools that can better represent transience and heterogeneity.

Addressing potential non-steady state conditions of groundwater flow and temperature over the period of investigation (1972–2018) is one of the biggest opportunities for improvement in this work; this includes evaluating the assumption that identified upflow zones have been active for a long enough duration to affect Rio Grande salinity. While geothermal discharge rates are commonly consistent for long periods of time in absence of geothermal development, the Mesilla is a developed and dynamic groundwater region. All but one of the upflow profiles were measured over a timespan of just 14 months (11 March 1979 through 10 May 1980), which favors consistent conditions during measurement; one upflow profile had an unknown collection date but was most likely measured in the late 1970s or early 1980s. It is important to keep in mind that the groundwater elevations shown in Figure 7 vary through time, thereby affecting the estimated flowpaths of upwelling waters through time as well. Generally, upflow zones that are nearest to the Mesilla Valley are the most likely to be affected by changing groundwater conditions, and their associated flux estimates and flowpaths are therefore more uncertain. Regions with small hydraulic gradients (i.e., slowly moving groundwater; Figure 7), such as the Mesilla interior, are less likely to have had time to transport salinity to the Rio Grande on short timescales. Additional thermal, geochemical, and groundwater elevation data could be collected to assess the consistency of the estimated fluxes and their corresponding flow history.

6. Conclusions

Evaluation of previously published flux estimates and 379 temperature profiles measured between 1972 and 2018 show the appreciable potential salinity contributions from upwelling geothermal waters to the shallow aquifer system and the Rio Grande within the Mesilla (United States)/Conejos-Médanos (Mexico) Basin. Upflow and/or warm lateral flow profiles were identified within the region's three known geothermal systems (Radium Springs, East Mesa, and East Potrillo).

Salinity flux analyses indicate that the East Mesa geothermal system may contribute about 36,700 tons of dissolved solids per year (t/y) to the shallow aquifer system, whereas the East Potrillo geothermal system may add around an additional 8500 t/y. Assuming these fluxes are steady through time and eventually enter the Rio Grande, these systems could account for a combined 22% (East Mesa = 18%, East Potrillo = 4%) of typical average annual Rio Grande salinity. These salinity proportions can be much greater in times of low streamflow and additional salinity contributions likely come from the Radium Springs geothermal system. Radium Springs flux estimates were not feasible in this study due to data coverage limitations but could be pursued in the future.

Regional water levels mapped in 2010 indicate upwelling brackish waters flow towards the Rio Grande and southern part of the Mesilla portion of the Basin, with some water intercepted by wells in Las Cruces and northern Chihuahua. These waters upwell from depths greater than 1 km with upflow being focused along fault zones, uplifted bedrock, and/or fractured igneous intrusions. This understanding may be used to guide future data collection efforts aimed at identifying additional upflow zones, particularly in areas that have limited thermal data coverage but adequate knowledge of faults and stratigraphy.

This work demonstrates the utility of using heat to identify regional and local sources of salinity and their associated fluxes and highlights the benefits of using thermal data in hydrologic studies. This effort could be improved upon by future research focused on improving data coverage and reducing uncertainties associated with transience and heterogeneity in the aquifer system. Overall, the results presented herein further inform stakeholders on the presence of several brackish upflow zones that could notably degrade the quality of international water supplies in this developed drought-stricken region.

Supplementary Materials: The following are available online at https://www.mdpi.com/article/10.3390/w14010033/s1, Figure S1: Plots of all 379 temperature profiles; Figure S2: Plots of smoothed profiles and derivatives for analyzed measurements made below estimated water table elevations; Table S1: Profile analysis details including water table depth estimates, curvature classifications, and analysis remarks.

Author Contributions: Conceptualization, J.D.P., A.J.R. and S.A.K.; methodology, J.D.P.; software, J.D.P.; validation, J.D.P.; formal analysis, J.D.P.; investigation, J.D.P., A.J.R. and S.A.K.; resources, J.D.P., A.J.R. and S.A.K.; data curation, S.A.K., J.D.P. and A.J.R.; writing—original draft preparation, J.D.P.; writing—review and editing, J.D.P., A.J.R. and S.A.K.; visualization, J.D.P.; supervision, A.J.R. and J.D.P.; project administration, A.J.R. and J.D.P.; funding acquisition, A.J.R. and S.A.K. All authors have read and agreed to the published version of the manuscript.

Funding: This work was partially funded by the U.S. Geological Survey (funding authorized by P.L. 109–448) for the Transboundary Aquifer Assessment Program. Shari A. Kelley received U.S. Department of Energy funding (NM-EE0002850) through the National Geothermal Database project to digitize and organize the thermal data used in this research.

Institutional Review Board Statement: Not applicable.

Informed Consent Statement: Not applicable.

Data Availability Statement: The data presented in this study are openly available in ScienceBase at https://doi.org/10.5066/P9FQ9BEN (accessed on 11 June 2019) [18]. USGS site identifiers may be used to extract additional data from the U.S. Geological Survey National Water Information System (NWIS) database [23].

Acknowledgments: The authors extend their gratitude to James Witcher for providing many of the original data documents that were then digitized and used in this research. This product has been peer reviewed and approved for publication consistent with U.S. Geological Survey Fundamental Science Practices (https://pubs.usgs.gov/circ/1367/ (accessed on 13 December 2021)). Any use of trade, firm, or product names is for descriptive purposes only and does not imply endorsement by the U.S. Government. Writings prepared by U.S. Government employees as part of their official duties, including this paper, cannot be copyrighted and are in the public domain.

Conflicts of Interest: The authors declare no conflict of interest. The funders had no role in the design of the study; in the collection, analyses, or interpretation of data; in the writing of the manuscript, or in the decision to publish the results.

References

1. National Oceanic and Atmospheric Administration (NOAA)—National Centers for Environmental Information—North American Drought Monitor. Available online: https://www.ncdc.noaa.gov/temp-and-precip/drought/nadm/ (accessed on 29 April 2021).
2. Alley, W.M. *Five-Year Interim, Report of the United States—Mexico Transboundary Aquifer Assessment Program: 2007–2012*; U.S. Geological Survey Open-File Report; U.S. Department of the Interior: Washington, DC, USA, 2013; 31p. [CrossRef]
3. Hogan, J.F.; Phillips, F.M.; Mills, S.K.; Hendrickx, J.M.H.; Ruiz, J.; Chesley, J.T.; Asmerom, Y. Geologic origins of salinization in semi-arid river: The role of sedimentary basin brines. *Geology* **2007**, *35*, 1063–1066. [CrossRef]
4. Driscoll, J.M.; Sherson, L.R. *Variability of Surface-Water Quantity and Quality and Shallow Groundwater Levels and Quality within the Rio Grande Project Area, New Mexico and Texas, 2009–2013*; U.S. Geological Survey Scientific Investigations Report; USGS: Reston, VA, USA, 2016; 33p. [CrossRef]
5. Phillips, F.M.; Mills, S.; Hendrickx, M.H.; Hogan, J. Environmental tracers applied to quantifying causes of salinity in arid-region rivers: Results from the Rio Grande Basin, Southwestern USA. *Dev. Water Sci.* **2003**, *50*, 327–334. [CrossRef]
6. Witcher, J.C.; King, J.P.; Hawley, J.W.; Kennedy, J.F.; Williams, J.; Cleary, M.; Bothern, L.R. Sources of Salinity in the Rio Grande and Mesilla Basin Groundwater; New Mexico Water Resources Institute Technical Completion Report; 2004; 168p. Available online: https://nmwrri.nmsu.edu/tr-330/ (accessed on 1 May 2021).
7. Moore, S.J.; Bassett, R.L.; Liu, B.; Wolf, C.P.; Doremus, D. Geochemical tracers to evaluate hydrogeologic controls on river salinization. *Ground Water* **2008**, *46*, 489–501. [CrossRef] [PubMed]
8. Szynkiewicz, A.; Witcher, J.C.; Modelska, M.; Borrok, D.M.; Pratt, L.M. Anthropogenic sulphate loads in the Rio Grande, New Mexico (USA). *Chem. Geol.* **2011**, *283*, 194–209. [CrossRef]
9. Moyer, D.L.; Anderholm, S.K.; Hogan, J.F.; Phillips, F.M.; Hibbs, B.J.; Witcher, J.C.; Matherne, A.M.; Falk, S.E. *Knowledge and Understanding of Dissolved Solids in the Rio Grande—San Acacia, New Mexico, to Fort Quitman, Texas, and Plan for Future Studies and Monitoring*; U.S. Geological Survey Open-File Report; USGS: Reston, VA, USA, 2013; 67p. [CrossRef]
10. Hiebing, M.; Doser, D.I.; Avila, V.M.; Ma, L. Geophysical studies of fault and bedrock control on groundwater geochemistry within the southern Mesilla Basin, western Texas and southern New Mexico. *Geosphere* **2018**, *14*, 1912–1934. [CrossRef]
11. Kubicki, C.; Carroll, K.C.; Witcher, J.C.; Robertson, A. An Integrated Geochemical Approach for Defining Sources of Groundwater Salinity in the Southern Rio Grande Valley of the Mesilla Basin, New Mexico and West Texas, USA; Desalination and Water Purification Research Program—Research and Development Office Report/New Mexico Water Resources Research Institute Technical Completion Report; 2021; 82p. Available online: https://nmwrri.nmsu.edu/tr-388/ (accessed on 1 May 2021).

12. Snyder, J.T. Heat Flow in the Southern Mesilla Basin with an Analysis of the East Potrillo Geothermal System, Dona Ana County, New Mexico. Master's Thesis, New Mexico State University, Las Cruces, NM, USA, May 1986.
13. Witcher, J.C. A Geothermal Resource Data Base—New Mexico. New Mexico State University—Southwest Technology Development Institute Report; 1995; 94p. Available online: https://digital.library.unt.edu/ark:/67531/metadc665986/m2/1/high_res_d/204690.pdf (accessed on 1 May 2021).
14. Lund, J.W. Radium Hot Springs. *GHC Bull.* **2002**, *23*, 48–49. Available online: https://www.osti.gov/servlets/purl/1209492 (accessed on 1 May 2021).
15. Witcher, J.C.; Cunniff, R.A. Geothermal Energy at New Mexico State University in Las Cruces. *GHC Bull.* **2002**, *23*, 30–36. Available online: https://www.osti.gov/servlets/purl/1209492 (accessed on 1 May 2021).
16. Witcher, J.C.; Lund, J.W. Masson Radium Springs Farm. *GHC Bull.* **2002**, *23*, 42–44. Available online: https://www.osti.gov/servlets/purl/1209492 (accessed on 1 May 2021).
17. Frenzel, P.F.; Kaehler, C.A.; Anderholm, S.K. *Geohydrology and Simulation of Ground-Water Flow in the Mesilla Basin, Dona Ana County, New Mexico, and El Paso County, Texas, with a Section on Water Quality and Geochemistry*; U.S. Geological Survey Professional Paper; USGS: Reston, VA, USA, 1992. [CrossRef]
18. Pepin, J.D.; Robertson, A.J.; Kelley, S.A. *Borehole Temperature Profiles Measured in the Mesilla Basin, New Mexico during the Period 1972 through 2018*; U.S. Geological Survey Data Release; USGS: Reston, VA, USA, 2019. [CrossRef]
19. Anderson, M.P. Heat as a ground water tracer. *Ground Water* **2005**, *43*, 951–968. [CrossRef] [PubMed]
20. Robertson, A.J.; Matherne, A.M.; Pepin, J.D.; Ritchie, A.; Sweetkind, D.S.; Teeple, A.P.; Granados-Olivas, A.; Garcia Vasquez, A.C.; Carroll, K.C.; Fuchs, E.H.; et al. Mesilla/Conejos-Médanos Basin: U.S.-Mexico Transboundary Water Resources and Research Needs. *Water* **2021**. under revision.
21. Teeple, A.P. *Geophysics- and Geochemistry-Based Assessment of the Geochemical Characteristics and Groundwater-Flow System of the U.S. Part of the Mesilla Basin/Conejos-Médanos Aquifer System in Doña Ana County, New Mexico, and El Paso County, Texas, 2010–2012*; U.S. Geological Survey Scientific Investigations Report; USGS: Reston, VA, USA, 2017; 183p. [CrossRef]
22. Kelley, R. New Mexico Geothermal Play Fairway Analysis from LANL. Geothermal Data Repository. 2015. Available online: https://gdr.openei.org/submissions/597 (accessed on 29 April 2021).
23. U.S. Geological Survey [USGS]. *USGS Water Data for the Nation*; U.S. Geological Survey National Water Information System Database; USGS: Reston, VA, USA, 2021. [CrossRef]
24. Chapin, C.E.; Seager, W.R. Evolution of the Rio Grande rift in the Socorro and Las Cruces areas. In *26th Annual Fall Field Conference Guidebook*; Seager, W.R., Clemons, R.E., Callender, J.F., Eds.; New Mexico Geological Society: Socorro, NM, USA, 1975; pp. 297–321.
25. Hawley, J.W. *Guidebook to the Rio Grande Rift in New Mexico and Colorado*; New Mexico Bureau of Mines and Mineral Resources Circular; New Mexico Bureau of Mines and Mineral Resources: Socorro, NM, USA, 1978; 241p.
26. Chapin, C.E.; Cather, S.M. Tectonic setting of the axial basins of the northern and central Rio Grande rift. In *Basins of the Rio Grande Rift—Structure, Stratigraphy, and Tectonic Setting*; Keller, G.R., Cather, S.M., Eds.; Geological Society of America: Boulder, CO, USA, 1994; pp. 5–25. [CrossRef]
27. Sweetkind, D.S. *Three-Dimensional Hydrogeologic Framework Model of the Rio Grande Transboundary Region of New Mexico and Texas, USA, and Northern Chihuahua, Mexico*; U.S. Geological Survey Scientific Investigations Report; USGS: Reston, VA, USA, 2017; 49p. [CrossRef]
28. Hoffer, J.M. *Geology of Potrillo Basalt Field, South-Central New Mexico*; New Mexico Bureau of Geology and Mineral Resources Circular; New Mexico Bureau of Mines and Mineral Resources: Socorro, NM, USA, 1976; 30p.
29. Servicio Geológico Mexicano. *Hydrogeological Activities in the Conejos-Medanos/Mesilla Basin Aquifer, Chihuahua, Phase 1*; Servicio Geológico Mexicano Report; Servicio Geológico Mexicano: Pachuca, Hidalgo, 2010; Volume 1, 467p.
30. LeRoy, S.; Garfin, G. *Climate Assessment for the Southwest (CLIMAS)—The Climate of Las Cruces*; University of Arizona: Tucson, NM, USA, 2017; 24p.
31. Hanson, R.T.; Ritchie, A.B.; Boyce, S.E.; Galanter, A.E.; Ferguson, I.A.; Flint, L.E.; Flint, A.; Henson, W.R. *Rio Grande Transboundary Integrated Hydrologic Model and Water-Availability Analysis, New Mexico and Texas, United States, and Northern Chihuahua, Mexico*; U.S. Geological Survey Scientific Investigations Report; USGS: Reston, VA, USA, 2020; 186p. [CrossRef]
32. Gross, J.T.; Giordano, T.H. The Las Cruces Geothermal Field—A flow model [Abstract]. *New Mex. Geol.* **1985**, *7*, 64.
33. Sorey, M.L. Measurement of vertical groundwater velocity from temperature profiles in wells. *Water Resour. Res.* **1971**, *7*, 963–970. [CrossRef]
34. Boyle, J.M.; Saleem, Z.A. Determination of recharge rates using temperature-depth profiles in wells. *Water Resour. Res.* **1979**, *15*, 1616–1622. [CrossRef]
35. Irvine, D.J.; Cartwright, I.; Post, V.E.A.; Simmons, C.T.; Banks, E.W. Uncertainties in vertical groundwater fluxes from 1-D steady state heat transport analyses caused by heterogeneity, multidimensional flow, and climate change. *Water Resour. Res.* **2016**, *52*, 813–826. [CrossRef]
36. Wang, X. fANCOVA: Nonparametric Analysis of Covariance. R Package Version 0.5-1. 2010. Available online: https://CRAN.R-project.org/package=fANCOVA (accessed on 13 December 2021).
37. R Core Team. R: A Language and Environment for Statistical Computing. R Foundation for Statistical Computing. 2019. Available online: https://www.R-project.org/ (accessed on 13 May 2021).

38. Bredehoeft, J.D.; Papadopulos, I.S. Rates of vertical groundwater movement estimated from the Earth's thermal profiles. *Water Resour. Res.* **1965**, *1*, 325–328. [CrossRef]
39. Kell, G.S. Density, thermal expansivity, and compressibility of liquid water from 0° to 150 °C: Correlations and tables for atmospheric pressure and saturation reviewed and expressed on 1968 temperature scale. *J. Chem. Eng. Data* **1975**, *20*, 97–105. [CrossRef]
40. Ingebritsen, S.E.; Sanford, W.E.; Neuzil, C.E. *Groundwater in Geologic Processes*, 2nd ed.; Cambridge University Press: New York, NY, USA, 2006; 536p.
41. U.S. Geological Survey (USGS) 3D Elevation Program (3DEP). National Elevation Dataset (NED) Seamless Layers. Available online: http://ned.usgs.gov/about.html (accessed on 18 May 2021).
42. Reiter, M.; Eggleston, R.E.; Broadwell, B.R.; Minier, J. Estimates of terrestrial heat flow from deep petroleum tests along the Rio Grande rift in Central and Southern New Mexico. *J. Geophys. Res.* **1986**, *91*, 6225–6245. [CrossRef]
43. Witcher, J.C.; Reiter, M.; Bland, D.; Barroll, M.W. Geothermal resources in New Mexico. *New Mex. Geol.* **1992**, 14–16.
44. Faulds, J.E.; Hinz, N.H. Favorable Tectonic and Structural Settings of Geothermal Systems in the Great Basin Region, Western USA: Proxies for Discovering Blind Geothermal Systems. In Proceedings of the World Geothermal Congress, Melbourne, Australia, 19–24 April 2015; p. 6. Available online: https://www.osti.gov/servlets/purl/1724082 (accessed on 13 December 2021).
45. Faulds, J.; Coolbaugh, M.; Bouchot, V.; Moek, I.; Oguz, K. Characterizing structural controls of geothermal reservoirs in the Great Basin, USA, and Western Turkey: Developing successful exploration strategies in extended terranes. In Proceedings of the World Geothermal Congress, Bali, Indonesia, 25–30 April 2010; pp. 1–11.

water

MDPI

Article

Investigation of the Origin of Hueco Bolson and Mesilla Basin Aquifers (US and Mexico) with Isotopic Data Analysis

Ana Cristina Garcia-Vasquez [1,*], Alfredo Granados-Olivas [2], Zohrab Samani [3] and Alexander Fernald [4]

1 Department of Water Science and Management Graduate Degree Program, New Mexico State University, Las Cruces, NM 88003, USA
2 Department of Civil and Environmental Engineering, Universidad Autónoma de Ciudad Juárez, Avenida del Charro 610 Norte, C.P. Chihuahua, Ciudad Juárez 32310, Mexico; agranados@uacj.mx
3 Department of Civil Engineering, New Mexico State University, Las Cruces, NM 88003, USA; zsamani@nmsu.edu
4 New Mexico Water Resources Research Institute, New Mexico State University, Las Cruces, NM 88003, USA; afernald@nmsu.edu
* Correspondence: gava@nmsu.edu

Abstract: An important tool to identify the origin of a groundwater resource is the use of isotopic signatures. Isotopic signatures give us the age of water and provide information as to the water's origin, potential transit at geologic structures, source of salinization, and possible recharge points. The purpose of this study was to collect and analyze well samples to evaluate isotopic tracers ($\delta^{18}O$ and tritium) in the transboundary Conejos-Médanos/Mesilla aquifer located between the US and Mexico. This new analyzed information was compared with the isotopic information available in the US Mesilla and US-MX Hueco basins generated by previous works, which described the common origin of the Hueco Bolson and Mesilla Basins aquifers. This study used isotopic analysis to validate the theory of the original formation and interconnectivity of both transboundary basins. This research presents new data of $\delta^{18}O$ and tritium, and a comparison with previous published data from other workers, versus the known global meteoric water line (GMWL) and the Rio Grande evaporation line (RGEL). Results show that the groundwater at the transboundary aquifer features an evaporated isotopic signal, which is consistent with referenced published data that discusses the geologic history of aquifer formations at the studied area. This study is important because isotopic studies from the area were nonexistent and because isotopic data can explain recharge scenarios that relate to groundwater quality.

Keywords: isotopes; transboundary aquifers assessment; Hueco Bolson; Mesilla Basin; Conejos-Médanos/Mesilla aquifer; groundwater

Citation: Garcia-Vasquez, A.C.; Granados-Olivas, A.; Samani, Z.; Fernald, A. Investigation of the Origin of Hueco Bolson and Mesilla Basin Aquifers (US and Mexico) with Isotopic Data Analysis. *Water* **2022**, *14*, 526. https://doi.org/10.3390/w14040526

Academic Editors: Sharon B. Megdal and Anne-Marie Matherne

Received: 16 August 2021
Accepted: 22 December 2021
Published: 10 February 2022

Publisher's Note: MDPI stays neutral with regard to jurisdictional claims in published maps and institutional affiliations.

1. Introduction

In the Paso del Norte (PdN) transboundary aquifers region, located between the United States and Mexico, where New Mexico, Texas, and Chihuahua meet, the climate is semiarid. Water is increasingly scarce due to surface supply reductions caused by drought and climate change, increased demands from growing regional populations, and municipal and industrial (M&I) expansion affecting availability for environmental demands. Based on these reductions, there is an urgent need for better understanding and management of the quantity and quality of the region's scarce water resources. In this binational region, groundwater is the main source for agriculture and M&I water demands; therefore, understanding the origin of groundwater recharge is critical for better management and long-term sustainability of the basin's groundwater [1,2]. Estimation of groundwater recharge can be made via different methods, such as the general water balance approach, field measurement, or isotopic studies. The evaluation is more accurate when isotope and geochemistry methods are combined [3,4]. Isotope and geochemistry methods are complementary tools that distinguish different water sources and provide information on the origin of groundwater, age of water, residence time, and recharge points [4–7].

1.1. Isotope Study

Isotopes in water molecules work as natural tracers. The isotopic composition of continental precipitation depends on the water's origin and pathway, which begins the moment it leaves the sea in the form of evaporation and ends when the sample is collected [8]. Additionally, isotopes exist in stable or unstable forms [5,6]. Stable isotopes for oxygen are ^{16}O, ^{17}O, ^{18}O, and for hydrogen are protium (^{1}H) and deuterium (^{2}H, D). When these isotopes are combined to form a water molecule, they also provide an isotopic composition that translates into a powerful hydrology tracer. A pair of isotopes commonly used in hydrology is the $\delta^{18}O$ combination, which is compared using the global meteoric water line (GMWL) to show the percentage of isotope present in the sample.

Another isotope used in hydrology is tritium, an unstable isotope of hydrogen (^{3}H or T). In the same manner as ^{14}C, tritium originates from neutrons (n) present in cosmic rays due to nuclear reactions with nitrogen present in the atmosphere; the following chemical reaction indicates this formation $^{14}N{+}n \rightarrow {}^{12}C + {}^{3}H$ [9,10]. After this reaction, the tritium joins the hydrological cycle in the atmospheric part [9,10]. In hydrology, tritium has been used to distinguish new waters from old waters, because of its short half-life of 12.3 years [5,7], and its predictable timing of origin during nuclear explosions in contact with the atmosphere.

In this research, we focus on the transboundary area formed by the Hueco Bolson and Conejos-Médanos/Mesilla Basin aquifers of the middle Rio Grande watershed. Our investigation includes isotopic and geochemical data collected from the Mexican portion of the Mesilla Basin aquifer referred to as the "Conejos-Médanos Aquifer" in Mexico. These data were obtained via a comprehensive field and laboratory analysis. The analysis was compared with a similar study on the US side of the Mesilla Basin [11]. In order to cover the entire transboundary area, we also included data from the Hueco Bolson Aquifer [12]. In the conclusion section of this work, we compare our results with the study reported by Hawley and Kottlowski (1969) [13], which indicates that the waters present in the Hueco Bolson and Conejos-Médanos/Mesilla Basin aquifer were part of a single aquifer before the formation of the *Sierra de Juárez* (Juarez Mountain Range).

1.2. Rio Grande

One of the most important rivers in the US is the Rio Grande, or the Rio Bravo as it is called in Mexico (Figure 1). The Rio Grande watershed has an area of approximately 924,300 miles^2 (2,394,000 km^2) and includes regions in both the US and Mexico [14]. With a length of about 1900 miles (3060 Km), it is the 20th longest river in the world, the 5th longest river in North America, and is the 2nd longest American river after the Mississippi [15]. The Rio Grande begins in the San Juan Mountains of southern Colorado, which are part of the Rocky Mountains, and flows through New Mexico and Texas. In the south, the Rio Grande marks the borderline between the US and Mexico [16]. In Mexico, the river runs through Chihuahua, Coahuila, Nuevo Leon, and Tamaulipas, finally ending in the Gulf of Mexico. The Rio Grande has two international dams, Falcon and La Amistad, that are managed by the International Boundary and Water Commission/Comisión Internacional de Limites y Agua (IBWC/CILA) [14]. Figure 1 shows the entire watershed of the Rio Grande from Colorado to the Gulf of Mexico.

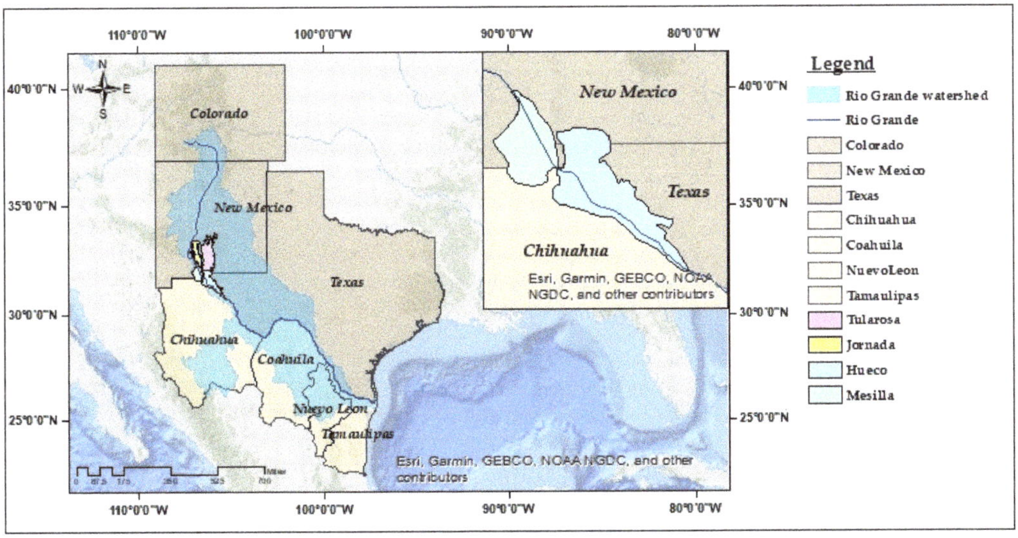

Figure 1. Map of Rio Grande watershed and river mainstem through the seven states in the US and Mexico.

1.3. Previous Studies

Starting in 1970, geomorphology, geophysics, hydrological prospecting, water quality, and isotopic studies have been carried out for various basins in the states of Texas and New Mexico (US) and Chihuahua (Mexico). These studies were conducted in Mexico by the Municipal Water and Sanitation Board (Junta Municipal de Agua y Saneamiento, JMAS) [17], the National Water Commission (Comisión Nacional del Agua, CONAGUA) [18], the Mexican Geological Service (Servicio Geológico Mexicano, SGM) [19], the Autonomous University of Juarez City (Universidad Autónoma de Ciudad Juárez, UACJ) [20,21], the Autonomous University of Chihuahua (Universidad Autónoma de Chihuahua, UACH) [20], the Comisión Internacional de Límites y Aguas-Mexican section (CILA) and International Border and Water Commission, US section (IBWC) [14,19,22]. On the US side, studies were conducted by El Paso Water Utilities [23], the New Mexico Water Resources Research Institute (NMWRRI) [24,25], New Mexico State University (NMSU) [26,27], Texas A&M AgriLife Research Center [28] and the Transboundary Aquifer Assessment Program (TAAP) [29].

From the above-mentioned studies, the ones using environmental tracers such as $\delta^{18}O$ and tritium as well as and basic physicochemical parameters were selected for our analysis. These studies also provided the spatial distribution that enabled us to cover the area between the Hueco Bolson and Conejos-Médanos/Mesilla aquifers.

1.4. Study Area

Of the various aquifers along the Rio Grande, this study focuses on one of the most important transboundary regions between the United States and Mexico: the cross-border area of Juárez, Chihuahua in Mexico and Las Cruces, NM and El Paso, TX in the US. In this Paso del Norte or PdN transboundary region, groundwater uses are mainly supported by two transboundary aquifers: the Hueco Bolson and the Conejos Médanos/Mesilla Basin aquifers (Figure 1). Several communities along the US-Mexico border in New Mexico, Texas and Chihuahua depend on these aquifers for domestic, agricultural, and industrial water use [30]. In this study, special attention was given to the Mexican side of the Conejos-Médanos Basin aquifer where isotopic studies that could explain recharge scenarios in the area and their relationship with groundwater quality were nonexistent.

Cliett (1969) [31] mentioned that the geology of the Conejos-Médanos Basin aquifer is comparable to the Hueco Bolson aquifer, both having similar depositional environments on the geological time scale of the aquifers. Despite these similarities, they differ in their lithology and groundwater qualities, with differing sediments from contemporary basin fill within the surface area of the aquifer. Additionally, Cliett (1969) [31] defined that the two sediment units are hydraulically connected, meeting the aquifer at an estimated average depth of 152.4 m (500 ft). Regarding water levels, in the case of the shallow Hueco Bolson aquifer, along the agricultural zone of the Valle de Juárez, static levels were on average 12.19 m (40 ft) and superficially at 3 m (10 ft).

Hawley et al. (2009) [32] developed a hydrogeological model based on reports and peer-reviewed research to promote the exchange of information to provide a better understanding of water problems and possible alternative solutions to address them. His group's hydrogeological model includes the area of the Mesilla aquifer, a section of the Rio Grande in north-central Chihuahua, Mexico, and parts adjacent to the south of the Jornada del Muerto Basin, where the contact between the strata is shown as well as the basin's sedimentary fill. The basement that represents the bedrock and the tectonic characteristics of the area are reflected not only in the composition of the sedimentary fill, but also in the groundwater flow and chemistry according to its time of residence. The source of sediment fill in this aquifer was the surrounding mountains, consisting largely of Paleozoic sedimentary rocks inclined on a base of Precambrian rocks; these mountains also contain Tertiary volcanic rocks [31].

Appendix A (see Figure A1) shows the sedimentary Santa Fe Group with the evolution and tectonic faults of the basins in the southern region of the Rio Grande. In the past 25 million years, this region has had a profound effect on the distribution of the groupings in the lithofacies (strata) of the Santa Fe Group [33]. Hawley and Lozinsky (1992) [34] subdivided the Santa Fe Group into three stratigraphic units: lower, middle, and upper. These units are defined based on the general lithological character, the depositional environments of the fill, and the characteristics related to the post-depositional history.

Hawley and Swanson (2022 in revision) [35], show that the hydrogeological framework controls on groundwater flow and chemistry in the transboundary—aquifers system west of the lower Mesilla Valley (MeV) and PdN transboundary aquifers systems in this area—are comprised of: 1) thick Santa Fe Group (SFG) rift-basin fill (as much as 600 m), and 2) the thin ($\leq$20 m) alluvial aquifers of the inner-river valley. They also recognized that at least the upper part of the SFG aquifer system was present in Chihuahua, located as far south as the Federal Highway 2 corridor west of the Juarez and Sapello mountain ranges in Mexico. In regard to groundwater quality in the transboundary Mesilla/Conejos-Médanos Basin aquifer, Hawley and Swanson (2022 in revision) [35] address that the ongoing research has demonstrated that very large quantities of fresh to slightly saline water are stored in the basin-fill aquifer system, where most groundwater in storage is at least 11ka and was recharged during the last glacial/pluvial stage of the Late Pleistocene Epoch (~29 to 11 ka).

2. Materials and Methods

The Conejos-Médanos Basin data were collected from the JMAS wells on the Mexican side of the Mesilla Basin aquifer. We collected sixteen samples (Figure 2a,b) on 9 and 10 June 2016. Sampling was conducted in collaboration with the JMAS team, Grupo CARSO, and the UACJ Environmental Engineering laboratory. The sixteen samples were analyzed for physicochemical and metallic parameters by Garcia-Vasquez in the UACJ Environmental Laboratory. A total of nine of these samples were analyzed for $\delta^{18}O$ and tritium isotopes in the Isotopic Hydrologic Laboratory at the Mexican Institute of Water Technology (IMTA) (Figure 3).

(**a**) (**b**)

Figure 2. Sampling with JMAS, Grupo CARSO, UACJ Environmental Laboratory: (**a**) sampling water in the Conejos Médanos from JMAS well set with the UACJ Environmental Laboratory, (**b**) sampling team members of CARSO, JMAS, this study, and UACJ Environmental Laboratory.

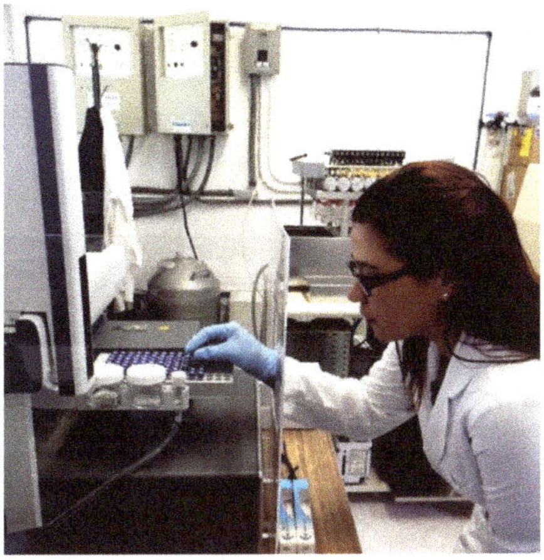

Figure 3. Analyzing stable isotopes ($\delta^{18}O$) samples at the Isotopic Hydrologic Laboratory (Laser analyzer Cavity Ringdown Spectrometer L2110-i Isotopic H_2O Picarro with high precision vaporizer A0211).

This study offers a significant contribution as it completes the characterization of the Conejos-Médanos/Mesilla Basin aquifer isotopic system by providing results from the Mexican side of the aquifer to the already existing data from the US side. To complete the system analysis in this region, we compared our results with similar previous research on the US side of the Mesilla Basin aquifer [11] and a study of the Hueco Bolson aquifer between the US and Mexican sides [12].

2.1. Mesilla Basin Aquifer Data

In 2010, Teeple (2017) [11] gathered 44 isotopic samples (Table 1) from four hydrologic units in the Mesilla Basin aquifer on the US side. He used the subdivision of the groundwater flow system outlined by Hawley and Lozinsky (1992) [34] to divide the study area. Subdivisions made by them were four hydrological units (Table 1) including the Rio Grande Alluvium, which is from a quaternary system and is part of the Santa Fe Group. The Santa Fe Group is a Tertiary system divided into three hydrogeologic units, the Upper, Middle, and Lower Santa Fe Group. The southern boundary in the study area of Teeple (2017) [11] was the border between the US and Mexico.

The aquifer was divided into four hydrogeological units based on the terrain stratigraphy and groundwater flow of the Mesilla aquifer as shown in Table 1.

Table 1. Samples in the Mesilla Basin aquifer by Teeple (2017) [11].

Area	Samples
Rio Grande alluvium (RGA)	3
Lower part of the Santa Fe Group (LSF)	4
Middle part of the Santa Fe Group (MSF)	24
Upper part of the Santa Fe Group (USF)	13
Total Samples	44

Teeple (2017) [11] gathered 44 samples from wells and sampled the same location at different depths from five sets of wells in different hydrologic units. For the first set of wells, TQ18, TQ19, TQ20, and TQ21, had depths of 55, 275, 280, and 200 ft, and hydrologic units of RGA, USF, MSF and LSF, respectively. For the second set of wells, TQ26, TQ27, TQ28, and TQ29, the depths were 47, 275, 275, and 280 ft, and the hydrologic units were RGA, USF, MSF and LSF, respectively. For the third set of wells, TQ31 and TQ32, the depths were 150 and 275 ft, and the hydrologic units were MSF and LSF, respectively. For the fourth set of wells, TQ34, TQ35, and TQ36, the depths were 135, 270 ft and one more unspecified, and the hydrologic units were USF, MSF and LSF, respectively. For the last set of wells, TQ40 and TQ41, the depths were 47 and 132, and the hydrologic units were USF and MSF, respectively. The coordinates for each set of wells are in Appendix B.

Tritium results shown by Teeple (2017) [11] were analyzed at the Menlo Park Tritium Laboratory in Menlo Park, CA under the procedures of Östlund and Werner (1962) [36] and Thatcher et al. (1977) [37].

The analyses for stable isotope ratios of δD and $\delta^{18}O$ in Teeple (2017) [11] were conducted at the USGS Stable Isotope Laboratory in Reston, Va. Under the described methods in Révész and Coplen (2008b) [38].

This study was carried out on the US side of the Mesilla aquifer in cooperation with the USGS, IBWC, NM WRRI, NMSU, Texas AgriLife Research, TWRI, and Texas A&M. The results from the 44 samples in the Teeple (2017) [11] study were predominantly Na-HCO3 or a Na-SO4-HCO3 geochemistry water groups. For tritium, the results indicate negative values, which means there was no tritium content because of the decay. Teeple (2017) [11] mentioned that results show groundwater flows are generally from the north to south-southeast and that there is a pattern of groundwater discharging in the PdN.

2.2. Hueco Bolson Aquifer Data

Previous studies of the Hueco Bolson aquifer on the Mexican side indicate an increasing trend of calcium and sulfate ions with total dissolved solids (TDS) of more than 750 mg/L. This shows a deterioration in water quality during the 1965–1999 period [39].

Eastoe et al. (2007) [12] conducted an analysis of the isotopic concentration in the Hueco Bolson. They made a subdivision of hydrologic units (Table 2). This subdivision encompasses the Hueco Bolson Aquifer in both the US and Mexico.

Table 2. Samples in Hueco Bolson Aquifer by Eastoe et al. (2007) [12].

Area	Samples
Hueco Bolson Aquifer, El Paso County, Texas	35
Hueco Bolson Aquifer, Chihuahua	31
Hueco Bolson Aquifer, Doña Ana and Otero Countries, New Mexico	5
Hueco Bolson Aquifer, Hudspeth County and east El Paso County, Texas	4
Total Samples	75

Eastoe et al. (2007) [12] gathered 75 samples of groundwater and precipitation. Groundwater was sampled from public and private wells; precipitation samples were from the Juarez region. Stable oxygen and hydrogen isotopes were measured with a gas source isotope radio-frequency mass spectrometer (Finnigan). The delta value was standardized with the Vienna Standard Mean Ocean Water (VSMOW). Liquid scintillation spectrophotometry was used for tritium analysis. The stable and unstable isotope analysis was carried out in the laboratory at the University of Arizona.

Results from stable isotope data showed four types of groundwater recharge. The authors identified two sources of recharge from the Rio Grande and another two sources of recharge from local precipitation.

Previous studies used to perform this assessment were selected as they have published the same type of analysis and data samples in different locations. Table 3 shows the data from the sources referred to in this study by the author.

Table 3. Data collected from different authors used in this investigation.

Source	Year	$\delta^{18}O$	Tritium	Coordinates	Aquifer
Eastoe et al.	2007				Hueco (US/MX.)
Teeple	2010				Mesilla (US)
This study	2015				Conejos Médanos (MX.)

Appendix B (see Table A1) contains a record of all the data used to perform the analysis. "ID" means the identification of the sample in this study; "Source" is the name of the well sampled; "Date" refers to the year when the sample was taken; "Latitude and longitude" mean the sample coordinates; "$\delta^{18}O$ and T" refer to the isotopic values obtained for oxygen, hydrogen, and tritium, respectively; and "Group," to the group previously named by the authors. Additionally, from Eastoe et al. (2007) [12], A = Rio Grande, B = Rio Grande near the Sierra de Juárez, C = Upper Hueco Bolson, D = South of the Hueco Bolson, and E = Middle Hueco Bolson. The other acronyms used are Upper Santa Fe (USF), Middle Santa Fe (MSF), Lower Santa Fe (LSF), and Rio Grande Alluvion (RGA) from Teeple (2017) [11]. In this study, the Conejos Médanos Basin is labeled (CM).

3. Results

Hydrogeochemical results show groundwater ions are predominantly $Cl+SO_4$ and HCO_3, throughout the area. There is a mixture of waters that have the main components Na^+, Cl^- and SO_4^- ions. Due to the type of sediment fill deposit around the Conejos Médanos aquifer, the presence of these ions throughout the aquifer was expected. Geochemically, this reflects the rock interaction that predominates in this area and reveals current rock deterioration through the mineralization of the waters throughout the region of the Conejos Médanos aquifer.

Figure 4 shows the Mesilla and Hueco aquifers and geographical locations of the samples collected by this study, Teeple (2017) [11], and Eastoe et al. (2007) [12].

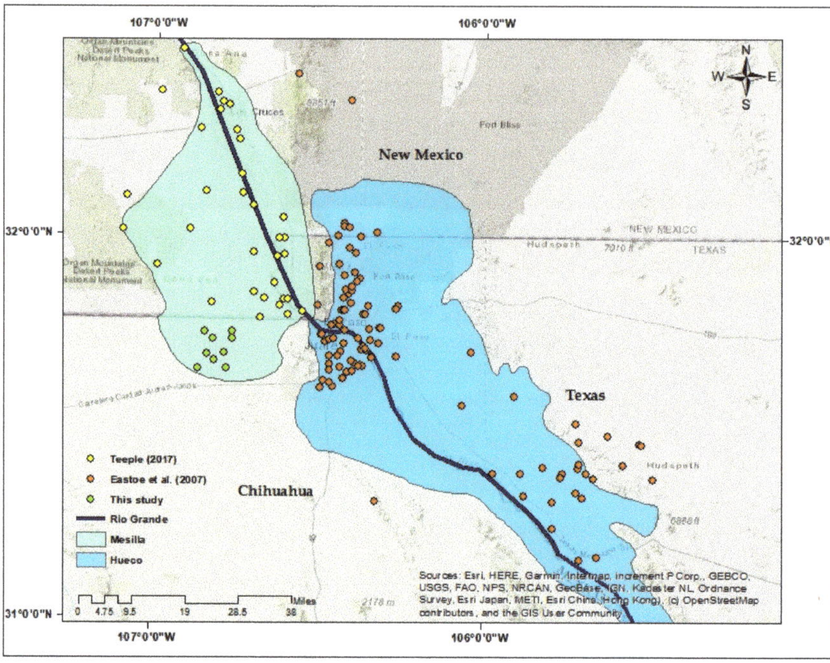

Figure 4. Location of samples collected in this study (green), Teeple (2017) [11] (yellow), and Eastoe et al. (2007) [12] (orange).

3.1. Tritium

The tritium results obtained in this study (Figure 4, green points) in the Conejos Médanos Basin varied from −0.70 to 0.58 Tritium Units (TU), which is a non-significant tritium content because the absence of tritium or values below <0.5 TU indicate that the age of waters is not greater than 50 years. This is an important finding because it indicates that the water present in this zone is not of recent origin, which demonstrates that there is no recharge in this zone. Furthermore, this study does not report any significant tritium concentrations in the Conejos Médanos Aquifer.

The Mesilla Basin aquifer results obtained by Teeple (2017) [11] indicate the presence of pre-boom waters, which refers to water recharged prior to 1950. Teeple (2017) [11] found high concentrations of tritium in two samples collected from wells in the Rio Grande Alluvium; the values were 4.6 TU (T Q18) and 7.5 TU (T Q26). In the Hueco Bolson, the highest concentrations followed the same path as the Mesilla Basin aquifer [12].

Figure 5 shows values over 2 TU for the samples taken by Eastoe et al. (2007) [12] near the Rio Grande Alluvium. These tritium concentration values range from 2.6 to 14.2 TU, which points to recharge points within the study's area. The area with recharge points and possible recharge near these points is in the alluvium of the Rio Grande, which is consistent with what other authors mentioned in their studies.

Recharge points in the Rio Grande, in the Conejos-Médanos/Mesilla Basin, and Hueco Bolson aquifers are present on the surface and exist mostly at the piedmont slopes of the mountains adjacent to the Rio Grande Alluvium. This indicates that in the Mexican portion of the Mesilla Basin, the water is old and does not have significant recharge areas. Thus, in the rest of the points with values <2 TU, there is no recharge, at least in the sampled points.

Data collection by the different authors occurred in 2006, 2010, and 2015. Although the collection of samples occurred at different times, for this analysis the variation in residence time from one sample to another is not significant because they are valid in time and space.

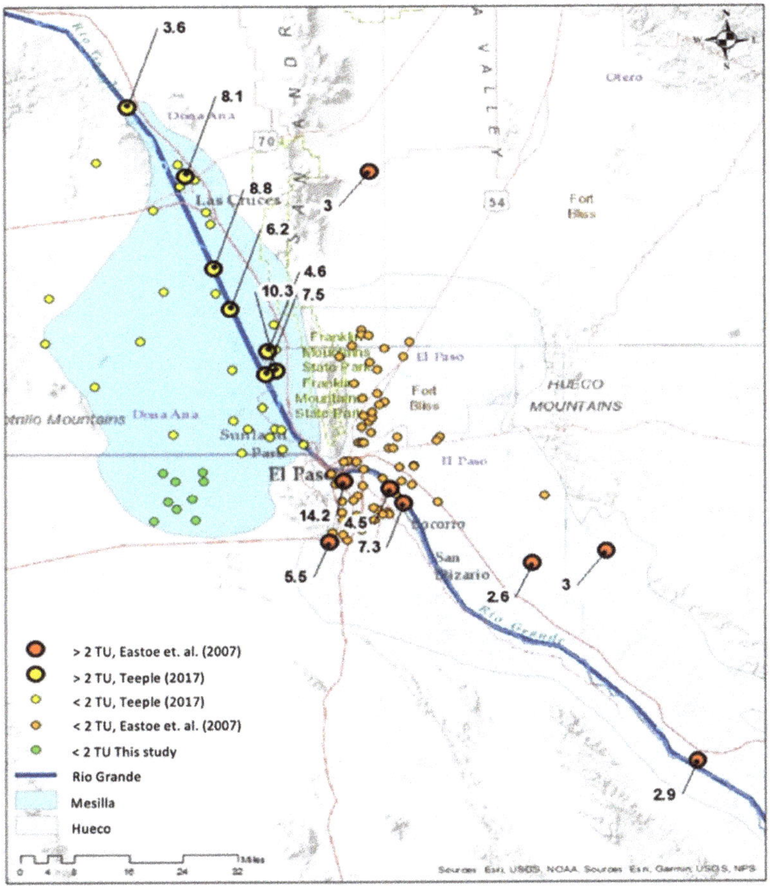

Figure 5. Tritium concentration values of more than 2 TU. The red points with black borders represent values more than 2 TU from Eastoe et al. (2007) [12]. The yellow points with black borders show the values with more than 2 TU from Teeple (2017) [11]. The orange points indicate values lower than 2 TU from Eastoe et al. (2007) [12]. The yellow points indicate values lower than 2 TU from Teeple (2017) [11]. The green points indicate values lower than 2 TU from this study. The Mesilla Basin aquifer is featured in blue, and the Hueco Bolson in light blue. The line in dark blue shows the Rio Grande mainstream.

3.2. Oxygen 18 ($\delta^{18}O$)

Figure 6 shows a compilation of the sample points. The samples are grouped into numbers and letters. The letters are given by the author and apply only to the samples taken by Eastoe et al. (2007) [12]. The data gathered from Eastoe et al. (2007) [12] are featured in orange squares (Group A), circles (Group B), and diamonds (Group C); each shape represents a different group given by the author. The data from this study are shown by green circles; and the data by Teeple (2017) [11] in yellow circles. The values of all points were compared with the GMWL and the RGEL to determine the changes in the water's isotopic composition, produced by different processes. A total of three groups were obtained.

Group 1 is in the GMWL and is made up of samples from group C. Some of these were taken by Eastoe et al. (2007) [12] from the Hueco Bolson (orange diamonds), while five samples came from the Teeple (2017) [11] study (yellow circles). Group C comes from the Franklin and Organ Mountains. Eastoe et al. (2007) [12] mentioned that similar water could be originating in the Juarez Mountains (Sierra de Juárez). On the other hand, the five samples from Teeple (2017) [11] (yellow points) are TQ12, TQ14, TQ16, TQ30, and TQ32 (See Appendix B). These samples were taken in the Mesilla Basin near the Rio Grande Alluvium, which means that water from the river is present in these locations. In Group 1, waters are located in or near the GMWL because no current depletion can be seen in the isotopes.

Group 2 results feature 14 samples close to the line while the rest are slightly above the line. Of those first fourteen samples, three (orange squares) are E1, E2, and E3 (See Appendix B); they are part of Group A and were taken by Eastoe et al. (2007) [12] in the Hueco Bolson aquifer in Chihuahua, near the Rio Grande. These three samples have an isotopic composition of $\delta^{18}O$, which varies slightly between −8.6 and −9.4. Another nine samples (yellow points) were TQ00, TQ03, TQ09, TQ13, TQ18, TQ23, TQ24, TQ25, and TQ36 (See Appendix B); they were taken by Teeple (2017) [11] and show an isotopic composition of $\delta^{18}O$ with a variation of −7.74 to −8.97. The last of the fourteen samples found in RGEL were taken by this study in the Conejos-Médanos set of wells of the JMAS; these featured an isotopic composition of $\delta^{18}O$ and a variation of −8.83. The rest of the Group 2 samples that are slightly above the RGEL were taken by this study and Teeple (2017) [11] in the Mesilla/Conejos-Médanos Basin.

The results of stable $\delta^{18}O$ isotopes in this study are not near the GMWL, but they are near the RGEL. According to Teeple (2017) [11], and Witcher et al. (2004) [24], these results could indicate that groundwater has a Rio Grande isotopic signature from the ancestral Rio Grande and this could be a sign of evaporated waters. In addition, they show that recharge sources include precipitation, bedrock fissure water, and irrigation return water. Finally, they also point to water evaporation.

Group 3 is made of three samples which are in or near the RGEL. This group is formed by three samples from Group A taken by Eastoe et al. (2007) [12] in the Hueco Bolson aquifer in Chihuahua near the Rio Grande. The group is made up of Group B (orange circles), taken by Eastoe et al. (2007) [12] and consisting of samples collected beneath the urban area of Juárez City and the Rio Grande floodplain in El Paso. The geographical area in which the samples were collected is a semi-arid area where evaporation processes occur; this phenomenon could have affected the process. This dataset falls below the GMWL, indicating that water has evaporated. Group 3 is also formed by samples taken by Teeple (2017) [11].

The study by Teeple (2017) [11] reports that values of less than −80.0 and −10.5 $\delta^{18}O/\delta D$ (‰) have an apparent age of less than 10,000 carbon-14 years before present (1950). Samples from this age are found near the Rio Grande Alluvion. Values greater than −80.0 and −10.5 $\delta^{18}O/\delta D$ (‰) have an age greater than 10,000 carbon-14 years before present (1950). Samples of this age are found in the southeast of the Mesilla Basin aquifer, near the Hueco Bolson and the Juarez Mountains. Such a group of results is consistent with results from this study in the Conejos-Médanos region and with those of Group C, from Eastoe et al. (2007) [12], which are marked as Group 3 in Figure 6.

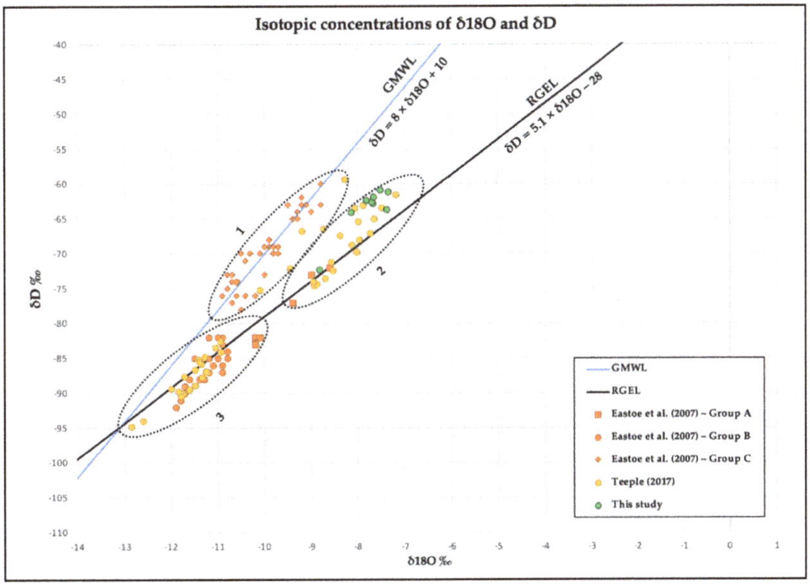

Figure 6. Plot of $\delta^{18}O/\delta D$ (‰) in groundwater from this study, Eastoe et al. (2007) [12] and Teeple (2017) [11] was compared to the global meteoric water line (GMWL) and Rio Grande evaporation line (RGEL). The graph was divided into three groups, these groups considered all the samples in Appendix B. Group 1) is formed by water samples from the Mesilla and Hueco basins taken by Teeple (2017) [11] and Eastoe et al. (2007) [12]. Group 2) consists of samples from the Hueco Bolson and the Mesilla/Conejos-Médanos Basin aquifers, and they are samples taken by this study, Teeple (2017) [11] and Eastoe et al. (2007) [12]. Group 3) contains samples from the Mesilla Basin aquifer and the Bolson del Hueco; the samples were taken by Teeple (2017) [11] and Eastoe et al. (2007) [12].

4. Conclusions

According to the age determined by the results of the isotopic concentration and the $\delta^{18}O/\delta D$ of the water, Group 2 is formed by old water. Occasionally an addition of ^{18}O is caused by dissolution processes, and this can increase with geothermal activity; having this geothermal change could have caused a movement to the right of the GMWL. This, in Figure 6, indicates that the "X" axis, which is ^{18}O, moved to the right, achieving a greater concentration of ^{18}O. On the contrary, the "Y" axis, which represents a ^{16}O concentration, decreased. This change to a concentration greater than ^{18}O and lower than ^{16}O results in an isotopically heavier $\delta^{18}O$ signature but without any change in the δ^2H signature [2,24]. Most of the groundwater samples that are plotted along the displaced GMWL represent isotopically lighter water, with δD values of less than −80.00 per thousand and $\delta^{18}O$ values of less than −10.50 per thousand [40]. This isotopic signature indicates that the samples in Group 2 probably underwent water recharge during the relatively humid and cool Pleistocene climate [40].

According to Witcher et al. (2004) [24] and Bumgarner (2012) [40], the GMWL in the studied area has been displaced and represents ancient groundwater and geothermal groundwater, from which ^{18}O of the rocks have been obtained. This was due to an exchange processes that typically occurs with the water-rock interaction and probable hydrothermal alteration. Such an alteration occurs when the oxygen present in the groundwater is exchanged due to the composition of the rock, temperature, texture, and length of contact [24].

The compilation of isotopic data provided by this article is important as it allows for the comparison of water samples from different locations in the US-Mexico borderland area of the Hueco Bolson and Mesilla Basin aquifers. The locations of the samples collected contribute to understanding the water origin of the studied area.

Hawley and Kottlowski (1969) [13] established that the Rio Grande flowed across the western area of the Juarez Mountains and that water from the Rio Grande drained into the Cabeza de Baca Ancient Lake, going through the sedimentary deposits which are presently part of the Mesilla Basin aquifer [3]. However, with the formation of the Juarez Mountains in the Quaternary period, the Rio Grande changed its course, carving its way through the El Paso Canyon over the course of recent geological times, flowing between the Franklin Mountains and the Juarez Mountains through the canyon that formed between the neighboring mountains [13].

As different authors mention, a primary source of recharge into the Mesilla Basin aquifer system is the Rio Grande Alluvium in the Mesilla Valley because of the seepage losses from the riverbed. From previous and new data evaluated, we conclude that the Conejos-Médanos Basin aquifer has the same source of water as the Hueco Bolson does from Group A of Eastoe et al. (2007) [12]. The Group A samples were taken near the Rio Grande at the foot mountain in the Juarez Mountains. Moreover, as was expected, the Group 1 samples collected by Teeple (2017) [11] at the south of the Mesilla Valley to the Conejos Médanos Basin aquifer signal the presence of the same type of water in this area.

In conclusion, the samples collected and analyzed by this study complete the description of the Hueco Bolson and the Mesilla/Conejos-Médanos Basin at the US-Mexico transboundary area. According to previous study results shown for Group 2, a stable isotope $\delta^{18}O$ concentration falls below the GMWL in the evaporated zone, which indicates that these are old waters that have undergone evaporation, horizontal infiltration, or dissolution processes. Moreover, groundwater values indicate that groundwater recharge sources include precipitation, bedrock fissure water, or both. Furthermore, results are consistent with findings by Eastoe et al. (2007) [12], Teeple (2017) [11], Hawley and Kottlowski (1969) [13], Witcher et al. (2004) [24], and Bumgarner (2017) [40], whose findings indicate that the groundwater is not recent and that it was recharged thousands of years ago when the climate was more humid, which could be the cause for the same isotopic content in the Hueco Bolson and Conejos-Médanos/Mesilla Basin aquifers near the Juarez Mountains.

Author Contributions: Conceptualization, A.C.G.-V., A.G.-O., Z.S., A.F.; methodology, A.C.G.-V., A.G.-O.; analysis, A.G.-O.; writing—original draft preparation, A.C.G.-V., A.G.-O., Z.S., A.F.; writing—review and editing, A.C.G.-V., A.G.-O., Z.S., A.F.; project administration, A.C.G.-V., A.G.-O., Z.S., A.F.; funding acquisition, A.C.G.-V., A.G.-O., Z.S., A.F. All authors have read and agreed to the published version of the manuscript.

Funding: The author's work has been supported by funding through the U.S. Geological Survey Award #G17AC00441 to New Mexico State University, entitled "Transboundary Aquifer Assessment Program".

Informed Consent Statement: Informed consent was obtained from all subjects involved in the study.

Data Availability Statement: The date presented in this study are available on request from the corresponding author.

Acknowledgments: We appreciate the Municipal Water and Sanitation Board (JMAS) of Juarez City for their help in accessing the water wells in Juarez. Likewise, we are grateful with María del Carmen Agüero and Miguel Hernandez for their help in taking the water samples and support with materials for conducting physicochemical analyzes at the Environmental Laboratory of Autonomous University of Juarez City (UACJ). In addition, we thank Luis González Hita and Eng. Blanca de Jesús Carteño for their support to carry out the isotope analysis at the Isotopic Hydrologic Laboratory in the Mexican Institute of Water Technology (IMTA).

Conflicts of Interest: The authors declare no conflict of interest.

Appendix A

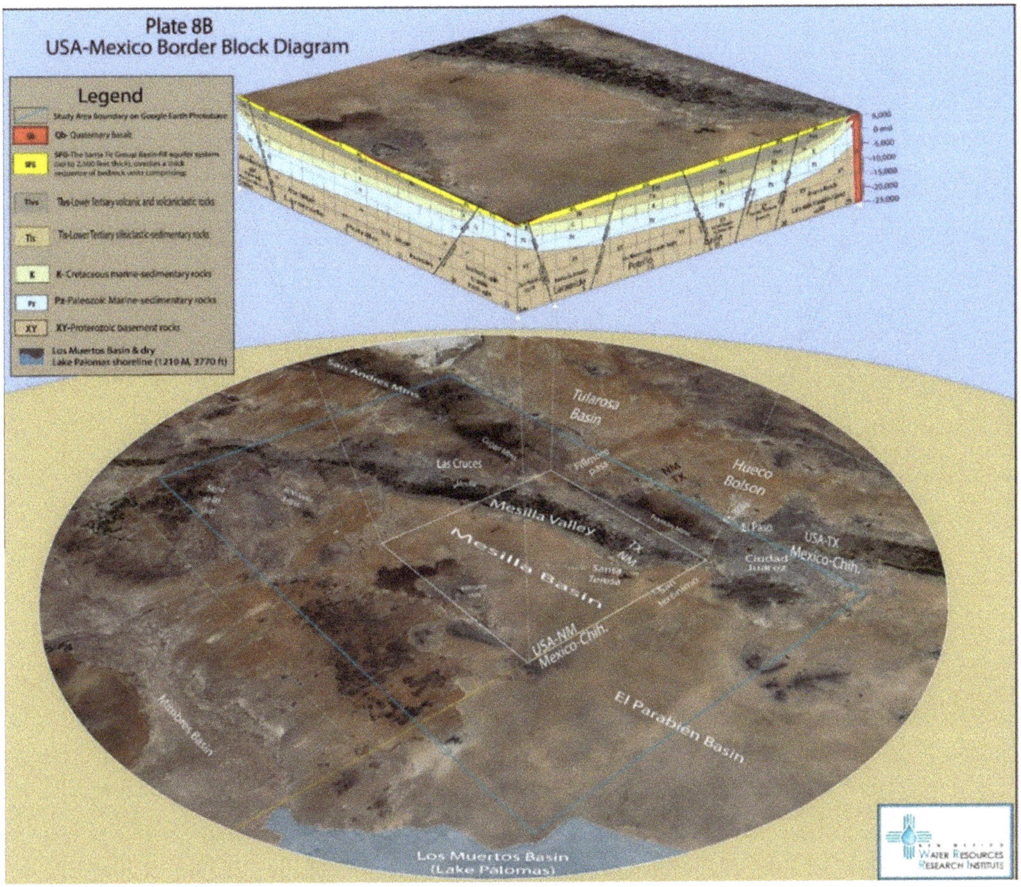

Figure A1. Northeast-facing block diagram of the southern Mesilla Basin, with its southern panel at the International-Boundary [35].

Appendix B

Details of data used for this study are in Table A1.

Table A1. Data used for this study.

ID	SOURCE	DATE	LATITUDE	LONGITUDE	$\delta^{18}O$	D	*T*	GROUP
E 1	JMAS well 3Z	2006	31.686	−106.339	−9.4	−77	7.3	A
E 2	JMAS well 9R	2006	31.745	−106.493	−8.6	−72		A
E 3	JMAS well 19R	2006	31.647	−106.415	−10.2	−83		A
E 4	JMAS well 53R	2006	31.606	−106.494	−9	−73	5.5	A
E 5	JMAS well 62	2006	31.745	−106.489	−10.2	−82		A
E 6	JMAS well 141	2006	31.701	−106.434	−10.1	−82		A
E 7	EPWU well 9	2006	31.772	−106.454	−11.5	−85	−0.5	B
E 8	EPWU well 14	2006	31.769	−106.463	−11.2	−85	1.2	B
E 9	EPWU well 408	2006	31.755	−106.421	−10.9	−82	1.6	B

Table A1. *Cont.*

ID	SOURCE	DATE	LATITUDE	LONGITUDE	$\delta^{18}O$	D	*T*	GROUP
E 10	EPWU well 414	2006	31.704	−106.356	−11.2	−82	−0.5	B
E 11	EPWU well 420	2006	31.735	−106.383	−10.6	−74	−0.6	B
E 12	JMAS well 1R	2006	31.725	−106.481	−10.8	−84		B
E 13	JMAS well 5	2006	31.61	−106.456	−10.9	−86	1.2	B
E 14	JMAS well 13RR	2006	31.625	−106.487	−10.9	−83	−0.5	B
E 15	JMAS well 17R	2006	31.731	−106.47	−10.9	−87		B
E 16	JMAS well 42R	2006	31.63	−106.426	−10.8	−85	1.4	B
E 17	JMAS well 47	2006	31.667	−106.374	−11.3	−88		B
E 18	JMAS well 50R	2006	31.66	−106.437	−11.7	−89	−0.4	B
E 19	JMAS well 56R	2006	31.662	−106.369	−11.6	−88		B
E 20	JMAS well 76	2006	32.357	−106.409	−11	−85	3	B
E 21	JMAS well 82R	2006	31.667	−106.467	−11.7	−89	−0.6	B
E 22	JMAS well 84	2006	31.651	−106.466	−11.9	−92		B
E 23	JMAS well 99R	2006	31.69	−106.443	−11.8	−91		B
E 24	JMAS well 115	2006	31.672	−106.394	−11.4	−88		B
E 25	JMAS well 120	2006	31.651	−106.4	−11.7	−90		B
E 26	JMAS well 130	2006	31.662	−106.381	−11.2	−87		B
E 27	JMAS well 134	2006	31.621	−106.466	−11	−82		B
E 28	JMAS well 142	2006	31.689	−106.468	−11.3	−85	−0.5	B
E 29	JMAS well 161	2006	31.735	−106.456	−11.1	−86		B
E 30	JMAS well 151	2006	31.706	−106.371	−11.8	−90	−0.7	B
E 31	JMAS well 165	2006	31.675	−106.402	−11.9	−92	−0.4	B
E 32	JMAS well 180	2006	31.731	−106.343	−11.7	−90	−0.9	B
E 33	JMAS well 183	2006	31.72	−106.424	−11.8	−91		B
E 34	JMAS well 186	2006	31.852	−106.41	−11.8	−90		B
E 35	JMAS well 193	2006	31.891	−106.378	−11.3	−88		B
E 36	West Windmill Bowen	2006	31.983	−106.473	−9.2	−63	1.2	C
E 37	LF4	2006	32	−106.377	−9.5	−63	−0.8	C
E 38	Vista Hills Blue well	2006	31.762	−106.317	−10.8	−75	−0.5	C
E 39	Well 2 Vista Hills	2006	31.761	−106.315	−10.8	−73	−0.6	C
E 40	Wheeler well #3B	2006	31.687	−106.265	−10.7	−77	−0.5	C
E 41	EPWU well 18	2006	31.769	−106.437	−10.9	−76	−0.8	C
E 42	EPWU well 20A	2006	31.841	−106.427	−9.3	−65	−0.6	C
E 43	EPWU well 25	2006	31.899	−106.423	−10	−69	−0.5	C
E 44	EPWU well 33	2006	31.957	−106.392	−9.3	−64	−0.5	C
E 45	EPWU well 42	2006	31.972	−106.409	−9.9	−68	−0.6	C
E 46	EPWU well 45	2006	31.798	−106.368	−10.3	−70	−0.9	C
E 47	EPWU well 52	2006	31.928	−106.442	−9.2	−62	−0.5	C
E 48	EPWU well 55	2006	31.862	−106.422	−9.9	−69	−0.6	C
E 49	EPWU well 63	2006	31.798	−106.361	−10.4	−71	0.5	C
E 50	EPWU well 69	2006	31.759	−106.347	−10.7	−73	−0.4	C
E 51	EPWU well 83	2006	31.715	−106.366	−10.2	−76	4.5	C
E 52	EPWU well 93	2006	31.819	−106.352	−10.7	−73	−0.7	C
E 53	EPWU well 519	2006	31.907	−106.392	−9.9	−68	−0.9	C
E 54	EPWU well 404	2006	31.722	−106.32	−10.7	−74	1.1	C
E 55	EPWU well 416	2006	31.709	−106.36	−10	−73	1.7	C
E 56	Well 2B Ft. Bliss	2006	31.829	−106.406	−10.1	−70	−0.5	C
E 57	Well 5A Ft. Bliss	2006	31.808	−106.432	−9	−64	−0.5	C
E 58	Well 6A Ft. Bliss	2006	31.808	−106.426	−8.8	−63	0.5	C
E 59	Well 7 Ft. Bliss	2006	31.808	−106.422	−9.8	−70	−0.4	C
E 60	Well 10 Ft. Bliss	2006	31.859	−106.403	−9.7	−70	0.5	C
E 61	Well 11 Ft. Bliss	2006	31.87	−106.403	−9.7	−69	−0.4	C
E 62	Well 12 Ft. Bliss	2006	31.885	−106.388	−9.8	−69	−0.5	C
E 63	Intl. Garment Proc. No.4	2006	31.82	−106.261	−10.4	−76	1.1	C
E 64	Intl. Garment Proc. No.1	2006	31.812	−106.267	−10.6	−76	1.5	C

Table A1. *Cont.*

ID	SOURCE	DATE	LATITUDE	LONGITUDE	δ18O	D	T	GROUP
E 65	Chaparral Edna	2006	32.036	−106.426	−9.9	−68	−0.5	C
E 66	Chaparral Sylvia	2006	32.028	−106.426	−9.1	−63	−0.8	C
E 67	Chaparral Rosencrans	2006	32.025	−106.41	−9.4	−65		C
E 68	Rinchem well	2006	32.004	−106.446	−10.7	−74	−0.7	C
E 69	Rhino pump well	2006	32.012	−106.325	−10.5	−70	−0.5	C
E 70	JMAS well 221	2006	31.73	−106.464	−10.5	−78	14.2	C
E 71	LF1	2006	31.983	−106.337	−8.5	−60	1	D
E 72	Esperanza PO	2006	31.16	−105.71	−6.3	−46	2.9	D
E 73	Indian Cliffs Ranch	2006	31.563	−106.066	−8.5	−67	2.6	E
E 74	Velarde	2006	31.587	−105.907	−6.8	−59	3	E
E 75	El Paso Lakes	2006	31.701	−106.038	−9.3	−69		E
T Q00	322320106551801	2010	32.48600	−106.9220	−8.53	−72.38	3.6	USF
T Q01	322233106590901	2010	32.37592	−106.98634	−11.26	−86.92	0	MSF
T Q02	322219106485001	2010	32.37200	−106.81400	−11.34	−87.71	0.3	MSF
T Q03	322054106475201	2010	32.34843	−106.79834	−8.71	−73.53	8.1	USF
T Q04	322024106463901	2010	32.34000	−106.77900	−11.25	−86.98	1.3	USF
T Q05	321934106482601	2010	32.32648	−106.80778	−11.79	−90.30	0.1	MSF
T Q06	321641106515401	2010	32.27800	−106.86500	−11.74	−90.06	−0.1	MSF
T Q07	321628106451501	2010	32.27426	−106.75417	−11.6	−89.46	0.3	MSF
T Q08	321501106443801	2010	32.25037	−106.74445	−11.49	−88.84	0.1	USF
T Q09	320939106441701	2010	32.16093	−106.73861	−8.95	−74.58	8.8	USF
T Q10	320654106504201	2010	32.11500	−106.84500	−11.71	−87.54	0	MSF
T Q11	320643106440401	2010	32.11181	−106.73448	−11.79	−90.41	0	MSF
T Q12	320604107051201	2010	32.10121	−107.08723	−8.75	−66.42	0	MSF
T Q13	320445106421001	2010	32.07927	−106.70333	−8.89	−74.40	6.2	USF
T Q14	320253106364001	2010	32.04800	−106.61100	−10.1	−75.16	0.1	USF
T Q15	320054106533901	2010	32.01510	−106.89473	−11.36	−85.8	0	USF
T Q16	320040107054601	2010	32.01121	−107.09668	−9.2	−66.71	−0.1	MSF
T Q17	315955106362201	2010	31.99649	−106.60694	−11.43	−85.18		MSF
T Q18	315940106372301	2010	31.99444	−106.62306	−8.04	−69.74	4.6	RGA
T Q19	315940106372302	2010	31.99444	−106.62306	−11.05	−83.41	0.2	USF
T Q20	315940106372303	2010	31.99444	−106.62306	−11.29	−84.76	0	MSF
T Q21	315940106372304	2010	31.99444	−106.62306	−11.39	−85.33	0	LSF
T Q22	315723106415201	2010	31.95677	−106.69833	−11.39	−85.6	0	MSF
T Q23	315712106361802	2010	31.95371	−106.60583	−7.97	−68.02	4.2	USF
T Q24	315712106361803	2010	31.95371	−106.60583	−8.96	−74.01	10.3	MSF
T Q25	315712106361804	2010	31.95371	−106.60583	−11.49	−86.65	0.9	LSF
T Q26	315646106374401	2010	31.94611	−106.62889	−8.57	−71.17	7.5	RGA
T Q27	315646106374402	2010	31.94611	−106.62889	−12.61	−93.96	−0.1	USF
T Q28	315646106374403	2010	31.94611	−106.62889	−12.85	−94.73	−0.1	MSF
T Q29	315646106374404	2010	31.94611	−106.62889	−11.84	−89.75	0	LSF
T Q30	315519106593101	2010	31.92200	−106.99200	−8.29	−59.36	0	MSF
T Q31	315245106380601	2010	31.87927	−106.63555	−12.0	−89.32	0	MSF
T Q32	315245106380602	2010	31.87927	−106.63555	−9.46	−72.09		LSF
T Q33	315114106414901	2010	31.85400	−106.69700	−10.93	−84.06	0	MSF
T Q34	315013106362601	2010	31.83705	−106.60777	−7.2	−61.57	−0.1	USF
T Q35	315013106362602	2010	31.83705	−106.60777	−7.51	−63.39		MSF
T Q36	315013106395301	2010	31.83705	−106.66527	−7.74	−67.08		MSF
T Q37	315006106354601	2010	31.83500	−106.59600	−7.67	−65.04		RGA
T Q38	314932106493401	2010	31.82594	−106.82527	−10.94	−82.65	0	MSF
T Q39	314908106371201	2010	31.81900	−106.62000	−7.89	−63.03	0.1	MSF
T Q40	314817106325801	2010	31.80483	−106.54999	−7.99	−65.41	1.3	USF
T Q41	314817106325802	2010	31.80483	−106.54999	−8.14	−68.74	0.1	MSF
T Q42	314746106353601	2010	31.79622	−106.59388	−8.38	−67.42	0.1	MSF
T Q43	314717106404401	2010	31.78800	−106.67900	−8.08	−63.5	0	MSF
TS 01	P1-CM-21	2015	31.65043	−106.8657	−7.39	−63.7	0.35	CM
TS 02	P3-CM-06	2015	31.68897	−106.8363	−7.69	−62.8	−0.16	CM

Table A1. *Cont.*

ID	SOURCE	DATE	LATITUDE	LONGITUDE	$\delta^{18}O$	D	*T*	GROUP
TS 03	P5-CM-24	2015	31.74661	−106.8465	−8.83	−72.3	−0.23	CM
TS 04	P7-CM-12	2015	31.7307	−106.820	−7.36	−61.2	−0.3	CM
TS 05	P9-CM-15	2015	31.7307	−106.820	−7.7	−62.6	−0.7	CM
TS 06	P11-CM-23	2015	31.65181	−106.7786	−7.68	−61.9	0.17	CM
TS 07	P12-CM-18	2015	31.69394	−106.7852	−7.54	−60.9	0.36	CM
TS 08	P16-CM-01	2015	31.7494	−106.7622	−8.15	−64.1	−0.23	CM
TS 09	P17-CM-14	2015	31.72955	−106.7593	−7.82	−62.3	0.44	CM

ID: Identification of samples, *E* by Eastoe et al. (2007) [12], *T* by Teeple (2017) [11], *TS* by This study.

References

1. Chowdhury, A.H.; Uliana, M.; Wade, S. Ground water recharge and flow characterization using multiple isotopes. *Groundwater* **2008**, *46*, 426–436. [CrossRef] [PubMed]
2. Wood, W.W.; Sanford, W.E. Chemical and isotopic methods for quantifying ground-water recharge in a regional, semiarid environment. *Groundwater* **1995**, *33*, 458–468. [CrossRef]
3. Allison, G.B. *A Review of Some of the Physical, Chemical, and Isotopic Techniques Available for Estimating Groundwater Recharge. Estimation of Natural Groundwater Recharge*; Simmers, I., Ed.; NATO ASI Series; Springer: Dordrecht, The Netherlands, 1988; Volume 222, pp. 49–72.
4. Clark, I. *Groundwater Geochemistry and Isotopes*; CRC Press: Boca Raton, FL, USA, 2015.
5. Gat, J. *Isotope Hydrology: A Study of the Water Cycle*; Series on Environmental Science and Management; Imperial College Press: London, UK, 2010; ISBN 978-1-86094-035-4.
6. Allegre, C.J. *Isotope Geology*; Cambridge University Press: Cambridge, UK, 2008.
7. White, W.M. *Isotope Geochemistry*; John Wiley & Sons: Hoboken, NJ, USA, 2015.
8. Fontes, J.; Fritz, P. Environmental Isotopes in Groundwater Hydrology. In *Handbook of Environmental Isotope Geochemistry*; Springer: Berlin/Heidelberg, Germany, 1980.
9. Libby, W.F. Atmospheric Helium Three and Radiocarbon from Cosmic Radiation. *Phys. Rev.* **1946**, *69*, 671. [CrossRef]
10. Gat, J.R.; Mook, W.G.; Meijer, H.A. *Environmental Isotopes in the Hydrological Cycle*; Principles and Applications UNESCO/IAEA: Paris, France, 2001; Volume 2, pp. 63–67.
11. Teeple, A.P. *Geophysics- and Geochemistry-Based Assessment of the Geochemical Characteristics and Groundwater-Flow System of the U.S. Part of the Mesilla Basin/Conejos-Médanos Aquifer System in Doña Ana County, New Mexico, and El Paso County, Texas, 2010–12*; Scientific Investigations Report; U.S. Geological Survey: Reston, VA, USA, 2017.
12. Eastoe, C.J.; Hibbs, B.J.; Olivas, A.G.; Hogan, J.F.; Hawley, J.; Hutchison, W.R. Isotopes in the Hueco Bolson Aquifer, Texas (USA) and Chihuahua (Mexico): Local and General Implications for Recharge Sources in Alluvial Basins. *Hydrogeol. J.* **2007**, *16*, 737–747. [CrossRef]
13. Hawley, J.W.; Kottlowski, F.E.; LeMone, D.V. Quaternary geology of the south-central New Mexico border region. In *Border Stratigraphy Symposium, El Paso, TX, USA, 30 April 1969*; New Mexico Bureau of Mines and Mineral Resources: Socorro, NM, USA, 1969; Volume 104, pp. 89–115.
14. International Boundary and Water Commission, U.S. Section Texas Clean Rivers Program. In *Basin Highlights Report for the Rio Grande Basin in Texas*; US International Boundary & Water Commn: El Paso, TX, USA, 2010.
15. McNeese, T. The Rio Grande. In *Rivers in World History*; Chelsea House: Philadelphia, PA, USA, 2005; ISBN 978-0-7910-8244-7.
16. Britannica. Rio Grande: River, United States-Mexico. Available online: https://www.britannica.com/place/Rio-Grande-river-United-States-Mexico (accessed on 7 June 2021).
17. Junta Municipal de Agua y Saneamiento (JMAS) de Ciudad Juárez. *Junta Municipal de Agua y Saneamiento Plan Maestro de JMAS Juárez 2010*; JMAS: Ciudad Juárez, Mexico, 2010.
18. Comisión Nacional del Agua (CONAGUA). *Comisión Nacional del Agua: Actualización de la Disponibilidad Media Anual del acuífero Conejos-Médanos (0823), Estado de Chihuahua 2015*; CONAGUA: Coyoacán, Mexico, 2015.
19. Servicio Geológico Mexicano (SGM). *Comisión Internacional de Límites y Aguas-Sección Mexicana*; Servicio Geológico Mexicano Actividades Hidrogeológicas Del Acuífero Conejos/Medanos, Estado de Chihuahua, I Etapa. 2010; SGM: Pachuca, Mexico, 2010.
20. Soto-Ontiveros, A.; Granados-Olivas, A.; Pinales-Munguía, A.; Saúl-Solís, S.; McConnel Heyman, J. Evolución temporal del flujo del agua subterránea en Ciudad Juárez, Chihuahua aplicando modelación geoespacial. *Tecnociencia Chih.* **2018**, *12*, 103–113.
21. Mayer, A.; Heyman, J.; Granados-Olivas, A.; Hargrove, W.; Sanderson, M.; Martinez, E.; Vazquez-Galvez, A.; Alatorre-Cejudo, L.C. Investigating Management of Transboundary Waters through Cooperation: A Serious Games Case Study of the Hueco Bolson Aquifer in Chihuahua, Mexico and Texas, United States. *Water* **2021**, *13*, 2001. [CrossRef]
22. International Boundary & Water Commission. About the Rio Grande. Available online: https://www.ibwc.gov/CRP/riogrande.htm (accessed on 3 June 2021).
23. Bredehoeft, J.; Ford, J.; Harden, B.; Mace, R.; Rumbaugh, J. *Review and Interpretation of the Hueco Bolson Groundwater Model*; El Paso Water Utilities: El Paso, TX, USA, 2004.

24. Witcher, J.C.; King, J.P.; Hawley, J.W.; Kennedy, J.F.; Cleary, M.; Bothern, L.R. *Sources of Salinity in the Rio Grande and Mesilla Basin Groundwater*; New Mexico Water Resources Research Institute: Las Cruces, NM, USA, 2004.
25. Updegraff, C.D.; Gelhar, L.W. *Parameter Estimation for a Lumped Parameter Groundwater Model of the Mesilla Valley, New Mexico*; New Mexico Water Resources Research Institute: Las Cruces, NM, USA, 1977.
26. Samani, Z.; Bawazir, A.S.; Bleiweiss, M.; Skaggs, R.; Longworth, J.; Tran, V.D.; Pinon, A. Using Remote Sensing to Evaluate the Spatial Variability of Evapotranspiration and Crop Coefficient in the Lower Rio Grande Valley, New Mexico. *Irrig. Sci.* **2009**, *28*, 93–100. [CrossRef]
27. Ward, F.A.; Hurd, B.; Rahmani, T.; Gollehon, N. Economic Impacts of Federal Policy Responses to Drought in the Rio Grande Basin. *Water Resour. Res.* **2006**, *42*, 1–13. [CrossRef]
28. Ahn, S.; Abudu, S.; Sheng, Z.; Mirchi, A. Hydrologic impacts of drought-adaptive agricultural water management in a semi-arid river basin: Case of Rincon Valley, New Mexico. *Agric. Water Manag.* **2018**, *209*, 206–218. [CrossRef]
29. Alley, W.M. *Five-Year Interim Report of the United States–Mexico Transboundary Aquifer Assessment Program: 2007–2012*; US Department of the Interior, US Geological Survey: Charleston, SC, USA, 2013.
30. Sanchez, R.; Eckstein, G. Groundwater management in the borderlands of Mexico and Texas: The beauty of the unknown, the negligence of the present, and the way forward. *Water Resour. Res.* **2020**, *56*, 1–20. [CrossRef]
31. Cliett, T. Groundwater occurrence of the El Paso area and its related geology. In *Guidebook of the Border Region, Chihuahua and the United States: New Mexico Geological Society Twentieth Field Conference*; New Mexico Bureau of Geology & Mineral Resources: Socorro, NM, USA, 1969; Volume 209, p. 214.
32. Hawley, J.W.; Kennedy, J.F.; Granados-Olivas, A.; Ortiz, M.A. *Hydrogeologic Framework of the Binational Western Hueco Bolson–Paso Del Norte Area, Texas, New Mexico, and Chihuahua: Overview and Progress Report on Digital-Model Development*; WRRI Technical Completion Report 349; New Mexico Water Resources Research Institute: Las Cruces, NM, USA, 2009.
33. Conover, C.S. *Ground-Water Conditions in the Rincon and Mesilla Valleys and Adjacent Areas in New Mexico*; US Government Printing Office: Washington, DC, USA, 1954.
34. Hawley, J.; Lozinsky, R.P. *Hydrogeologic Framework of the Mesilla Basin in New Mexico and Western Texas*; New Mexico Bureau of Mines and Mineral Resources: Socorro, NM, USA, 1992.
35. Hawley, J.W.; Swanson, B.H. Conservation of shared groundwater resources in the binational Mesilla Basin-El Paso Del Norte region–A hydrogeological perspective. In *Hydrological Resources in Transboundary Basins between Mexico and the United States: El Paso Del Norte and the Binational Water Governance*; Granados Olivas, A., Ed.; Universidad Autónoma de Ciudad Juarez (UACJ) and Universidad Autónoma de Chihuahua (UACH): Chiyoda City, Tokyo, 2022.
36. Révész, K.; Coplen, T.B. *Determination of the δ (2H/1H) of water: RSIL Lab Code 1574*; Révész, K., Coplen, T.B., Eds.; U.S. Geological Survey: Reston, Virginia, 2008.
37. Östlund, H.G.; Werner, E. The electrolytic enrichment of tritium and deuterium for natural tritium measurements. In *Tritium in the Physical and Biological Sciences Symposium*; International Atomic Energy Agency: Vienna, Austria, 1962.
38. Thatcher, L.L.; Janzer, V.J.; Edwards, K.W. *Methods for Determination of Radioactive Substances in Water and Fluvial Sediments*; US Government Printing Office: Washington, DC, USA, 1977; Volume 5, p. 95.
39. Del Hierro Ochoa, J.C. Comportamiento Hidrogeoquímico del Bolsón del Hueco en el área urbana de Ciudad Juárez, Chihuahua, México, 1965–1999. Ph.D. Thesis, Universidad Autónoma de Ciudad Juárez (UACJ), Ciudad Juárez, Mexico, 2009.
40. Bumgarner, J.R.; Stanton, G.P.; Andrew, P.T.; Thomas, J.V.; Payne, J.D.; Musgrove, M. *A conceptual model of the hydrogeologic framework, geochemistry, and groundwater-flow system of the Edwards-Trinity and related aquifers in the Pecos County Region, Texas*; U.S. Department of the Interior: Washington, DC, USA; U.S. Geological Survey: Reston, VA, USA, 2012.

water

MDPI

Review

Mesilla/Conejos-Médanos Basin: U.S.-Mexico Transboundary Water Resources

Andrew J. Robertson [1,*], Anne-Marie Matherne [1], Jeff D. Pepin [1], Andre B. Ritchie [1], Donald S. Sweetkind [2], Andrew P. Teeple [3], Alfredo Granados-Olivas [4], Ana Cristina García-Vásquez [5], Kenneth C. Carroll [6], Erek H. Fuchs [7] and Amy E. Galanter [1]

1 U.S. Geological Survey, New Mexico Water Science Center, Albuquerque, NM 87113, USA; matherne@usgs.gov (A.-M.M.); jpepin@usgs.gov (J.D.P.); abritchie@usgs.gov (A.B.R.); agalanter@usgs.gov (A.E.G.)
2 U.S. Geological Survey, Geosciences and Environmental Change Science Center, Denver, CO 80225, USA; dsweetkind@usgs.gov
3 U.S. Geological Survey, Oklahoma-Texas Water Science Center, Austin, TX 78754, USA; apteeple@usgs.gov
4 Instituto de Ingeniería y Tecnología, Ciudad Juárez, Universidad Autónoma de Ciudad Juárez, Ciudad Juárez 32500, Mexico; agranados@uacj.mx
5 Water Science & Management, New Mexico State University, Las Cruces, NM 88003, USA; gava@nmsu.edu
6 Plant & Environmental Sciences Department, New Mexico State University, Las Cruces, NM 88003, USA; kccarr@nmsu.edu
7 Elephant Butte Irrigation District, Las Cruces, NM 88005, USA; efuchs@ebid-nm.org
* Correspondence: ajrobert@usgs.gov; Tel.: +1-505-830-7975

Abstract: Synthesizing binational data to characterize shared water resources is critical to informing binational management. This work uses binational hydrogeology and water resource data in the Mesilla/Conejos-Médanos Basin (Basin) to describe the hydrologic conceptual model and identify potential research that could help inform sustainable management. The Basin aquifer is primarily composed of continuous basin-fill Santa Fe Group sediments, allowing for transboundary through-flow. Groundwater flow, however, may be partially or fully restricted by intrabasin uplifts and limited recharge. The shallow groundwater in the Rio Grande alluvium receives recharge from the Rio Grande and responds to changes in water supply and demand. About 11% of Rio Grande alluvial groundwater volume is recharged annually, an amount that is less than recent withdrawals. Potentially recoverable fresh to slightly brackish groundwater was estimated at 82,600 cubic hectometers in the U.S. portion of the Basin and 69,100 cubic hectometers in the Mexican portion. Alluvial groundwater geochemistry is governed by the evaporative concentration of the Rio Grande and agricultural diversions, whereas deeper groundwater geochemistry is governed by mixing and geochemical processes. Continued refinements to storage estimates, the water budget, and deep groundwater extent and geochemistry can improve estimates of sustainable use and inform alternative water sources.

Keywords: transboundary; water resources; Rio Grande; conceptual model; hydrogeology; geochemistry

Citation: Robertson, A.J.; Matherne, A.-M.; Pepin, J.D.; Ritchie, A.B.; Sweetkind, D.S.; Teeple, A.P.; Granados-Olivas, A.; García-Vásquez, A.C.; Carroll, K.C.; Fuchs, E.H.; et al. Mesilla/Conejos-Médanos Basin: U.S.-Mexico Transboundary Water Resources. *Water* **2022**, *14*, 134. https://doi.org/10.3390/w14020134

Academic Editor: Aizhong Ye

Received: 31 August 2021
Accepted: 18 December 2021
Published: 6 January 2022

Publisher's Note: MDPI stays neutral with regard to jurisdictional claims in published maps and institutional affiliations.

1. Introduction

As in many arid regions, increasing water demand due to increasing population and agricultural uses within the Mesilla/Conejos-Médanos Basin (Basin), combined with multi-year drought conditions [1], has resulted in diminished surface-water supplies and increased reliance on groundwater withdrawals. The Basin, which is largely defined by the presence of basin-fill sediments and bounding uplifts, is home to a regionally important transboundary aquifer underlying portions of Texas and New Mexico (U.S.), as well as Chihuahua (Mexico) (Figure 1). The Rio Grande, which flows through the northeastern portion of the Basin (Figure 1), along with groundwater, provides water for the residents and industries of Las Cruces, New Mexico; El Paso, Texas; and Ciudad Juárez, Chihuahua, along with numerous smaller communities. The Basin is also home to one of the largest agricultural producing regions in the state of New Mexico [2]. The high demands on the water resources, coupled with decades of reduced streamflow, are resulting in reduced groundwater supplies. The New Mexico Universities Working Group on Water Supply Vulnerabilities (2015) reports that the groundwater resources in the Basin " ... may no longer have the capacity to provide a reliable, supplemental supply during extended drought conditions and with the current levels of intensive use of groundwater." [3].

Figure 1. Location of the Mesilla/Conejos-Médanos Basin in Texas and New Mexico (U.S.), as well as Chihuahua (Mexico). Base terrain and geographic references from U.S. Geological Survey (2021) [4], Teeple (2017) [5] and Driscoll and Sherson (2016) [6]. The geographic coordinate system is the North American Datum of 1983.

The transboundary and interstate nature of the Basin aquifer adds to the complexity of resolving competing demands. The research needs for a scarce conjunctive-use resource that is managed by multiple governments benefits from not only understanding of the physical characteristics and limits of the resource but cooperation and common objectives of stakeholders. Surface water of the Rio Grande (known as the Rio Bravo in Mexico) is allocated according to a series of legal agreements, including the tri-state (Colorado, New Mexico, and Texas) Rio Grande Compact, the federal Rio Grande Project, and the 1906 U.S.-Mexico treaty. In contrast to the complex legal structures apportioning surface water, there are no state compacts or binational treaties to govern the apportionment of groundwater. Pending (2022) lawsuits (State of Texas v. State of New Mexico and State of Colorado [No. 141, Orig.] and the Rio Grande adjudication, New Mexico v. Elephant Butte Irrigation District, et al. [No. CV 96-888]) add uncertainty to the existing legal framework and future allocation of the limited resources, putting increased emphasis and importance on the Basin and resource characterization to support sound management decisions.

The objective of this work is to summarize the state of the hydrologic science in the Basin, describe the working hydrologic conceptual model and water budget, and identify potential research that would add key information to assist sustainable binational management. The U.S. portion of the Basin is often referred to as the Mesilla Basin and will be referred to as the Mesilla portion of the Basin in this report. The Mexican portion of the Basin is often referred to as the Conejos-Médanos aquifer and will be referred to as the Conejos-Médanos portion of the Basin in this report. Much of the work in the Basin has focused on either the U.S. or Mexican portion, and interpretative study areas generally do not extend across the international border. Most of the published work available and cited in this article is from the U.S. side of the Basin and primarily focuses on the Mesilla Valley (Figure 1). This work seeks to synthesize that research and describe it in the context of the international resource. Specifically, Basin characteristics, surface water, groundwater, and water-chemistry information are used to (1) determine the extent of the regional aquifer and estimate the amount of groundwater exchange between Mesilla and Conejos-Médanos portions of the aquifer; (2) evaluate prior estimates of groundwater storage in the regional aquifer and develop a revised estimate using updated information, (3) determine the extent and amount of present-day recharge, and (4) summarize the source and mechanisms of salinity near Paso del Norte, where the Rio Grande and groundwater flow out of the Basin.

To bridge the disparate research and gain a more complete understanding of the regional aquifer system, new estimates of storage and water-budget components are introduced, as well as a binational water-level map. This work introduces new storage estimates for the Mesilla portion of the Basin using a newly constructed digital hydrogeologic framework [7], as well as for the Conejos-Médanos portion of the Basin, for which storage has not been previously estimated. An international water-level map that combines groundwater elevation data from both sides of the international border was developed in order to visualize potential groundwater flow through the Basin. Finally, a water-budget approach is presented that includes a new estimate of groundwater flow across the international border. Three versions of the water budget are developed under different assumptions by using data for the period of record, a dry period, and a wet period to compare differences resulting from surface-water availability.

This paper is organized into conventional water resource assessment sections, and the following narrative provides a roadmap for how each these sections address the objective and research objectives listed above. To quantify the amount of water available in the Basin, Section 3 summarizes the hydrogeology of the Basin sediments that make up the aquifer and bounding features, along with descriptions of faults and sub-basins that may affect groundwater flow. Streamflow characteristics and variability are described in Section 4.1. Estimates of the aquifer dimensions and hydrologic properties are then combined with groundwater levels in Section 4.2 to estimate the amount of groundwater in storage in the Basin. Using the supporting evidence of groundwater fluctuations in Section 4.2 and water chemistry in Section 5, the extent and fraction of present-day recharge from surface water is estimated in the water budgets presented in Section 6. Potential contributors to salinity and their extent are introduced in Section 4 and further developed in Section 5. Uncertainty and gaps in data and understanding are noted throughout the paper and are summarized in Section 7.

2. Materials and Methods

The information and research summarized in this paper were gathered from published data and interpretations. Several previous and current federal, state, and local programs are responsible for the large body of work contributing to our understanding of the Basin. Key works used in this paper to estimate groundwater storage and to delineate groundwater flow direction include previous descriptions of geologic structure, basin-fill sediments, and bounding features by Hawley and Kennedy (2004) [8] and Sweetkind (2017) [7] for the Mesilla portion of the Basin, as well as Jimenez and Keller (2000) [9] and Servicio Geológico Mexicano (SGM) (2011) [10] for the Conejos-Médanos portion of the Basin. Previous storage estimates are limited to the Mesilla portion of the Basin and have been reported by Hawley and Kennedy (2004) [8], Wilson and colleagues (1981) [11], and Balleau (1999) [12]. Water-chemistry data reported by numerous authors, but primarily Witcher and colleagues (2004) [13] and Teeple (2017) [5], were used to describe the extent of recharge and groundwater flow, as well as salinity increases at the terminus of the Basin, near Paso del Norte (Figure 1). To understand the extent and quantity of present-day recharge as a contribution to the water budget, data were compiled from Wilson et al. (1981) [11], the numerical modeling efforts of Frenzel and Kaehler (1992) [14] and S.S. Papadopulos and Associates, Inc. (SSPA; 2007) [15], from Hanson and colleagues (2020) [16], and the associated data release by Ritchie et al. (2018) [17]. Interpretation of the present groundwater/surface-water interactions was supported with long-term data-collection records available from the U.S. Geological Survey (USGS; 2021) [18].

The total volume of groundwater stored in the aquifer system of the Mesilla portion of the Basin and a small part of the Conejos-Médanos portion was estimated using the method documented by the Texas Water Development Board (TWDB), which differentiates between groundwater storage in unconfined and confined aquifers [19,20]. Groundwater storage for unconfined aquifers (S_u) is equal to the volume of water that would be released by dewatering the entire saturated thickness of the aquifer and is calculated as follows:

$$S_u = A\ (b_s)\ S_y \tag{1}$$

where A is the area of the saturated aquifer, b_s is the saturated thickness (equal to the groundwater elevation minus the elevation of the bottom of aquifer), and S_y is the specific yield. Specific yield is the ratio of the volume of water that drains from a saturated rock by gravity to the total volume of the saturated aquifer.

Groundwater storage in confined aquifers (S_c) consists of the volume of water that would be released by dewatering the entire aquifer and the elastic properties of the aquifer (expansion of water and deformation of aquifer solids), calculated as:

$$S_c = [A\ (b_s)\ S_y] + [A\ (h)\ Ss\ (b_a)] \tag{2}$$

where b_a is the aquifer thickness (equal to the elevation of the top of the aquifer minus the elevation of the bottom of aquifer), h is the hydraulic head (equal to the groundwater elevation minus the elevation of the top of the aquifer), and Ss is the specific storage. Specific storage is the amount of water released from or taken into storage per unit volume of a porous medium per unit change in hydraulic head. Other terms are as defined for Equation (1).

The elevations of the top and bottom of the aquifer units were estimated by using the three-dimensional hydrogeologic framework model developed by Sweetkind (2017) [7]. Sweetkind (2017) assigned a thickness of about 1.5 meters (m)to a hydrogeologic framework model unit where it was absent within the stratigraphic sequence [7]; these areas were excluded from the total volume estimates. The groundwater elevation was estimated from the groundwater-elevation surface presented in Section 4. Equation (1) was used to calculate the S_u where the groundwater elevation was equal to or below the top of the aquifer unit, and Equation (2) was used to calculate the S_c where the groundwater elevation was above the top of the aquifer unit. An S_y of 0.1 and an Ss of 0.00001 per foot were assumed for all aquifer units. These values were chosen to be consistent with reported values and previous storage estimates [8].

The three-dimensional hydrogeologic framework model used to estimate the volume of the Mesilla portion of the Basin represents about 45% of the total area of the Mesilla/Conejos-Médanos Basin, as depicted in Teeple (2017) [5] (Figure 1). Thus, only a small portion the northern part of the Conejos-Médanos portion of the Basin, roughly corresponding to the southernmost active extent of the Rio Grande Transboundary Integrated Hydrologic Model [16], was included in the volume estimates for the Mesilla portion of the Basin. In addition, areas with a high standard error from kriging the groundwater elevations were excluded from the total volume estimates.

Different spatial resolutions were used for the three-dimensional hydrogeologic framework model, a 200-m grid-cell size, and for the groundwater-elevation surface, a 23-m grid-cell size. To combine the different spatial resolutions, the ArcGIS Zonal Statistics tool [21] was used to calculate the mean groundwater elevation within each hydrogeologic framework model grid.

A binational water-level map was constructed by interpolating median groundwater-elevation data from measurements made in 2010 within the basin-fill sediments [10,18]. Measurements made in Mexico and reported in the National Geodetic American Vertical Datum of 1929 were converted to reference the North American Vertical Datum of 1988 (NAVD88) by using the correction factors from Carrera-Hernández (2020) [22] so that measurements within both countries were relative to a common datum. Data that were not measured within the basin-fill sediments or that were noted as being affected by nearby pumping, surface water, or other factors were omitted from the analysis. The final dataset included a total of 217 measurement locations, with 108 in Mexico and 109 in the United States [23]. Median values from these locations were contoured by using universal kriging interpolation techniques in R (version 3.5.3), an open-source programming language [24,25]. The kriging configuration used a 2nd-order trend with an interaction term, a spherical variogram model (range = 61,134 m; sill = 592 square meters (m^2); nugget = 0 m^2), and a 23-m grid-cell size. Portions of the map that were associated with relatively high interpolation uncertainty, as determined by the kriging standard error and largely due to sparce data coverage, were omitted from the final map [23]. Elevations referenced throughout this paper are reported relative to the NAVD88.

3. Geologic Setting and Hydrogeology

The semiarid Mesilla/Conejos-Médanos Basin (Basin) covers 8290 square kilometers (km^2), of which about 5960 km^2 are in Chihuahua, Mexico [8,10]. Physiographically, the Mesilla portion of the Basin can be divided into the West Mesa, the incised Mesilla Valley that contains the Rio Grande and associated sediments, and the East Bench (Figure 1). The Conejos-Médanos portion of the Basin is divided into the lowlands and dunes of Laguna El Barreal to the southwest and the mesa features that may be considered as the southern extension of the West Mesa in New Mexico (Figure 1). Basin elevations range from about 1100 m near the Paso del Norte to greater than 2700 m in the Organ Mountains east of Las Cruces (Figure 1) [16]. Average annual precipitation in Las Cruces, New Mexico is about 21.3 centimeters (cm) per year [26], and average annual precipitation from four climate stations in and near the Conejos-Médanos portion of the Basin is about 16.8 cm per year [27]. The average annual estimated reference evapotranspiration ranges from about 187 cm/year along the Rio Grande Valley to about 79 cm/year in the surrounding mountains [16]. The Rio Grande is an important surface-water feature flowing through the northern Mesilla portion of the Basin, whereas ephemeral surface waters in the Conejos-Médanos portion of the Basin are considered to be in closed basins and not connected to the Rio Grande.

3.1. Geologic Setting

The Basin is located in the southern part of the Rio Grande rift, a tectonic feature that is characterized by a series of generally north-south-trending structural extensional basins. The still-active Rio Grande rift has been evolving for more than 25 million years through episodic crustal extension and basin subsidence [8]. The Basin is bound by volcanic highlands and fault-block (horst) ranges that expose tilted Paleozoic and Early Cretaceous carbonate and siliciclastic rocks and include some Tertiary igneous intrusions [8,28–30]. The eastern margin of the Mesilla portion of the Basin is bound by the Organ and Franklin Mountains, and the western margin by fault blocks and volcanic uplands of the East Potrillo Mountains and West Potrillo Mountains. The Robledo and Doña Ana Mountains define the northern end of the Basin, except in the northeast, where it transitions to the Jornada del Muerto Basin (Figure 1) and where interbasin groundwater flow is reported to occur [5,8]. The Conejos-Médanos portion of the Basin is bound to the east by low-elevation (<1490 m) mountains and hills, including Sierra de Juárez, Cerro El Picacho, and Sierra de Samalayuca; to the south by the Cenozoic outcrops around Sierra La Candelaria; and to the southwest by the hills and mountains west of Laguna El Barreal, such as Sierra El Malpais and Sierra La Nariz [31].

Tectonic deformation and faulting within the Basin formed a series of structural sub-basins and uplifts (Figure 2) [7,8,30,32]. These interbasin and intrabasin structures play a major role in groundwater flow. The Mesilla portion of the Basin is divided into sub-basins by a normal-fault bounded horst block, the Mid-Basin Uplift. The Mid-Basin Uplift does not extend through the entire saturated thickness but may result in the division of deeper groundwater flow [8]. To the north, the entire saturated thickness of the southern portion of the Jornada Del Muerto Basin (Figure 2) is separated from the Basin by sub-crops resulting from a complex series of faults [7,8]. In the Conejos-Médanos portion of the Basin, a series of northwest—southeast trending sub-basins, Los Muertos, El Parabien, and Southcentral (or Conejos-Médanos), are defined by mostly buried uplifts [9,33]. Previous groundwater-level maps indicate the potential for groundwater throughout the Conejos-Médanos portion of the Basin to flow north [10,27]; however, these buried uplifts may restrict northerly groundwater flow to only a small portion near the international border.

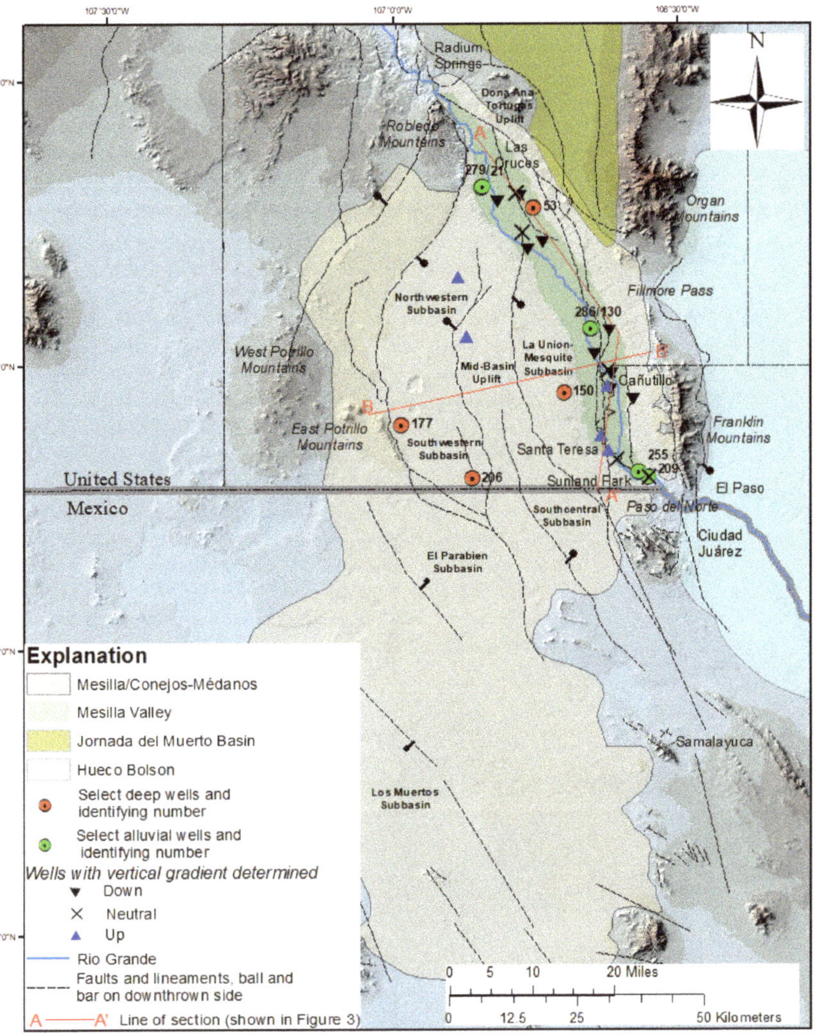

Figure 2. Major faults and structural features of the Mesilla/Conejos-Médanos Basin. Also included are the locations of nested wells, where vertical gradients have been determined, and select deep and shallow wells with long-term groundwater-level records. Base terrain and geographic references are from [5,6,34]. Coordinates are the geographic coordinate system relative to the North American Datum of 1983.

3.2. Hydrogeology

Aquifers in the Rio Grande rift, like the Basin, are made up of saturated recent valley -fill deposits and older thick intermontane sedimentary basin-fill deposits. While climate regimes influenced the geomorphology of the fluvial deposits in the Basin, tectonics were—and are—the primary controls for the aggradation from the Rio Grande [35,36]. The Basin aquifer system is made up of as much as 900 m of partially consolidated eolian, alluvial, lacustrine, and fluvial (ancestral Rio Grande) deposits that comprise the Pliocene to Pleistocene Santa Fe Group [7,8,10]. A thin layer (<25 m) of unconsolidated Quaternary alluvial and fluvial deposits, known as the Rio Grande alluvium, overlies the Santa Fe Group [8]. The Santa Fe Group unconformably overlies a succession of older consolidated basement rocks that include, in descending order: lower-to-middle Tertiary and volcaniclastic rocks [7]; up to 760 m of Upper Cretaceous marine and non-marine sandstone and shale; 900 m to over 1500 m of Paleozoic dolomite, limestone, and sandstone with interbedded shale and gypsum beds [8,10]; and a complex of Proterozoic granite and metamorphic rocks.

The Santa Fe Group is often divided into three hydrostratigraphic units that roughly correspond to the stages of basin filling: the upper Santa Fe, the middle Santa Fe, and the lower Santa Fe hydrostratigraphic units (Figure 3), which are correlative with the Camp Rice and Palomas, the Rincon Valley, and the Hayner Ranch formations, respectively [7,8]. Deposition of the lower Santa Fe unit occurred between 10 and 25 million years ago (Ma) in a broad, shallow basin that predated the uplifts of the flanking mountain blocks [8]. The lower Santa Fe unit is primarily fine-grained and partly consolidated with some calcium-sulfate and sodium-sulfate evaporites and cementation, which was deposited in a closed basin. The middle Santa Fe unit is composed of eolian dune sequences up to 610 m thick that intertongue with alluvial deposits near the bounding mountains and fluvial and playa-lake deposits in the inner Basin [8]. The middle Santa Fe unit was deposited about 4 to 10 Ma, when rift tectonics were most active. Rapid aggradation in the Basin during this time, caused by subsidence of the central basin blocks relative to the surrounding uplifts [8], resulted in deposition of coarse clastic alluvial deposits derived from the uplift of the ranges bounding both sides of the basin [8]. The upper Santa Fe unit is about 3 to 4 Ma and consists of fluvial deposits from a large, braided river of the ancestral Rio Grande, with channel sands and gravels from as far north as the mountains in southern Colorado and alluvial fan deposits derived from basin-bounding highlands (Figure 1) [8]. The fluvial system discharged into the playa-lake plains of the eastern Hueco Bolson and the southern Conejos-Médanos portion of the Basin.

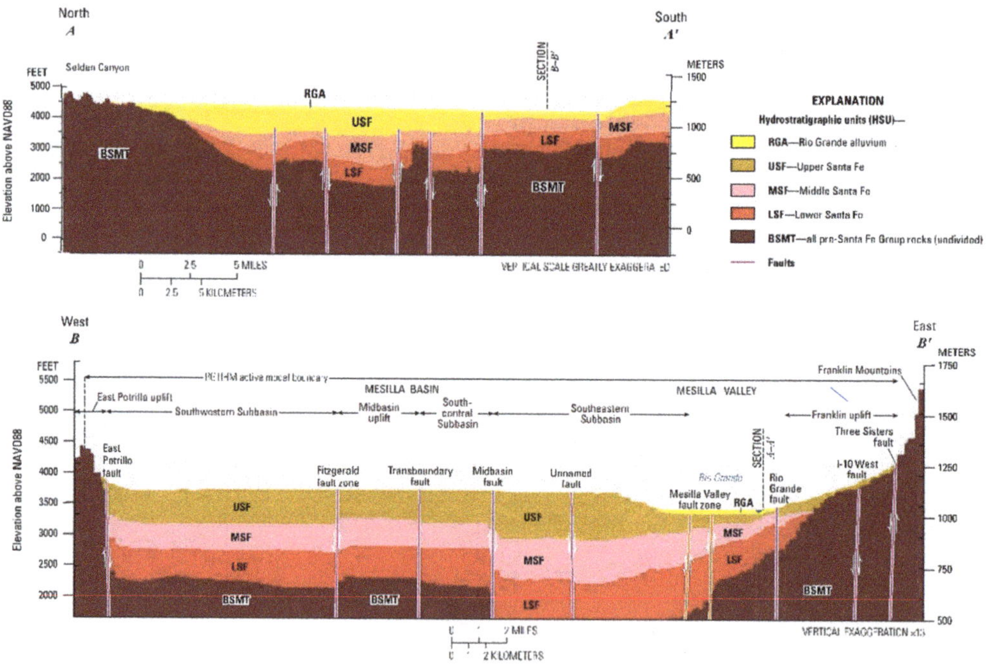

Figure 3. Cross sections showing the hydrostratigraphic units of the Mesilla/Conejos-Médanos Basin (adapted from [16]). Location of cross sections shown in Figure 2.

The Quaternary Rio Grande alluvium and upper Santa Fe unit were deposited by the through-flowing Rio Grande [16]. The alluvial sequence was produced by multiple episodes of valley entrenchment during glacial stages and aggradation during interglacial (interpluvial) stages. The Rio Grande alluvium is composed of river-channel and overbank depositional facies ranging in texture from sand and gravel to silt and clay that are generally 15 to 38 m thick, respectively [11]. A basal-channel gravel and sand layer, up to 9 to 12 m thick, was deposited during the interval of maximum valley incision near the end of the late Pleistocene ice age [37].

Deposition of basin-fill sediments as a result of both tectonics and varying climate regimes has produced the sediment fill structure that has affected groundwater conditions through time. Prior to incision by the Rio Grande, starting about 700,000 years ago, the Basin was a broad plain with water tables in the West Mesa portion of the Basin up to 110 m above present-day (2022) groundwater levels [8]. The lacustrine deposits of the ancestral Lake Palomas in the Conejos-Médanos portion of the Basin indicate lake levels were about 60 m above present-day (2022) groundwater levels [8]. Geologic evidence indicates that the Rio Grande once flowed through what is now the Jornada del Muerto Basin and discharged to the Hueco Bolson through Filmore Pass [35], indicating the potential for past and present interbasin groundwater flow. The present course of the Rio Grande probably did not become established until the middle Pliocene [35], allowing lacustrine deposits to form in the southern portions of the Basin [8].

Groundwater in the Santa Fe Group generally occurs under leaky-confined conditions as a result of interbedded clays. These clay layers result in horizontal conductivity exceeding vertical conductivity [11]. Hydraulic conductivities generally decrease with depth and with increasing distance from the source of sediments [8]. The fluvial sediments of the Rio Grande alluvium and Santa Fe Group generally represent a fining-upward sequence of sediments [16]. A broad range of aquifer-specific capacity, transmissivity, and hydraulic-conductivity values have been estimated for this aquifer system, as supported by the observed and inferred lithofacies [8]. Specific capacities of 2 to 45 liters per second per meter (L/s/m) are reported for the coarse-grained deposits in the upper Santa Fe unit and Rio Grande alluvium, but specific capacity estimates of the middle Santa Fe are usually less than 8 L/s/m and between 0.2 to 2 L/s/m for the lower Santa Fe [8]. Based on aquifer tests of about 50 wells and test holes completed in the Santa Fe Group, the average well yield is about 95 L/s [11]. Transmissivities of the upper 300 m of the aquifer in the West Mesa average around 0.01 square meters per second (m^2/s) and range from 0.01 to 0.04 m^2/s in the Mesilla Valley [8,11] Transmissivities in the Conejos-Médanos portion of the Basin range from about 0.002 to 0.004 m^2/s [10]. Vertical hydraulic conductivity measured in West Mesa wells ranged from about 10^{-6} to 10^{-5} m per second (m/s) for the entire thickness of confining layers [14]. The storage coefficients from aquifer tests range from 0.001 to 0.00003 [11,38]. The horizontal-to-vertical hydraulic conductivity anisotropy ratio of basin-fill aquifer systems in the Rio Grande rift may range from 200:1 to 1000:1 [8]. Horizontal hydraulic conductivities of the Rio Grande alluvium range from about 4×10^{-4} to 1.2×10^{-3} m/s, and the average estimated specific yield is 0.2 [37], with reported estimates ranging from 0.1 to 0.3 [39–41].

Faults throughout the Basin may allow upward leakage of saline water from older Mesozoic units into the Santa Fe Group and downward or lateral leakage of fresher water from the Rio Grande valley fill into the Santa Fe Group [34,42]. Cross faults, not mapped in the Texas portion of the Basin, have been interpreted from geophysical investigations and may produce sharp divides between deep and shallow basin structures in addition to the upflow zones [43]. Differences in thickness and lithology of the Basin sediments are observed across certain faults. For example, the Santa Fe Group on the horst near the Robledo Mountains (Figure 2) is reported to be mainly composed of clay facies, in contrast to alternating layers of sands, gravels, and clays in the graben to the south. The high groundwater gradient, lithologic data, and water-yield data associated with the horst are indicative of low transmissivity values relative to other portions of the Basin [44]. The Santa Fe Group in this area is also thinner, with Permian bedrock units found at depths as shallow as 350 m, whereas thickness of the Santa Fe Group to the south has been observed at over

730 m [11]. The West Mesa (Figure 1) also contains numerous volcanic features, including a line of cinder cones, lava flows, and a series of maars. Igneous dikes associated with these centers are often aligned with groups of faults [7]. Several wells have encountered basaltic rock at depth [8], with reports of varying degrees of groundwater productivity.

4. Water Availability

Water resources in the Mesilla/Conejos-Médanos Basin (Basin) occur both as surface water and groundwater in the U.S. portion of the Basin, whereas the Mexican portion of the Basin relies entirely on groundwater. Both comprise a substantial source of water for the semiarid region, but drought and increasing demands are reducing water availability.

4.1. *Surface Water*

The Rio Grande enters the Basin through Selden Canyon (about 1210 m elevation) and leaves through a gap known as Paso del Norte (about 1130 m elevation) that separates the Franklin Mountains from Sierra de Juárez to the south near El Paso, Texas (Figure 1). The Upper Rio Grande Impact Assessment [45] reports that supplies from all native water sources to the Rio Grande are projected to decrease, on average, by one third overall by the end of the of the 21st century. Other projections indicate increased variability in monthly and annual flows, and climate change modeling for the region indicates earlier snowmelt runoff and warmer average temperatures, leading to increased variability in the magnitude, timing, and spatial distribution of streamflow [45].

4.1.1. Streamflow

Except for a few perennial seeps and springs that flow for short distances in the surrounding mountains, the Rio Grande is the only perennial surface-water body in the Basin. The Rio Grande's drainage basin above Caballo Dam is about 71,700 km^2 (Figure 1), and about 60 to 75% of Rio Grande streamflow is derived from seasonal snowpack in the high-elevation headwaters [46,47]. While surface-water availability in the Rio Grande is largely driven by regional climate and landscape patterns and upstream water use, the local climate can affect both supply and demand in the Basin [16].

Streamflow in the Rio Grande through the Basin is controlled by Rio Grande Project (Project) demands. Surface water in the Rio Grande is apportioned by the 1906 international treaty, which apportions 74.0 cubic hectometers per year (hm^3/year) to Mexico, as well as the 1938 Rio Grande Compact that divides streamflow among the states of Colorado, New Mexico, and Texas based on measured streamflow. The Project releases water from the upstream Elephant Butte and Caballo Reservoirs (Figure 1) during the growing season (irrigation season) to provide water to the Elephant Butte Irrigation District (EBID) in New Mexico, the El Paso County Water Improvement District No. 1 (EP No. 1) in Texas, and the Juárez Valley Irrigation District 009 in Mexico. During the winter season, when no dam releases occur (non-irrigation season), the riverbed is often dry for long stretches within the Basin [6]. However, this is not only a recent phenomenon. Conover (1954) [39] reported that even prior to construction of Elephant Butte Dam, the river would sometimes dry for months at a time. Recent (2022) surface-water conditions in the Basin represent a reduced supply that has persisted since about 2000 [16]. Average annual streamflow for 1940 to 2018 from releases out of Caballo Reservoir was approximately 791 hm^3/year (standard deviation 316 hm^3/year) or about 25 m^3/s (Figure 4) [16]. In contrast, the average annual streamflow during the period of persistent drought from 2000 to 2018 was only about 663 hm^3/year (standard deviation 221 hm^3/year) or 21 m^3/s [47]. At El Paso (the outlet of the Basin), the average annual streamflow between 1940 and 2018 was 487 hm^3/year [17].

Based on precipitation and geologic records, Rio Grande streamflow in the Pleistocene would have been far greater than present (2022) [8,35]. Flood flows during that time likely covered the floodplain for weeks at a time [11]. In the decades since agricultural development in the Mesilla Valley, upstream reservoirs have greatly reduced flood flows. Following the construction of Elephant Butte Dam in 1915 and the controlled releases beginning in

1916, there was a substantial drop in peak streamflow [5]. For example, annual peak flows of 280 m^3/s, including several annual peak flows of more than 430 m^3/s, were recorded before 1915, but since 1915, the annual peak flow has not exceeded 260 m^3/s [5]. Where the Rio Grande leaves the Basin, a flow of 680 m^3/s was measured at El Paso during a flood in 1905, but no flow over 230 m^3/s has been measured since [8].

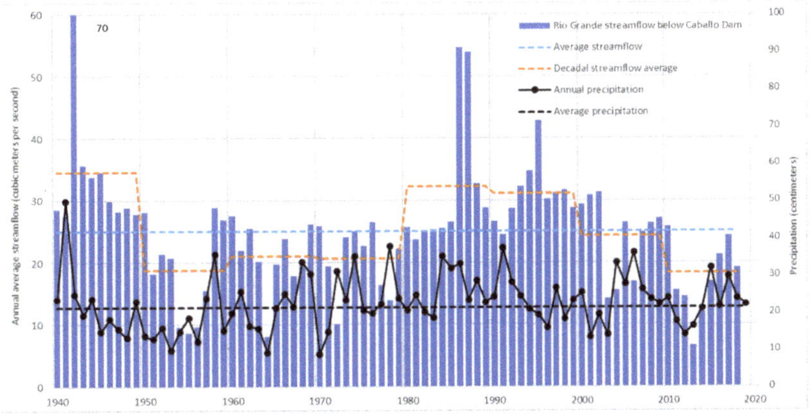

Figure 4. Annual Rio Grande streamflow below Caballo Reservoir from 1940 to 2018 [18,48], average annual streamflow for the period of record, decadal average streamflow, annual precipitation from 1940 to 2019 at the New Mexico State University COOP station [26], and average precipitation for that period of record.

As a primary source of recharge to the groundwater, the Rio Grande, through much of its 100-km length in the Basin, loses water by seepage to the aquifer [49–53]. When the river has water in it, seepage to the aquifer can maintain groundwater levels. However, when the river is dry, the water table may drop below the riverbed. When streamflow commences, the initial seepage rate is believed to be equal to the infiltration rate of the riverbed. After several weeks of streamflow, the groundwater levels rebound, and the Rio Grande hydraulically reconnects to the water table; seepage rates are then proportional to the hydraulic conductivity of the aquifer and the hydraulic gradient between the river (surface-water elevation) and the water table [38].

4.1.2. Irrigated Agriculture

Elephant Butte Dam was completed in 1916, and Caballo Dam was completed in 1938 in order to provide a more predictable supply of irrigation water to downstream users [16]. Elephant Butte Reservoir is the primary means of storage for the Project, and Caballo Reservoir, located immediately downstream from Elephant Butte on the Rio Grande, serves to regulate Project seasonal releases and electricity generation. There is about 370 km^2 of farmland in the Mesilla Valley, but currently (2022) only about 300 km^2 is in active cultivation in New Mexico and about 50 km^2 in Texas (out of about 450 km^2 of total valley area) [48,54]. A full annual allotment of water for EBID irrigators is about 338 hm^3, or about 0.9 hm^3/km^2, and was delivered each year, on a pro rata basis, between 1979 and 2002 [55]. However, drought conditions starting in the late 1990s have reduced surface-water availability in the Rio Grande and therefore reduced irrigation deliveries in the first two decades of the 21st century. As a conjunctively managed system, irrigators in the Basin offset surface-water deficits with groundwater pumping, resulting in increased pumping when surface-water supplies are limited. Starting in the early 1950s, because of drought and reduced surface-water availability, irrigators in the Mesilla Valley began rapidly developing the groundwater resources [11]. This was limited primarily to the shallow alluvium during early development but has since expanded to include many wells completed in the deeper

Santa Fe Group. Municipal, industrial, and domestic dependence on groundwater also expanded during this time, but irrigation remains the dominant use of groundwater [56].

Water released from Project reservoirs first flows thorough the Palomas Basin, where some water is diverted to irrigate fields. Water for irrigation in the Mesilla Valley is diverted from the Rio Grande at Leasburg Dam near the mouth of Selden Canyon and at the Mesilla Dam south of Las Cruces (Figure 1). The diverted water then enters a network of irrigation canals and laterals that convey and deliver surface water to farm headgates for irrigation use. Historically, the canals and laterals were unlined earthen structures, but an increasing number have been lined or converted to pipe to reduce conveyance and evaporative losses [16].

Agriculture is both the primary user of surface water and a substantial contributor to groundwater recharge. Groundwater recharge from irrigated agriculture occurs as deep percolation (water that infiltrates below the root zone) because of the infiltration of applied water on crop lands and infiltration through the conveyance structures [14,15]. Historically, about 40 to 50% of the headgate diversions made it to the fields [11], but that percentage has increased with efficiencies in operations and canal improvements. Operational and measurement errors, noted in the past, have been minimized in part because drought conditions have sensitized the farming community to the value of water and because operational and maintenance programs and engineering design improvements have evolved substantially. The losses in the system can include unintended discharge back to the river following diversion into the canal system, ditch breaks, and evaporation and transpiration. Losses also occur as channel seepage into the groundwater [57]. In order to reduce water-table elevations from channel seepage and deep percolation following an increase in irrigation, drains were constructed in the 1910s and 1920s. Approximately 320 km of drains were constructed to maintain optimal growing conditions by preventing waterlogged soils [16,39]. The drains remove excess irrigation water from the groundwater and return it to the Rio Grande at various locations. During surface-water shortfalls, groundwater levels may fall below the drain elevation and reduce or eliminate flows [14]. Based on average annual discharges in selected drains in the Mesilla Valley, Wilson and others (1981) estimated the discharge from these drains to the Rio Grande at just under 123 hm^3/year [11]. Presently (2022), however, the average water-table elevation throughout most of the Mesilla Valley has fallen below the design grade of the drains, rendering most of the drains dry.

4.1.3. Salinity

Water-quality impairment can lead to reduced water availability. A major source of water-quality impairment is salinity, often expressed as the concentration of dissolved solids (DS) [58]. Salinization of aquifers and surface water in arid regions is a growing concern because of increasing water use resulting from population growth and increasing agricultural demands [59]. The Rio Grande has been shown to have experienced substantial increases in salinity as it flows from north to south, in particular at the terminuses of the rift basins [60–66]. In the Basin, salinity increases in the Rio Grande have been documented by numerous researchers, and the source of the salinity has been the subject of multiple works [5,6,13,42,59,63,64,67]. Major salinity sources have been attributed to irrigation [68], evaporite mineral dissolution [67], and shallow evaporite brines [65,66]. Some have also suggested that topographically driven groundwater flow, a concept originally described by Toth [69], is responsible for the higher-salinity groundwater located near the Paso del Norte [8,13,63,64,70]. Topographically driven groundwater flow has also been shown in other Rio Grande rift basins where terminal bedrock constrictions force deep groundwater upward toward the surface and into the Rio Grande [64,70].

The two major sources of salinity increases in the Rio Grande are likely the leaching of salts from soils by irrigation and a deep, higher-salinity groundwater source discharging to the river [5]. Increased irrigation in the late 1910s resulted in more irrigation water recharging the shallow groundwater system, causing the water table to rise and salts to accumulate in the soils [14]. As these salts were leached from the soils by the application of irrigation water that seeped past the root zone, the salinity in the shallow groundwater system increased [71]. Wilson and others (1981) and Phillips and others (2003) indicated that increasing DS concentrations from Caballo Dam to El Paso are primarily due to evapotranspiration of irrigation water that percolated to the water table and is removed as drain flow [11,70]. As added evidence, they also reported that during irrigation season, the salinity of drain flow can be two to three times greater than salinity in the river [11,70]. Szynkiewicz and colleagues (2011) used sulfur isotopes to conclude that sulfate concentrations in the Rio Grande were consistent with fertilizers and not a geologic evaporite source [59]. However, the sulfate ratios in the groundwater in the Mesilla Valley indicate multiple sources of sulfate. Bedrock dissolution was a more likely source of sulfate in deeper portions of the aquifer and in groundwater near Paso del Norte. In later work, Szynkiewicz and others (2015) attributed up to 60% of the non-irrigation sulfate concentrations in the Rio Grande near Paso de Norte to evaporative brines, which was reduced to about 20% of the sulfate concentrations during the irrigation season [65,66].

Other investigations have focused on natural processes of salinity increases, such as groundwater recharge associated with Paleozoic sedimentary rocks, the upwelling of deep basinal waters, and geothermal outflows [13,62–64,70,71]. Based on the mass-balance modeling of Mills (2003) [61] and an isotopic mixing model by Hogan and colleagues (2007) [63], groundwater near Paso del Norte has been estimated to contribute approximately 6500 to 9750 tons of chloride per year to the Rio Grande, or between 10 to 15% of the annual chloride load in the Rio Grande at El Paso. Helicopter frequency-domain electromagnetic data showed a marked decrease in relative resistivity in the middle of the Mesilla Valley, near Fillmore Pass (Figure 2) [5]. This resistivity change, along with DS concentrations from groundwater wells, indicates a sharp increase in the spatial distribution of DS concentrations in the groundwater and a potential source of salinity for the Rio Grande [5]. Hawley and Kennedy (2004) suggested the possibility of preferential pathways for deeply circulating fluids in the Paleozoic and Cretaceous carbonates that are exposed and shallowly buried near Paso del Norte [8]. This suggestion is supported by the presence of extensive fractures associated with fault zones and observed dissolution features in outcrops. Higher salinity in shallow groundwater and the Rio Grande near Paso del Norte is also reported to be caused by structurally forced upwelling of brackish and saline water from deep groundwater and by the upflow of geothermal water from shallow bedrock structures and bedrock boundaries [13,42]. Teeple (2017) reported lower resistivity at depth, indicating saline water upwelling though fractures in the bedrock [5].

4.2. Groundwater

The Mesilla portion of the Mesilla/Conejos-Médanos Basin (Basin) has been reported to host up to an estimated 80,200 hm^3 of recoverable fresh and moderately brackish water (<3000 milligrams per liter of dissolved solids) [8]. However, most of that groundwater was recharged tens of thousands of years ago during the cooler and wetter parts of the Pleistocene [8]. In this section, groundwater flow patterns, revised storage estimates, and long-term groundwater levels are described to identify potential recharge sources and to estimate the amount of groundwater potentially available.

4.2.1. Groundwater Flow

The distribution of the groundwater resources in the Basin is controlled by the location of recharge and discharge, aquifer properties, and hydraulic gradient and may be interrupted by local gradient changes and hydraulic barriers, such as faulting and subcropping bedrock highs. Groundwater levels in the Rio Grande alluvium are shallow and unconfined and generally decrease from north to south at an average gradient of about 0.8 to 1.1 m/km [38], closely following the topographic gradient. Groundwater flow directions are influenced locally by hydraulic stresses, such as leakage to or from the Rio Grande, drains and canals, groundwater pumping, and infiltration from heavily irrigated fields [5,37]. Groundwater elevations within the Santa Fe Group decrease from the Basin margins to the

Basin outlet at the Paso del Norte (Figures 2 and 5). The hydraulic gradient in the Santa Fe Group ranges from about 19 m/km in the northwestern part of the Basin (near the Robledo Mountains) to less than 0.4 m/km near Paso del Norte [38]. Overall groundwater flow in the Mesilla portion of the Basin is southeasterly, toward the Paso del Norte (Figure 5), and may be partially restricted by the Mid-Basin Uplift [5]. The general groundwater flow direction in the Conejos-Médanos portion of the Basin is north and northeast (Figure 5) [10]. Groundwater pumping locally alters groundwater flow by creating cones of depression in the central parts of the Mesilla Basin near Las Cruces, New Mexico and Cañutillo, Texas, as well as at the Juárez well field (which supplies water to Ciudad Juárez, Chihuahua) near the international border [5,10] (Figure 5). Groundwater flow may also be affected by faults in the area, but 2010 (Figure 5) and previous groundwater-elevation maps have not indicated substantial gradient changes across mapped faults except where they are associated with large topographic uplifts. Water-level elevations also indicate hydraulic connection between the saturated sediments underlying the West Mesa and those in the Mesilla Valley [8,11]. Water-level contours interpreted by SGM in 2007 [10] and those presented here (Figure 5) indicate largely uninterrupted groundwater flow over the entire Conejos-Médanos portion of the Basin. However, the presence of intrabasin uplifts between structural sub-basins [9] and the lack of substantial recent recharge [8] could interrupt groundwater flow in the Santa Fe Group between the Conejos-Médanos sub-basins.

Despite reports of groundwater mixing [5,13,72] and lack of a single direction of flow (Figure 5), the groundwater system along the Mesilla Valley appears to be adequately characterized by an active shallow system, up to about 30 m deep, primarily in the Rio Grande alluvium, as well as long flow paths in the deeper Santa Fe Group. The shallow groundwater flows occur as relatively fast flow paths (<5 years), as evidenced by seasonal and annual groundwater-level fluctuations and age-dating tracers [5,14,38]. Deep groundwater elevations do not notably change in response to seasonal or annual climate or surface-water changes (except near production wells) [14]. Outside the Mesilla Valley, the Basin structure and groundwater elevations indicate a groundwater system in which recharge occurs primarily as mountain-front recharge and flows are at depth, with mixing occurring only with other deep groundwater flow paths and geothermal upwelling. Most groundwater naturally discharges through hydraulic upwelling to the river and shallow groundwater system and is lost to evaporation and transpiration near Paso del Norte at the southern end of the Mesilla Valley, where a thin alluvial veneer overlies a bedrock high [11].

Groundwater Recharge

Groundwater flow can occur when recharge at a location increases groundwater elevations, creating a hydraulic gradient. Previous studies have identified multiple sources of recharge; however, all conclude that the predominant source of recharge is water from the Rio Grande. Investigations into recharge include evaluations of groundwater gradients, seepage measurements, water-budget estimates, water chemistry (in particular the stable isotopes of water), and seasonal groundwater fluctuations.

Many studies conclude that the Basin groundwater in the Mesilla Valley is recharged from surface water associated with Rio Grande streamflow and irrigated agriculture [5,8,11,13,14,16,38,53]. Infiltration of the Rio Grande to the aquifer can exceed evapotranspiration. As a result, recharge for the Basin primarily occurs as vertical flow from the surface-water system (the Rio Grande, canals, laterals, drains, and irrigated cropland [11]. Previous seepage investigations conducted during steady, low-flow conditions indicate that the Rio Grande is often a losing stream along most of the 100-km reach in the Mesilla Valley but with gains documented at the mouth of Selden Canyon and near the Paso del Norte [38,53]. Surface-water seepage measurements from the Rio Grande have typically been collected in the winter and range from about 0.2 to 1.4 m^3/s between Selden Canyon and Paso del Norte [11,15,39,49–52]. Increases in streamflow in the Rio Grande are rapidly followed by increases in groundwater elevations in the shallow subsurface near the Rio Grande, and recent groundwater declines throughout the shallow aquifer are closely related to reductions in surface-water availability [11]. In addition to channel seepage, much of the streamflow through the Basin is diverted for use in irrigated agriculture, and recharge results from both the direct application of water to fields and seepage from the valley-wide irrigation conveyance system (laterals and canals). Stable isotopes of water measured along the Rio Grande and irrigation structures were observed to become increasingly enriched in the heavier hydrogen and oxygen isotopes in the direction of flow, indicating the cumulative effects of evaporation [70]. Subsequent research has shown that shallow groundwater in much of the Rio Grande alluvium has isotopic signatures that are similar to surface water in the Rio Grande, enriched by evaporation, indicating a large component of recent Rio Grande recharge [5,13,42]. In contrast, the stable isotope compositions of water measured in deeper wells in the Basin are more depleted than in the shallower wells, indicating a large component of recharge from the Rio Grande that occurred during the cooler periods of the Pleistocene [5,13,42].

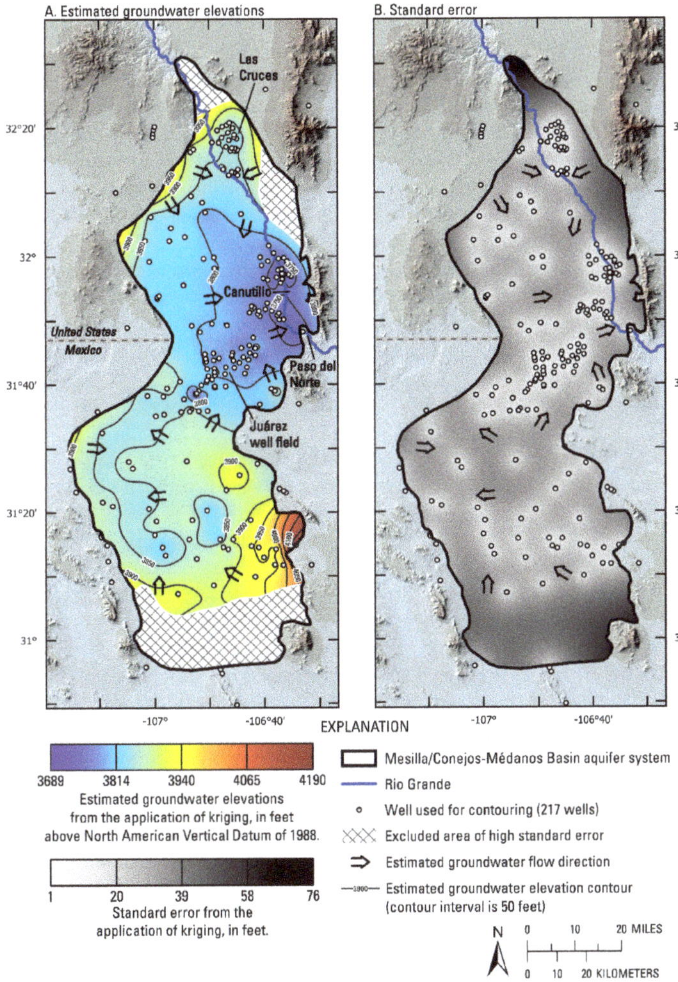

Figure 5. Estimated groundwater elevations (**A**) and standard error values (**B**) from application of kriging in the Mesilla/Conejos Médanos Basin [23]. Base terrain and geographic references are from [5,6,34]. Coordinates are the geographic coordinate system relative to the North American Datum of 1983.

Despite the large surface area underlain by the Basin aquifer system, evaporation and transpiration by desert vegetation creates large negative soil pressures that prevent the deep percolation of precipitation beneath the root zone over most of the Basin [14]. Near-surface caliche deposits also limit the potential recharge from precipitation [10,38]. For these reasons, recharge in arid lands is often the result of focused recharge from ephemeral streams or ponded water [73]. Estimates of recharge from precipitation on the West Mesa range from 0 to 0.34 hm^3/year [14,15] and up to 3.0 hm^3/year for the Conejos-Médanos portion of the Basin [10,14]. Although recharge from precipitation over much of the Basin may be considered negligible [38], several models have simulated the effective precipitation for agriculture and concluded that amount of deep percolation to be between 5 and 64 hm^3/year (about 6 to 10% of the total amount of water applied) [14,15].

Higher groundwater elevations at the boundaries of the Basin indicate that mountain-front recharge is a source of groundwater inflow (Figure 5) [11]. Based on a geochemical analysis, Teeple (2017) identified groundwater at the margins of the Basin as having a substantial amount of mountain-front recharge because the water-chemistry characteristics were distinct from other groundwater types [5]. The age-dating tracers associated with those samples indicate, however, that mountain-front recharge flows slowly along deeper flow paths in the aquifer. Most estimates of mountain-front recharge assume that 2% of the annual precipitation infiltrates to the aquifer over the area of the uplands [8,13]. Frenzel and Kaehler (1992) estimated that mountain-front recharge accounts for about 12 hm^3/year to the Mesilla portion of the Basin [14], whereas Hanson and colleagues (2020) estimate about half that amount using the basin characterization model [16,74]. A majority of the mountain-front recharge in the Mesilla portion of the Basin is thought to originate from the Organ and Franklin Mountains, while a smaller portion is thought to originate from the East and West Potrillo Mountains [14,15]. Groundwater elevations in the Conejos-Médanos portion of the Basin also indicate recharge is occurring in the surrounding highlands (Figure 5). The Comisión Nacional del Agua (CONAGUA) (2020) estimates this mountain-front recharge to be about 5.2 hm^3/year [27]. Stable isotopes of water measured in Organ Mountain springs [75], high-elevation bedrock wells in the San Andres Mountains [76], and wells in the Jornada del Muerto (Jornada) Basin [77] represent mountain-front recharge and plot close to the Global Meteoric Water Line [78], indicating minimal evaporation. Mountain-front recharge is also reported as the primary recharge source of the adjacent Jornada Basin [77,79], and in work in the neighboring Hueco Bolson, Eastoe and colleagues (2008) reported similar stable isotope compositions in a group they characterized as the "Organ and Franklin Mountain group" [80]. With a few exceptions near the boundaries of the Basin, however, these isotopic compositions are not evident in groundwater samples within the Basin.

The extent of interbasin flow is unknown, but groundwater discharge into or out of the Basin is limited by the surrounding uplifts. Interbasin flow has been estimated, or speculated, between the Basin and the Palomas Basin to the north, the Jornada Basin to the north and northeast, and the Hueco Bolson at Fillmore Pass and Paso Del Norte (Figure 1). The thin alluvial sediments in Selden Canyon limit the groundwater inflow from the Palomas Basin to the Mesilla [38]. Recharge to the Basin is also reported to occur from the southern Jornada Basin, where remnants of the ancestral Rio Grande fluvial plain exist [5,8]. The previously hypothesized flow paths in the Santa Fe Group sediments overlying the bedrock high are possible [5,8], but work by Langman and Ellis (2013) [77] and Witcher and colleagues (2004) [13] indicates that most, if not all, of the Jornada Basin groundwater entering the Mesilla portion of the Basin travels through deeper and more tortuous flow paths within the buried Tertiary intrusions that divide the two basins. These deeper flow paths are present likely because of the prevalent faults in the area. In these deeper flow paths, the composition of Jornada Basin groundwater is altered because of the local geothermal effect [77]. Interbasin flow has also been speculated between the Mesilla portion of the Basin and Hueco Bolson through Fillmore Pass (Figure 2). Fillmore Pass once hosted an ancestral Rio Grande channel and has been reported to still be a potential route for interbasin flow. In a numerical model of the Hueco Bolson, Orr and Risser (1992) assigned a constant flux of about 0.32 hm^3/year into the Hueco Bolson from the Basin [81]. However, hydraulic gradients between the Mesilla Basin and Hueco Bolson provide little evidence of interbasin fluxes [8,11,14,18]. CONAGUA (2020) and SGM (2011) report interbasin flow to the Conejos-Médanos portion of the Basin primarily from the Laguna de Santa Maria aquifer in the south (Figure 1) and estimate inflow to be about 11 hm^3/year [10,27].

Historical and modern water levels indicate that groundwater moves from the Conejos-Médanos portion of the Basin into the Mesilla portion of the Basin as throughflow. Based on mapped lacustrine-sediment elevations and paleo-groundwater elevations, there was reportedly a substantial (but unquantified) amount of "paleo-groundwater flux" originating from the pluvial Lake Palomas in the Conejos-Médanos portion of the Basin, moving north to the Mesilla portion of the Basin [8]. Present groundwater elevations (Figure 5) provide evidence that groundwater continues this throughflow. Based on area estimates from cross sections developed by Hawley and Kennedy (2004) [8], hydraulic conductivity estimates by Frenzel and Kaehler (1992) [14], and groundwater hydraulic gradients developed herein (Figure 5), the median estimate of throughflow from the Conejos-Médanos portion of the Basin to the Mesilla portion of the Basin is about 22.8 hm^3/year, with a minimum discharge estimated to be about 4.7 hm^3/year and a maximum discharge of about 82.5 hm^3/year (Table 1). These estimates represent about 30%, 140%, and 520% of the estimated total annual recharge (15.8 hm^3/year) to the Conejos-Médanos portion of the Basin, respectively [27]. This northerly groundwater flow is supported by stable isotope compositions of water measured in groundwater samples collected from the Juárez well field, which indicates a distinct isotopic signature from Rio Grande, mountain-front, or geothermal waters [82] that is also present in some groundwater samples collected near the U.S./Mexico border.

Table 1. Values used in groundwater throughflow estimates between the Conejos-Médanos portion of the Mesilla/Conejos Médanos Basin (Basin), moving north to the Mesilla portion of the Basin. Area estimates are determined from cross-section K-K′ by Hawley and Kennedy (2004) [8], hydraulic conductivities are reported values from Frenzel and Kaehler (1992) [14], and hydraulic gradients are estimated from Figure 5. [m^2, square meters; m/day, meters per day; m/m, meters per meter; m^3/day, cubic meters per day; hm^3/yr, cubic hectometers per year; USF, upper Santa Fe unit; MSF, middle Santa Fe unit; LSF, lower Santa Fe unit].

Formation	Area Estimates	Hydraulic Conductivities (m/Day)			Hydraulic Gradients (m/m)		
	(m^2)	Minimum	Maximum	Median	Minimum	Maximum	Median
USF	3,252,000	13	34	21	1.89×10^{-4}	9.47×10^{-4}	4.73×10^{-4}
MSF	7,860,000	2.7	13	6.7	1.89×10^{-4}	9.47×10^{-4}	4.73×10^{-4}
LSF	6,190,000	0.61	4.3	1.5	1.89×10^{-4}	9.47×10^{-4}	4.73×10^{-4}

Formation	Discharge (m^3/Day)			Discharge (hm^3/Year)		
	Minimum	Maximum	Median	Minimum	Maximum	Median
USF	8100	103,000	32,900	2.96	37.60	12.01
MSF	4100	98,000	25,000	1.50	35.77	9.13
LSF	700	25,000	4500	0.26	9.13	1.64
Sum	12,900	226,000	62,400	4.71	82.49	22.78

Geothermal waters that are associated with the region's known geothermal systems upwell from depths of greater than 1 km to recharge shallower groundwater supplies. The brackish (1800 to more than 4800 milligrams per liter (mg/L) DS concentrations) waters are evidence of deep groundwater circulation and ascend in the northernmost portion of the Basin at Radium Springs, in the west near the East Potrillo Mountains, and along the East Bench along the eastern border zone of the Mesilla Valley [83,84]. Discharge estimates for the East Bench range from 0.01 to 19 hm^3/year and between 0.5 and 1.6 hm^3/year for the East Potrillo Mountains [83–85]. No estimates are available for the Radium Springs system [83]. These inflows represent localized upflow within the geothermal systems and tend to be focused along faults, uplifted bedrock horst blocks, and fractured igneous intrusions [83].

Vertical Groundwater Flow

Groundwater in the Rio Grande alluvium generally occurs under unconfined conditions, while groundwater in the Santa Fe Group is typically semi-confined [38]. Due to the presence of interbedded gravels, sands, and clays in the alluvium, horizontal permeability usually exceeds vertical permeability by several orders of magnitude [38]. The vertical gradient potential is generally downward in shallower wells (<150 m) but becomes smaller and, at some locations, reverses with depth, potentially allowing for mixing of deep waters where clay layers may be discontinuous [11,17]. This condition is evident in several nested wells in the Mesilla Valley and in two paired deep wells on the West Mesa (Figure 2). Locations where upward vertical gradients are persistent and throughout the water column [5,17] are near the U.S./Mexico border, where groundwater gradients (Figure 5) indicate a confluence of horizontal flow directions where several faults occur (Figure 2) [60,84]. The location of the upward gradients may also result from the thinning aquifer (Figure 3), in contrast to farther south, where vertical gradients are neutral and the aquifer thickness remains relatively uniform (Figure 3) [17,72]. Large production well fields, like the one near Cañutillo, Texas (Figure 2), can also induce large changes in the vertical gradient by locally lowering groundwater elevations in vertically adjacent aquifer units.

Groundwater Age

Age-dating tracers (tritium and carbon-14) in samples collected from groundwater wells in the Basin generally indicate increasing age of residence with depth, from modern waters near the surface to older waters with increasing depth. Indications of age from carbon-14 analysis in samples from the Basin groundwater tend to be grouped into modern (near 100 percent modern carbon; pmc), moderately old (between 50 and 100 pmc), and old (<17 pmc) [5]. Age-dating tracers also indicate that recent recharge in quantities large enough to dilute older groundwater is occurring primarily in the Rio Grande alluvium [5,13]; however, Teeple (2017) identified a few locations where modern recharge may be mixing with older waters in the Santa Fe Group [5]. Two of those locations are near production wells that have been documented to create cones of depression, and the third is near the terminus of the Basin, where the aquifer thins. These findings support the conceptual model of groundwater flow in which there are locally nested flow systems in the Mesilla Valley that overlie more extensive and deep flow paths throughout the Basin.

Groundwater Discharge

Most of the discharge from the Rio Grande alluvium occurs through pumping for irrigation and seepage of shallow groundwater to the surface-water drain system [16,37]. Discharge from the deeper Santa Fe hydrostratigraphic units is primarily through groundwater pumping for municipal and industrial use, a small amount of upward leakage [37], and a small amount of groundwater discharge to the neighboring Hueco Bolson through the Paso del Norte, as indicated by groundwater elevations (Figure 5).

Groundwater is withdrawn for agricultural, municipal, industrial, and domestic uses. Except for the Juárez well field, located just south of the U.S./Mexico border, the majority of irrigation and municipal pumping occurs in the Mesilla Valley. Before 1951, the number of agricultural withdrawals from groundwater was small due to adequate surface-water supplies [41]. However, in 1964, when there was a record low surface-water allotment of only 0.10 hm^3/km^2, Richardson and others (1972) estimated that irrigation pumpage exceeded 255 hm^3/year [57]. During the wet years of the 1980s, groundwater withdrawals for irrigation in Doña Ana County decreased to about 70 hm^3/year [16]. Drought conditions returned during the late 1990s and early 2000s, and agricultural pumping in Doña Ana County increased to about 120 hm^3/year [16]. In recent years (2016–2019), agricultural pumping in the Mesilla Valley has ranged from about 150 to 210 hm^3/year [86–88].

The largest municipal users of groundwater in the Basin are the cities of Las Cruces, New Mexico; El Paso, Texas; Ciudad Juárez, Chihuahua; and the Camino Real Regional Utility Authority (CRRUA), which supplies Santa Teresa and Sunland Park, New Mexico (Figure 2) [15]. The population of these communities has grown rapidly in the last 60 years. For example, the population of the city of Las Cruces has grown from 12,300 in 1950 to about 104,100 in 2021, and the Ciudad Juárez/El Paso urban center has grown from a population of about 255,000 to over 2.2 million over that same time [89]. Las Cruces Utilities began supplying water to Las Cruces through a series of groundwater wells in the 1920s [15,90]. In recent years (2016–2019), municipal pumping for the City of Las Cruces has ranged from about 25 to 27 hm^3/year [86–88]. El Paso Water Utility operates a well field in Cañutillo, Texas, which began production in the 1950s [15,91]. Annual groundwater pumping at the Cañutillo well field increased steadily, starting from 3.7 hm^3/year in 1952 and recently supplying El Paso with between 31 and 43 hm^3/year [16,91–93]. Developments in Santa Teresa and Sunland Park along the New Mexico/Texas border rely on groundwater supply from the CRRUA through a network of groundwater wells that were drilled in the early 1970s [15]. Annual groundwater withdrawal from these wells has increased steadily, from 2.5 hm^3 in 1973 to more than 6.9 hm^3 in 2003 [15]. In 2007, water withdrawal from the Conejos-Médanos portion of the Basin was about 1.6 hm^3, mainly for domestic and livestock use [10]. However, in 2010, in order to meet growing demand in Ciudad Juárez and the increasing stress on the Hueco Bolson aquifer, the city began supplementing municipal water with about 31 hm^3/year of groundwater extracted from 23 wells located near the international border (Figure 5) [10].

4.2.2. Storage

Changes in groundwater storage can occur year to year, depending on land use, pumping, and climatic conditions [16]. Using an integrated hydrologic model, Hanson and colleagues (2020) showed interannual variability in depletion and replenishment of groundwater storage within the Mesilla portion of the Basin and that the largest annual groundwater storage depletions corresponded to increased groundwater pumping [16]. Inflows to the groundwater system, predominately recharge from the Rio Grande and infiltration of irrigation water, resulted in groundwater storage replenishment in portions of the aquifer in years with decreased groundwater pumping.

Previous Groundwater Storage Estimates

Previous estimates of groundwater storage are limited to parts of the Mesilla portion of the Basin. The approximate thickness of saturated freshwater sediments ranges from 120 m in the north and south to almost 910 m in the central portion of the Basin [8,11]. The thickest saturated sediments containing freshwater generally coincide with the present course of the Rio Grande. Wilson and colleagues (1981) estimated that there is approximately 24,700 hm^3 of recoverable freshwater in the Rio Grande alluvium and Santa Fe Group sediments underlying the Mesilla Valley north of Cañutillo, Texas, based on a specific yield (S_y) of 15% and assuming that 60% of the sediments are sands and gravels [11]. Wilson and others (1981) also estimated the freshwater storage beneath the West Mesa to be about 41,900 hm^3 [11]. Hawley and Kennedy (2004) estimated the "most productive" portion of the aquifer system (the Rio Grande alluvium, upper Santa Fe unit, and middle Santa Fe unit) to hold about 17,300 hm^3 of available freshwater (<1000 mg/L DS) [8]. This estimate was made using an average saturated thickness of 61 m over a 2600 km^2 area and an S_y of 0.1. This is similar but not spatially exclusive to the 16,000 hm^3 estimated for the West Mesa area by Balleau (1999) [12]. Hawley and Kennedy (2004) went on to estimate that there could be as much as 61,700 hm^3 of fresh to slightly brackish (1000 to 3000 mg/L

DS) water in the deeper parts of the Basin, by assuming a thickness of about 300 m over 1940 km^2 and an S_y of 0.1 [8]. There are no reported estimates of the storage beneath the Conejos-Médanos portion of the Basin.

New Estimates of Groundwater Storage

The recent construction of a digital hydrogeologic framework [7] allows for a new estimate of groundwater storage using a more detailed aquifer volume. From Equations (1) and (2) (in Section 2), the estimates presented herein are derived using values of S_y and specific storage; we will therefore refer to the estimated volume as "potentially recoverable groundwater." We define this term as the volume of groundwater that could potentially be removed by pumping to completely drain the aquifer and excepting the groundwater retained by capillary forces. The Rio Grande alluvium fills the incised Rio Grande Valley floodplain (Figure 1), which, in places, is as much as 8-km wide. In the digital hydrogeologic framework model for the Basin, the base of the Rio Grande alluvium was defined as 24 m below land surface [7]. Using this assumption, the total volume of saturated sediments in the Rio Grande alluvium (equivalent to the river-channel hydrostratigraphic unit of Sweetkind (2017) [7]) was estimated to be 8600 hm^3. Assuming an S_y of 0.1, the volume of potentially recoverable water calculated using Equation (1) is about 860 hm^3. Within the portions of the Basin included in this analysis, the total volume of saturated sediment in the upper Santa Fe unit was estimated to be about 333,000 hm^3, the total volume of saturated sediments in the middle Santa Fe unit was estimated to be about 456,000 hm^3, and the total volume of saturated sediments in the lower Santa Fe unit was estimated to be about 530,000 hm^3. Using Equation (2) and assuming an S_y of 0.1 and an Ss of 0.00001 per foot, the total volumes of potentially recoverable groundwater in the upper, middle, and lower Santa Fe units were estimated to be about 33,300, 48,100, and 58,000 hm^3, respectively. The total storage estimate of 141,000 hm^3 is almost double the Hawley and Kennedy (2004) estimated volume of about 80,200 hm^3 [8]. However, this estimate includes all of the lower Santa Fe unit, whereas Hawley and Kennedy [8] used a lower cutoff depth of about 300 m, and therefore, their estimate does not include most of the lower Santa Fe unit. Based on reports of low yields and higher salinities of water from the lower Santa Fe unit relative to the upper units, we believe that a more realistic estimate of recoverable groundwater is limited to the Rio Grande alluvium and the upper and middle Santa Fe units. Removing the lower Santa Fe unit volume yields an estimate of recoverable groundwater of 82,600 hm^3.

In order to estimate the amount of groundwater in storage in the Conejos-Médanos portion of the Basin, we assume the same S_y as Hawley and Kennedy (2004) [8] (0.1) and calculate an aquifer volume by using 87% of the reported area, 5180 of the 5960 km^2, and a uniform saturated thickness of 133 m. The smaller area accounts for the thinning aquifer at the margins of the Basin and several igneous intrusions that likely reduce the amount of aquifer sediments. The saturated thickness is assumed from reported thicknesses of fresh and slightly brackish groundwater cited in SGM (2011) [10] and interpreted from several geoelectric units (geologic units with similar electrical properties) that were grouped into three horizons: 20–40 m of low-permeability sediments with freshwater, 40–150 m of slightly brackish water, and a deeper unit with higher salinity [10]. Based on these assumptions, the amount of recoverable fresh or slightly brackish water stored in the Conejos-Médanos portion of the Basin is estimated to be about 69,100 hm^3.

4.2.3. Aquifer Dynamics

Seasonal and annual recharge and discharge in the shallow Rio Grande alluvial aquifer and long flow paths in the deeper Santa Fe Group aquifer are recorded by changes in groundwater levels [16]. Hydrographs from selected wells with long-term records in the Basin are displayed in Figure 6, using winter measurements to reduce the effect of seasonal pumping on the record. Well locations are shown in Figure 2. Data from nearby wells with similar screen depths are included to add to, and in some cases extend, the period of record for a given well. Figure 6a shows depths to water for Rio Grande alluvial wells.

The full dataset is included in the background (gray shaded points and lines) to show the seasonality in alluvial-well groundwater levels. Figure 6b shows groundwater elevations in wells screened in the Santa Fe Group sediments to show hydrographs and groundwater levels in relation to one another.

The shallow alluvial wells include, in downstream order, 21 (USGS Site ID 321853106504001), 130 (USGS Site ID 320403106390401) and 209 (USGS Site ID 314817106325801) (and their associated nearby wells, 279 [USGS Site ID 321859106503101], 286 [USGS Site ID 320404106385801], and 255 [USGS Site ID 314854106340101], respectively) (Figure 2) [18]. Fluctuations in shallow groundwater levels indicate a dynamic surface-water/groundwater interaction in the northern portion of the Basin [6,38]. Smaller groundwater level variations near wells 209 and 255 compared to other wells for the same point in time (Figure 6a) are consistent with previous indications of a reduction in vertical gradients near the terminus of the Basin (Paso del Norte). Long-term groundwater-level data in wells 21, 279, 130, and 286 show multiyear declines in the Rio Grande alluvium during the drought of the 1950s and the current (2022) drought, starting around 2000 (Figure 6a). These declines are far less discernable in the more southerly wells, 209 and 255. Reductions in streamflow due to drought not only limit the amount of surface water available for irrigation and recharge but also lead to an increase in groundwater withdrawals [16]. The transition to groundwater pumping, reflected in multiyear declines in the depth to water, reflects the reduced availability of surface water in the Basin.

Annual groundwater elevations in the Santa Fe Group sediments typically vary by less than 15 cm per year unless they are near areas of large extractions [15] (Figure 6b). The hydrograph of well 53 (USGS Site ID 321650106451201), located in Las Cruces, New Mexico, for example, shows the largest annual changes and a steady decline in groundwater elevations due to groundwater pumping. The highest groundwater elevation (1174.1 m) in well 53 was measured in 1964, and the lowest (1161.4 m) was in 2015. Well 177 (USGS Site ID 315349106585701), located away from any large pumping areas, has varied by just over 30 cm since measurements began in 1962. Groundwater elevations in well 206 (USGS Site ID 314810106513601), near the U.S./Mexico border, have typically varied by only about 3 to 6 cm per year between 1983 and 2010. However, a decline of almost 1.37 m occurred between 2010 and 2020, likely resulting from the start of pumping at the Juárez well field in 2010 (Figure 6b). Groundwater elevations in the Conejos-Médanos portion of the Basin did not change substantially between 1987 and 2007 [10], prior to pumping in the Juárez well field. The decline in groundwater elevations in well 206 following the start of pumping also coincides with water-level declines in alluvial wells (for example, wells 21, 279, 130, and 286) that can be attributed to reduced surface-water availability, and therefore, care is warranted when attributing the groundwater elevation declines. A similar decline is observed in the central Basin at well 150 (USGS Site ID 315720106415601).

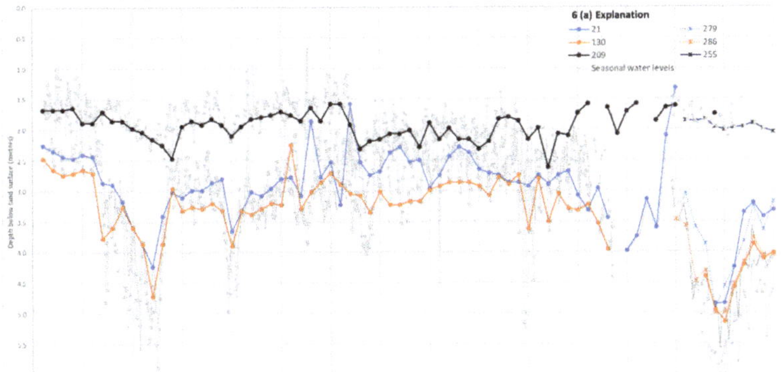

Figure 6. *Cont.*

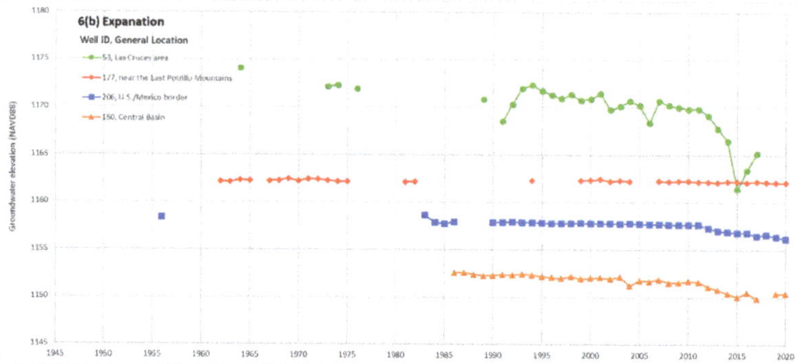

Figure 6. Groundwater levels, 1945–2020. (**a**) Depth to water measured in piezometers screened in the Rio Grande alluvium; (**b**) groundwater elevation measured in wells screened in the Santa Fe Group sediments at depths greater than 90 m below land surface. Well locations are shown in Figure 2. Data from USGS (2021) [18].

5. Water Chemistry

Spatial patterns and relations between water chemical and isotopic data are useful for determining recharge sources, direction of flow, and geochemical processes [5,13]. As stated previously, much of the recharge in the Basin originates from the Rio Grande and, to a lesser extent, local precipitation. Once the surface water has infiltrated, water-rock interactions and groundwater mixing can change the chemical composition of the water [93]. The geochemistry of shallow groundwater throughout the Basin is governed primarily by the evaporative concentration of solutes in Rio Grande streamflow and agricultural diversions, while deeper groundwater geochemistry results from multiple processes, including mixing between geothermal and nongeothermal groundwater, dissolution and mineral precipitation reactions, and ion exchange [13].

5.1. Groundwater Chemistry

Water chemistry in the Rio Grande alluvium generally reflects the chemistry of the surface-water system, whereas water chemistry in the Santa Fe Group depends on the source and time of recharge and water-rock interactions along the flow path. Although groundwater chemistry varies greatly with location, Wilson and colleagues (1981), examining specific conductance data between 1953 and 1976, reported that groundwater chemistry changed little over time in the wells they sampled [11].

5.1.1. Dissolved Solids

Dissolved solids (DS) concentration is a measure of the mineral content of water and is a conservative property, meaning that concentration is not expected to change as water moves downgradient unless it mixes with water from a different source or interacts with a different rock or sediment type [94]. Because of this property, DS can be used to identify areas of similar water types and can provide evidence of groundwater flow and mixing [5]. The mineral content in the Basin groundwater is extremely variable. Land (2016) reports DS concentrations ranging from 234 to 30,800 milligrams per liter (mg/L) from 408 records in New Mexico [95], Teeple (2017) reports a range of 161 to 31,000 mg/L from 239 wells along the Texas/New Mexico border [5], and SGM (2011) reports a range of 72 to 11,370 mg/L from 91 wells in the Conejos-Médanos portion of the Basin [10]. The average DS concentrations for these datasets are similar but reflect increasing salinity from north to south, with average values of just over 1200 mg/L from Land (2016) [95], 1500 mg/L from Teeple, (2017) [5], and 1700 mg/L from SGM (2011) [10]. These average values are skewed by the presence of samples with high mineral content. This bias is reflected in median DS concentrations of 693 mg/L [95], 857 mg/L [5], and 817 mg/L [10], which are lower than their respective mean concentrations. For comparison, DS concentrations in the Rio Grande are reported to range from 400 to 1200 mg/L, with higher concentrations observed in the downstream part of the river [96].

Water chemistry in the upper Santa Fe unit is similar to water chemistry of the shallow Rio Grande alluvium. Wells less than 30-m deep in the Mesilla Valley that are farther from the Rio Grande have slightly brackish water (DS concentrations 1000 to 2000 mg/L), with predominant ions of calcium, sodium, and sulfate [11,38]. This relation is consistent throughout the Mesilla Valley and has been attributed to the effects of surface-irrigation practices and evapotranspiration [11]. Individual wells with high salinity or iron concentrations may tap abandoned river channels or ancestral swamp or bog deposits [43]. Wells slightly deeper (between 45 and 245 m) are typically fresh (DS concentrations < 1000 mg/L) and are not characterized by specific dominant ions. Groundwater in the middle Santa Fe hydrostratigraphic unit typically has lower DS concentrations than in the overlying units, and the DS concentrations tend to be greater in groundwater samples from wells screened in the lower Santa Fe unit. The presence of fresher water between the upper Santa Fe unit and the lower Santa Fe unit throughout most of the Basin supports the conceptual flow model in which the deep and shallow groundwater have limited interactions [5].

Spatial variability of water chemistry throughout the Mesilla portion of the Basin may result from numerous factors, such as the source of recharge, groundwater mixing, presence of evaporites, and geothermal inputs. Groundwater in the northern part of the Basin is generally fresher, with concentrations of trace elements that are less than regulatory water quality criteria. In contrast, water from wells on the east side of the Mesilla Valley and near Paso del Norte often have DS concentrations greater than 1000 mg/L. The DS concentrations in West Mesa wells become more brackish at shallower depths farther west and south, with increases in the relative amounts of chloride and sulfate [38]. Similar major ion concentrations indicate good connection between groundwater from wells on the west side of the Mesilla Valley and groundwater from the West Mesa. At the southern end of the Mesilla Valley, however, DS concentrations can exceed 10,000 mg/L [37].

Several drilling records note increased DS concentrations near the bottom of the borehole [97], but data are too sparse to determine the extent of brackish water in the deep lower Santa Fe unit. Similarly, there are very few reports of wells screened in the underlying consolidated bedrock. Leggatt and others (1963) described one bedrock well that "flowed salty water," whereas the description from a well screened in Cretaceous rocks is that of "moderate supply and satisfactory for industrial use" [41].

5.1.2. Temperature

The average temperature from 154 groundwater wells in the Basin was about 26 degrees Celsius (°C; median was also 26 °C) and ranged from 14 to 37 °C [5,10,13]. Groundwater temperatures generally increase with depth at a rate of about 35 °C per km in the Basin but can be affected by groundwater advection in places [98]. For example, temperature gradients are much higher within known geothermal areas, where waters ascend from depths of over 1 km (e.g., East Bench, East Potrillo Mountains, and Radium Springs) (Figure 1) [83,84]. Localized groundwater upflow zones within these geothermal areas can have temperatures exceeding 55 °C, and geothermal waters at Radium Springs are close to 100 °C [83,99]. Temperature gradients are sometimes lower or negative within the Mesilla Valley, where surface-water recharge or horizontal groundwater flow affect temperatures [83].

5.1.3. Major Ions

The majority of the groundwater in the Mesilla portion of the Basin is sodium-cation-dominant and a chloride-sulfate-anion type [5]. The dominant water type measured in the Conejos-Médanos portion of the Basin is sodium-bicarbonate-sulfate, with increasing chloride as groundwater flows north [10]. Except for calcium, all the major ions increase as DS concentrations increase [5], indicating that general mineral dissolution and cation exchange processes are occurring. Many studies report that there is an evolution of calcium-dominated water to sodium-dominated water [5,13] and a general shift from bicarbonate recharge waters toward chloride and sulfate waters in the direction of flow in the southern Mesilla Valley. The transition in water types occurring along a given flow path indicates increased influence from evaporite dissolution and cation exchange on groundwater chemistry [42]. Evaporite minerals (halite, gypsum, and anhydrite) are present in sediments in parts of the aquifer and contribute sodium, chloride, sulfate, and calcium into solution [67]. Isometric log plots [100] of Mesilla Valley groundwater provide supporting evidence of this and tend toward equilibrium with evaporite minerals as the DS concentrations increase, particularly halite in the southern Mesilla Valley [42]. However, most samples were enriched with sodium relative to halite dissolution, indicating additional processes, such as the dissolution of silicate minerals, calcite dissolution, cation-exchange, or a combination of each, are occurring [5,13,42].

5.2. End Members

Geochemical and isotopic attributes are commonly used by researchers to classify groundwater [5,13,94]. Several studies have identified distinct geochemical water types in the Basin's groundwater to answer a variety of hydrologic questions [5,13,63,64,76]. The conceptual model of the Basin characteristics, such as groundwater flow, sources of salinity, and geochemical evolution, are supported and further developed using these tracers [5,13,42]. Based on characterizations by Teeple (2017), we describe four groundwater end members that have distinct geochemical and isotopic compositions that can be attributed to different sources and water-rock interactions [5]. We acknowledge the presence of other groundwater types and groundwater that contains a mixture of different end members; however, by focusing on the most prevalent types, we believe we can adequately describe the prevailing processes that influence groundwater geochemical attributes. The end members include (1) ancestral Rio Grande (pre-Holocene) geochemical group, (2) modern Rio Grande geochemical group, (3) mountain-front geochemical group, and (4) deep-groundwater upwelling geochemical group. The groups identified include freshwater likely sourced from the ancestral Rio Grande, as well as younger groundwater from the modern Rio Grande seepage, older groundwater near the western highlands sourced from mountain-front recharge, and brackish to saline water from deeper sources.

5.2.1. Ancestral Rio Grande (Pre-Holocene)

Groundwater belonging to this end member is postulated to have recharged into the system as seepage from the ancestral Rio Grande. This ancestral Rio Grande groundwater type occurs throughout the Basin and is found primarily in the middle Santa Fe unit. Based on overall groundwater flow patterns, groundwater of this geochemical group is moving south and southeast, before flowing laterally east toward the Paso del Norte (Figure 5). Modern-day flow for this paleo-recharged groundwater is likely driven by small amounts of recharge at the Basin margins and by withdrawals at large groundwater pumping centers. This geochemical group is characterized by low DS concentrations (average = 415 mg/L; median = 360 mg/L), low amounts of radiocarbon, and depleted stable isotopes of water [5]. Despite the long flow path, the small gains in DS concentrations along the flow path of this groundwater type indicate that the aquifer lacks large amounts of soluble minerals, possibly because of the depositional environment, as well as being depleted in soluble minerals by large volumes of groundwater flow in the past. The general geochemical evolution along the flow path is from a calcium-sodium-bicarbonate to a sodium-sulfate-bicarbonate water type. The low radiocarbon and depleted stable isotope composition reflect the recharge conditions before the Holocene.

5.2.2. Modern Rio Grande (Holocene)

Groundwater belonging to this geochemical group is found primarily in the Rio Grande alluvium and, to a smaller extent, the upper Santa Fe unit in the Mesilla Valley. High concentrations of age-dating tracers, tritium, and carbon-14 indicate that shallow groundwater was recharged from the Rio Grande and agricultural infrastructure within the past 100 years [5]. When there are ample surface-water supplies, water may enter and leave the groundwater system within a season, as occurs with water conveyed by the drains. Stable isotope compositions [5,13] support evaporated Rio Grande water as the source of this groundwater group, with sample results plotting on the Rio Grande evaporation line developed by Phillips and others (2003) [70]. Groundwater temperatures also indicate rapid recharge of surface water, with groundwater temperatures (averaging about 19 °C) near the average annual temperature in Las Cruces (11 °C) [5,13]. This end member ranges from calcium-sulfate water type in the northern Mesilla Valley to sodium-chloride-sulfate water type in the southern Mesilla Valley [5]. DS concentrations in the shallow alluvium average about 1950 mg/L (the median is about 1600 mg/L). Evaporative processes can lead to elevated DS concentrations (>2000 mg/L) along agricultural drains, and recharge from these drains often leads to elevated concentrations of dissolved ions in the shallow subsurface [13,72].

5.2.3. Mountain-Front Recharge

In mountain-front recharge areas, precipitation and high-elevation springs and wells are characterized by low DS concentrations (<250 mg/L) and typically have dominant ions of calcium, magnesium, and bicarbonate [13,75,76]. The stable isotope composition of these epigenic waters plot on or near the Global Meteoric Water Line [78], unlike the stable isotope composition of Rio Grande water. Groundwater sampled near the boundaries of the Basin show mixing patterns between the epigenic sources and several different endogenic sources. The lack of samples with a large epigenic component from large portions of the Basin aquifer supports previous assertions that mountain-front recharge is presently a minor component of recharge to the Basin aquifer [14,16]. Wells near the Basin boundaries that are considered to have a substantial component of mountain-front recharge generally have low concentrations of age-dating tracers, indicating that the water likely recharged under wetter and cooler conditions than at present (2022). Water belonging to this group is somewhat mineralized, with average DS concentrations around 780 mg/L (median of about 710 mg/L) [5].

5.2.4. Deep Upwelling Groundwater

The presence of high-temperature groundwater, faults, and highly mineralized water indicates that groundwater from depth is upwelling to shallower zones in portions of the aquifer [83,101,102]. Vertical leakage from deep-seated regional groundwater flow systems, including geothermal systems, may also be a substantial source for salinity increases in the shallow aquifers and the Rio Grande near Paso del Norte [13,70,83]. Waters with elevated concentrations of trace ions are observed in wells on the West Mesa, near the East Potrillo Mountains, at Radium Springs, and on the East Bench (Figures 1 and 2) [5,13,71,94]. The deep upwelling groundwater has been characterized as having high concentrations of chloride, arsenic, potassium, silica, aluminum, iron, and lithium and may be originating from the Paleozoic and Cretaceous carbonate bedrock [5,42]. Szynkiewicz and colleagues (2011), using sulfate isotopes and principal component analysis, concluded that a substantial amount of major ion concentrations in the groundwater could be attributed to groundwater flow from the bedrock [59]. The original source of this deep groundwater is unknown but may include paleo-mountain-front recharge in the southern and western portions of the Basin and interbasin flow from the Jornada Basin along the East Bench [13,14,77]. Temperature analyses indicate that geothermal waters in the Mesilla portion of the Basin ascend from depths of at least 1 km, with geothermal waters at Radium Springs coming from depths upwards of 2 km [83].

6. Water Budget

The water budget for the Mesilla portion of the Basin is approached by balancing inflows and outflows between three interrelated components of water supply and use:

surface water, groundwater, and agriculture. Only the Mesilla portion of the Basin was selected because of the available data and the presence of the Rio Grande. This budget relies on the throughflow estimates from the Conejos-Médanos portion of the Basin made in Section 4.2.1. The budget terms were considered and refined by compiling the available estimates for three periods: the entire time period for which there are reliable and publicly available data (generally 1940 to 2014), the dry period between 1950 and 1969, and a wetter period between 1980 and 1999. Dry periods and wetter periods were classified by Hanson and colleagues (2020) [16].

The surface-water component of the water-budget estimate includes only the Rio Grande and the water that flows into or is diverted from it. Between 1940 and 2014 the flow measured entering the Basin above Leasburg Dam averaged about 759 hm^3/year (+/− 310 hm^3/year; median value 772 hm^3/year), and streamflow measured leaving the Basin averaged about 491 hm^3/year (+/− 279 hm^3/year; median value 465 hm^3/year) [17]. The difference indicates that, on average, there was about 268 hm^3/year of water lost from the Rio Grande in the Basin. As the largest of the inflows and outflows, these budget terms agree well with other reported estimates and provide a solid starting point for adding other budget items [11,14,15]. The largest diversions of surface water in the Basin are made for irrigating crops. Between 1940 and 2014 the average amount of water diverted in the Basin was 433 hm^3/year (+/− 152 hm^3/year; median value 461 hm^3/year) or about 57% of the Rio Grande inflow at Leasburg Dam and ranged from about 60 hm^3/year in 2012 to almost 680 hm^3/year in 1945 [17]. These diversions are measured, and the values considered reliable within appropriate gaging errors. The amount of water returned to the Rio Grande from canal return flow is estimated to be about 10% of the diversions [11,17,39], and return flow from the drains was typically about 33% of the diversions for the available data between 1940 and 1978 [11,17]. The total return flow over that period ranged from 9% to 56% of the diversions [11,17]. Between 1940 and 1992, drain flows varied from just under 5.6 hm^3/year in 1956 to a little over 310 hm^3/year in 1944 and averaged about 143 hm^3/year (+/− 80 hm^3/year; median value 148 hm^3/year) [17]. Wastewater returns to the Rio Grande increased from about 6.2 to 15 hm^3/year between 1976 and 2014, and the average was about 12 hm^3/year [17]. Based on the average precipitation in the area (21 cm) and the estimated surface area of the Rio Grande (7.8 km^2), the average amount of precipitation falling directly on the river is about 1.6 hm^3/year. The annual evaporation from the river surface was estimated to be 13 hm^3/year by Frenzel and Kaehler (1992) [14] and was used for this budget term. Finally, surface-water seepage measurements from the Rio Grande have typically been collected in the winter and range from about 1.2 to 80 hm^3/year [11,15,39,49–52]. A value of 31 hm^3/year was selected to balance the annual surface-water budget for the three periods and the selection of that value is supported by the frequency of seepage measurements that were reported close to that value. For comparison the simulated stream seepage at the end of the Frenzel and Kaehler (1992) model was about 68 hm^3/year [14].

The groundwater component of the water budget consists of the entire aquifer and does not differentiate between geologic units. The largest depletion of groundwater is by pumping and is distinguished between municipal and industrial pumping and agricultural pumping. Groundwater for municipal use is metered and has generally increased in the Basin through time [15,16]. Estimates of municipal and industrial water use have ranged from about 49 hm^3/year in 1975 [11,14] to 86 hm^3/year in 2005 [15]. For this water-budget component, we assigned the low estimate of 49 hm^3/year to the early (1950 to 1969) period and the high estimate of 86 hm^3/year for the later period (1980 to 1999). Hanson and colleagues' (2019) estimate of 67 hm^3/year [16] was used for the period of record. Metered agricultural groundwater withdrawals have only recently become available, and groundwater withdrawals for irrigated agriculture have been estimated using a variety of techniques. Because of the conjunctive use of water for irrigation, agricultural pumping amounts vary dramatically year to year, with annual estimates ranging from less than 12 hm^3/year to over 370 hm^3/year [15,16,47]. Previously reported average agricultural pumping estimates cover different time periods and often include areas outside the Mesilla Valley [14–16]. For this analysis, the period of record average annual pumping rate of 110 hm^3/year was selected from estimates by Frenzel and Kaehler (1992) and SSPA (2007) [14,15]. However, because those pumping estimates include the Palomas Basin, the estimate was reduced by using the fraction of total irrigated lands that are farmed in the Mesilla Valley (about 84%) [48] to 94 hm^3/year. In recent years (2010–2018), agricultural pumping to supplement reduced surface-water deliveries has exceeded municipal and industrial pumping by about 3:1, withdrawing over 250 hm^3/year [47]. Sources of recharge include deep percolation of applied irrigation water as a fraction of diversions, seepage from the Rio Grande (31 hm^3/year), and mountain-front recharge (14 hm^3/year). The majority of recharge to the aquifer since the 1940s is estimated to be from the infiltration of water through the irrigation conveyance system and the surface application of irrigation water [8,11,38]. The amount of groundwater recharge through deep percolation was estimated to be 13% of the total applied water (both surface water and groundwater), which is lower than the 39% of applied water estimated by SSPA (2007) and 24% by Hanson and colleagues (2020) [15,16]. This discrepancy is because the models route this groundwater to both groundwater recharge and drain flow, while this estimate assumes all deep percolation goes to recharge and the drain flow is a separate function of the applied water. Mountain-front recharge was estimated to be about 14 hm^3/year based on previous estimates, which have ranged from about 1.2 to 17 hm^3/year [14,15,17]. Groundwater flow from the Conejos-Médanos Basin towards the Basin outlet was estimated in the previous section to be about 23 hm^3/year. Finally, Slichter (1905) estimated groundwater flow from the Mesilla portion of the Basin to the Hueco Bolson at Paso del Norte to be about 3 L/s, or about 0.1 hm^3/year [103].

The agricultural component of the water budget includes the diversions, conveyance structures, fields, and drains of the EBID in New Mexico, as well as EP No. 1 in Texas. The primary inflow to agriculture is the surface water diverted from the Rio Grande. As reported above, the average amount of water diverted in the Basin was about 433 hm^3/year. The effective precipitation falling on agricultural lands in the Basin has been estimated to be between 50% [15] and 90% [14]. For this budget estimate, 50% of the average annual precipitation (21 cm) was applied over 330 km^2 of irrigated lands to yield about 36 hm^3/year of precipitation available for agriculture. Water leaving the system includes return flow through canals and drains, which are estimated as a percentage of surface-water diversions (about 10% and 33%, respectively). Evapotranspiration from irrigated fields in the Basin has been estimated to be about 10 m^3 per km^2 for lower water-use crops, such as cotton, and about 20 m^3 per km^2 for higher water-use crops, such as pecans. While more acreage is being converted to higher water-use crops [15,16], for this estimate, 15 m^3/km^2 over 330 km^2 of irrigated lands was used to estimate the water use by crops at 308 hm^3/year, which is similar to previous estimates of about 302 hm^3/year by Frenzel and Kaehler (1992) and 305 hm^3/year by SSPA (2007) [14,15]. Groundwater pumping generally increases with reductions in surface-water deliveries, although historical pumping records are sparse. Estimated groundwater pumping for agricultural use is described above.

Several items can be noted from these water-budget estimates (Table 2). The first is that present-day average groundwater recharge from the Rio Grande and irrigated agriculture accounts for about 73% of the recharge to the Mesilla portion of the Basin and 66% of the recharge to the entire Basin (15.8 hm^3/year of recharge reported for the Conejos-Médanos portion of the Basin [27]). Additionally, present-day recharge accounts for about 11% of Rio Grande alluvium groundwater volume annually, an amount that is lower than recent withdrawals. Second, for the period of record, the total loss from storage is estimated to be about 24.8 hm^3/year. This estimate is substantially lower than the 53 hm^3/year estimated by Hanson and colleagues (2020), and the difference may be due to the fact that their estimate includes both the Palomas Basin and the Mesilla portion of the Basin, as well as different estimates of throughflow from the Conejos-Médanos portion of the Basin [16]. While the estimates presented in this report may provide an estimate of the magnitude of the total loss over the period of record, it can be noted that much of the variability in the system is lost by selecting single values and percentages in order to simplify the computations. The average values and unadjusted percentages are sure to include biases and therefore warrant using with caution and will be increasingly less reliable for shorter time periods.

Table 2. Water-budget components (in cubic hectometers) and the sum of inflows and outflows for the Mesilla portion of the Basin for three time periods. Parentheses and red font used in the table for negative numbers.

Budget Items	Average Amounts for the Period of Record (1940 to 2014)			Average Amounts for the Dry Period 1950 to 1969			Average Amounts for the Wet Period 1980 to 1999		
	Surface Water	Ground-Water	Agri-Culture	Surface Water	Ground-Water	Agri-Culture	Surface Water	Ground-Water	Agri-Culture
Rio Grande inflow Leasburg Dam	759			586			944		
Irrigation diversions	(433)		433	(384)		384	(516)		516
Precipitation	1.6		36	1.6		36	1.6		36
Return flow (canals)	43		(43)	38		(38)	52		(52)
Drain return flow	143		(143)	127		(127)	170		(170)
Deep percolation		68	(68)		66	(66)		74	(74)
Wastewater returns	12			3.7			12		
Evapotranspiration	(13)		(308)	(13)		(308)	(13)		(308)
Agricultural pumping		(94)	94		(121)	121		(53)	53
Municipal and industrial pumping		(67)			(49)			(86)	
Groundwater outflow near El Paso		(0.1)			(0.1)			(0.1)	
Rio Grande seepage	(31)	31		(31)	31		(31)	31	
Conejos-Médanos throughflow		23			23			23	
Mountain-front recharge		14			14			14	
Rio Grande outflow at El Paso	(491)			(319)			(617)		
Inflows	958	136	562	756	133	541	1179	141	604
Outflows	(968)	(160)	(563)	(748)	(170)	(539)	(1,176)	(139)	(604)
Balance	(9.41)	(24.9)	(0.79)	8.86	(37.6)	1.38	2.44	1.54	0.16
Basin surface water loss	268			266			327		
Total agricultural use			527			505			569

7. Potential Future Research Directions

The research needs for a scarce conjunctive-use resource that is managed by multiple governments would benefit from not only understanding of the physical characteristics and limits of the resource but cooperation and common objectives of stakeholders. This discussion is provided as a starting point to acknowledge the existing data gaps and to provide strategies for how to fill gaps. Perhaps the most important concern facing managers of any finite resource is the sustainability and resiliency of the resource and a management strategy that reduces uncertainty. To help address this concern, several data gaps were identified in this work that may benefit from future research, including (1) high-resolution storage, inflow, and use estimates; and (2) expanded water-chemistry data.

7.1. High-Resolution Water Storage, Inflow, and Use Estimates

Groundwater storage estimates provide an understanding of the water available at a given time and allow for estimates of resource capacity under different withdrawal scenarios. Storage is estimated from the volume and capacity of the aquifer. These values are highly variable and often generated from point data, where data density may be reasonably high or sparingly low, particularly at depth. Continued refinement of the underlying geology and aquifer properties of the Basin would allow for more accurate estimates of the amount of groundwater while also providing a better understanding of its distribution. Added drilling and geophysical data could provide better estimates of the aquifer thicknesses under the West Mesa and in the Conejos-Médanos portion of the Basin. Aquifer tests and geophysical methods for determining aquifer characteristics could improve understanding, particularly in the lower Santa Fe unit and throughout the Conejos-Médanos portion of the Basin. In addition, quantification of the changes in hydraulic gradients with continuous monitoring between parts of the West Mesa and the Conejos-Médanos portion of the Basin to the Mesilla Valley would provide a better understanding of the effect that pumping in the Mesilla Valley may have on groundwater flow and support improved resource-management decisions.

Groundwater storage allows managers to maintain use of water resources in times of supply shortages, but sustainable use would require that withdrawals not exceed inputs over the long-term. Therefore, additional refinements of the inflow and use estimates would be useful to determine withdrawal thresholds and associated effects on the inflow and outflow balance. New agricultural groundwater metering would be valuable to get better estimates of agricultural extractions, delivery and on-farm efficiencies, and recharge. Analysis of the reported data trends following multiple years of collection could define patterns and trends in water use. Continued collection of streamflow data at Leasburg and El Paso, along with upgraded gage sites to provide a higher-quality record, would allow managers to monitor surface-water trends within the Mesilla portion of the Basin. Water-level contours interpreted by SGM (2011) [10] and those presented here (Figure 5) indicate connected groundwater flow over the entire Conejos-Médanos portion of the Basin. However, the presence of intrabasin uplifts between structural sub-basins and the lack of substantial recent recharge indicates that groundwater flow in large areas of the Conejos-Médanos portion of the Basin may be disconnected. Additional groundwater-elevation, geochemical and isotopic data around the Los Muertos sub-basin, particularly near the northern extent, would provide further evidence as to whether groundwater in the southern Conejos-Médanos portion of the Basin is flowing northward. Additional flux estimates for groundwater throughflow between the Conejos-Médanos portion and the Mesilla portion of the Basin would help to refine the recharge estimates in the Conejos-Médanos portion of the Basin, the overall water budget for groundwater in the Basin, and the quantity and limitations of the brackish groundwater resource. This could be addressed by installation of nested wells and perhaps flux-meter installations. Finally, by incorporating and validating groundwater flow models with geochemical and isotopic tracers, inflow estimates (such as mountain-front recharge and deep upwelling) could be further supported and refined.

Refinement of the Basin water budget would allow managers to quantify threshold values and make management decisions to maintain the resource for multiple uses. Actionable limits could include maximum drawdown thresholds to prevent irreversible subsidence, decreased well performance, water-quality degradation, or a combination of these factors.

7.2. Water Chemistry

The amount of available water is also dependent on the quality of the water. The quality of the water is important not only for determining the amount of water available for consumption without treatment but also for the feasibility of treating brackish water to use-specified standards. In order to understand the potential for further salinization of the resource and, if needed, its mitigation, future research could include drilling several deeper wells (screened in the bedrock) near the Paso del Norte to improve current understanding of the bedrock-groundwater contribution to groundwater flow and salinity. If groundwater was derived from these bedrock units along a regional flow path, an upwelling geothermal signature would likely be observed in these deeper wells. Additional research may also improve current understanding of the relation between groundwater-age signatures and groundwater flow paths. Such research could improve our current understanding of the relation between groundwater age and salinity in the Mesilla Valley, which appears to be correlated in certain circumstances [5,42]. Furthermore, given that most of the drains throughout the Mesilla Valley have not been functioning over the last decade, future research could be beneficial to determine whether flushing thresholds exist at which salt accumulations at damaging levels may be reached in farm fields.

In order to deal with a persistent lack of sufficient water, researchers have looked to the potential of brackish groundwater as a resource. Concerns unique to inland desalination projects include uncertainty regarding the size of the resource, issues relating to water treatment for constituents in groundwater, such as silica, and disposal of brine concentrate [104]. Previous works [37,105] have suggested that substantial resources of slightly brackish water are present in the Mesilla Basin. However, there are limited records for the deeper portions of the Basin, with an average well depth in the Basin of only 100 m [95]. In addition to mapping the quantity of the brackish resource, a geochemical characterization of the reservoir would be useful to further assess desalination favorability.

8. Summary

The Mesilla/Conejos-Médanos Basin (Basin) is located in the southern part of the Rio Grande rift, a tectonic feature that is characterized by generally north-south-trending structural extensional basins. The Basin's aquifer is composed of the basin-fill sediments of the Santa Fe Group within the Basin-bounding uplifts. Interbasin uplifts and subcrops separating the sub-basins within the Basin likely restrict deeper flow in the Mesilla portion of the Basin and may restrict all flow between the Los Muertos and El Parabien sub-basins in the Conejos-Médanos portion of the Basin.

Using a newly developed hydrogeologic framework, a new international water-level map, and previously reported aquifer property assumptions, the amount of potentially recoverable fresh to slightly brackish groundwater in the Mesilla portion of the Basin is estimated to be about 82,600 hm^3. This new estimate is largely in agreement with previous estimates. A new estimate of storage for the Conejos-Médanos portion of the Basin is also presented in this work. Based on areal-extent and saturated-thickness assumptions, the amount of recoverable fresh or slightly brackish water stored in the Conejos-Médanos portion of the Basin is estimated to be about 69,100 hm^3. The majority of groundwater stored in the Basin is thousands to tens of thousands of years old and was recharged during cooler and wetter parts of Quaternary glacial-pluvial cycles [8,106,107]. This water is very slowly being displaced at the boundaries by mountain-front recharge and near pumping centers, where vertical gradients are increased by large withdrawals from groundwater pumping.

The Rio Grande flows through the Mesilla Valley in the Mesilla portion of the Basin, which contains recent deposits of the Rio Grande alluvium. Relatively dynamic surface-water/groundwater interactions in the Mesilla Valley between the Rio Grande and the Rio Grande alluvium result in groundwater levels that are responsive to annual and seasonal changes in water supply and demand, in contrast to the deeper groundwater levels that remain stable or show a gradual response. This concept of groundwater movement that includes short groundwater flow paths in the shallow aquifer and groundwater flow paths that increase in length with depth is supported by the hydrologic and water-chemistry data presented in this report. Based on evidence presented in various sections of this report, the Rio Grande alluvium is the only unit currently receiving substantial amounts of recharge from the Rio Grande, and the amount of groundwater in the Rio Grande alluvium represents a little less than 0.6% of the entire regional aquifer. Approximately 11% of the volume of the Rio Grande alluvium is estimated to be recharged annually from Rio Grande seepage and deep percolation of agricultural water, but that amount is often offset by pumping.

The geochemistry of shallow groundwaters in the Mesilla Valley portion of the Basin is governed by the evaporative concentration of Rio Grande streamflow and agricultural diversions, while deeper groundwater geochemistry results from multiple processes, including the mixing of geothermal and non-geothermal groundwater, dissolution and precipitation reactions, and ion exchange [13]. As such, water chemistry in the Rio Grande alluvium generally reflects the chemistry of the surface-water system, with DS concentrations ranging from about 500 to over 1000 mg/L, whereas water chemistry in the Santa Fe Group depends on the source and time of recharge and the water-rock interactions along the flow path. Generally, groundwater in the middle Santa Fe unit has some of the lowest DS concentrations in the Basin, perhaps indicating large groundwater recharge and fluxes in pre-Holocene time, reducing the amount of soluble minerals in the solid matrix. DS concentrations are reported to increase in the lower Santa Fe unit and, in particular, at the basement rock contact. Brackish and saline groundwater in the Mesilla portion of the Basin are reported, where upflow areas associated with faults and bedrock features are most common. The extent of brackish groundwater in the Conejos-Médanos portion of the

Basin is not known, but records of lacustrine evaporites and increasing salinity with depth indicate a substantial brackish reservoir.

Continued refinements to the storage estimates and water-budget items with well installation, geophysics, and monitoring would allow for better estimates of sustainable use limits. Further characterization of the deep groundwater extent and geochemistry would inform development of alternative water sources, such as desalination.

Author Contributions: Conceptualization, A.J.R. and A.-M.M.; methodology, A.J.R., J.D.P. and A.B.R.; software, A.J.R. and A.B.R.; validation, A.J.R., J.D.P., A.B.R., E.H.F. and A.E.G.; formal analysis, A.J.R., J.D.P. and A.B.R.; investigation, A.J.R., A.-M.M., J.D.P., A.B.R. and D.S.S.; resources, A.J.R.; data curation. A.J.R., J.D.P. and A.E.G.; writing—original draft preparation, A.J.R.; writing—review and editing, A.J.R., A.-M.M., J.D.P., A.B.R., D.S.S., A.P.T., A.G.-O., A.C.G.-V., K.C.C., E.H.F. and A.E.G.; visualization, A.J.R. and J.D.P.; supervision, A.J.R.; project administration, A.J.R. and A.-M.M.; funding acquisition, A.J.R. and A.-M.M. All authors have read and agreed to the published version of the manuscript.

Funding: This work was partially funded by the U.S. Geological Survey (funding authorized by P.L. 109–448) for the Transboundary Aquifer Assessment Program.

Data Availability Statement: Groundwater elevations used in the construction of the groundwater contour map are available in the following data release: Pepin, J.D., and Robertson, A.J., 2021, Estimated groundwater elevations and standard errors from the application of kriging to median groundwater elevation data from 2010 in the Mesilla/Conejos-Médanos Basin, United States and Mexico: U.S. Geological Survey data release, https://doi.org/10.5066/P98Z25JG accessed on 16 December 2021.

Acknowledgments: This product has been peer-reviewed and approved for publication consistent with U.S. Geological Survey Fundamental Science Practices (https://pubs.usgs.gov/circ/1367/, accessed on 16 December 2021). Any use of trade, firm, or product names is for descriptive purposes only and does not imply endorsement by the U.S. Government. Writings prepared by U.S. Government employees as part of their official duties, including this paper, cannot be copyrighted and are in the public domain.

Conflicts of Interest: The authors declare no conflict of interest. The funders had no role in the design of the study; in the collection, analyses, or interpretation of data; in the writing of the manuscript, or in the decision to publish the results.

References

1. National Drought Mitigation Center United States Drought Monitor. New Mexico Tabular Data Archive. Available online: http://droughtmonitor.unl.edu/Data/DataTables.aspx (accessed on 7 June 2019).
2. U.S. Department of Agriculture. *2012 Census of Agriculture, United States Summary and State Data*; U.S. Department of Agriculture: North Bend, WA, USA, 2014; Volume 1, 586p.
3. New Mexico Universities Working Group on Water Supply Vulnerabilities. Final report to the Interim Committee on Water and Natural Resources, 2015. In *New Mexico Bureau of Geology and Mineral Resources Open File Report*; New Mexico Geological Society: Socorro, NM, USA, 2015; Volume 577, p. 22.
4. U.S. Geological Survey, The National Map. Available online: https://www.usgs.gov/core-science-systems/national-geospatial-program/national-map (accessed on 30 June 2021).
5. Teeple, A.P. Geophysics- and geochemistry-based assessment of the geochemical characteristics and groundwater-flow system of the U.S. part of the Mesilla Basin/Conejos-Médanos aquifer system in Doña Ana County, New Mexico, and El Paso County, Texas, 2010–2012. In *U.S. Geological Survey Scientific Investigations Report 2017–5028*; U.S. Department of the Interior: Washington, DC, USA; U.S. Geological Survey: Reston, VA, USA, 2017; 183p. [CrossRef]
6. Driscoll, J.M.; Sherson, L.R. Variability of surface-water quantity and quality and shallow groundwater levels and quality within the Rio Grande Project area, New Mexico and Texas, 2009–2013. In *U.S. Geological Survey Scientific Investigations Report 2016–5006*; U.S. Geological Survey: Reston, VA, USA, 2016.
7. Sweetkind, D.S. Three-dimensional hydrogeologic framework model of the Rio Grande transboundary region of New Mexico and Texas, USA, and northern Chihuahua, Mexico. In *U.S. Geological Survey Scientific Investigations Report 2017-5060*; U.S. Geological Survey: Reston, VA, USA, 2017. [CrossRef]
8. Hawley, J.W.; Kennedy, J.F. Creation of a digital hydrogeologic framework model of the Mesilla Basin and Southern Jornada del Muerto Basin. In *WRRI Technical Completion Report*; New Mexico Water Resources Research Institute, New Mexico State University: Las Cruces, NM, USA, 2004.

9. Jimenez, A.J.; Keller, G.R. Rift Basin structure in the border region of Northern Chihuahua. In *New Mexico Geological Society, 51st Annual Field Conference Guidebook*; New Mexico Geological Society: Socorro, NM, USA, 2000; pp. 79–83.
10. Servicio Geológico Mexicano. *Hydrogeological Activities in the Conejos-Médanos Aquifer, State of Chihuahua, Phase I*; International Boundary and Water Commission: El Paso, TX, USA, 2011; Volume 1, 109p. Available online: https://nmwrri.nmsu.edu/wp-content/uploads/2015/publish/TAAP/Final_report_English_Mesilla_ConejosMedanos_Study.pdf (accessed on 16 December 2021).
11. Wilson, C.A.; White, R.R.; Orr, B.R.; Roybal, R.G. Water resources of the Rincon and Mesilla Valleys and adjacent areas, New Mexico. In *New Mexico State Engineer Technical Report 43*; New Mexico State Engineer: Santa Fe, NM, USA, 1981; 514p.
12. Balleau, W.P. Groundwater modeling in the lower Rio Grande. In Proceedings of the 43rd Annual New Mexico Water Conference: Water Challenges on the Lower Rio Grande, Las Cruces, NM, USA, 22–23 October 1998; Ortega Klett, C.T., Ed.; New Mexico Water Resources Research Institute: Las Cruces, NM, USA, 1999. Report No. 310. pp. 38–46.
13. Witcher, J.C.; King, J.P.; Hawley, J.W.; Kennedy, J.F.; Williams, J.; Cleary, M.; Bothern, L.R. Sources of salinity in the Rio Grande and Mesilla Basin groundwater. In *New Mexico Water Resources Research Institute Technical Completion Report*; New Mexico State University: Las Cruces, NM, USA, 2004; no. 330.
14. Frenzel, P.F.; Kaehler, C.A. Geohydrology and simulation of ground-water flow in the Mesilla Basin, Doña Ana County, New Mexico, and El Paso County, Texas. In *US Geological Survey Professional Paper 1407-C*; US Government Printing Office: Washington, DC, USA, 1992. [CrossRef]
15. S.S. Papadopulos and Associates, Inc. (SSPA). Draft—Groundwater flow model for administration and management in the Lower Rio Grande Basin. In *Draft Reports for the New Mexico Office of the State Engineer*; S.S. Papadopulos and Associates, Inc.: Rockville, MD, USA, 2007; 58p.
16. Hanson, R.T.; Ritchie, A.B.; Boyce, S.E.; Galanter, A.E.; Ferguson, I.A.; Flint, L.E.; Henson, W.R. Rio Grande transboundary integrated hydrologic model and water-availability analysis, New Mexico and Texas, United States, and Northern Chihuahua, Mexico. In *US Geological Survey Scientific Investigations Report 2019-5120*; U.S. Geological Survey: Reston, VA, USA, 2020. [CrossRef]
17. Ritchie, A.B.; Hanson, R.T.; Galanter, A.E.; Boyce, S.E.; Damar, N.A.; Shephard, Z.M.; Tillman, F.D. Digital hydrologic and geospatial data for the Rio Grande transboundary integrated hydrologic model and water-availability analysis, New Mexico and Texas, United States, and Northern Chihuahua, Mexico. In *U.S. Geological Survey Data Release*; U.S. Geological Survey: Reston, VA, USA, 2018. [CrossRef]
18. U.S. Geological Survey. USGS water data for the Nation. In *U.S. Geological Survey National Water Information System Database*; U.S. Geological Survey: Reston, VA, USA, 2020. [CrossRef]
19. Kohlrenken, W. GAM Task 15-006: Total estimated recoverable storage for aquifers in Groundwater Management Area 1. Texas Water Development Board. 2015. Available online: https://www.twdb.texas.gov/groundwater/docs/GAMruns/Task15-006.pdf (accessed on 2 June 2021).
20. Thompson, J.C.; Kreitler, C.W.; Young, M.H. Exploring groundwater recoverability in Texas, maximum economically recoverable storage. *Texas Water J.* **2020**, *11*, 152–171. [CrossRef]
21. ArcGIS Pro Zonal Statistics (Spatial Analyst). Available online: https://pro.arcgis.com/en/pro-app/latest/tool-reference/spatial-analyst/zonal-statistics.htm (accessed on 16 March 2021).
22. Carrera-Hernández, J.J. Vertical datum transformation grids for Mexico. *Sci. Data* **2020**, *7*, 10. [CrossRef]
23. Pepin, J.D.; Robertson, A.J. Estimated groundwater elevations and standard errors from the application of kriging to median groundwater elevation data from 2010 in the Mesilla/Conejos-Médanos Basin, United States and Mexico. In *U.S. Geological Survey Data Release*; U.S. Geological Survey: Reston, VA, USA, 2021. [CrossRef]
24. Fisher, J.C. Optimization of water-level monitoring networks in the eastern Snake River Plain aquifer using a kriging-based genetic algorithm method. In *U.S. Geological Survey Scientific Investigations Report 2013-5120*; U.S. Geological Survey: Reston, VA, USA, 2013; 74p. [CrossRef]
25. R Core Team. R: A Language and Environment for Statistical Computing. In *R Foundation for Statistical Computing*; R Core Team: Vienna, Austria, 2019. Available online: https://www.R-project.org/ (accessed on 1 June 2021).
26. Western Regional Climate Center. Annual Totals from State University, NM (290131). Available online: https://wrcc.dri.edu/cgi-bin/cliMAIN.pl?nm0131 (accessed on 12 February 2021).
27. CONAGUA. Actualization de la Disponibilidad Media Anual de Agua en el Acuifero Conejos-Medanos (0823) Estado de Chihuahua. In *Subdireccion General Tecnica Gerencia de Aguas Subterraneas*; Comisión Nacional del Agua: Mexico City, Mexico, 2020; 36p. Available online: https://sigagis.conagua.gob.mx/gas1/Edos_Acuiferos_18/chihuahua/DR_0823.pdf (accessed on 1 July 2021).
28. Chapin, C.E.; Seager, W.R. Evolution of the Rio Grande rift in the Socorro and Las Cruces areas. In *New Mexico Geological Society, 26th Annual Field Conference Guidebook*; Seager, W.R., Clemons, R.E., Callender, J.F., Eds.; NM Bureau of Geology & Mineral Resources: Socorro, NM, USA, 1975; pp. 297–321.
29. Hawley, J.W. Guidebook to the Rio Grande Rift in New Mexico and Colorado. In *New Mexico Bureau of Mines and Mineral Resources Circular 163*, 1st ed.; New Mexico Bureau of Geology & Mineral Resources: Socorro, NM, USA, 1978; 241p.
30. Chapin, C.E.; Cather, S.M. Tectonic setting of the axial basins of the northern and central Rio Grande Basin, New Mexico. In *Basins of the Rio Grande Rift—Structure, Stratigraphy, and Tectonic Setting*; Keller, G.R., Cather, S.M., Eds.; Geological Society of America: McLean, VA, USA, 1994; pp. 5–25.

31. Instituto Nacional de Estadística. Geografía e Informática. Ciudad Juarez, 1:250000, H 13-1. 1983. Available online: https://www.inegi.org.mx/contenidos/productos/prod_serv/contenidos/espanol/bvinegi/productos/geografia/tematicas/Geologia_hist/1_250_000/702825675530.pdf (accessed on 16 December 2021).
32. Sellepack, B.P. The Stratigraphy of the Pliocene–Pleistocene Santa Fe Group in the Southern Mesilla Basin. Master's Thesis, University of Texas at El Paso, El Paso, TX, USA, 2003; 268p.
33. Khatun, S.; Doser, D.I.; Imana, E.C.; Keller, G.R. Locating faults in the southern Mesilla bolson, West Texas and southern New Mexico, using 3-D modeling of precision gravity data. *J. Environ. Eng. Geophys.* **2007**, *12*, 149–161. [CrossRef]
34. Kelley, R. New Mexico Geothermal Play Fairway Analysis from LANL. In *Geothermal Data Repository*; Los Alamos National Laboratory: Los Alamos, NM, USA, 2015. Available online: https://gdr.openei.org/submissions/597 (accessed on 29 April 2021).
35. Mack, G.H.; Seager, W.R.; Leeder, M.R.; Perez-Arlucea, M.; Salyards, S.L. Pliocene and Quaternary history of the Rio Grande, the axial river of the southern Rio Grande rift, New Mexico, USA. *Earth-Sci. Rev.* **2006**, *79*, 141–162. [CrossRef]
36. Seager, W.R.; Morgan, P. Rio Grande rift in southern New Mexico, west Texas, and northern Chihuahua. In *Rio Grande Rift: Tectonics and Magnetism*; Riecker, R.E., Ed.; American Geophysical Union: Washington, DC, USA, 1979; pp. 87–106.
37. Hawley, J.W.; Kennedy, J.F.; Creel, B.J. The Mesilla Basin Aquifer System of New Mexico, West Texas, and Chihuahua—An overview of its hydrogeologic framework and related aspects of groundwater flow and chemistry. In *Aquifers of West Texas*; Mace, R.E., Mullican, W.F., III, Angle, E.S., Eds.; Texas Water Development Board: Austin, TX, USA, 2001; pp. 76–94.
38. Nickerson, E.L.; Myers, R.G. Geohydrology of the Mesilla ground-water Basin, Doña Ana County, New Mexico, and El Paso County, Texas. In *U.S. Geological Survey Water-Resources Investigations Report 92-4156*; U.S. Geological Survey: Reston, VA, USA, 1993; p. 23.
39. Conover, C.S. Ground-water conditions in the Rincon and Mesilla Valleys and adjacent areas in New Mexico. In *U.S. Geological Survey Water-Supply Paper 1230*; U.S. Geological Survey: Reston, VA, USA, 1954; 200p.
40. Gates, J.S.; White, D.E.; Leggat, E.R. Preliminary study of the aquifers of the Lower Mesilla Valley in Texas and New Mexico by model simulation. In *U.S. Geological Survey Water-Resources Investigations Report 84-4317*; U.S. Geological Survey: Reston, VA, USA, 1984. [CrossRef]
41. Leggat, E.R.; Lowry, M.E.; Hood, J.W. Ground-water resources of the lower Mesilla Valley Texas and New Mexico. In *U.S. Geological Survey Water-Supply Paper 1669-AA*; U.S. Geological Survey: Reston, VA, USA, 1963. [CrossRef]
42. Kubicki, C.; Carroll, K.C.; Witcher, J.C.; Robertson, A.J. An integrated geochemical approach for defining sources of groundwater salinity in the southern Rio Grande Valley of the Mesilla Basin, New Mexico. In *New Mexico Water Resources Research Institute Technical Completion Report*; New Mexico State University: Las Cruces, NM, USA, 2021; 40p.
43. Arunshankar, B.N. Use of Earth Resistivity Method for Monitoring Saline Groundwater Movement in Aquifers. Master's Thesis, University of Texas at El Paso, El Paso, TX, USA, 1993; 79p.
44. Myers, R.G.; Orr, B.R. Geohydrology of the aquifer in the Santa Fe Group, northern West Mesa of the Mesilla Basin near Las Cruces, New Mexico. In *U.S. Geological Survey, Water-Resources Investigations Report 84-4190*; U.S. Geological Survey: Reston, VA, USA, 1986; 37p.
45. Llewellyn, D.; Vaddey, S. *West-Wide Climate Risk Assessment: Upper Rio Grande Impact Assessment*; U.S. Bureau Reclamation, Upper Colorado Region: Albuquerque, NM, USA, 2013; 138p.
46. Rango, A. Snow: The Real Water Supply for the Rio Grande Basin: New Mexico. *J. Sci.* **2006**, *44*, 99–118. Available online: https://jornada.nmsu.edu/files/bibliography/06-035.pdf (accessed on 16 December 2021).
47. New Mexico State Office of the Engineer, Lower Rio Grande Water Master Reports. Available online: https://www.ose.state.nm.us/WM/WMdistrict4.php (accessed on 3 February 2021).
48. Bureau of Reclamation, Supplemental Environmental Assessment—Implementation of Rio Grande Project Operating Procedures, New Mexico and Texas: Bureau of Reclamation. 2013. Available online: https://www.usbr.gov/uc/albuq/envdocs/ea/riogrande/op-Proced/Supplemental/Final-SuppEA.pdf (accessed on 10 April 2021).
49. Crilley, D.M.; Matherne, A.M.; Thomas, N.; Falk, S.E. Seepage investigations of the Rio Grande from below Leasburg Dam, Leasburg, New Mexico, to above American Dam, El Paso, Texas, 2006–2013. In *Geological Survey Open-File Report 2013-1233*; U.S. Geological Survey: Reston, VA, USA, 2013. Available online: http://pubs.usgs.gov/of/2013/1233/ (accessed on 16 December 2021).
50. Briody, A.C.; Robertson, A.J.; Thomas, N. Seepage Investigation of the Rio Grande From below Leasburg Dam, Leasburg, New Mexico, to above American Dam, El Paso, Texas, 2015. In *U.S. Geological Survey Scientific Investigations Report 2016-5011*; U.S. Geological Survey: Reston, VA, USA, 2016. [CrossRef]
51. Briody, A.C.; Robertson, A.J.; Thomas, N. Seepage Investigation of the Rio Grande From Below Leasburg Dam, Leasburg, New Mexico, to Above American Dam, El Paso, Texas, 2014. In *U.S. Geological Survey Scientific Investigations Report 2016-5010*; U.S. Geological Survey: Reston, VA, USA, 2016. [CrossRef]
52. Ball, G.P.; Robertson, A.J.; Medina Morales, K. Seepage investigation of the Rio Grande from below Leasburg Dam, Leasburg, New Mexico, to above El Paso, Texas, 2018. In *U.S. Geological Survey Scientific Investigations Report 2019–5140*; U.S. Geological Survey: Reston, VA, USA, 2020. [CrossRef]
53. Ikard, S.; Teeple, A.; Humberson, D. Gradient self-potential logging in the Rio Grande to identify gaining and losing reaches across the Mesilla Valley. *Water* **2021**, *13*, 1331. [CrossRef]
54. Fuchs, E.H.; King, J.P.; Carroll, K.C. Quantifying disconnection of groundwater from managed-ephemeral surface water during drought and conjunctive agricultural use. *Water Resour. Res.* **2019**, *55*, 5871–5890. [CrossRef]

55. King, P. Balancing Agricultural Use of Surface Water and Groundwater Along the Rio Grande. In Proceedings of the Abstract for the 60th Annual New Mexico Water Conference, Taos, NM, USA, 23 September 2015. Available online: https://nmwrri.nmsu.edu/wp-content/uploads/Conference60/SpeakerAbstracts/King.pdf (accessed on 5 February 2021).
56. New Mexico Water Rights Reporting System: New Mexico Office of the State Engineer. Available online: http://nmwrrs.ose.state.nm.us/nmwrrs/index.html (accessed on 1 June 2021).
57. Richardson, G.L.; Gebhard, T.G., Jr.; Brutsaert, W.F. Water table investigation in the Mesilla Valley. In *New Mexico State University Engineering Experiment Station Technical Report*; Engineering Experiment Station, New Mexico State University: Las Cruces, NM, USA, 1972; Volume 76.
58. U.S. Geological Survey, Chloride, Salinity, and Dissolved Solids. Available online: https://www.usgs.gov/mission-areas/water-resources/science/chloride-salinity-and-dissolved-solids?qt-science_center_objects=0#qt-science_center_objects (accessed on 5 May 2021).
59. Szynkiewicz, A.; Witcher, J.C.; Modelska, M.; Borrok, D.M.; Pratt, L.M. Anthropogenic sulfate loads in the Rio Grande, New Mexico (USA). *Chem. Geol.* **2011**, *283*, 194–209. [CrossRef]
60. Bexfield, L.M.; Anderholm, S.K. Water-quality assessment of the Rio Grande Valley, Colorado, New Mexico, and Texas—groundwater quality in the Rio Grande flood plain, Cochiti Lake, New Mexico, to El Paso, Texas, 1995. In *U.S. Geological Survey Water-Resources Investigations Report*; U.S. Geological Survey: Reston, VA, USA, 1997; Volume 93, pp. 96–4249.
61. Mills, S.K. Quantifying Salinization of the Rio Grande Using Environmental Tracers. Master's Thesis, New Mexico Institute of Mining and Technology, Socorro, NM, USA, 2003.
62. Hibbs, B.; Merino, M. Discovering a geologic salinity source in the Rio Grande aquifer. *Southwest Hydrol.* **2007**, *6*, 20–23.
63. Hogan, J.F.; Phillips, F.M.; Mills, S.K.; Hendrickx, J.M.H.; Ruiz, J.; Chelsey, J.T.; Asmerom, Y. Geologic origins of salinization in a semiarid river—The role of sedimentary basin brines. *Geology* **2007**, *35*, 1063–1066. [CrossRef]
64. Moore, S.J.; Bassett, R.L.; Liu, B.; Wolf, C.P.; Doremus, D. Geochemical tracers to evaluate hydrogeologic controls on river salinization. *Ground Water* **2008**, *46*, 489–501. [CrossRef]
65. Szynkiewicz, A.; Borrok, D.M.; Skrzypek, G.; Rearick, M.S. Isotopic studies of the Upper and Middle Rio Grande. Part 1—Importance of sulfide weathering in the riverine sulfate budget. *Chem. Geol.* **2015**, *411*, 323–335. [CrossRef]
66. Szynkiewicz, A.; Borrok, D.M.; Ganjegunte, G.K.; Skrzypek, G.; Ma, L.; Rearick, M.S.; Perkins, G.B. Isotopic studies of the Upper and Middle Rio Grande. Part 2—Salt loads and human impacts in south New Mexico and west Texas. *Chem. Geol.* **2015**, *411*, 336–350. [CrossRef]
67. Hiebing, M.; Doser, D.I.; Avila, V.M.; Ma, L. Geophysical studies of fault and bedrock control on groundwater geochemistry within the southern Mesilla Basin, western Texas and southern New Mexico. *Geosphere* **2018**, *14*, 1912–1934. [CrossRef]
68. Sheng, Z.; Darr, M.; King, J.P.; Bumgarner, J.; Michelsen, A. Mesilla Basin/Conejos-Médanos section of the Transboundary Aquifer Assessment Program, in Five-year interim report of the United States-Mexico transboundary aquifer assessment program—2007–2012. In *U.S. Geological Survey Open-File Report 2013–1059*; Alley, W.M., Ed.; U.S. Geological Survey: Reston, VA, USA, 2013. [CrossRef]
69. Toth, J. A theoretical analysis of groundwater flow in small drainage basins. *J. Geophys. Res.* **1963**, *68*, 4795–4812. [CrossRef]
70. Phillips, F.M.; Hogan, J.F.; Mills, S.K.; Hendrickx, J.M.H. Environmental tracers applied to quantifying causes of salinity in arid-region rivers: Results from the Rio Grande Basin, Southwestern USA. *Dev. Water Sci.* **2003**, *50*, 327–334.
71. Anderholm, S.K. Water quality and geochemistry of the Mesilla Basin. In *Geohydrology and Simulation of Ground-Water Flow in the Mesilla Basin, Doña Ana County, New Mexico, and El Paso County, Texas*; Frenzel, P.F., Kaehler, C.A., Eds.; U.S. Geological Survey: Denver, CO, USA, 1992; Volume 1407–C, pp. 64–75.
72. CH2MHill, Distal Mesilla conceptual model. In *Technical Report to the U.S. Army Corps of Engineers*; U.S. Army Engineer: Washington, DC, USA, 2013; 224p.
73. Cuthbert, M.O.; Acworth, R.I.; Andersen, M.S.; Larsen, J.R.; McCallum, A.M.; Rau, G.C.; Tellam, J.H. Understanding and quantifying focused, indirect groundwater recharge from ephemeral streams using water table fluctuations. *Water Resour. Res.* **2016**, *52*, 827–840. [CrossRef]
74. Flint, L.E.; Flint, A.L. California Basin Characterization Model: A dataset of historical and future hydrologic response to climate change. In *U.S. Geological Survey Data Release*; U.S. Geological Survey: Reston, VA, USA, 2014; Volume 1. [CrossRef]
75. Gross, J. A Hydrogeological Investigation of the Las Cruces Geothermal Field. Master's Thesis, New Mexico State University, Las Cruces, NM, USA, 1988.
76. Langman, J.B.; Robertson, A.J.; Bynum, J.; Gebhardt, F.E. Geochemical trends and natural attenuation of RDX, nitrate, and perchlorate in the Hazardous Test Area fractured-granite aquifer, White Sands Missile Range, New Mexico, 1996–2006. In *U.S. Geological Survey Scientific Investigations Report*; U.S. Geological Survey: Reston, VA, USA, 2008; Volume 5157.
77. Langman, J.B.; Ellis, A.S. Geochemical indicators of interbasin groundwater flow within the southern Rio Grande Valley, southwestern USA. *Environ. Earth Sci.* **2013**, *68*, 1285–1303. [CrossRef]
78. Harmon, C. Isotopic variations in meteoric waters. *Science* **1961**, *133*, 1702–1703.
79. Hibbs, B. Hydrogeologic and water quality issues along the El Paso/ Juárez corridor. *Environ. Eng. Geosci.* **1999**, *1*, 27–39. [CrossRef]

80. Eastoe, C.J.; Hibbs, B.J.; Olivas, A.G.; Hogan, J.F.; Hawley, J.; Hutchinson, W.R. Isotopes in the Hueco Bolson aquifer, Texas (USA) and Chihuahua (Mexico)—Local and general implications for recharge sources in alluvial basins. *Hydrogeol. J.* **2008**, *16*, 737–747. [CrossRef]
81. Orr, B.R.; Risser, D.W. Geohydrology and potential effects of development of freshwater resources in the northern part of the Hueco Bolson, Doña and Otero Counties, New Mexico, and El Paso County, Texas. In *U.S. Geological Survey Water-Resources Investigations Report 91-4082*; U.S. Geological Survey: Reston, VA, USA, 1992. [CrossRef]
82. Garcia-Vasquez, A.C. Caracterización Hidrogeoquímica e Isotópica del acuífero de Conejos-Médanos, Ciudad Juárez, Chihuahua. Master's Thesis, Universidad Autónoma de Ciudad Juárez, Juárez, Chihuahua, Mexico, 2017.
83. Pepin, J.D.; Robertson, A.J.; Kelley, S.A. Salinity contributions from geothermal waters to the Rio Grande and shallow aquifer system in the transboundary Mesilla Basin (United States)/Conejos-Médanos (Mexico). *Water* **2021**, *14*, 33. [CrossRef]
84. Witcher, J.C. *A Geothermal Resource Data Base—New Mexico*; Southwest Technology Development Institute, New Mexico State University: Las Cruces, NM, USA, 1995. [CrossRef]
85. Snyder, J.T. Heat Flow in the Southern Mesilla Basin with an Analysis of the East Potrillo Geothermal System, Doña Ana County, New Mexico. Master's Thesis, New Mexico State University, Las Cruces, NM, USA, May 1986.
86. Serrano, R.J. Lower Rio Grande Water Master Annual Report 2016 Accounting Year. State of New Mexico Office of the State Engineer. 2016. Available online: https://www.ose.state.nm.us/WM/LRG/Reports/2016%20LRG%20WM%20REPORT.pdf (accessed on 17 March 2021).
87. Serrano, R.J. Lower Rio Grande Water Master Annual Report 2017 Accounting Year. State of New Mexico Office of the State Engineer. 2017. Available online: https://www.ose.state.nm.us/WM/LRG/Reports/2017%20WM%20REPORT%20-%20FINAL%20DRAFT.pdf (accessed on 17 March 2021).
88. Serrano, R.J. Lower Rio Grande Water Master Annual Report 2019 Accounting Year. State of New Mexico Office of the State Engineer. 2019. Available online: https://www.ose.state.nm.us/WM/LRG/Reports/2019%20WM%20REPORT%20-%20FINAL.pdf (accessed on 17 March 2021).
89. World Population Review. Available online: https://worldpopulationreview.com/world-cities (accessed on 11 June 2021).
90. Petronis, L.H.; Finch, S.T., Jr.; Shomaker, J.W. *Ground-Water Flow and Solute Transport Model for the Griggs and Walnut Superfund Site*; Albuquerque, N. Mex., John Shomaker and Associates, Inc.: Las Cruces, NM, USA, 2006; 53p. Available online: https://donaanacounty.org/sites/default/files/pages/FSRappendixA.pdf (accessed on 16 December 2021).
91. Hutchison, W.R. Conceptual Evaluation of Surface Water Storage in El Paso County—Final Report. El Paso Water Utilities Hydrogeology Report 08–02. 2008. Available online: https://www.twdb.texas.gov/publications/reports/contracted_reports/doc/0704830690_RegionE/SurfaceWaterStorage-EPCounty-revisedfinal.pdf (accessed on 16 December 2021).
92. White, D.E. Summary of hydrologic information in the El Paso, Texas area, with emphasis on ground-water studies, 1903–1980. In *U.S. Geological Survey Open-File Report 83–775*; U.S. Geological Survey: Reston, VA, USA, 1983. [CrossRef]
93. Faure, G. *Principles of Isotope Geology*, 2nd ed.; Wiley: New York, NY, USA, 1986.
94. Plummer, L.N.; Bexfield, L.M.; Anderholm, S.K.; Sanford, W.E.; Busenberg, E. Geochemical characterization of ground-water flow in the Santa Fe Group aquifer system, Middle Rio Grande Basin, New Mexico. In *U.S. Geological Survey Water-Resources Investigation Report 03–4131*; U.S. Geological Survey: Reston, VA, USA, 2004; 414p. [CrossRef]
95. Land, L. Overview of Fresh and Brackish Water Quality in New Mexico. In *New Mexico Bureau of Geology and Mineral Resources Open-File Report*; New Mexico Bureau of Geology & Mineral Resources: Socorro, NM, USA, 2016; p. 583.
96. Moyer, D.L.; Anderholm, S.K.; Hogan, J.F.; Phillips, F.M.; Hibbs, B.J.; Witcher, J.C.; Matherne, A.M.; Falk, S.E. Knowledge and understanding of dissolved solids in the Rio Grande–San Acacia, New Mexico, to Fort Quitman, Texas, and plan for future studies and monitoring. In *U.S. Geological Survey Open-File Report 2013–1190*; U.S. Geological Survey: Reston, VA, USA, 2013. Available online: http://pubs.usgs.gov/of/2013/1190/ (accessed on 16 December 2021).
97. Nickerson, E.L. Selected geohydrologic data for the Mesilla ground-water basin, 1987 through 1992 water years, Doña Ana County, New Mexico and El Paso County, Texas. In *U.S. Geological Survey Open-File Report*; U.S. Geological Survey: Reston, VA, USA, 1995; Volume 111.
98. Reiter, M.; Eggleston, R.E.; Broadwell, B.R.; Minier, J. Estimates of terrestrial heat flow from deep petroleum tests along the Rio Grande rift in Central and Southern New Mexico. *J. Geophys. Res.* **1986**, *91*, 6225–6245. [CrossRef]
99. Witcher, J.C.; Lund, J.W. Masson Radium Springs Farm. *GHC Bull.* **2002**, *23*, 42–44. Available online: https://www.osti.gov/servlets/purl/1209492 (accessed on 1 May 2021).
100. Engle, M.A.; Rowan, E.L. Interpretation of Na–Cl–Br systematics in sedimentary basin brines: Comparison of concentration, element ratio, and isometric log-ratio approaches. *Math. Geosci.* **2012**, *45*, 87–101. [CrossRef]
101. Swanberg, C.A. Detection of geothermal components in groundwaters of Doña Ana County, southern Rio Grande rift, New Mexico. In *26th Annual Field Conference Guidebook*; Seager, W.R., Clemons, R.E., Callender, J.F., Eds.; New Mexico Geological Society: Socorro, NM, USA, 1975; pp. 175–180.
102. Decker, E.R.; Cook, F.A.; Ramberg, I.B.; Smithson, S.B. Significance of geothermal and gravity studies in the Las Cruces area. In *26th Annual Field Conference Guidebook*; Seager, W.R., Clemons, R.E., Callender, J.F., Eds.; New Mexico Geological Society: Socorro, NM, USA, 1975; pp. 251–259.
103. Slichter, C.S. Observations on groundwater of the Rio Grande valley. In *U.S. Geological Survey Water-Supply Paper*; U.S. Geological Survey: Reston, VA, USA, 1905; Volume 141. [CrossRef]

104. Scruggs, C.E.; Thomson, B.M. Opportunities and Challenges for Direct Potable Water Reuse in Arid Inland Communities. *J. Water Resour. Plan. Manag.* **2017**, *143*, 4017064. [CrossRef]
105. Hawley, J.W. Challenges and opportunities for brackish groundwater resource development in New Mexico–prediction hydroscience from an octogenarian hydrogeologist's perspective. In Proceedings of the Urban Land Institute-New Mexico Section, Albuquerque, NM, USA, 28 April 2016. Available online: https://aquadoc.typepad.com/files/uli-nm_whitepaper.pdf (accessed on 28 April 2016).
106. Plummer, L.N.; Bexfield, L.M.; Anderholm, S.K.; Sanford, W.E.; Busenberg, E. Geologic and hydrologic framework: C. Geochemical characterization of groundwater flow in parts of the Santa Fe Group aquifer system, Middle Rio Grande Basin, New Mexico. In *U.S. Geological Survey Middle Rio Grande Basin Study—Proceedings of the Fourth Annual Workshop, Albuquerque, NM, USA, 15–16 February 2000*; U.S. Geological Survey Open File Report; Cole, J.C., Ed.; U.S. Geological Survey: Denver, CO, USA, 2000; Volume 00-488, pp. 7–12. [CrossRef]
107. Scanlon, B.R.; Darling, B.K.; Mullican, W.F., III. Evaluation of groundwater recharge in basins of Trans-Pecos Texas. In *Aquifers of West Texas*; Mace, R.E., Mullican, W.F., III, Angle, E.S., Eds.; Texas Water Development Board: Austin, TX, USA, 2001; Volume 356, pp. 26–40.

water

MDPI

Article

Transboundary Aquifers between Baja California, Sonora and Chihuahua, Mexico, and California, Arizona and New Mexico, United States: Identification and Categorization

Rosario Sanchez [1,*] **and Laura Rodriguez** [1,2]

[1] Texas Water Resources Institute, Texas A&M University, College Station, TX 77843, USA
[2] Water Management and Hydrologic Science Program, Texas A&M University, College Station, TX 77845, USA; lmrodriguezlo@tamu.edu
* Correspondence: rosario@tamu.edu

Abstract: In 2016, research suggested there might be up to 36 transboundary aquifers located along the border between Mexico and the U.S. The main contribution of this study was to put together the available segments already existent in the literature without considering the validity of the criteria used to define the boundaries of those segments. In 2018, updated research reported 33 hydrogeological units (HGUs) crossing the boundaries between Mexico and Texas. This later analysis included the homogenization of geological nomenclatures, standardization of geological and hydrogeological criteria, using a specific methodology to correlate, identify, and delineate each HGU. The purpose of this paper is to use this latter methodology and expand the same analysis to include the transboundary aquifers between Baja California/California, Sonora/Arizona, and Chihuahua/New Mexico. Results of this study indicate that a total of 39 HGUs have been identified in this region which accounts for an approximate shareable land of 135,000 km^2 where both countries share half of the area. From the total shareable area, around 40% reports good to moderate aquifer potential and water quality, of which 65% is in the U.S. and 35% on the Mexico side. Border-wide, the total number of HGUs in the border region between Mexico and the United States is 72, covering an approximate area of 315,000 km^2 (180,000 km^2 on the U.S. side and 135,000 km^2 on the Mexico side). The total area that reports good to moderate aquifer potential as well as good to regular water quality ranges between 50 and 55% (of which approximately 60% is in the U.S. and the rest in Mexico).

Keywords: transboundary; aquifers; Mexico; United States; border; groundwater

Citation: Sanchez, R.; Rodriguez, L. Transboundary Aquifers between Baja California, Sonora and Chihuahua, Mexico, and California, Arizona and New Mexico, United States: Identification and Categorization. *Water* **2021**, *13*, 2878. https://doi.org/10.3390/w13202878

Academic Editors: Sharon B. Megdal and Anne-Marie Matherne

Received: 27 June 2021
Accepted: 9 October 2021
Published: 14 October 2021

1. Introduction

In 2016, Sanchez et al. [1] suggested there might be up to 36 transboundary aquifers located along the border between Mexico and the U.S. This first assessment attempted to represent the first draft of all aquifers across the frontier between the two countries. However, this initial step was only able to put together the available information already existent without considering the validity of the criteria used to define the boundaries of those aquifers. Two years later, Sanchez et al. [2] reported 33 hydrogeological units (HGUs) crossing the boundaries between Mexico and Texas. This later analysis included the homogenization of geological nomenclatures and the standardization of geological and hydrogeological criteria to define aquifer boundaries, and it used a methodology to correlate, identify, and delineate each HGU based primarily on geological parameters. Although this methodology might differ from other transboundary studies in the border region [3–7], it did provide for the first time important physical elements that highlighted the transboundary nature of groundwater at the border at a regional scale. In fact, apart from the available studies performed under the umbrella of the Transboundary Aquifer Assessment Program (TAAP), which include de San Pedro/San Pedro, Santa Cruz/Santa Cruz (including Nogales/Nogales), Valle de Juarez/Hueco Bolson, Conejos-Medanos/Mesilla

Bolson, and Allende-Piedras Negras transboundary aquifers, there are limited references to physical studies of transboundary aquifers at a regional and even transboundary scale. There are some additional projects led by the U.S. Geological Survey (USGS) that have studied the Lower Rio Grande Basin, the Tijuana River basin, Mimbres/Las Palmas Aquifer, and the Lower Colorado River Basin [8]; however, their analysis tends to be limited to the U.S side of the aquifers. Therefore, the rest of the aquifers or shared areas in the border region remain to be explored. As of 2018, from the 33 HGUs identified by Sanchez et al. [2] between Mexico and Texas, only four aquifers have reported some type of assessment at transboundary level.

The purpose of this paper is to use the methodology applied in the border region between Mexico and Texas from Sanchez et al. [2] and expand the analysis to the remaining border region between Mexico and the United States. This study will report on the existing set of transboundary aquifers reported by Sanchez et al. [2] and include Baja California/California, Sonora/Arizona, and Chihuahua/New Mexico. Therefore, the overall result will be a border-wide assessment of transboundary aquifers utilizing one unique methodology that identifies, delineates, and initially assesses the physical conditions of all the hydrogeological units (HGUs) east of the Rio Grande/Rio Bravo across both countries. This information will serve as the basis for future assessments and prioritization analysis of transboundary aquifers in the border region between Mexico and the United States.

Results indicate that a total of 39 HGUs have been identified in the border region between California, Arizona, and New Mexico on the U.S. side and Baja California, Sonora, and Chihuahua on the Mexico side. This region accounts for an approximate shareable area of 135,000 km^2 where both countries share half of the area (65,000 km^2 Mexico and 69,000 km^2 the U.S.) From the total shareable area, around 40% reports good to moderate aquifer potential and water quality, of which 65% is located in the U.S. and 35% on the Mexico side.

From a statewide perspective, the border between Baja California, Mexico, and California, U.S., reports a total of 5 HGUs, from which 3 (Tijuana-San Diego Aq., Valle de Mexicali-San Luis Rio Colorado/Yuma-Imperial Valley and a great portion of the Quaternary deposits of Laguna Salada Aq./Coyote Wells Valley) report good to moderate aquifer potential and generally good to moderate water quality. Available data on water quality varies across the Valle de Mexicali-San Luis Rio Colorado/Yuma-Imperial Valley from good to poor (limited data included), particularly in the southern portions where saline water intrusion has been reported. In the case of Sonora and Arizona, 25 HGUs have been identified, with at least 7 HGUs (Nogales-Rio Santa Cruz Aq./Upper Santa Cruz Basin, Rio San Pedro Aq./Upper San Pedro Basin, Rio Agua Prieta Aq./Douglas Basin, Rio Altar Aq., San Simon Wash, Sonoyta-Puerto Peñasco Aq., and La Abra Plain) with generally good to moderate aquifer potential and good to moderate water quality. Variability of water quality for Sonoyta-Puerto Peñasco Aq., and San Simon Wash is also reported [9]. Additionally, 4 HGUs reported good to moderate aquifer potential but poor water quality with uncertainty considering the data limitations. Those include Cerro Colorado Numero 3 Valley, Lukeville-Sonoyta Valley, The Great Plain, and Arroyo Seco Aq. In the border region between Chihuahua and New Mexico, good aquifer potential and good water quality were identified in at least 3 out of the 8 HGUs reported. These are Janos Aq./Playas Basin, Ascension Aq./Hachita-Moscos Basin, and Las Palmas Aq./Mimbres Basin. Potrillo Mountains also report good aquifer potential but limited data on water quality.

Border-wide and adding the HGUs previously reported by Sanchez et al. [2] between Texas and Mexico, the total number of HGUs in the border region between Mexico and the United States totals 72, covering an approximate area of 315,000 km^2 (180,000 km^2 on the U.S. side and 135,000 km^2 on the Mexico side). The total area that reports good to moderate aquifer potential as well as good to regular water quality ranges between 50 and 55% (of which approximately 60% is in the U.S. and the rest in Mexico).

The first part of this paper presents the geological correlation of formations along the border across California, Arizona, and New Mexico on the U.S. side and Baja California,

Sonora, and Chihuahua on the Mexico side. The second part focuses on integrating and delineating the identified HGUs according to hydrological, lithological, topographical, surficial, and structural geology criteria. The third part of the paper shows the classification of the geological units within the boundaries of the corresponding HGUs according to aquifer potential and water quality. This study uses the same methodology and criteria developed by its predecessor, Sanchez et al. [2], with its corresponding limitations and adaptations considering the differences in geological characterization and data availability described below.

2. Materials and Methods

2.1. Geological Correlation

The basis of the analysis is to perform the geological correlation among units across the border. First, it was necessary to develop a review of available literature of geological units between Mexico, California, Arizona, and New Mexico, along with an extensive visualization and analysis of geographical information using ArcMap 10.5. Geological data, and maps from Mexico were downloaded from the federal agency Servicio Geologico Mexicano (SGM) [10] in shapefile format at 1:250,000 scale: Cartas Geologico mineras Tijuana I11-11, Mexicali I11-12, I12-10 Los Vidrios, Ensenada H11-2, San Felipe H11-3, Puerto Peñasco H12-1, Nogales H12-2, Agua Prieta H12-3, Cananea H12-5, Nacozari H12-6, Ciudad Juarez H13-1, and Nuevo Casas Grandes H13-4 [11–22]. For the states of California, Arizona, and New Mexico, geological data and maps were downloaded from the USGS online spatial data website, which covers the entire states [23,24] in shapefile format at 1:100,000 scale. The map scales were selected according to data availability on both sides of the border.

To address the issues related to differences of geological equivalence across the border, we first correlated the geological units by comparing the ages and stratigraphic lexicons and matched the geological units with their corresponding equivalent on the other side of the border. We used the lexicons available on the SGM website since they offer detailed lithological descriptions and geologic ages of the units across the border. After identifying the geological age ranges, name, and description of the units in Mexico, we correlated them with their equivalents using the USGS lexicons as reference.

Once the geological correlation process was performed, a geological structural and stratigraphic analysis (vertical geology) was developed using the geologic map profiles and well lithology descriptions to identify and delineate the boundaries of the formations. The physical continuity of geological units can be truncated by folds, lineaments, or faults, and in other cases, several formations were clustered together considering their lithological and hydrogeological similarities. A challenging issue was the igneous and metamorphic bodies outcropping at different regions. Due to their uneven distribution as outcrops, it is not possible to confirm their continuity underground, in contrast to the sedimentary rocks that are usually distributed as tabular masses and whose continuity across the border is easily traceable. Therefore, the criteria are that only geologic units outcropping on the international border (boundary formations) or crossing the border (transboundary formations) are considered in the analysis of classification of HGUs. Though there is no evidence of geological continuity across the border of the boundary formations, they are considered in the analysis as they constitute important geological and hydrological pieces within their corresponding HGUs. They appear in bold (Mex or U.S.) in the legends of the maps. As for the geologic units that outcrop only on one side of the border but do not appear close to the international border, they are considered in the geological correlation analysis and in the maps for visualization purposes but do not appear in bold in the map legends. This criterion was applied to most of the igneous and metamorphic rocks.

2.2. Delimitation of HGUs

As in Sanchez et al. 2018, this paper uses the term hydrogeological unit or HGU to refer to any soil or rock unit or zone that by virtue of its hydraulic properties has a distinct influence on the storage or movement of groundwater [25]. Therefore, considering the different hydrogeological conditions among units, some may or may not be categorized as aquifers.

The delimitation of the boundaries of the HGUs was the product of the aggrupation of geological units with common lithological features (such as high porosity) from other units where the impermeable rocks dominate. An important methodological criterion that was added as compared to Sanchez et al. [2] was topography. We integrated this variable because it was significantly important in those areas where the surficial geology was not enough to identify the limits of the unit, or the geologic heterogeneity of several units did not provide enough elements to draw a surficial boundary. For these cases, the geological maps were overlapped with the topographic applications of StreamStats from USGS [26], and SIATL from INEGI [27] which provided lineaments and slope changes to complement the HGUs' delineation. If the topography was still not definitive to identify a specific portion of the boundaries, we reviewed the available literature to confirm or adjust the boundary delineation for each case. Well lithological descriptions were also useful as indicative of aquifer features (aquifer potential) since rocks can have different conditions on the surface as compared to underground, which may modify the capacity of the aquifer to yield water. Therefore, this criterion was also added to the analysis of the HGU delimitation as compared to Sanchez et al. [2].

Another different criterion was the one applied to several HGUs where their delimitations included outcrops of crystalline igneous and metamorphic rocks with low porosity capabilities that appear as isolated hills in the topographic maps. Considering that available information about these hills does not provide enough confidence to discriminate them from the area covered by the corresponding HGU, this study included them within the boundaries of the corresponding HGU, pending further research to clarify if these crystalline rocks have an interaction with the rest of the area of the HGU.

Lastly, we assigned names to the HGUs based on preexistent aquifer names reported in the area on either side of the border. If there were no aquifers identified in previous studies, we used geographical marks, such as mountains, valleys, or towns to assign a name to the corresponding HGU.

2.3. Classification of Geological Formations

The last task was the classification of geological units (boundary and transboundary formations), which is based on hydrogeological features (aquifer potential) and water quality data, according to the same criteria used by Sanchez et al. [2].

"Aquifer potential" is defined as the potential that a geological unit, a group of geological units, or part of a geological unit contains sufficient saturated permeable material to yield significant quantities of water for wells and springs [28]. The criteria used to define aquifer potential considers mainly lithological features, permeability, porosity, hydraulic conductivity, and transmissivity (Table 1). Because the natural complexity and heterogeneity across the units and the different methods that are used to characterize units on both sides of the border, a combination of criteria had to be used to classify aquifer potential as "good", "moderate" or "poor". This study uses geological and lithological descriptions of the units, porosity and hydraulic conductivity when available, or standardized values according to the predominant lithology [29]. We also used permeability reports and assessments from the National Water Commission (CONAGUA), and technical reports from the New Mexico Water Resources Research Institute (NMWRRI), the Arizona Department of Water Resources (ADWR), California Division of Mines and Geology, and the USGS. We obtained data from federal, state, and local agencies, as well as from technical, academic, and scientific reports. The common criterion used in the literature for water quality was TDS (total dissolved solids), which were available for almost the complete border region.

Table 1. Geological formations classified into five groups according to aquifer potential (Good, Moderate, Poor) and water quality (Good, Regular, Poor). The unit of water quality is Total Dissolved Solids (TDS). The colors represent an ID later used in the classification of the units and on the maps. (Adapted from Sanchez et al. [2].)

Geological Formation			Water Quality			
			Good	**Regular**	**Poor**	**No Info**
			<1000 ppm	**1000–3000 ppm**	**>3000 ppm**	
			1	**2**	**3**	**4**
Aquifer Potential	Good	A	A1	A2	A3	A4
	Moderate	B	B1	B2	B3	B4
	Poor	C	C1	C2	C3	C4
	Aquitard	D	D1	D2	D3	D4
	No Info	E	E1	E2	E3	E4

Following the methodology of Sanchez et al. [2], we used the TDS ranges from the Texas Water Development Board [30] to classify groundwater quality: freshwater, less than 1000 mg/L; slightly saline (usually called "brackish water"), 1000–3000 mg/L; moderately saline, 3000–10,000 mg/L; very saline, 10,000–35,000 mg/L; and brine, over 35,000 mg/L. Some studies refer to "parts per million" (ppm), where 1 ppm is equivalent to 1 mg/L; ppm are the units used in this study. The categories defined in Table 1 for water quality consider freshwater as "good", slightly saline as "regular", and moderately saline with very saline are combined into one category as "poor". Table 1 shows how the formations will be classified into five groups according to aquifer potential for each one and its corresponding reported water quality.

3. Results and Discussion

3.1. Geological Correlation between Mexico (Baja California, Sonora, and Chihuahua) and the U.S. (California, Arizona, and New Mexico)

This section covers the geological features of the formations identified and correlated between Baja California, Sonora, and Chihuahua in Mexico, and California, Arizona, and Nuevo Mexico in the U.S. which are described in detail in Table 2. Geological formations in Table 2 are listed according to their geological age (oldest first), and if their names differ across countries, the first name listed corresponds to what is reported in Mexico and then in the U.S. Table 2 also includes hydrological features available and the reported names of those geological units that have been referred by the literature aquifers.

As in Sanchez et al. [2], there are formations that have been identified only on one side of the border (therefore not crossing to the other side); those formations are identified as boundary formations with a parenthetical (USA) or (MEX) after their name. Boundary and transboundary formations (the formations that cross the border) are the ones subject to classification analysis in this study and are highlighted in bold in the figures. Figures 1–4 list all the identified geological units with their reported names from both sides (Mex/U.S., even if they are the same). Other geological units located in the area but not outcropping the border are not considered in the analysis but are included in the maps and legends (not in bold) for visualization purposes. It is worth mentioning that in comparison to our antecessor, the geological maps in this study include geological faults and main topographic and hydrologic references that were not included in Sanchez et al. [2].

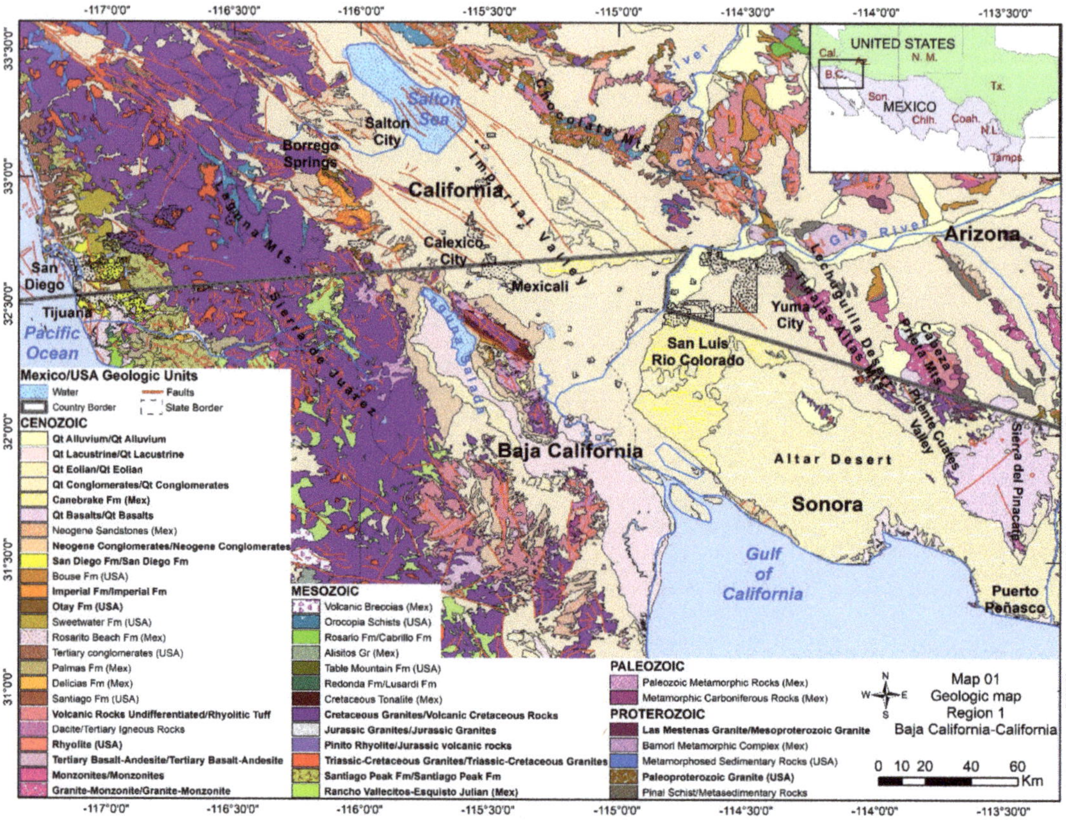

Figure 1. Geologic map, Baja California—California.

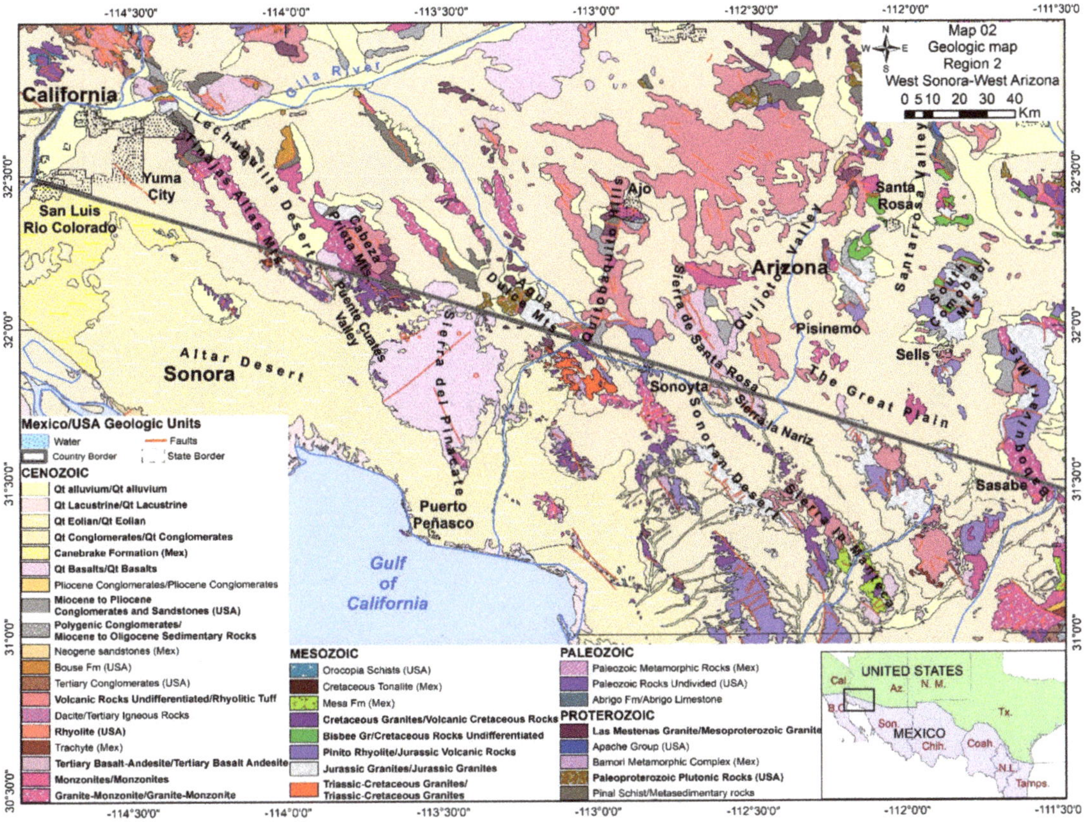

Figure 2. Geologic map, West Sonora—West Arizona.

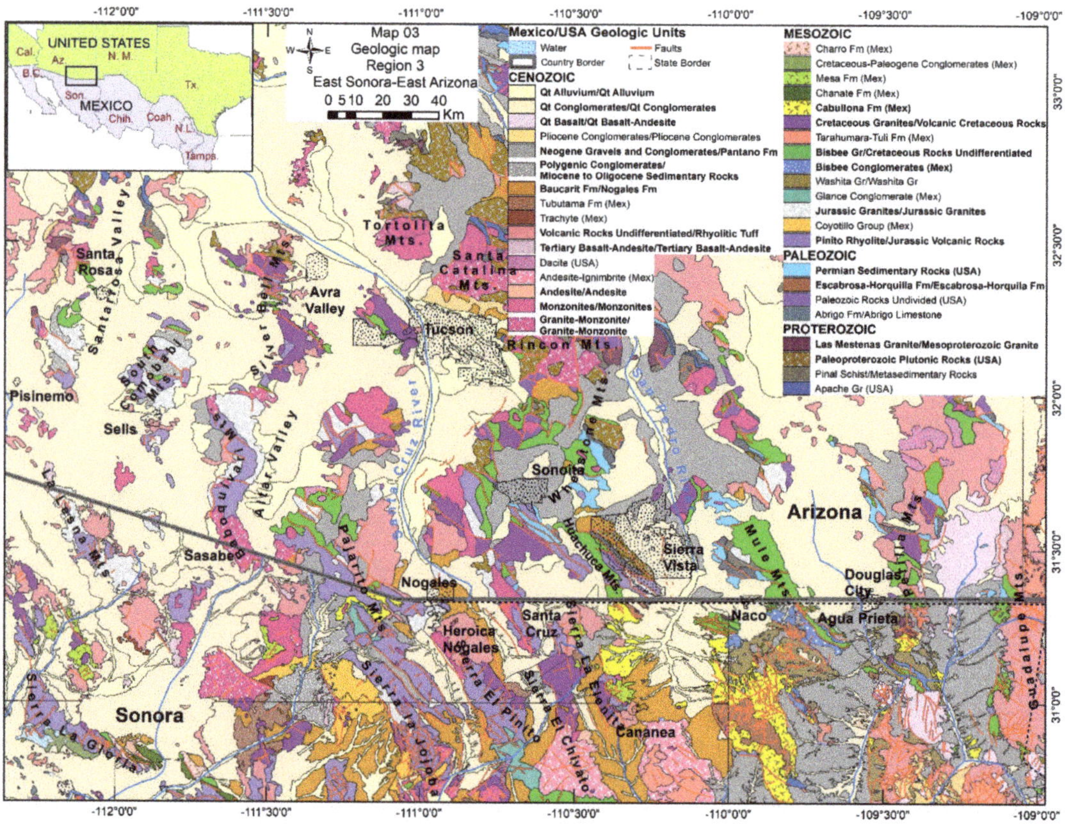

Figure 3. Geologic map, East Sonora—East Arizona.

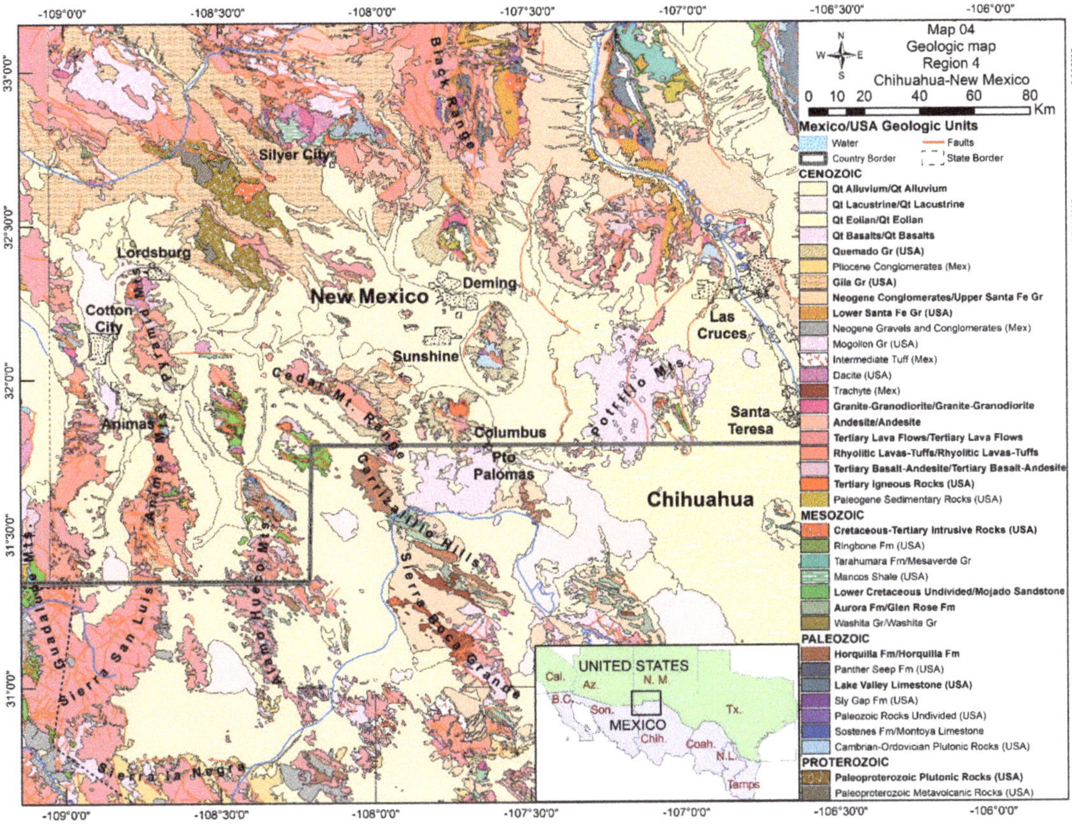

Figure 4. Geologic map, Chihuahua—New Mexico.

Table 2. Geological correlation/equivalence and hydrological features.

Unit Name (Mex/USA)	Age	Location (Mex/USA)	Lithological Description	Hydrological Features	Aquifer
Pinal Schist/ Metasedimentary Rocks (Figures 1–3).	Paleoproterozoic.	Western Chihuahua and Northern Sonora/ Central New Mexico, Southern Arizona, Southern California.	Gray quartzite-sericite schists and quartzites [31].	Low to nonexistent permeability [32]. Local secondary porosity due to fracturing [7].	San Pedro Aq. [7].
Paleoproterozoic Plutonic Rocks (USA) (Figures 1–4).	Paleoproterozoic.	Central New Mexico/ Southern Arizona.	Granitic Gneisses and foliated granites [33].	-	-
Bamori Metamorphic Complex (Mex) (Figures 1 and 2).	Paleoproterozoic.	Northeast Sonora.	Muscovite, biotite and quartz schists [34].	-	-
Apache Gr (USA) (Figures 2 and 3).	Mesoproterozoic.	Central and Southern Arizona.	Quartzites, with few shales and conglomerates [35].	Local secondary porosity due to fracturing [7].	San Pedro Aq. [7].
Las Mestenas Granite/ Mesoproterozoic Granite (Figures 1–3).	Mesoproterozoic.	Northern Sonora/ Southern Arizona.	Granites with coarse- grained holocrystalline texture. Some areas show gneiss texture [36].	Local secondary porosity due to fracturing [7,37]	San Pedro Aq. [7]. Los Vidrios Aq. [37].
Abrigo Fm/ Abrigo Limestone (Figures 2 and 3).	Cambrian.	Northeast Sonora/ Southern Arizona.	Gray-blue limestones, stratified in thin layers [31].	-	-
Cambrian-Ordovician Plutonic Rocks (USA) (Figure 4).	Cambrian-Ordovician.	Southwest New Mexico.	Granites and syenites [38].	Dense, impermeable rocks [39].	-
Sostenes Fm/ Montoya Limestone (Figure 4).	Ordovician.	Central and Northern Chihuahua/ West Texas, Southern New Mexico.	Gray dolomites, some layers are recrystallized limestones. Few mudstones to the top [40].	Montoya Dolomites have not been developed, but some evidence indicates that they may be capable of yielding water to wells [41].	Mimbres Basin [41].
Paleozoic Rocks Undivided (USA) (Figures 2 and 3).	Devonian-Carboniferous.	Southeastern New Mexico, Arizona.	Nodular, silty limestones that locally contain limy shales and siltstones [41].	Few wells yielded water from fractures [7,41].	Mimbres Basin [41]. San Pedro Aq. [7].
Paleozoic Metamorphic Rocks (Mex) (Figures 1 and 2).	Devonian-Carboniferous.	Northern Baja California/ Northern Sonora.	Biotite schists and slates [42].	Local secondary porosity due to fracturing [37].	Los Vidrios Aq. [37].

Table 2. *Cont.*

Unit Name (Mex/USA)	Age	Location (Mex/USA)	Lithological Description	Hydrological Features	Aquifer
Lake Valley Limestone (USA) (Figure 4).	Carboniferous.	Western Texas, southern New Mexico.	Gray limestones filled with nodular cherts [43].	-	-
Panther Seep Fm (USA) (Figure 4).	Carboniferous.	Western Texas, Central New Mexico.	Interbedded dark gray shales and calcareous siltstones [44].	-	-
Horquilla Fm/ Horquilla Fm (Figures 3 and 4).	Carboniferous.	Northern Chihuahua, Northern Sonora/ Southern Mexico, Arizona.	Thin pink limestone layers to the top. Gray massive limestone layers to the top [45].	Massive rocks which constitute limits of adjacent younger deposits [46].	Josefa Ortiz de Dominguez Aq. [46].
Escabrosa Fm/ Escabrosa Fm (Figure 3).	Carboniferous.	Northeast Sonora/ Southeast Arizona.	White to dark gray limestones, coarse stratification [31].	-	-
Permian Sedimentary Rocks (USA) (Figure 3).	Permian.	Southeastern Arizona.	Thick bedded limestones with layers of shales and sandstones [47].	Local secondary porosity due to fracturing [7].	San Pedro Aq. [7].
Triassic-Cretaceous Granites/ Triassic-Cretaceous Granites (Figures 1 and 2).	Triassic-Cretaceous.	Northern Baja California, Northwestern Sonora/ Southern California.	Granodiorites, Tonalites [15].	Poor porosity, secondary porosity on the surface due to alteration [48].	Tecate Aq. [48].
Jurassic Granites/ Jurassic Granites Figures 1–3).	Triassic-Jurassic.	Northern Sonora/ Southern Arizona.	Plutonic and volcanic rocks, with local occurrences of redbeds [47].	-	-
Pinito Rhyolite/ Jurassic Volcanic Rocks (Figures 1–4).	Jurassic.	Northern Sonora/ Southern Arizona.	White to light gray rhyolites and rhyodacites [49].	Low primary permeability but moderate secondary permeability [50,51].	Santa Cruz Aq. [50]. Rio Alisos Aq. [51].
Bisbee Gr/ Bisbee Gr (Figures 2 and 3).	Late Jurassic-Early Cretaceous.	Northern Sonora/ Southeast Arizona.	Conglomerates, sandstones, and argilites [52,53].	-	-
Washita Gr/ Washita Gr	Early Cretaceous.	Northern Chihuahua.	Limestones interbedded with clays [54].	-	-
Cretaceous Tonalite (Mex) (Figures 1 and 2).	Aptian.	Northwest Sonora, Northern Baja California.	Tonalites and Granites	-	-

Table 2. *Cont.*

Unit Name (Mex/USA)	Age	Location (Mex/USA)	Lithological Description	Hydrological Features	Aquifer
Aurora Fm/ Glen Rose Fm (Figure 4).	Albian.	Northeast Chihuahua/ Western Texas.	Limestone layers, sandy limestones with interbedded sandy clays, sandstone and marl [55].	Secondary porosity due to fracturing [56].	Palomas-Guadalupe Victoria Aq. [56].
Lower Cretaceous Undivided/Mojado Sandstone (Figure 4).	Albian-Cenomanian.	Northern Chihuahua/ Central New Mexico.	Quartz sandstone interbedded with gray shales [57]	Local secondary porosity due to fracturing [7,58].	San Pedro Aq. [7]. Arroyo San Bernardino Aq. [58]
Cretaceous Granites/ Volcanic Cretaceous Rocks (Figures 1–3)	Late Cretaceous.	Northern Sonora, Northern Baja California/ Southern Arizona, Southern California.	Rhyolitic to andesitic volcanic rocks, locally associated sedimentary and subvolcanic intrusive rocks [47].	These crystalline rocks are dense and contain only small amounts of water in fractures and weathered zones [59].	Mexicali-Rio Colorado Valley [59]. Santa Cruz Aq. [50].
Mancos Shale (USA) (Figure 4).	Turonian-Coniacian.	Central New Mexico.	Shales with local siltstones, sandstones, and bentonite [60].	-	-
Cabullona Fm (Mex) (Figure 3)	Santonian-Maastrichtian.	Northern Sonora.	Conglomerates, sandstones, shales, and tuffs [61].	-	-
Ringbone Fm (USA) (Figure 4).	Campanian.	Southwestern New Mexico.	Dark shales with conglomerates at the bottom, tuffaceous sandstone at the top [62].	-	-
Rosario Fm/Cabrillo Fm (USA) (Figure 1).	Maastrichtian.	Southwestern California.	Massive medium-grained sandstone with thin siltstone beds and conglomerate lenses [63].	-	-
Orocopia Schists (USA) (Figures 1 and 2).	Late Cretaceous-Paleogene.	Western Arizona, Southeastern California.	Gray quartz-feldspar schists, peridotites, schistose amphibolite, metachert, siliceous marble [64].	-	-
Cretaceous-Tertiary Intrusive Rocks (USA) (Figure 4).	Late Cretaceous-Paleogene.	Southern New Mexico.	Granodiorite, quartz monzonite, monzonite porphyry dikes [65].	-	-
Granite-Monzonite/ Granite-Monzonite (Figures 1–3).	Campanian-Eocene.	Northern Sonora, Northern Baja California/ Southern Arizona.	Muscovite granites with garnets, monzonites [66].	Poor porosity [50]. Secondary porosity [67].	Santa Cruz [50]. San Diego Aq. [67].

Table 2. *Cont.*

Unit Name (Mex/USA)	Age	Location (Mex/USA)	Lithological Description	Hydrological Features	Aquifer
Monzonites/ Monzonites (Figures 1–3).	Paleocene-Eocene.	Northern Sonora/ Southern Arizona.	Monzonites and quartz monzonites [19].	Local secondary porosity due to fracturing [7].	San Pedro Aq. [7].
Paleogene Sedimentary Rocks (USA) (Figure 4).	Paleocene-Oligocene.	Southern New Mexico.	Calcareous sandstones, gray limestones [68].	-	-
Delicias Fm (Mex) (Figure 1).	Eocene.	Northwestern Baja California.	Green shales and dark yellow sandstones [69].	-	-
Tertiary Igneous Rocks (USA) (Figure 4).	Eocene-Pliocene.	Southwestern New Mexico.	Monzonites to granites, andesites, dacites [65].	-	-
Rhyolitic Lavas-Tuffs/ Rhyolitic Lavas-Tuffs (Figure 4).	Oligocene.	Northern Chihuahua/ Western New Mexico.	Rhyolitic tuffs. Tuffaceous and silty sandstones [41].	Rocks rarely developed for groundwater production [41].	Mimbres Basin [41].
Tertiary Basalt-Andesite/ Tertiary Basalt-Andesite (Figures 1–4).	Oligocene.	Northern Baja California, Northern Sonora/ Southern California, Southern Arizona, Southern New Mexico.	Basaltic-andesitic sequence, pyroclastic rocks of silicic to intermediate composition ranging from soft pumiceous ashfall tuff to densely welded ash-flow tuff [70].	Locally fractured rocks allow secondary permeability between 18 and 25% [71].	Mimbres Basin [41].
Tertiary Lava Flows/ Tertiary Lava Flows (Figure 4).	Oligocene.	Northern Chihuahua and Sonora/ Southern New Mexico.	Locally erupted lavas, rhyolitic pyroclastic flows and tuffs [65].	-	-
Andesite/Andesite (Figures 3 and 4).	Oligocene.	Northern Chihuahua/ Southern New Mexico.	Andesitic and rhyolitic rocks [72].	Moderate to good secondary porosity [58].	Arroyo San Bernardino Aq. [58].
Granite-Granodiorite/ Granite-Granodiorite (Figure 4).	Oligocene.	Northern Chihuahua/ Southern New Mexico.	Quartz-Monzonite porphyry and granodiorites [72].	Locally fractured rocks allow secondary permeability between 18 and 25% [71,73].	Mimbres Basin [41]. Santa Cruz Aq. [73].
Trachyte (Mex) (Figures 2–4).	Oligocene.	Northern Chihuahua, Northern Sonora.	Trachyte and volcanic felsic flows [11].	-	-
Rhyolite (USA) (Figures 1 and 2).	Oligocene-Miocene	Southern California.	Volcanic rhyolitic flows and tuffs [74].	-	-

Table 2. *Cont.*

Unit Name (Mex/USA)	Age	Location (Mex/USA)	Lithological Description	Hydrological Features	Aquifer
Neogene gravels and conglomerates/Pantano Fm (Figures 3 and 4)	Oligocene-Miocene	Northern Sonora and Chihuahua/Eastern Arizona.	Well consolidated fine to coarse-grained alluvial fan and playa deposits, volcanic flows, and rock-avalanche beds [75].	Provides water to wells and alluvial deposits [75].	Tucson AMA [75].
Polygenic conglomerates/Miocene to Oligocene sedimentary rocks (Figures 2 and 3)	Oligocene-Miocene	Northern Sonora/Central and western Arizona.	Conglomerates, sandstones and mudstones undifferentiated [19,66].	-	-
Lower Santa Fe Gr (USA) (Figure 4)	Oligocene-Miocene	South-central New Mexico.	Coarse sandstones and alluvial fan deposits [76].	Source of fresh water in the bolsons area [77].	Mesilla Bolson [77].
Palmas Fm (Mex) (Figure 1).	Oligocene-Pleistocene.	Northwestern Baja California.	Polymictic conglomerates with few sandstones and claystones [14].	-	-
Dacite (USA) (Figures 1–4).	Miocene.	Southern New Mexico, Southern Arizona.	Rhyolite and Dacite flows [65].	-	-
Volcanic Rocks Undifferentiated/ Rhyolitic Tuff (Figures 1–3).	Miocene.	Northern Sonora, Northern Baja California/ Southern Arizona.	Flows of rhyolites and rhyolitic tuffs [21].	Poorly water bearing rocks, may form the boundaries of the groundwater reservoir [59].	Mexicali-Rio Colorado Valley [59].
Tertiary Conglomerates (USA) (Figures 1 and 2).	Miocene.	Southern California.	Coarse grained non-marine deposits [59].	These deposits are capable of yielding moderate amounts of fresh groundwater [59].	Mexicali-Rio Colorado Valley [59].
Sweetwater Fm (USA) (Figure 1).	Miocene.	Southwestern California.	Gritty sandstones and red claystones [78].	-	-
Dacite (USA) (Figures 1–4).	Miocene.	Southern New Mexico, Southern Arizona.	Rhyolite and Dacite flows [65].	-	-
Otay Fm (USA) (Figure 1)	Miocene	Southwestern California.	Conglomerates and sandstones with few mudstones and bentonites [79].	Water bearing unit [80].	San Diego Aq. [80].

Table 2. *Cont.*

Unit Name (Mex/USA)	Age	Location (Mex/USA)	Lithological Description	Hydrological Features	Aquifer
Imperial Fm/ Imperial Fm (Figure 1).	Miocene-Pliocene.	Southern California.	Siltstones and conglomerates [81].	Generally low permeability [59].	Yuma Aq. [59].
Bouse Fm (USA) (Figure 2)	Miocene-Pliocene.	Southern Arizona and California.	Basal limestones and a distinctive tuff, interbedded clay, silt, and sandstones [82].	The lower part of the formation is generally poorly permeable, but the upper part is fairly permeable where sand is more abundant [59].	Lower Colorado River and Salton Sea basins [59].
Baucarit Fm/Nogales Fm (Figure 3)	Miocene-Pliocene	Northern Sonora/Southeastern Arizona	Volcanic conglomerates, sandstones and clays with local thin basalt flows [83].	Effective porosity ranges from 16 to 42% and hydraulic conductivity from 4 to 57 cm per day [84]. Some layers from Baucarit Fm. work as confining units [58].	Santa Cruz Aq. [85]. Rio San Pedro Aq. [86]
Gila Gr (USA) (Figure 4).	Miocene-Pliocene.	Central New Mexico/ Central Arizona.	Conglomerates with calcareous cement. Interbeddings of sandstones [87].	Very low hydraulic conductivities and storage coefficients, indicative of semiconfined to confined hydraulic conditions [88].	Mimbres, Hachita, Playas, Animas basins [89]. Mimbres Basin [41].
Pliocene Conglomerates/Pliocene Conglomerates (Figures 2 and 3)	Pliocene	Northern Baja California, Northern Sonora/ Southern New Mexico, Southern Arizona, Southern California.	Conglomerates and sandstones [22].	-	-
Neogene Conglomerates/ Upper Santa Fe Gr (Figure 4).	Pliocene-Holocene.	Northern Chihuahua, Northern Sonora/Southwestern New Mexico.	Semiconsolidated polymictic conglomerates [11]. Fluvial cemented deposits [76].	Major source of fresh water [77].	Mesilla Bolson [77].
Quemado Gr (USA) (Figure 4).	Pliocene-Pleistocene.	Southern New Mexico.	Light brown friable sandstones and gravels [90].	-	-
San Diego Fm/San Diego Fm (Figure 1).	Pliocene-Pleistocene.	Northern Baja California/Southern California.	Sandstones and conglomerates with thin beds of bentonite [91].	Moderate hydraulic conductivity [92].	San Diego aq. [92].

Table 2. *Cont.*

Unit Name (Mex/USA)	Age	Location (Mex/USA)	Lithological Description	Hydrological Features	Aquifer
Canebrake Fm (Mex) (Figures 1 and 2).	Pliocene-Pleistocene.	Northern Baja California.	Gray and conglomerates with few layers of unconsolidated sandstones [93].	-	-
Neogene Sandstones (Mex) (Figure 1).	Pliocene-Holocene.	Northwestern Sonora.	Unconsolidated sands and gravels [17].	-	-
Qt Conglomerates/ Qt Conglomerates (Figures 1–3).	Pleistocene-Holocene.	Northern Sonora/ Southern Arizona.	Terrace deposits of coarse sand and gravel [41].	Deposits saturated with saline water [94]. Moderate porosity [95].	Mimbres Basin [41]. Arroyo Seco Aq. [95]. Puerto Peñasco Aq. [94].
Qt Eolian/ Qt Eolian (Figures 1–4).	Pleistocene-Holocene.	Northern Chihuahua, northwestern Sonora/ Southern California.	Unconsolidated sands [59].	-	Animas Aq. [72]. Yuma Aq. [59]. Laguna Salada Aq. [96].
Qt Lacustrine/ Qt Lacustrine (Figures 1–4).	Pleistocene-Holocene.	Northern Chihuahua, Northwestern Sonora, Northeastern Baja California/ Southern New Mexico.	Unconsolidated gray clay, red shales and bentonite [41].	Deposits with low permeability [37]. The deposits are part of swamps near the coast [97].	Sonoyta-Puerto Peñasco Aq. [97].
Qt Alluvium/ Qt Alluvium (Figures 1–4).	Pleistocene-Holocene.	Chihuahua, Sonora, Baja California/ New Mexico, Arizona, California.	Conglomerates cemented with calcium carbonate in southern New Mexico [39]. The alluvium consists of permeable lenses of sand and gravel interbedded with clay and silt in southeastern Arizona [98]. Clean medium to coarse sand in California [59].	The valley floor is underlain by permeable alluvium, capable of producing large amounts of ground water at Avra Valley [98]. High K, n, and S at San Diego Aq. [67]. High K in unconsolidated deposits at San Pedro Aq. [7].	Avra and Altar Valleys [98], Yuma and Mexicali-Rio Colorado Valley [59]. Mimbres Basin [41]. San Diego Aq. [67]. San Pedro Aq. [7].

3.2. Geologic Transboundary and Boundary Formation Limits

The geologic limits of the formations in the borderland are shown in Figures 5–8. These figures represent a more detailed identification and delimitation of transboundary and boundary geological units. Examples of boundary units in Figures 5–8 are Canebrake (Mex), Rhyolite (USA), Rancho Vallecitos-Esquisto Julian (Mex), Paleoproterozoic Granite (USA), Upper Santa Fe Gr (USA), Gila Gr (USA), Lake Valley Limestone (USA), Cabullona Fm (Mex), and Bisbee Conglomerates (Mex), among others. Though these formations seem to appear only on one side of the border at the surface, they could be continuous across the other side. However, limited information on these geologic units does not allow for further conclusions.

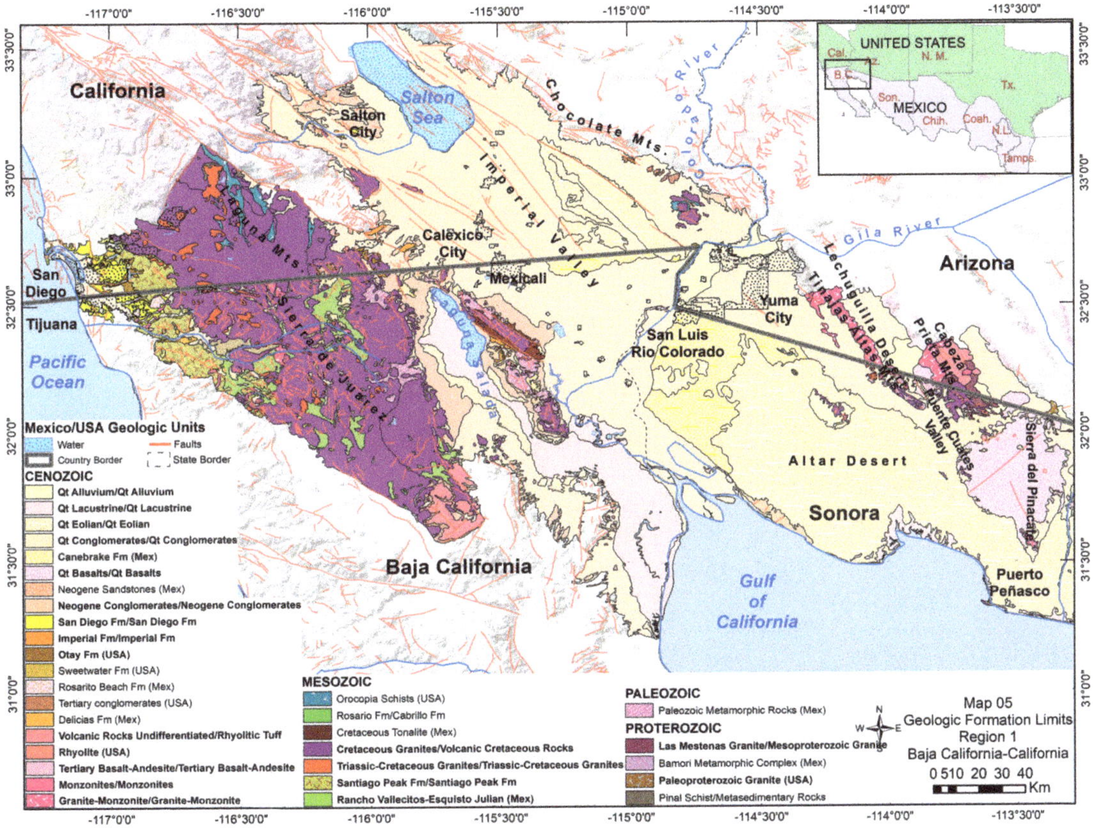

Figure 5. Geologic formation limits, Baja California—California.

The extension limits of the transboundary formations (crossing the border) were defined according to lithology and regional structural geology, such as faults, folds, and lineaments. Additionally, topography and hydrological features were also used to complement the analysis. The geological extensions shown in Figures 5–8 were defined mainly by deformation due to transpressive regimes, which originated the lineament systems known as the Walper Lineament and the Mojave-Sonora Megashear [99]. These lineament systems cross Baja California, Sonora, and Chihuahua in Mexico. In the northeastern part of the study area, the Texas Lineament defined the geological boundaries of most of New Mexico and Texas on the U.S. side [100]. Steep faults with orientation NW-SE formed as a response to the movement on the Mojave-Sonora Megashear, developing pull apart basins, which later filled with sediments originating most of the HGUs identified in this study area [99]. We will expand on the lithologic/structural boundaries on the individual descriptions of the HGUs in the following section.

As it has been mentioned before, there are formations that perform as extent limits of the boundary formations (those units that do not seem to cross the borderland) or that occur as igneous inclusions within, surrounding, or adjacent to the boundary formations. Analyses of these formations was not included in the current study but are included in the figures for mapping and visualization purposes. They are also listed in the corresponding legends of the figures (not highlighted in bold).

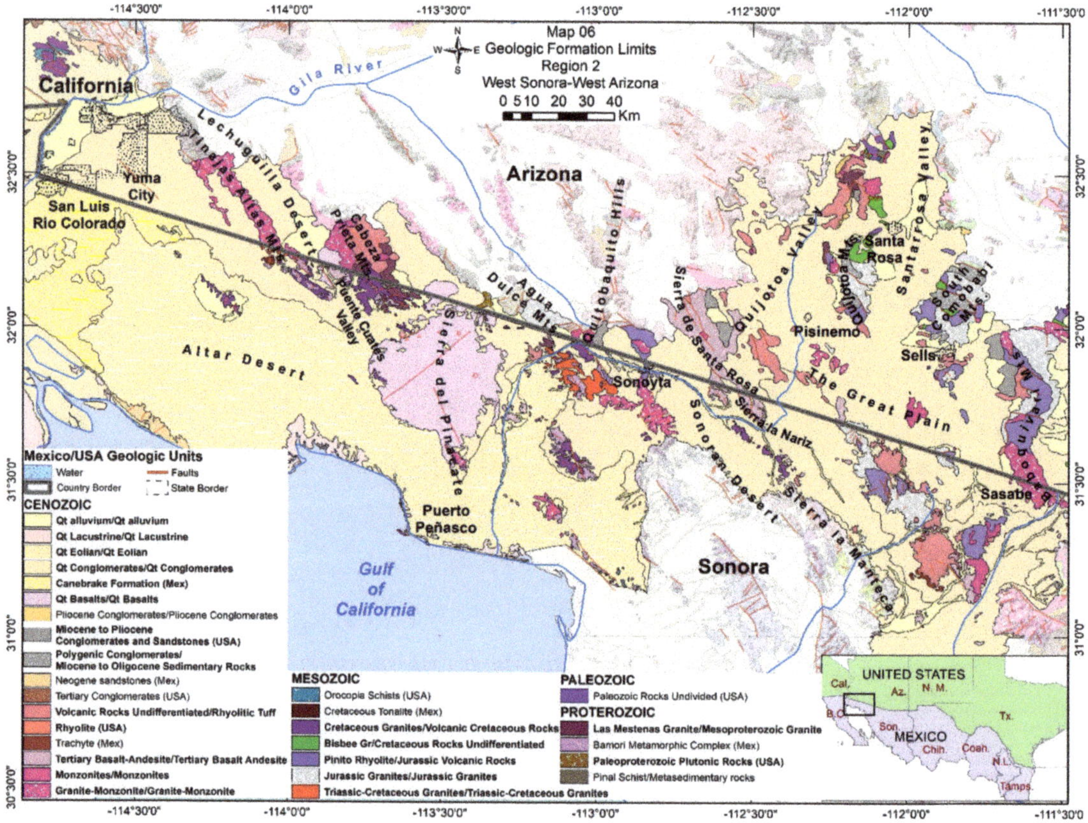

Figure 6. Geologic formation limits, West Sonora—West Arizona.

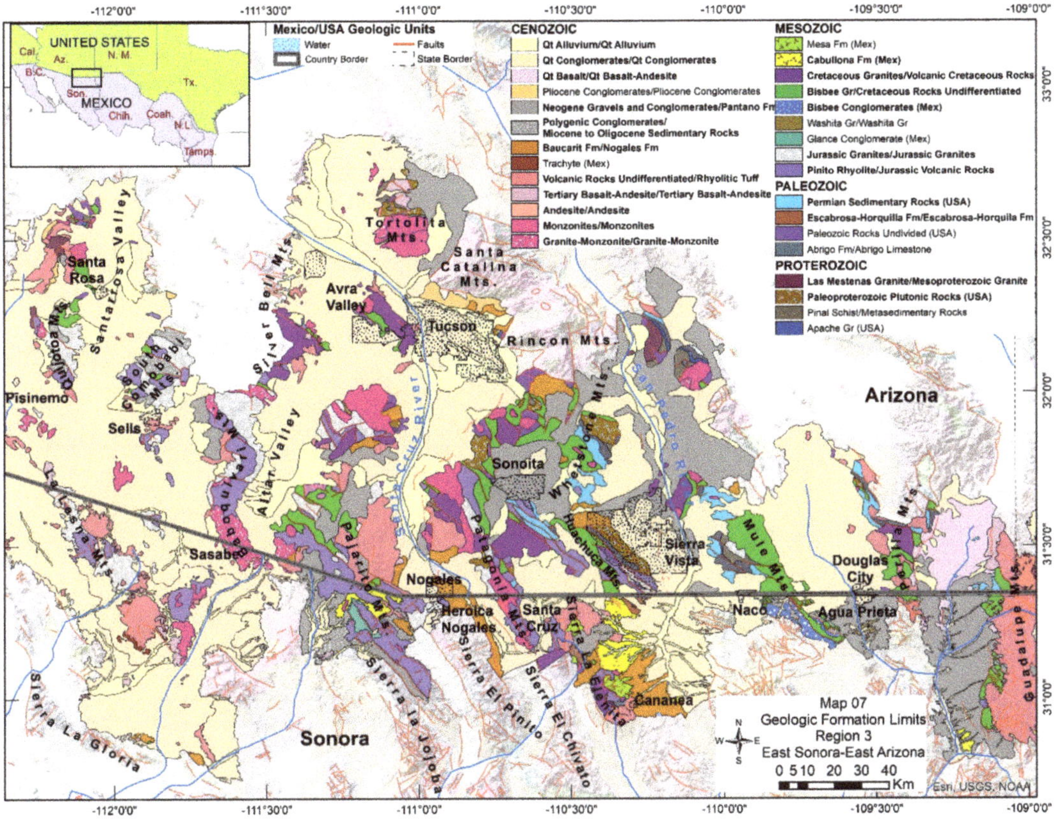

Figure 7. Geologic formation limits, West Sonora—West Arizona.

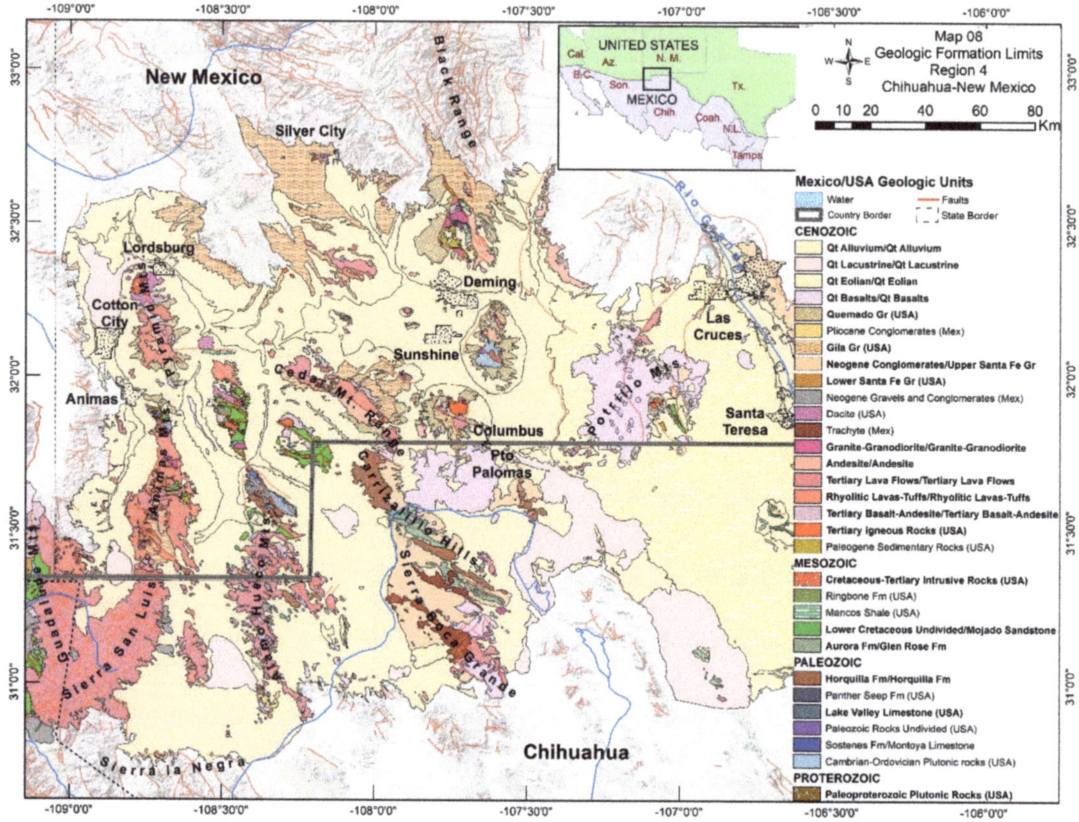

Figure 8. Geologic formation limits, West Sonora—West Arizona.

3.3. Delineation of Transboundary Hydrogeological Units (HGUs)

Figures 9–12 show the transboundary and boundary geological formations grouped into HGUs. This clustering of geological formations represents a refined delineation of transboundary geological formations considering lithological features, hydrogeological linkages and boundary limitations described in Table 2. As it has been mentioned earlier, they are referred to as "hydrogeological units" or "HGUs" (instead of aquifers) considering the different hydrogeological conditions among units that may or may not be categorized as aquifers. This section will cover how this clustering was integrated for each identified HGU.

The physical limits of the HGUs located across Baja California and California (Figure 9) are a combination of structural and lithological variations. The physical limits on the northern portion of Baja California have a stronger structural component. The Tijuana-San Diego Aquifer northern and southeastern boundaries are defined by the contact with volcanic rocks of local secondary permeability to non-existent permeability characteristics. According to the Internationally Shared Aquifer Resources Management (ISARM) [101], the official reported boundaries of this aquifer on the U.S. side match with the quaternary deposits shown in Figure 9; however, we extended the boundaries to include neighboring Neogene rocks, since groundwater flows from the recharge zone on the Otay Reserve towards the coast [80]. On the Mexico side, aquifer boundaries are delineated according to administrative criteria [1], and therefore, the physical boundaries presented in this study will mostly not coincide with those recognized officially by the CONAGUA.

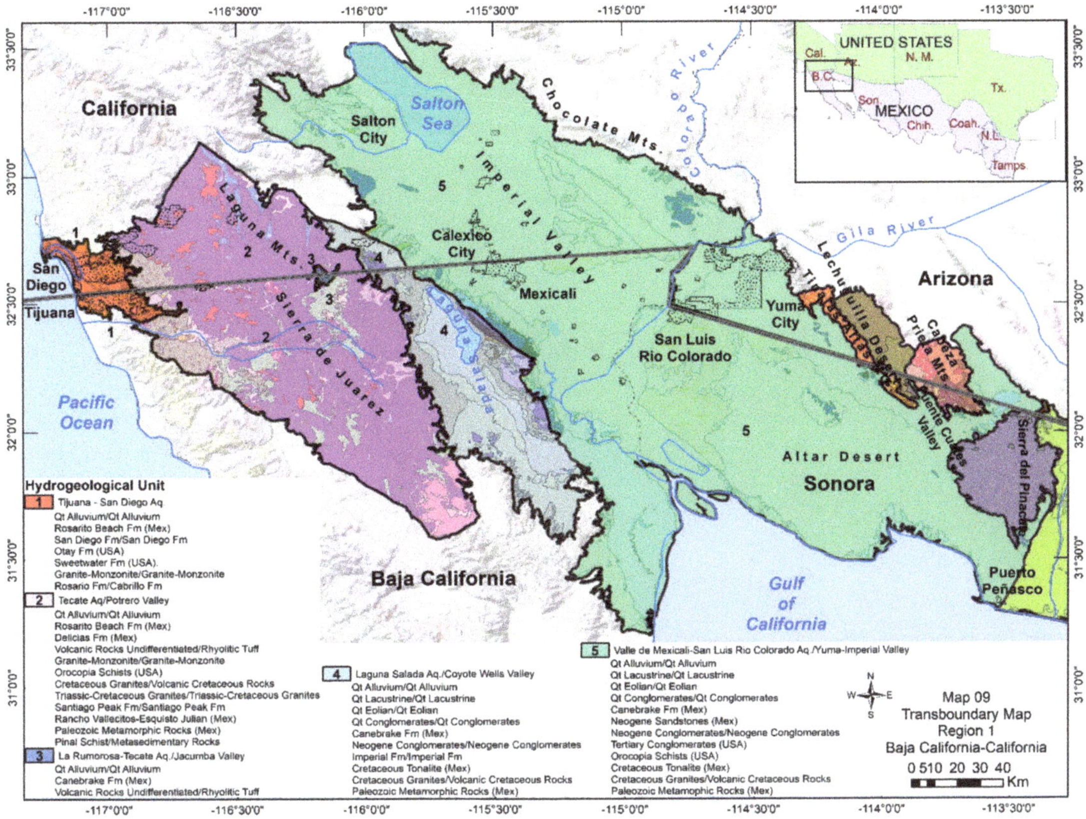

Figure 9. Transboundary map, Baja California—California.

The Tecate Aquifer/Potrero Valley is formed mainly by crystalized Triassic to Neogene igneous bodies with secondary permeability. This HGU has a strong structural control: the northern and eastern boundaries are defined by the Elsinore Fault Zone [102], and the southern boundary is defined by the San Miguel-Vallecitos Fault [14]. La Rumorosa-Tecate Aquifer/Jacumba Valley comprises quaternary deposits accumulated in a depression surrounded by impermeable granitic and metamorphic rocks of Neogene and Mesozoic age [103], and therefore, the physical limits are exclusively lithologic. Laguna Salada Aquifer/Coyote Wells Valley has a predominant structural control with the Sierra Juarez Fault to the west and the Laguna Salada Fault to the north-northeast. The southern limit is defined by Neogene volcanic rocks outcropping on Sierra Las Tinajas [96].

Moving towards the western side of Arizona and Sonora, the Valle de Mexicali-San Luis Rio Colorado Aq./Yuma-Imperial Valley HGU (Figures 9 and 10) western limit is defined by Sierra Cucapa, where granitic rocks of Cretaceous age and the Cucapa Fault comprise this side of the boundary. The crystalline rocks of Mesozoic age configure the northeastern boundary of the HGU at Chocolate Mountains [104], which together with the Salton Sea comprise the northern boundary in California [59]. The eastern boundary is defined by differences in lithology between the quaternary deposits forming this HGU and the Mesozoic-Neogene granites and Quaternary Basalts that formed the neighboring HGUs of Tinajas Altas Mountains and Los Vidrios Aquifer. The southern boundary is defined by the extension of the Rio Colorado deltaic deposits into the Gulf of California which constitutes a physical rather than lithological feature. The northern and eastern boundaries of the Valle de Mexicali-San Luis Rio Colorado Aq./Yuma-Imperial Valley on the U.S. side appear to be based on lithological differences [105], which are very similar to the boundaries presented in this study. The southern and western boundaries on the Mexico side are defined as well by lithology and match the official reports [105]; however, the eastern boundary does not coincide with official reports as they seem to respond to an administrative boundary [105,106].

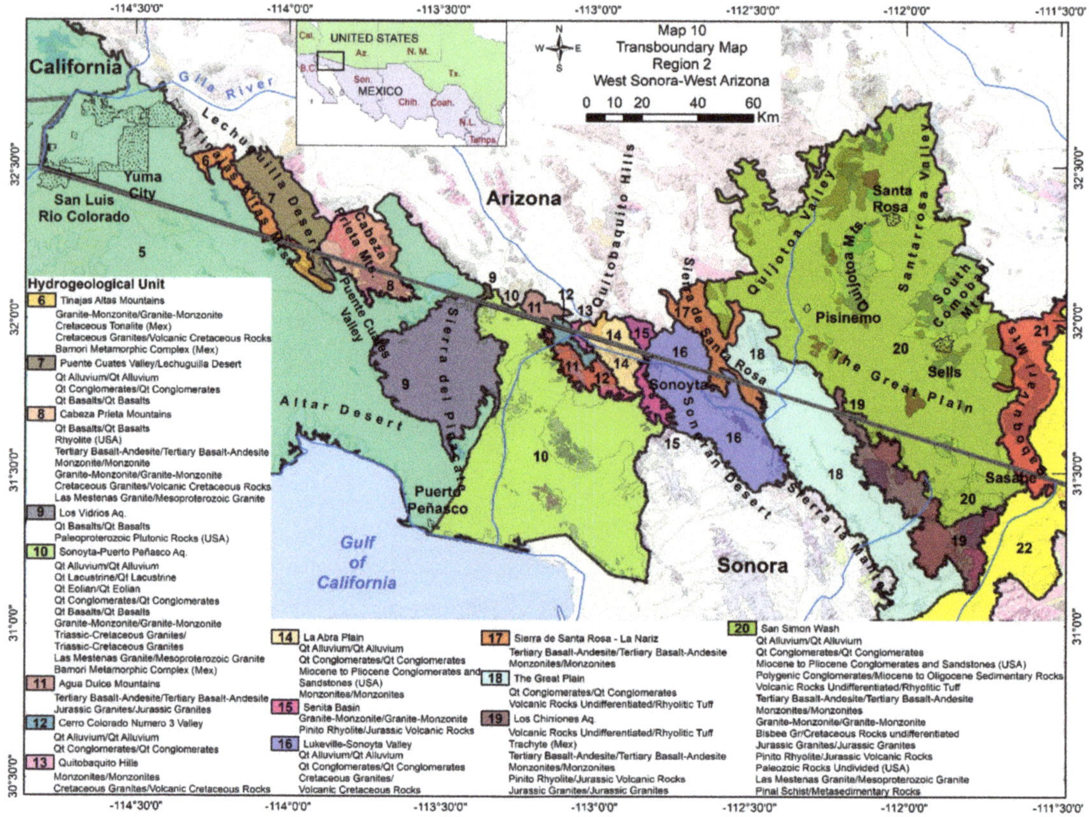

Figure 10. Transboundary map, West Sonora—West Arizona.

The geological limits of the western side of the state of Sonora and Arizona (Figure 10) are based on a combination of lithological variations. Boundaries are mostly defined by contrasting quaternary deposits in contact with old crystalline rocks with limited to non-existent permeability. These older units work as a basement for the identified HGUs in this area.

The Tinajas Altas Mountains, Puente Cuates Valley, Cabeza Prieta Mountains, and Sonoyta-Puerto Peñasco Aquifers have a strong structural component, since the boundaries are defined by pull apart basins associated with the Mojave-Sonora Megashear [99]. Due to this structural feature, it is possible to identify a sequence of Precambrian to Mesozoic crystalline rocks outcropping as mountains, with depressions filled with recent quaternary deposits. The exception to this structural feature is the Los Vidrios Aquifer, which is the product of recent quaternary volcanic activity, and it is located in an area where the volcanic outcrops work as a boundary between the Valle de Mexicali-San Luis Rio Colorado Aq./Yuma-Imperial Valley and the Sonoyta-Puerto Peñasco Aquifer.

The HGUs located between Agua Dulce Mountains and Baboquivari Mountains are the result of a similar structural environment related to the Mojave-Sonora Megashear, where Jurassic rocks intruded through the thrust faults, originating a series of volcanic and metamorphic belts intercalated with depressions filled by Quaternary deposits across the Sonoran Desert [107]. Topography also plays an important role in defining the northern boundaries of these HGUs. We identified these depressions in Figure 6 and integrated them into their corresponding HGUs as shown on Figure 10. The HGUs that are worth mentioning due to some degree of aquifer potential are Cerro Colorado Numero 3 Valley, Quitobaquito Hills, La Abra Plain, Lukeville-Sonoyta Valley, the Great Plain, and San Simon Wash. Initially, the USGS [9] used the term San Simon Wash to refer to the San Simon River watershed and the Papago Indian Reservation; however, in this report, the USGS also stated that the boundary of the San Simon Wash was "arbitrary". ISARM [105] also identified the Sonoyta-Papagos TBA (Transboundary Aquifer), which includes the San Simon Wash area in the U.S. as well as the administrative boundaries of the Sonoyta-Puerto Peñasco aquifer on the Mexico side.

The eastern side of Sonora/Arizona (Figure 11) consists of a combination of small faults and lithological changes in the north, as well as topography and drainage features particularly in southern Arizona. The differences in lithology are the predominant feature that this study used in the northern region of Sonora to define the HGUs' boundaries. The Arroyo Seco Aquifer and the Nogales-Rio Santa Cruz Aq./Upper Santa Cruz Basin consist of two parallel north–south trending alluvial basins, separated by blockfaulted mountains formed by Jurassic to Cretaceous igneous rocks. The first one outcrops at the Baboquivari and Silver Bell mountains on the west of Arroyo Seco Aquifer [108]. The second one is the mountain chain between the Tortolita Mountains and Pajarito Mountains that separates Arroyo Seco Aquifer and Nogales-Rio Santa Cruz Aq./Upper Santa Cruz Basin. The mountain chain between Santa Catalina and Huachuca Mountains defines the boundaries on the eastern side of Nogales-Rio Santa Cruz Aq./Upper Santa Cruz Basin [85]. The northern boundary of the system in this paper does not align with the official reports [105,109] mainly because we use a geological approach, and the published reports are based on watershed and management delimitations. The southern boundaries of the Nogales-Rio Santa Cruz Aq./Upper Santa Cruz Basin rely on a natural barrier formed by Sierras El Pinito and El Chivato, where crystalline volcanic rocks are abundant. The eastern boundaries are defined by the Whetstone and Huachuca Mountains that comprise the surroundings of Upper Sonoita Creek which is a basin fill alluvial aquifer that constitutes an important tributary of the Upper Santa Cruz feeding the underlying sediments [110,111]. The Rio Altar Aquifer is formed by the interaction of Neogene and Quaternary deposits, limited on the north by the Pajarito Mountains as well.

The Rio San Pedro Aq./Upper San Pedro Basin is limited on the west by the Rincon, Whetstone, Huachuca Mountains, and Sierra La Elenita, where volcanic and metamorphic rocks from Precambian to Neogene age outcrop, working as a barrier between this aquifer and the Nogales-Rio Santa Cruz Aq./Upper Santa Cruz Basin. The eastern boundary is defined by sedimentary crystalline rocks of Paleozoic to Cretaceous age with limited permeability that outcrop on the Mule Mountains (Figure 11). These natural barriers minimize groundwater connections with adjacent aquifers, even in the northern portion of the HGU [7]. The northern boundaries that we defined for this aquifer are close to those reported by ISARM [105], but they extend beyond what Callegary et al. [109] reports as the northern boundary. As in the case of Nogales-Rio Santa Cruz Aq./Upper Santa Cruz Basin, slight differences rely on our geology-based approach as compared to the watershed approach used by published official reports.

It should also be noted that the slight differences in extent presented here for both the Nogales-Rio Santa Cruz Aq./Upper Santa Cruz Basin and the Rio San Pedro Aq./Upper San Pedro Basin, as compared to those reported by TAAP, might also be related to the administrative and regulatory boundaries on the Arizona side (e.g., the Santa Cruz Active Management Area (AMA) jurisdiction). Nevertheless, the main geological features according to the ADWR for both San Pedro and Santa Cruz aquifers are in close agreement with our study [112,113].

The Rio Agua Prieta Aquifer/Douglas Basin and the Arroyo San Bernardino Aq./San Bernardino Valley are depressions filled by Neogene to Quaternary deposits and separated by the Perilla Mountains, where there are volcanic and old sedimentary rocks with limited permeability outcrop. The eastern boundary of the Arroyo San Bernardino Aq./San Bernardino Valley consists of half a graben structure located on the piedmont of the Guadalupe Mountains [114]. This HGU is locally covered by fractured Quaternary Basalts, which have the potential to work as aquifers or as confining layers.

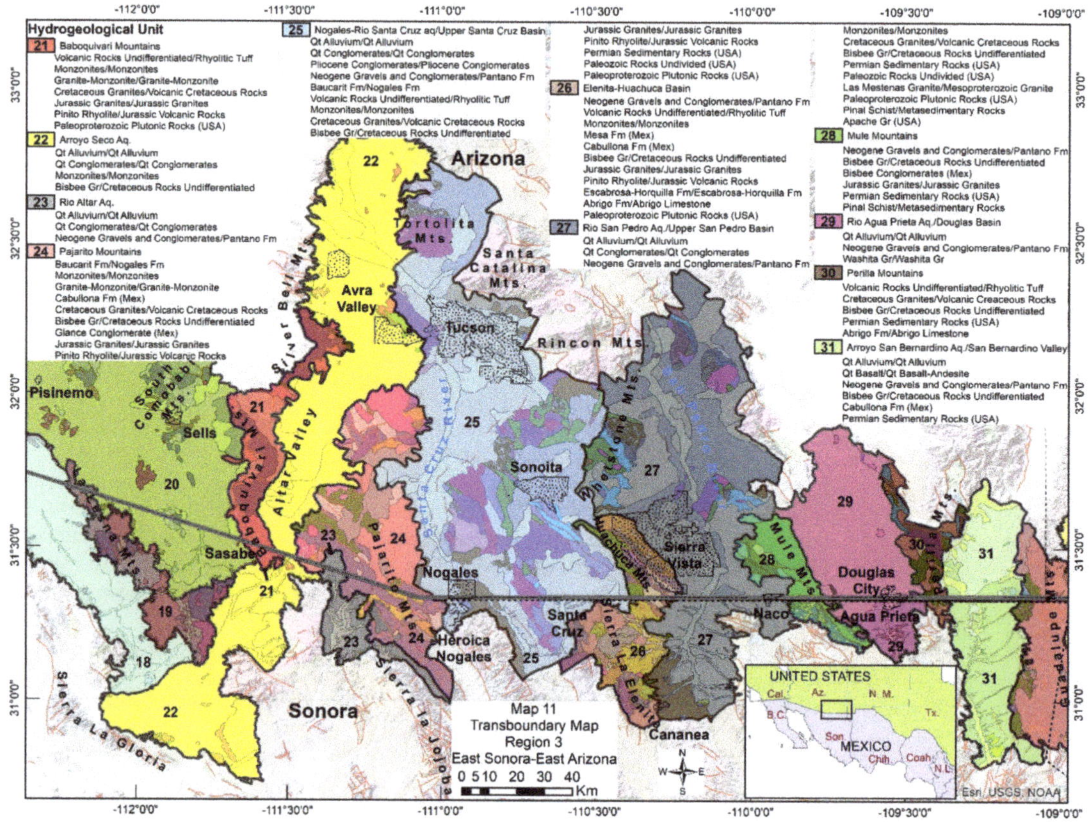

Figure 11. Transboundary map, East Sonora—East Arizona.

Figure 12 shows the formation limits between Chihuahua and New Mexico. Changes in geologic structures and lithology were definitive in delineating the boundaries of the units, and topography to a lesser extent. This area is dominated by graben structures associated to the Rio Grande Rift, and each individual graben is commonly bounded by steep faults, where old sediments and crystalline rocks outcrop [115], and the depressions or bolsons are filled with unconsolidated and coarse grain sized sediments [114]. The Continental Water Divide, located at the northern end of the area of interest works as a groundwater divide as well [39,114]; therefore, we used this topographic feature to define the northern boundary of some of the HGUs between Chihuahua and New Mexico. We defined the southern boundaries based on contrasting lithologic differences between crystalline rocks and the unconsolidated bolson-like sediments.

The Animas Basin is bounded on the west by the Guadalupe Mountains, comprising mostly igneous crystalline and volcanic rocks with limited permeability. The northern limit follows the surface seepage and groundwater flow divide between the Gila River Basin and the Animas Basin. The eastern boundaries are comprised by the Continental Water Divide, the Pyramid Mountains, and the mountain chain between Animas Mountains in the U.S. and Sierra San Luis in Mexico [114]. The latter mountain chains also work as the western boundary for the Janos Aq./Playas Basin. Sierra La Negra in Mexico bounds the Janos Aq./Playas Basin to the south. The Alamo Hueco Mountains separate the Janos Aq./playas Basin from the Ascension Aquifer/Hachita Moscos Basin, restraining the water flows between these two HGUs. The northern boundary of the Ascension Aq./Hachita Moscos is defined by the Continental Water Divide [39], and the cedar Mountain Range to the east, or what we have named as the Josefa Ortiz de Dominguez Aquifer, where Neogene volcanic rocks with limited to nonexistent permeability configure this HGU.

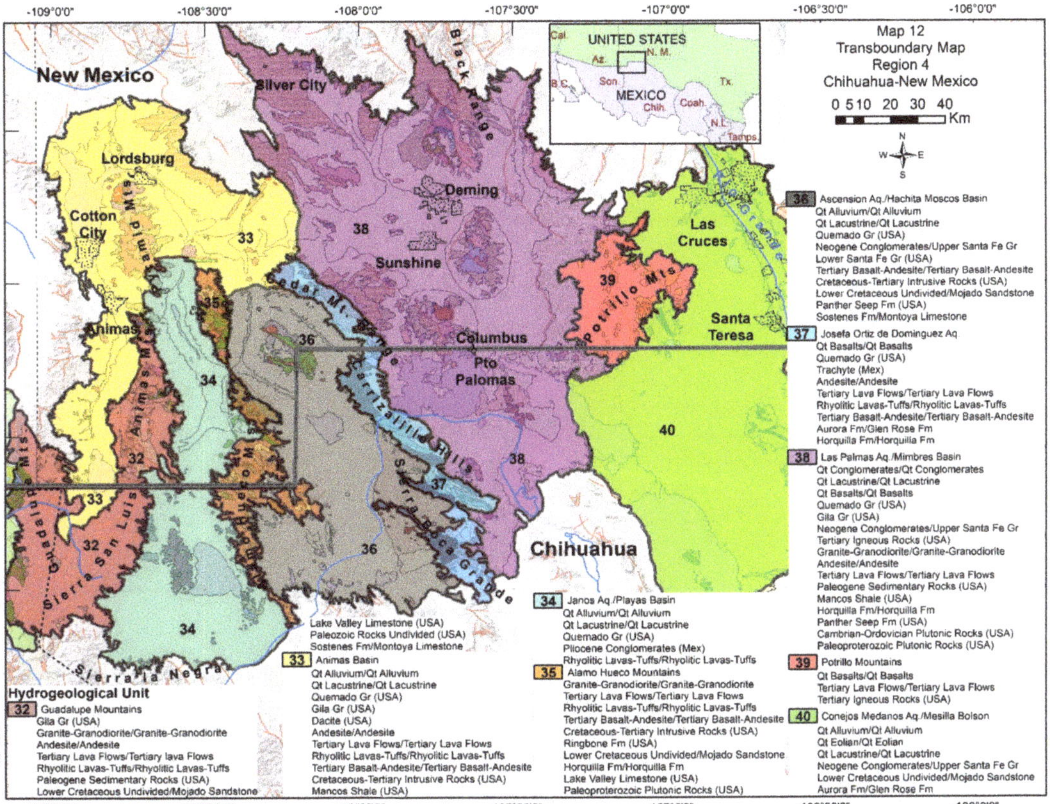

Figure 12. Transboundary map, Chihuahua—New Mexico.

The Cedar Mountain Range and Carrizalillo Hills are part of the Cedar Arc [39], which is one of several complex basin and range Province fault-block systems, and work as the western limit of Las Palmas Aq./Mimbres Basin. The Sierra Boca Grande in Mexico represents a similar echelon fault-block system that forms part of the southwestern boundary of the Mimbres Basin [41]. This HGU is bounded to the north by the Continental Water Divide and the Black Range [114] and to the east by the Potrillo Mountains, where fractured Quaternary Basalts occur. The southern limit is defined according to lithological and topographic differences with the Conejos-Medanos Aq./Mesilla Bolson which was already addressed in Sanchez et al. [2], but it is included in the maps for visualization purposes.

3.4. Classification of Geological Formations/Aquifers

According to the geological description and hydrogeological features noted in previous sections and in Table 2, boundary and transboundary formations within each HGU were classified with an ID value (color) with similar characteristics of aquifer potential and water quality. Table 1 shows the grouping and the corresponding ID value for each group. According to aquifer potential and water quality parameters, Group 1 (dark green), the most important units/formations in terms of groundwater potential and water quality, corresponds to A1, A2, B1, and B2. Group 2 (light green) includes those units/formations that have good to moderate aquifer potential but poor water quality or with limited water quality information on that area (A3, A4, B3, and B4). Group 2 constitutes a second level of priority areas because they could represent future resource development as water treatment options become more feasible. Group 3 (orange) includes those units with poor aquifer potential or aquitards with good to moderate water quality (C1, C3, D1, and D2). This group is considered the third level of priority due to the limited aquifer potential but is still useful for small communities and because the water quality is good to moderate. Group 4 (light maroon) is the lowest-priority group: this units report poor aquifer potential and poor water quality, or alternatively, they report limited information on water quality in that area (C3, C4, D3, and D4). Group 5 (gray) includes those units/formations with lack of information on both aquifer potential and water quality; therefore, their priority is undefined (E1, E2, E3, and E4).

The classification shown in Table 3 is based on the predominant hydrogeological conditions of the formations based on the available data. The formations (boundary and transboundary) are organized and listed within the limits of their corresponding HGU/aquifer described in the previous section. Therefore, the first column contains the corresponding name of the HGU or the reported Aquifer name according to Section 3.3, followed by the formations that integrate each HGU and the specific ID value for each one according to aquifer potential and water quality. Figures 13–16 show the HGUs colored according to the classification of each formation that integrates them, therefore showing the predominant ID value for each HGU. According to Table 3, a total of 39 boundary and transboundary formations were identified in the region that cover an approximate shareable area of 135,000 km^2 of which both countries share almost half (65,000 km^2 Mexico and 69,000 km^2 the U.S.). From the total shareable area, around 40% reports good to moderate aquifer potential and water quality, of which 65% is in the U.S. and 35% on the Mexico side.

In the area between Baja California and California, the HGUs with predominant good aquifer potential and good to moderate water conditions are Laguna Salada Aq./Coyote Wells Valley (Qt Alluvium, Qt Eolian, Qt Conglomerates and Neogene Conglomerates), followed by variable water quality conditions in the Tijuana-San Diego Aq. (Qt. Alluvium, Imperial Fm./Imperial Fm.) and the mostly overall extension of the Valle de Mexicali-San Luis Rio Colorado Aq./Yuma-Imperial Valley (Qt Alluvium, Qt Eolian, Qt Conglomerates and Neogene Conglomerates) covering an important area across California, Baja California, and West of Arizona and Sonora (Figure 13). The latter area of this HGU is well known for its high dependency on surface and groundwater, particularly, for intensive and extensive irrigated agriculture on both sides of the border, and also for the connectivity of the surface–groundwater systems from which native ecosystems and endangered species are equally dependent [1]. This HGU also encompasses the area of what is referred to as the Yuma Aquifer (which is also shared by Arizona and Sonora) that is subject to the only agreement between Mexico and the United States that has established pumping limitations and binational monitoring on both extraction rates and salinity levels (Minute 242 of the International Boundary and Water Commission, IBWC) [116]. The Tijuana-San Diego Aq. is the main water supply for the sister cities of Tijuana and San Diego and has good aquifer potential but has important salinity issues that are recurrent in the whole borderland between California and Baja California and that also expand into the western side of Arizona and Sonora.

Table 3. Classification of geological formations (within their corresponding HGU) in the border region between California, Arizona, and New Mexico, U.S., and Baja California, Sonora, and Chihuahua, Mexico, according to aquifer potential and water quality (*T = Transmissivity m^2/d, K = Hydraulic conductivity m/d, n = porosity %). The colors represent the differences among geological units. It is based on Table 1.

HGU/Aquifer	Boundary (*) and Transboundary Formations	Aquifer Potential	Hydrogeologic Features	Water Quality	TDS (ppm)	ID
(1) Tijuana–San Diego Aq.	Qt Alluvium/Qt Alluvium.	Good.	$T = 3024\ m^2/d$ K = 190 m/d	Fresh to Saline.	600–4900	A1-A3
	*Rosarito Beach Fm (Mex).	Good.	$T = 346\ m^2/d$ K = 0.17 m/d	Fresh to Saline.	600–4900	A1-A3
	San Diego Fm/San Diego Fm	Moderate.	K = 0.02 m/d	Poor.	-	B3
	*Otay Fm (USA).	Good	-	Poor		B4
	*Sweetwater Fm (USA).	Unknown.		Unknown.		E4
	Granite–Monzonite/Granite–Monzonite.	Moderate–Poor.		Unknown.		B4-C4
	Rosario Fm/Cabrillo Fm (USA).	Unknown.		Unknown.		E4
(2) Tecate Aq./ Potrero Valley.	Qt Alluvium/Qt Alluvium.	Moderate.	$T = 2074\ m^2/d$ K = 81.6 m/d	Good.	300–900	B1
	*Rosarito Beach Fm (Mex).	Moderate.	$T = 55\ m^2/d$ K = 0.82 m/d	Good.	300–900	B1
	*Delicias Fm (Mex).	Unknown.		Unknown.		E4
	Volcanic Rocks Undifferentiated/Rhyolitic Tuff.	Poor.		Unknown.		C4
	Granite–Monzonite/Granite–Monzonite.	Moderate–Poor.		Unknown.		B4-C4
	*Orocopia Schists (USA).	Unknown.		Unknown.		E4
	*Cretaceous Granites/Volcanic Cretaceous Rocks.	Moderate–Poor.		Unknown.		B4-C4
	Triassic–Cretaceous Granites/Triassic–Cretaceous Granites.	Moderate–Poor.		Unknown.		B4-C4

Table 3. *Cont.*

HGU/Aquifer	Boundary (*) and Transboundary Formations	Aquifer Potential	Hydrogeologic Features	Water Quality	TDS (ppm)	ID
	*Santiago Peak Fm/Santiago Peak Fm.	Unknown.		Unknown.		E4
	*Rancho Vallecitos–Esquisto Julian (Mex).	Unknown.		Unknown.		E4
	*Paleozoic Metamorphic Rocks (Mex).	Unknown.		Unknown.		E4
	*Pinal Schist/Metasedimentary Rocks.	Poor.		Poor.		C3
(3) La Rumorosa–Tecate Aq./ Jacumba Valley.	Qt Alluvium/Qt Alluvium.	Good.	T = 52–95 m^2/d	Moderate to good.	1184–1236	A2
	*Canebrake Fm (Mex).	Good.		Moderate to good.	1184–1236	A2
	Volcanic Rocks Undifferentiated/Rhyolitic Tuff.	Poor.		Unknown.		C4
(4) Laguna Salada Aq./ Coyote Wells Valley.	Qt Alluvium/Qt Alluvium.	Good.	T = 43–173 m^2/d K = 0.05–22 m/d	Brackish.	1180	A2
	Qt Lacustrine/Qt Lacustrine.	Poor.		Brackish.	1180	C2
	Qt Eolian/Qt Eolian.	Good.	T = 43–173 m^2/d K = 0.05–22 m/d	Brackish.	1180	A2
	Qt Conglomerates/Qt Conglomerates.	Good.	T = 43–173 m^2/d. K = 0.05–22 m/d	Brackish.	1180	A2
	*Canebrake Fm (Mex).	Moderate.		Brackish.	1180	B2
	Neogene Conglomerates/Neogene Conglomerates.	Moderate.		Brackish.	1180	B2
	Imperial Fm/Imperial Fm.	Low.		Brackish.	1180	C2
	*Cretaceous Tonalite (Mex).	Poor.		Unknown.		C4
	*Cretaceous Granites/Volcanic Cretaceous Rocks.	Poor.		Unknown.		C4
	*Paleozoic Metamorphic Rocks (Mex).	Poor.		Unknown.		C4
(5) Valle de Mexicali–San Luis Rio Colorado Aq./ Yuma–Imperial Valley.	Qt Alluvium/Qt Alluvium.	Good.	N = 28% T = 4300–30,200 m^2/d	Fresh to Saline.	498–7280	A1-A3
	Qt Lacustrine/Qt Lacustrine.	Poor.		Brackish.	498–7280	C2-C3
	Qt Eolian/Qt Eolian.	Good.	N = 28% T = 4300–30,200 m^2/d	Fresh to Saline.	498–7280	A1-A3
	Qt Conglomerates/Qt Conglomerates.	Good.	N = 28% T = 4300–30,200 m^2/d	Fresh to Saline.	498–7280	A1-A3

Table 3. *Cont.*

HGU/Aquifer	Boundary (*) and Transboundary Formations	Aquifer Potential	Hydrogeologic Features	Water Quality	TDS (ppm)	ID
	*Canebrake Fm (Mex).	Moderate.		Brackish.	498–7280	B1-B3
	*Neogene Sandstones (Mex).	Moderate.		Brackish.	498–7280	B1-B3
	Neogene Conglomerates/Neogene Conglomerates.	Moderate.		Brackish.	498–7280	B1-B3
	*Tertiary Conglomerates (USA).	Unknown.		Unknown.		E4
	*Orocopia Schists (USA).	Unknown.		Unknown.		E4
	*Cretaceous Tonalite (Mex).	Poor.		Unknown.		C4
	*Cretaceous Granites/Volcanic Cretaceous Rocks.	Poor.		Unknown.		C4
	*Paleozoic Metamorphic Rocks (Mex).	Poor.		Unknown.		C4
(6) Tinajas Altas Mountains.	Granite–Monzonite/Granite–Monzonite.	Poor.		Unknown.		C4
	*Cretaceous Tonalite (Mex).	Poor.		Unknown.		C4
	*Cretaceous Granites/Volcanic Cretaceous Rocks.	Poor.		Unknown.		C4
	*Bamori Metamorphic Complex (Mex).	Poor.		Unknown.		C4
(7) Puente Cuates Valley/Lechuguilla Desert.	Qt Alluvium/Qt Alluvium.	Good.		Unknown.		A4
	Qt Conglomerates/Qt Conglomerates.	Good.		Unknown.		A4
	Qt Basalts/Qt Basalts.	Poor.		Unknown.		C4
(8) Cabeza Prieta Mountains.	Qt Basalts/Qt Basalts.	Poor.		Unknown.		C4
	*Rhyolite (USA).	Poor.		Unknown.		C4
	Tertiary Basalt–Andesite/Tertiary Basalt–Andesite.	Poor.		Unknown.		C4
	Monzonites/Monzonites.	Poor.		Unknown.		C4
	Granite–Monzonite/Granite–Monzonite.	Poor.		Unknown.		C4
	*Cretaceous Granites/Volcanic Cretaceous Rocks.	Poor.		Unknown.		C4
	Las Mestenas Granite/Mesoproterozoic Granite.	Poor.		Unknown.		C4
(9) Los Vidrios Aq.	Qt Basalts/Qt Basalts	Poor.		Unknown.		C4
	*Paleoproterozoic Plutonic Rocks (USA).	Poor.		Unknown.		C4

Table 3. *Cont.*

HGU/Aquifer	Boundary (*) and Transboundary Formations	Aquifer Potential	Hydrogeologic Features	Water Quality	TDS (ppm)	ID
(10) Sonoyta–Puerto Peñasco Aq.	Qt Alluvium/Qt Alluvium.	Good.	T = 4400 m^2/d	Fresh to Saline.	353–25,076	A1-A3
	Qt Lacustrine/Qt Lacustrine.	Poor.	T = 1550 m^2/d	Fresh to Saline.	353–25,076	C1-C3
	Qt Eolian/Qt Eolian.	Good.	T = 4400 m^2/d	Fresh to Saline.	353–25,076	A1-A3
	Qt Conglomerates/Qt Conglomerates.	Good.	K = 302 m/d	Fresh to Saline.	353–25,076	A1-A3
	Qt Basalts/Qt Basalts.	Poor.		Unknown.		C4
	Granite–Monzonite/Granite–Monzonite.	Poor.		Unknown.		C4
	Triassic–Cretaceous Granites/Triassic–Cretaceous Granites.	Poor.		Unknown.		C4
	Las Mestenas Granite/Mesoproterozoic Granite.	Moderate–Poor.		Unknown.		B4-C4
	*Bamori Metamorphic Complex (Mex).	Moderate–Poor.		Unknown.		B4-C4
(11) Agua Dulce Mountains.	Tertiary Basalt–Andesite/Tertiary Basalt–Andesite.	Poor.		Unknown.		C4
	Jurassic Granites/Jurassic Granites.	Poor		Unknown		C4
(12) Cerro Colorado Numero 3 Valley.	Qt Alluvium/Qt Alluvium	Good.		Unknown.		A4
	Qt Conglomerates/Qt Conglomerates.	Good.		Unknown.		A4
(13) Quitobaquito Hills.	Monzonites/Monzonites.	Moderate–Poor.		Good.	662–783	B1-C1
	*Cretaceous Granites/Volcanic Cretaceous Rocks.	Moderate–Poor.		Good.	662–783	B1-C1

Table 3. *Cont.*

HGU/Aquifer	Boundary (*) and Transboundary Formations	Aquifer Potential	Hydrogeologic Features	Water Quality	TDS (ppm)	ID
(14) La Abra Plain.	Qt Alluvium/Qt Alluvium.	Good.	K = 15–30 m/d	Slightly saline.	1500	A2
	Qt Conglomerates/Qt Conglomerates.	Good.		Slightly saline.	1500	A2
	*Miocene to Pliocene Conglomerates and Sandstones (USA)	Poor.		Unknown.		C4
	Monzonites/Monzonites.	Poor.		Unknown.		C4
(15) Senita Basin.	Granite–Monzonite/Granite–Monzonite.	Poor.		Unknown.		C4
	Pinito Rhyolite/Jurassic Volcanic Rocks.	Poor.		Unknown.		C4
(16) Lukeville–Sonoyta Valley.	Qt Alluvium/Qt Alluvium.	Good.		Unknown.		A4
	Qt Conglomerates/Qt Conglomerates.	Good.		Unknown.		A4
	*Cretaceous Granites/Volcanic Cretaceous Rocks.	Poor.		Unknown.		C4
(17) Sierra de Santa Rosa–La Nariz.	Tertiary Basalt–Andesite/Tertiary Basalt–Andesite.	Poor.		Unknown.		C4
	Monzonites/Monzonites.	Poor.		Unknown.		C4
(18) The Great Plain.	Qt Conglomerates/Qt Conglomerates.	Good.		Poor.	4880	A3
	Volcanic Rocks Undifferentiated/Rhyolitic Tuff.	Poor.		Unknown.		C4
(19) Los Chirriones Aq.	Volcanic Rocks Undifferentiated/Rhyolitic Tuff.	Poor.		Unknown.		C4
	*Trachyte (Mex).	Poor.		Unknown.		C4
	Tertiary Basalt–Andesite/Tertiary Basalt–Andesite.	Poor.		Unknown.		C4
	Monzonites/Monzonites.	Poor.		Unknown.		C4
	Pinito Rhyolite/Jurassic Volcanic Rocks.	Poor.		Unknown.		C4
	Jurassic Granites/Jurassic Granites.	Poor.		Unknown.		C4
(20) San Simon Wash.	Qt Alluvium/Qt Alluvium.	Good.		Fresh to Saline.	180–4900	A1-A3
	Qt Conglomerates/Qt Conglomerates.	Good.		Fresh to Saline.	180–4900	A1-A3
	*Miocene to Pliocene Conglomerates and Sandstones (USA)	Unknown.		Unknown.		E4
	Polygenic Conglomerates/Miocene to Oligocene Sedimentary Rocks	Unknown		Unknown		E4
	Volcanic Rocks Undifferentiated/Rhyolitic Tuff.	Poor.		Unknown.		C4
	Tertiary Basalt–Andesite/Tertiary Basalt–Andesite.	Poor.		Unknown.		C4
	Monzonites/Monzonites.	Poor.		Unknown.		C4
	Granite–Monzonite/Granite–Monzonite.	Poor.		Unknown.		C4

Table 3. *Cont.*

HGU/Aquifer	Boundary (*) and Transboundary Formations	Aquifer Potential	Hydrogeologic Features	Water Quality	TDS (ppm)	ID
	Bisbee Gr/Cretaceous Rocks Undifferentiated	Poor.		Unknown.		C4
	Jurassic Granites/Jurassic Granites.	Poor.		Unknown.		C4
	Pinito Rhyolite/Jurassic Volcanic Rocks.	Poor.		Unknown.		C4
	*Paleozoic Rocks Undivided (USA).	Poor.		Unknown.		C4
	Las Mestenas Granite/Mesoproterozoic Granite.	Poor.		Unknown.		C4
	*Pinal Schist/Metasedimentary Rocks.	Poor.		Poor.		C3
(21) Baboquivari Mountains.	Volcanic Rocks Undifferentiated/Rhyolitic Tuff	Poor.		Unknown.		C4
	Granite–Monzonite/Granite–Monzonite.	Poor.		Unknown.		C4
	Monzonites/Monzonites.	Poor.		Unknown.		C4
	*Cretaceous Granites/Volcanic Cretaceous Rocks	Poor.		Unknown.		C4
	Jurassic Granites/Jurassic Granites.	Poor.		Unknown.		C4
	Pinito Rhyolite/Jurassic Volcanic Rocks.	Poor.		Unknown.		C4
	*Paleoproterozoic Plutonic Rocks (USA).	Poor.		Unknown.		C4
(22) Arroyo Seco Aq.	Qt Alluvium/Qt Alluvium.	Good.		Unknown.		A4
	Qt Conglomerates/Qt Conglomerates.	Moderate.	$T = 86\ m^2/d$	Unknown.		B4
	Monzonites/Monzonites.	Poor.		Unknown.		C4
	Bisbee Gr/Cretaceous Rocks Undifferentiated.	Poor.		Unknown.		C4
(23) Rio Altar Aq.	Qt Alluvium/Qt Alluvium.	Good.		Good.	243–640	A1
	Qt Conglomerates/Qt Conglomerates.	Good.		Good.	243–640	A1
	Pantano	Moderate.		Unknown.		B4
(24) Pajarito Mountains.	Baucarit Fm/Nogales Fm..	Moderate.		Unknown.		B4
	Granite–Monzonite/Granite–Monzonite.	Moderate–Poor.		Unknown.		B4-C4
	Monzonites/Monzonites.	Moderate–Poor.		Unknown.		B4-C4
	*Cabullona Fm (Mex).	Poor.		Unknown.		C4
	Bisbee Gr/Cretaceous Rocks Undifferentiated.	Poor.		Unknown.		C4
	*Glance Conglomerate (Mex).	Poor.		Unknown.		C4

Table 3. *Cont.*

HGU/Aquifer	Boundary (*) and Transboundary Formations	Aquifer Potential	Hydrogeologic Features	Water Quality	TDS (ppm)	ID
	*Cretaceous Granites/Volcanic Cretaceous Rocks.	Moderate–Poor.		Unknown.		B4-C4
	Jurassic Granites/Jurassic Granites.	Moderate–Poor.		Unknown.		B4-C4
	Pinito Rhyolite/Jurassic Volcanic Rocks.	Moderate–Poor.		Unknown.		B4-C4
(25) Nogales–Rio Santa Cruz aq./Upper Santa Cruz Basin.	Qt Alluvium/Qt Alluvium.	Good.	K = 1–90 m/d	Good.	500	A1
	Qt Conglomerates/Qt Conglomerates.	Good.		Good.	500	A1
	Pliocene Conglomerates/Pliocene Conglomerates	Moderate		Good	500	A1
	Neogene Gravels and Conglomerates/Pantano Fm	Moderate.		Unknown.		B4
	Baucarit Fm/Nogales Fm.	Moderate.	K = 0.1–1 m/d	Good.	500	A1
	Volcanic Rocks Undifferentiated/Rhyolitic Tuff.	Poor.		Unknown.		C4
	Monzonites/Monzonites.	Poor.		Unknown.		C4
	*Cretaceous Granites/Volcanic Cretaceous Rocks.	Poor.		Unknown.		C4
	Bisbee Gr/Cretaceous Rocks Undifferentiated.	Poor.		Unknown.		C4
	Jurassic Granites/Jurassic Granites.	Poor.		Unknown.		C4
	Pinito Rhyolite/Jurassic Volcanic Rocks.	Moderate.		Unknown.		B4
	*Permian Sedimentary Rocks (USA).	Poor.		Unknown.		C4
	*Paleozoic Rocks Undivided (USA).	Poor.		Unknown.		C4
	*Paleoproterozoic Plutonic Rocks (USA).	Poor.		Unknown.		C4
(26) Elenita-Huachuca Basin.	Neogene Gravels and Conglomerates/Pantano Fm	Moderate.		Unknown.		B4
	Monzonites/Monzonites.	Poor.		Unknown.		C4
	Volcanic Rocks Undifferentiated/Rhyolitic Tuff.	Poor.		Unknown.		C4
	Bisbee Gr/Cretaceous Rocks Undifferentiated.	Poor.		Unknown.		C4
	*Mesa Fm (Mex).	Unknown.		Unknown.		E4
	*Cabullona Fm (Mex).	Poor.		Unknown.		C4
	Jurassic Granites/Jurassic Granites.	Poor.		Unknown.		C4
	Pinito Rhyolite/Jurassic Volcanic Rocks.	Moderate.		Unknown.		B4
	Escabrosa–Horquilla Fm/Escabrosa–Horquilla Fm.	Poor.		Unknown.		C4
	*Abrigo Fm/Abrigo Limestone.	Unknown.		Unknown.		E4
	*Paleoproterozoic Plutonic Rocks (USA).	Poor.		Unknown.		C4

Table 3. *Cont.*

HGU/Aquifer	Boundary (*) and Transboundary Formations	Aquifer Potential	Hydrogeologic Features	Water Quality	TDS (ppm)	ID
(27) Rio San Pedro Aq./Upper San Pedro Basin.	Qt Alluvium/Qt Alluvium.	Good.	K = 12.5–7.5 m/d	Good.	229–751	A1
	Qt Conglomerates/Qt Conglomerates.	Good.	K = 12.5–7.5 m/d	Good.	229–751	A1
	Neogene Gravels and Conglomerates/Pantano Fm.	Moderate.	K = 3.5 m/d	Good.	229–751	B1
	Monzonites/Monzonites.	Poor.	K = 0.006 m/d	Unknown.		C4
	*Cretaceous Granites/Volcanic Cretaceous Rocks.	Poor.	K = 0.018 m/d	Unknown.		C4
	Bisbee Gr/Cretaceous Rocks Undifferentiated.	Poor.	K = 0.039 m/d	Unknown.		C4
	*Paleozoic Rocks Undivided (USA).	Poor.	K = 0.039 m/d	Unknown.		C4
	*Permian Sedimentary Rocks (USA).	Poor.	K = 0.039 m/d	Unknown.		C4
	*Apache Gr (USA).	Poor.	K = 0.072 m/d	Unknown.		C4
	Las Mestenas Granite/Mesoproterozoic Granite.	Poor.	K = 0.006 m/d	Unknown.		C4
	*Pinal Schist/Metasedimentary Rocks.	Poor.	K = 0.006 m/d	Poor.		C3
	*Paleoproterozoic Plutonic Rocks (USA).	Poor.		Unknown.		C4
(28) Mule Mountains.	Neogene Gravels and Conglomerates/Pantano Fm.	Moderate–Poor.		Unknown.		B4-C4
	Bisbee Gr/Cretaceous Rocks Undifferentiated.	Poor.		Unknown.		C4
	*Bisbee Conglomerates (Mex)	Unknown.		Unknown.		E4
	Jurassic Granites/Jurassic Granites.	Poor.		Unknown.		C4
	*Permian Sedimentary Rocks (USA).	Poor.		Unknown.		C4
	*Pinal Schist/Metasedimentary Rocks.	Poor.		Poor.		C3
(29) Rio Agua Prieta Aq./Douglas Basin.	Qt Alluvium/Qt Alluvium.	Good.	T = 147 m^2/d K = 10 m/d	Good.	344–552	A1
	Neogene Gravels and Conglomerates/Pantano Fm.	Moderate–Poor.		Unknown.		B4-C4
	Washita Gr/Washita Gr.	Unknown.		Unknown.		E4
(30) Perilla Mountains.	Volcanic Rocks Undifferentiated/Rhyolitic Tuff.	Poor.		Unknown.		C4
	*Cretaceous Granites/Volcanic Cretaceous Rocks.	Poor.		Unknown.		C4
	Bisbee Gr/Cretaceous Rocks Undifferentiated.	Poor.		Unknown.		C4
	*Permian Sedimentary Rocks (USA).	Poor.		Unknown.		C4
	*Abrigo Fm/Abrigo Limestone.	Unknown.		Unknown.		E4

Table 3. *Cont.*

HGU/Aquifer	Boundary (*) and Transboundary Formations	Aquifer Potential	Hydrogeologic Features	Water Quality	TDS (ppm)	ID
(31) Arroyo San Bernardino Aq./San Bernardino Valley.	Qt Alluvium/Qt Alluvium.	Good.	$T = 100\ m^2/d$ $K = 43\ m/d$	Good.	<1000	A1
	Neogene Conglomerates/Neogene Conglomerates.	Moderate–Good.		Unknown.		A4-B4
	Qt Basalts/Qt Basalts.	Good.		Unknown.		A4
	Bisbee Gr/Cretaceous Rocks Undifferentiated.	Moderate–Poor.		Unknown.		B4-C4
	*Cabullona Fm (Mex).	Moderate–Poor.		Unknown.		B4-C4
	*Permian Sedimentary Rocks (USA).	Poor.		Unknown.		C4
(32) Guadalupe Mountains.	*Gila Gr (USA).	Good.		Unknown.		A4
	Tertiary Lava Flows/Tertiary Lava Flows.	Poor.		Unknown.		C4
	Granite–Granodiorite/Granite–Granodiorite.	Poor.		Unknown.		C4
	*Andesite/Andesite.	Good.		Unknown.		A4
	*Paleogene Sedimentary Rocks (USA).	Unknown.		Unknown.		E4
	Rhyolitic Lavas–Tuffs/Rhyolitic Lavas–Tuffs.	Poor.		Unknown.		C4
	Bisbee Gr/Cretaceous Rocks Undifferentiated.	Moderate–Poor.		Unknown.		B4-C4
	*Paleozoic Rocks Undivided (USA).	Poor.		Unknown.		C4
	*Lake Valley Limestone (USA).	Unknown.		Unknown.		E4
	*Sostenes Fm/Montoya Limestone.	Unknown.		Unknown.		E4
(33) Animas Basin.	Qt Alluvium/Qt Alluvium.	Good.	$T = 273–3055\ m^2/d$	Unknown.		A4
	Qt Lacustrine/Qt Lacustrine.	Poor.		Unknown.		C4
	*Quemado Gr (USA).	Unknown.		Unknown.		E4
	*Gila Gr (USA).	Good.		Unknown.		A4
	*Dacite (USA).	Unknown.		Unknown.		E4
	*Andesite/Andesite.	Poor.		Unknown.		C4
	Tertiary Lava Flows/Tertiary Lava Flows.	Poor.		Unknown.		C4
	Rhyolitic Lavas–Tuffs/Rhyolitic Lavas–Tuffs.	Poor.		Unknown.		C4
	Tertiary Basalt–Andesite/Tertiary Basalt–Andesite.	Poor.		Unknown.		C4
	*Cretaceous–Tertiary Intrusive Rocks (USA).	Unknown.		Unknown.		E4
	*Mancos Shale (USA).	Unknown.		Unknown.		E4

Table 3. *Cont.*

HGU/Aquifer	Boundary (*) and Transboundary Formations	Aquifer Potential	Hydrogeologic Features	Water Quality	TDS (ppm)	ID
(34) Janos Aq./Playas Basin.	Qt Alluvium/Qt Alluvium.	Good.	T = 345 m^2/d	Good.	250–500	A1
	Qt Lacustrine/Qt Lacustrine.	Poor.		Unknown.		C4
	*Quemado Gr (USA).	Unknown.		Unknown.		E4
	Pliocene Conglomerates (Mex)	Unknown.		Unknown.		E4
	Rhyolitic Lavas–Tuffs/Rhyolitic Lavas–Tuffs.	Poor.		Unknown.		C4
(35) Alamo Hueco Mountains.	Granite–Granodiorite/Granite–Granodiorite.	Poor.		Unknown.		C4
	Tertiary Lava Flows/Tertiary Lava Flows.	Poor.		Unknown.		C4
	Rhyolitic Lavas–Tuffs/Rhyolitic Lavas–Tuffs.	Poor.		Unknown.		C4
	Tertiary Basalt–Andesite/Tertiary Basalt–Andesite.	Poor.		Unknown.		C4
	*Cretaceous–Tertiary Intrusive Rocks (USA).	Unknown.		Unknown.		E4
	*Ringbone Fm (USA).	Unknown.		Unknown.		E4
	Lower Cretaceous Undivided/Mojado Sandstone.	Moderate–Poor.		Unknown.		B4-C4
	Horquilla Fm/Horquilla Fm	Poor.		Unknown.		C4
	*Lake Valley Limestione (USA).	Unknown.		Unknown.		E4
	*Paleoproterozoic Plutonic Rocks (USA).	Poor.		Unknown.		C4
(36) Ascension Aq./Hachita Moscos Basin.	Qt Alluvium/Qt Alluvium.	Good.	T = 346 m^2/d K = 0.17 m/d	Good.	500	A1
	Qt Lacustrine/Qt Lacustrine.	Moderate.		Unknown.		B4
	*Quemado Gr (USA).	Unknown.		Unknown.		E4
	Neogene Conglomerates/Upper Santa Fe Gr	Good.		Good.	250–1000	A1
	Lower Santa Fe Gr.	Moderate.		Unknown.		B4
	Tertiary Basalt–Andesite/Tertiary Basalt–Andesite.	Moderate.		Unknown.		B4
	*Cretaceous–Tertiary Intrusive Rocks (USA).	Unknown.		Unknown.		E4
	Lower Cretaceous Undivided/Mojado Sandstone.	Moderate–Poor.		Unknown.		B4-C4
	*Panther Seep Fm (USA).	Unknown.		Unknown.		E4
	*Sostenes Fm/Montoya Limestone.	Unknown.		Unknown.		E4

Table 3. *Cont.*

HGU/Aquifer	Boundary (*) and Transboundary Formations	Aquifer Potential	Hydrogeologic Features	Water Quality	TDS (ppm)	ID
(37) Josefa Ortiz de Dominguez Aq.	Qt Basalts/Qt Basalts.	Poor.		Unknown.		C4
	*Quemado Gr (USA).	Unknown.		Unknown.		E4
	*Trachyte (Mex).	Poor.		Unknown.		C4
	*Andesite/Andesite.	Poor.		Unknown.		C4
	Tertiary Lava Flows/Tertiary Lava Flows.	Poor.		Unknown.		C4
	Rhyolitic Lavas–Tuffs/Rhyolitic Lavas–Tuffs	Poor.		Unknown.		C4
	Tertiary Basalt–Andesite/Tertiary Basalt–Andesite.	Poor.		Unknown.		C4
	Aurora Fm/Glen Rose Fm.	Moderate–Poor.		Unknown.		B4-C4
	Horquilla Fm/Horquilla Fm.	Poor.		Unknown.		C4
(38) Las Palmas Aq./Mimbres Basin.	Qt Conglomerates/Qt Conglomerates.	Good.	T = 35–330 m^2/d K = 0.7 m/d	Fresh.	340	A1
	Qt Lacustrine/Qt Lacustrine.	Poor.	K = 0.4 m/d	Fresh.	340	C1
	Qt Basalts/Qt Basalts.	Good.	K = 0.24–0.36 m/d	Fresh.	340	A1
	*Quemado Gr (USA).	Unknown.		Unknown.		E4
	Neogene Conglomerates/Upper Santa Fe Gr	Moderate.		Fresh to slightly saline.	120–1400	B1-B2
	*Gila Gr (USA).	Good.		Fresh.	200–380	A1
	*Tertiary Igneous Rocks (USA).	Good.		Unknown.		A4
	Granite–Granodiorite/Granite–Granodiorite.	Good.		Unknown.		A4
	*Andesite/Andesite.	Good.	n = 18%–25%	Unknown.		A4
	Tertiary Lava Flows/Tertiary Lava Flows.	Good.		Fresh.	260–560	A1
	*Paleogene Sedimentary Rocks (USA).	Unknown.		Unknown.		E4
	*Mancos Shale (USA).	Unknown.		Unknown.		E4
	Horquilla Fm/Horquilla Fm.	Poor.		Slightly saline.		C2
	*Panther Seep Fm (USA).	Unknown.		Unknown.		E4
	*Cambrian–Ordovician Plutonic Rocks (USA).	Good.		Unknown.		A4
	*Paleoproterozoic Plutonic Rocks (USA).	Good.		Fresh.	500	A1
(39) Potrillo Mountains.	Qt Basalts/Qt Basalts.	Good.	n = 18%–25%	Unknown.		A4
	Tertiary Lava Flows/Tertiary Lava Flows.	Good.	n = 18%–25%	Unknown.		A4
	*Tertiary Igneous Rocks (USA).	Good.	n = 18%–25%	Unknown.		A4

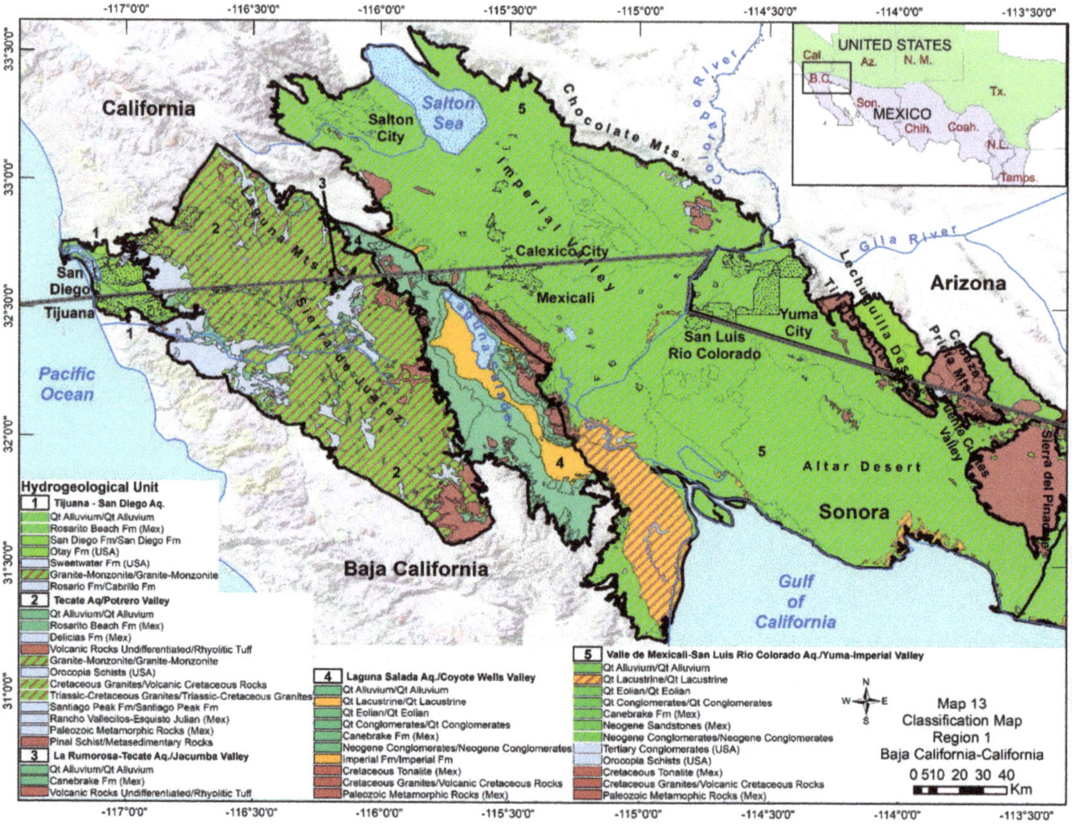

Figure 13. Classification map, Baja California—California.

Figure 14 shows the classification of the formations within their corresponding HGUs across the remaining western side of Arizona and Sonora. This region is also characteristic of good aquifer potential formations but with moderate to poor or unknown water quality conditions. There is an important presence of aquitard conditions in several of the identified HGUs that characterize the area such as the cases of Tinajas Altas Mountains, Cabeza Prieta Mountains, Los Vidrios Aq., Agua Dulce Mountains, Senita Basin, and Los Chirriones Aq. These HGUs are conformed primarily by Quaternary Basalts, Granite-Monzonites, Jurassic Granites, and Volcanic Cretaceous Rocks. Some of these geologic characteristics are also present but to a lesser extent in San Simon Wash and Sonoyta-Puerto Peñasco Aq., where moderate water quality conditions can be found. Puerto Cuates Valley/Lechugilla Desert, Lukeville-Sonoyta Valley, and The Great Plain report good aquifer conditions, but there is limited information related to water quality. As in the Baja California–California region, this region relies heavily on groundwater for agriculture and domestic use considering the limited availability of surface water. Figure 15 shows the eastern part of the border between Arizona and Sonora. Good aquifer potential and good levels of water quality are present to a greater extent in this region as compared to the westernmost side. The Nogales-Rio Santa Cruz Aq./Upper Santa Cruz Basin, the Rio San Pedro Aq./Upper San Pedro Basin, and the Rio Agua Prieta Aq./Douglas Basin, all recognized transboundary aquifers at binational level, show good aquifer potential and good water quality. These aquifers have been categorized as high priority given the level of groundwater dependence for domestic use and population growth and therefore the vulnerability of the aquifer to overexploitation and contamination. The Pajarito Mountains, Arroyo Seco Aq., and Arroyo San Bernardino Aq./San Bernardino Valley also show good aquifer potential, but there is limited information on water quality.

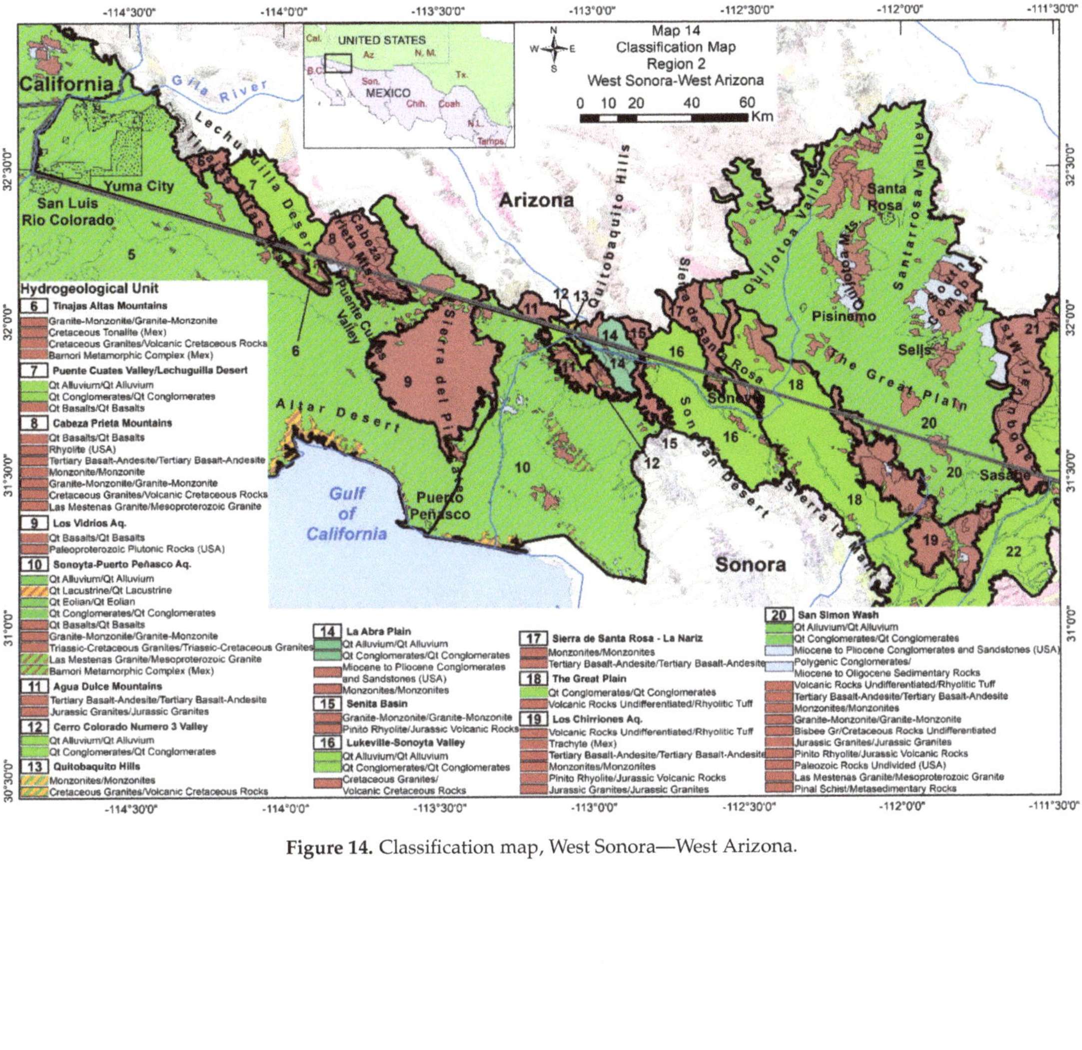

Figure 14. Classification map, West Sonora—West Arizona.

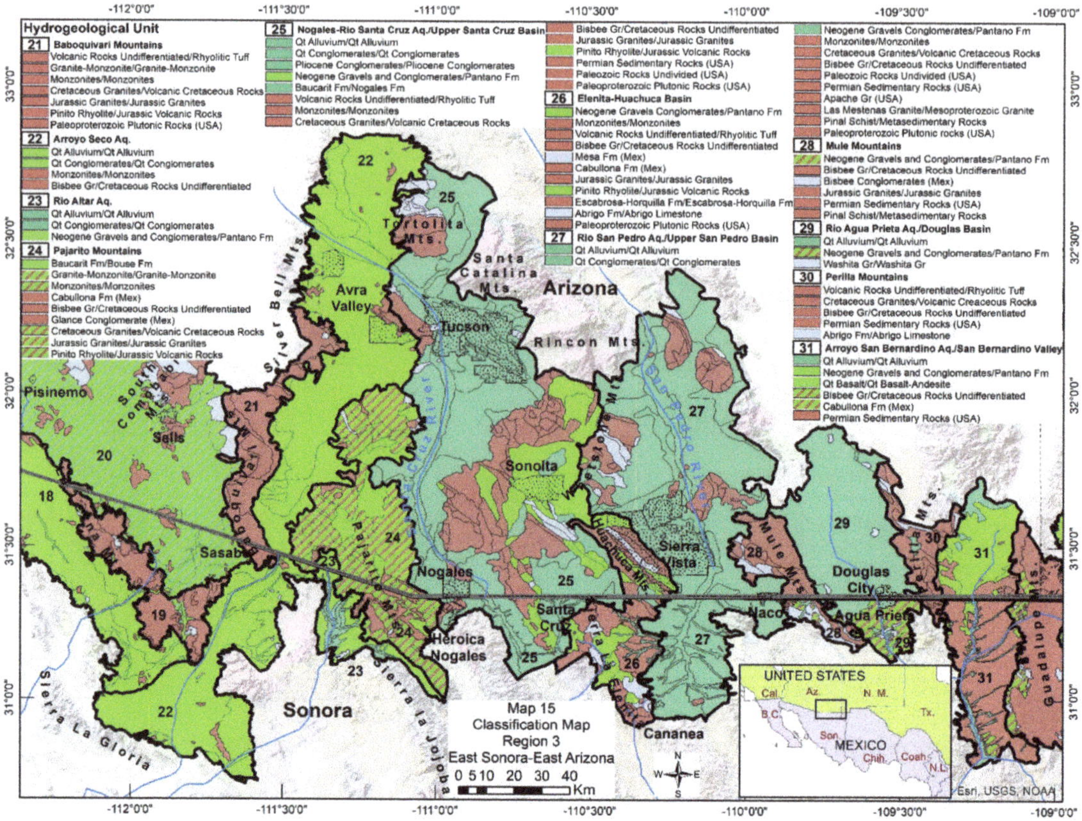

Figure 15. Classification map, East Sonora—East Arizona.

Figure 16 shows the classification of units between Nuevo Mexico and Chihuahua. Janos Aq./Playas Basin, Ascension Aq./Hachita Moscos Basin, and Las Palmas Aq./Mimbres Basin show the highest levels of aquifer potential and water quality, followed by Conejos-Medanos Aq./Mesilla Bolson, and Animas Basin, which report poor to moderate water quality. The Mimbres Basin is an officially recognized transboundary aquifer according to ISARM databases; however, the delineation officially reported is an undefined line in the area, meaning more research is required to confirm the delineation of this aquifer at transboundary level [101]. Over-pumping has been reported around the Columbus-Palomas region as well as high levels of salinity associated with mining activities [1]. It is worth mentioning that, from the total shareable land in this region, approximately 85 percent reports good aquifer potential and water quality. Small communities in the border region rely on these aquifers for potable and local agricultural use, and therefore, the strategic value for this area for future sources of water in the region is one of the highest in the U.S.–Mexico border region.

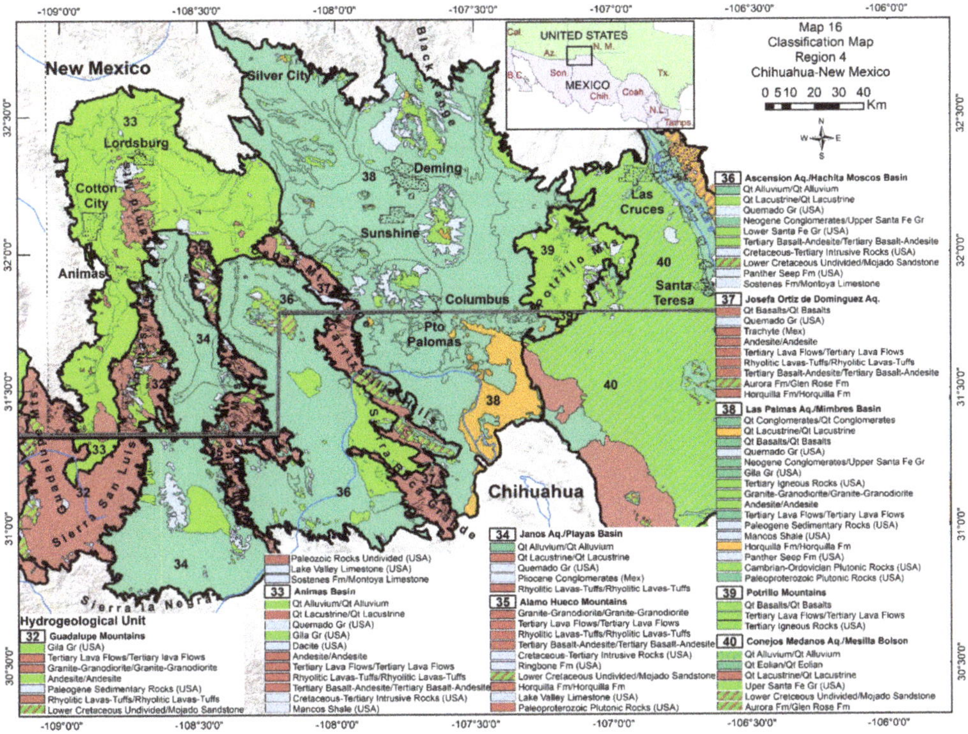

Figure 16. Classification map, Chihuahua—New Mexico.

4. Conclusions

Results indicate that a total of 39 HGUs have been identified in the border between California, Arizona, and New Mexico on the U.S. side and Baja California, Sonora, and Chihuahua on the Mexico side. This region accounts for an approximate shareable area of 135,000 km^2 where both countries share half of the area (65,000 km^2 Mexico and 69,000 km^2 the U.S). From the total shareable area, around 40% reports good to moderate aquifer potential and water quality, of which 65% is in the U.S. and 35% on the Mexico side. It should be noted that approximately 15% of the shareable land that reports good aquifer conditions also reports unknown or limited data on water quality conditions; therefore, this could mean that estimations of good aquifer conditions and water quality along the region might be underestimated.

Border-wide and adding the HGUs previously reported by Sanchez et al. [1] between Texas and Mexico, the total number of HGUs in the border region between Mexico and the United States is 72, covering an approximate area of 315,000 km^2 (180,000 km^2 in the U.S. and 135,000 km^2 on the Mexico side). The total area considered to have good to moderate aquifer potential as well as good to regular water quality ranges between 50 and 55% (of which approximately 60% is in the U.S. and the rest in Mexico).

From a statewide perspective, the border between Baja California and California reports a total of 5 HGUs, from which 3 (Tijuana-San Diego Aq., Valle de Mexicali-San Luis Rio Colorado/Yuma-Imperial Valley, and a great portion of the Quaternary deposits of Laguna Salada Aq./Coyote Wells Valley) report good to moderate aquifer potential and generally good to moderate water quality. Available data on water quality vary across the Valle de Mexicali-San Luis Rio Colorado/Yuma-Imperial Valley from good to poor (included limited information), particularly in the southern portions where saline intrusion has been reported. In the case of Sonora and Arizona, 25 HGUs have been identified, with at least 7 HGUs (Nogales-Rio Santa Cruz Aq./Upper Santa Cruz Basin, Rio San Pedro Aq./Upper San Pedro Basin, Rio Agua Prieta Aq./Douglas Basin, Rio Altar Aq., San Simon Wash, Sonoyta-Puerto Peñasco Aq., and La Abra Plain) with generally good to moderate aquifer potential and good to moderate water quality. Variability in water quality for Sonoyta-Puerto Peñasco Aq. and San Simon Wash is also reported. Additional 4 HGUs reported good to moderate aquifer potential but poor water quality with also uncertainty considering the data limitations. Those include Cerro Colorado Numero 3 Valley, Lukeville-Sonoyta Valley, The Great Plain, and Arroyo Seco Aq. In the border region between Chihuahua and New Mexico, good aquifer potential and good water quality were identified in at least 3 out of the 8 HGUs reported. These HGUs are Janos Aq./Playas Basin, Ascension Aq./Hachita-Moscos Basin, and Las Palmas Aq./Mimbres Basin. Potrillo Mountains also report good aquifer potential but limited data on water quality.

Figure 17 shows the complete map of the HGUs/aquifers identified in this paper from California through New Mexico and their corresponding southern border states in Mexico. This is the first ever recorded map that shows the geological continuity across the border between both countries in the complete study area and, along with that reported by Sanchez et al. [2], that covers the border between Texas and Mexico, constituting the first geological assessment on this scale for the complete border region between Mexico and the United States. Further research must incorporate new data particularly on vertical geology, water quality, three-dimensional distribution of HGUs, evidence of groundwater flow systems, isotope assessments for residence times and so on. This new scientific information will support the potential discussions of transboundary groundwater management possibilities towards a more sustainable groundwater use in the border region.

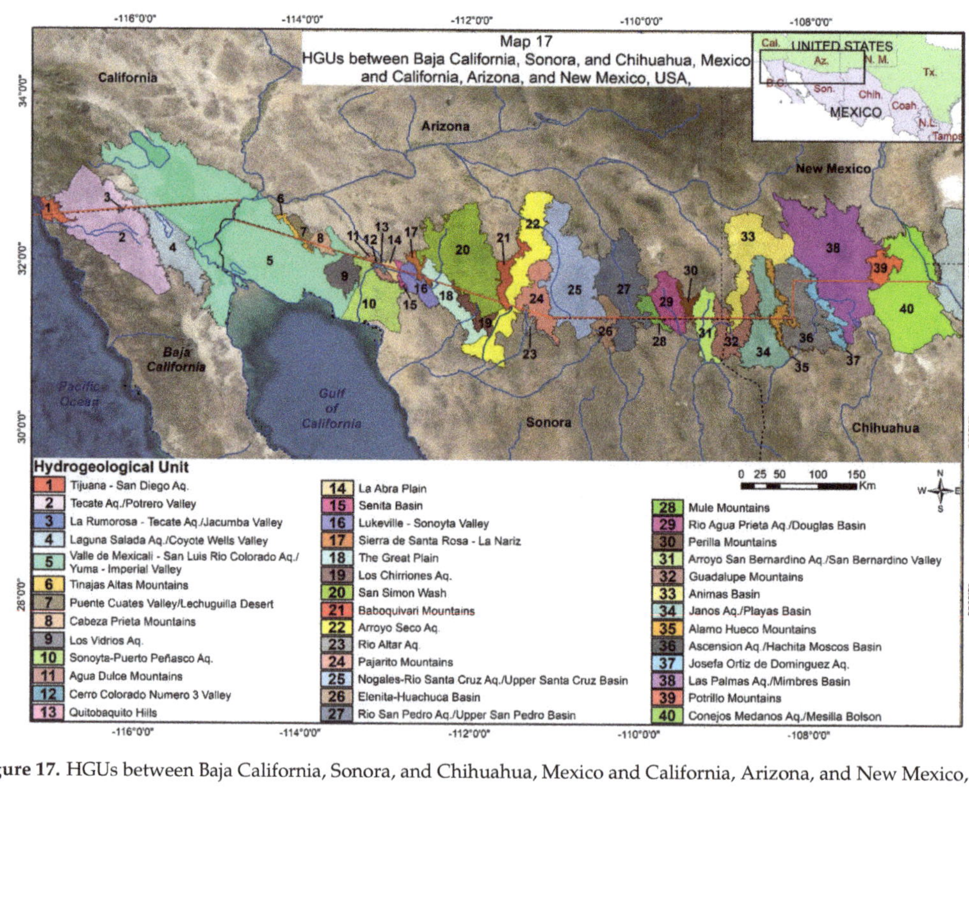

Figure 17. HGUs between Baja California, Sonora, and Chihuahua, Mexico and California, Arizona, and New Mexico, USA.

Author Contributions: Conceptualization, R.S.; methodology, R.S., L.R.; software, L.R.; validation, R.S.; formal analysis, R.S., L.R.; investigation, L.R.; resources, R.S.; data curation, L.R.; writing—original draft preparation, R.S., L.R.; writing—review and editing, R.S.; visualization, L.R.; supervision, R.S.; project administration, R.S.; funding acquisition, R.S. All authors have read and agreed to the published version of the manuscript.

Funding: This research was funded by the United States Geological Service, Identifying and Assessing the Conditions of Transboundary Aquifers Between Texas and Mexico. Grant number USGS M1800330.

Data Availability Statement: Data are available in a publicly accessible repository that does not issue DOIs. Publicly available datasets were analyzed in this study. This data will be found here after publication: Home | Transboundary Water Portal (tamu.edu).

Acknowledgments: We are grateful for the support of the U.S. Geological Service and for the funding for this project. We also want to specially recognize the revisions made by reviewer who made an extensive and detailed revision of our paper which significantly impacted the quality of this manuscript and its contribution to the overall literature on the topic. Thank you!

Conflicts of Interest: The authors declare no conflict of interest. The funders had no role in the design of the study; in the collection, analyses, or interpretation of data; in the writing of the manuscript; or in the decision to publish the results.

References

1. Sanchez, R.; Victoria, L.; Eckstein, G. Identifying and Characterizing Transboundary Aquifers Along the Mexico–Us Border: An Initial Assessment. *J. Hydrol.* **2016**, *535*, 101–119. [CrossRef]
2. Sanchez, R.; Rodriguez, L.; Tortajada, C. Transboundary Aquifers between Chihuahua, Coahuila, Nuevo Leon and Tamaulipas, Mexico, and Texas, USA: Identification and Categorization. *J. Hydrol. Reg. Stud.* **2018**, *20*, 74–102. [CrossRef]
3. Tapia-Villaseñor, E.M.; Megdal, S.B. The Us-Mexico Transboundary Aquifer Assessment Program as a Model for Transborder Groundwater Collaboration. *Water* **2021**, *13*, 530. [CrossRef]
4. Hanson, A.T.; Schmid, W.; Knight, J.; Maddock, T., III. Integrated Hydrologic Modeling of a Transboundary Aquifer System—Lower Rio Grande. In Proceedings of the MODFLOW and More 2013—Translating Science into Practice, Golden, CO, USA, 2–5 June 2013.
5. Erwin, G. *Groundwater Flow Model of the Santa Cruz Active Management Area Microbasins International Boundary to Nogales International Wastewater Treatment Plant Santa Cruz County, Arizona*; Arizona Department of Water Resources: Phoenix, AZ, USA, 2007.
6. Callegary, J.B.; Megdal, S.B.; Tapia Villaseñor, E.M.; Petersen-Perlman, J.D.; Minjárez Sosa, I.; Monreal, R.; Gray, F.; Grijalva Noriega, F. Findings and Lessons Learned from the Assessment of the Mexico-United States Transboundary San Pedro and Santa Cruz Aquifers: The Utility of Social Science in Applied Hydrologic Research. *J. Hydrol. Reg. Stud.* **2018**, *20*, 60–73. [CrossRef]
7. Callegary, J.B.; Minjárez Sosa, I.; Tapia Villaseñor, E.M.; Dos Santos, P.; Monreal Saavedra, R.; Grijalva Noriega, F.J.; Huth, A.K.; Gray, F.; Scott, C.A.; Megdal, S.B. Binational Study of the Transboundary San Pedro Aquifer. In *El Paso TX*; 2016; 1, pp. 1–164. Available online: http://www.ibwc.gov/Files/Binational_Study_Transboundary_San_Pedro_Aquifer.pdf (accessed on 10 September 2020).
8. Sanchez, R.; Brena, A.; Rivera, A.; Hanson, R.T.; Hernandez-Espriu, A.; Hogeboom, R.; Benavides, J.; Milman, A.; Pedrozo-Acuna, A.; Soriano-Monzalvo, J.C.; et al. Perspectives on Building Water Security and a Resilient Capacity for Transboundary Groundwater Systems in the Borderlands of Mexico and the United States. *Water Int.* **2021**, in press.
9. Hollet, K.J. *Maps Showing Ground-Water Conditions in the San Simon Wash Area, Papago Indian Reservation, Arizona-1979*; Geological Survey (US): Reston, VS, USA, 1981. [CrossRef]
10. Servicio Geologico Mexicano. Geoinfomex. El Banco De Datos Del Sgm. Available online: https://mapasims.sgm.gob.mx/GeoInfoMex_gobmx/ (accessed on 11 January 2021).
11. Servicio Geologico Mexicano. *Carta Geologico Minera Ciudad Juarez, H13-1, El Porvenir H13-2, Chihuahua, Escala 1:250,000*; Servicio Geológico Mexicano, 2003. Available online: http://mapserver.sgm.gob.mx/Cartas_Online/geologia/33_H13-1_GM.pdf (accessed on 15 October 2020).
12. Servicio Geologico Mexicano. *Carta Geologico Minera Agua Prieta, H12-3 Sonora, Chihuahua Escala 1:250,000*; Servicio Geológico Mexicano, 2003. Available online: http://mapserver.sgm.gob.mx/Cartas_Online/geologia/25_H12-3_GM.pdf (accessed on 15 October 2020).
13. Servicio Geologico Mexicano. *Carta Geologico Minera Ensenada, H11-2, Baja California, Escala 1:250,000*; Servicio Geológico Mexicano, 2003. Available online: http://mapserver.sgm.gob.mx/Cartas_Online/geologia/2_H11-2_GM.pdf (accessed on 15 October 2020).
14. Servicio Geologico Mexicano. *Carta Geologico Minera Tijuana, I11-11, Baja California, Escala 1:250,000*; Servicio Geológico Mexicano, 2003. Available online: http://mapserver.sgm.gob.mx/Cartas_Online/geologia/1_I11-11_GM.pdf (accessed on 15 October 2020).

15. Servicio Geologico Mexicano. *Carta Geologico Minera Mexicali I11-12, Baja California Y Sonora, Escala 1:250,000*; Servicio Geológico Mexicano, 2003. Available online: http://mapserver.sgm.gob.mx/Cartas_Online/geologia/3_I11-12_GM.pdf (accessed on 15 October 2020).
16. Servicio Geologico Mexicano. *Carta Geologico Minera Nueva Casas Grandes, H13-4 Chihuahua, Escala 1:250,000*; Servicio Geológico Mexicano, 2002. Available online: http://mapserver.sgm.gob.mx/Cartas_Online/geologia/34_H13-4_GM.pdf (accessed on 15 October 2020).
17. Servicio Geologico Mexicano. *Carta Geologico Minera Puerto Peñasco H12-1, Sonora, Escala 1:250,000*; Servicio Geológico Mexicano, 2002. Available online: http://mapserver.sgm.gob.mx/Cartas_Online/geologia/10_H12-1_GM.pdf (accessed on 15 October 2020).
18. Servicio Geologico Mexicano. *Carta Geologico Minera Los Vidrios, I12-10, Sonora, Escala 1:250,000*; Servicio Geológico Mexicano, 2002. Available online: http://mapserver.sgm.gob.mx/Cartas_Online/geologia/9_I12-10_GM.pdf (accessed on 15 October 2020).
19. Servicio Geologico Mexicano. *Carta Geologico Minera Nogales, H12-2 Sonora, Escala 1:250,000*; Servicio Geológico Mexicano, 2000. Available online: http://mapserver.sgm.gob.mx/Cartas_Online/geologia/16_H12-2_GM.pdf (accessed on 15 October 2020).
20. Servicio Geologico Mexicano. *Carta Geologico Minera Cananea H12-5, Sonora, Escala 1:250,000*; Servicio Geológico Mexicano, 1999. Available online: http://mapserver.sgm.gob.mx/Cartas_Online/geologia/17_H12-5_GM.pdf (accessed on 15 October 2020).
21. Servicio Geologico Mexicano. *Carta Geologico Minera San Felipe, H11-3 Baja California Y Sonora, Escala 1:250,000*; Servicio Geológico Mexicano, 1999. Available online: http://mapserver.sgm.gob.mx/Cartas_Online/geologia/4_H11-3_GM.pdf (accessed on 15 October 2020).
22. Servicio Geologico Mexicano. *Carta Geológico-Minera Nacozari H12-6 Sonora Y Chihuahua, 1:250000: Pachuca*; Servicio Geológico Mexicano, 1998. Available online: http://mapserver.sgm.gob.mx/Cartas_Online/geologia/26_H12-6_GM.pdf (accessed on 15 October 2020).
23. Stoeser, D.B.; Green, G.N.; Morath, L.C.; Heran, W.D.; Wilson, A.B.; Moore, D.W.; Van Gosen, B.S. *Preliminary Integrated Geologic Map Databases for the United States Central States: Montana, Wyoming, Colorado, New Mexico, Kansas, Oklahoma, Texas, Missouri, Arkansas, and Louisiana,-the State of Texas*; Geological Survey (US): Reston, VS, USA, 2005. [CrossRef]
24. Ludington, S.; Moring, B.; Miller, R.; Flynn, K.; Hopkins, M.; Stone, P.; Bedford, D.; Haxel, G. *Preliminary Integrated Databases for the United States-Western States: California, Nevada, Arizona, Washington, Oregon, Idaho, and Utah*; Geological Survey (US): Reston, VS, USA, 2005. [CrossRef]
25. Poland, J.F.; Lofgren, B.E.; Riley, F.S. *Glossary of Selected Terms Useful in Studies of the Mechanics of Aquifer Systems and Land Subsidence Due to Fluid Withdrawal*; Geological Survey (US): Reston, VS, USA, 1972.
26. Ries, K.G., III; Newson, J.K.; Smith, M.J.; Guthrie, J.D.; Steeves, P.A.; Haluska, T.; Kolb, K.R.; Thompson, R.F.; Santoro, R.D.; Vraga, H.W. *Streamstats, Version 4*; US Geological Survey: Reston, VS, USA, 2017.
27. INEGI. Siatl (Simulador De Flujos De Agua De Cuencas Hidrográficas). Available online: https://antares.inegi.org.mx/analisis/red_hidro/siatl/ (accessed on 30 August 2021).
28. U.S. Geological Survey. Water Science Glossary of Terms. *Water Encycl.* **2005**, *2*, 541–548. [CrossRef]
29. Hiscock, K.M. *Hydrogeology: Principles and Practice*; John Wiley & Sons: Hoboken, NJ, USA, 2009.
30. Hutchison, W.R. Conceptual Model Report: Lower Rio Grande Valley Groundwater Transport Model. 2016. Available online: https://www.twdb.texas.gov/publications/reports/contracted_reports/doc/1548301854_GSI_ConceptualReport.pdf (accessed on 20 April 2017).
31. Ransome, F.L. *The Geology and Ore Deposits of the Bisbee Quadrangle, Arizona*; US Government Printing Office: Washington, DC, USA, 1904; Volume 21.
32. CONAGUA. Actualizacion De La Disponibilidad Media Anual De Agua En El Acuifero Cumpas (2661), Estado De Sonora. Mexico, DF: Subdireccion General Tecnica, CONAGUA. 2018. Available online: https://sigagis.conagua.gob.mx/gas1/Edos_Acuiferos_18/sonora/DR_2661.pdf (accessed on 20 April 2017).
33. Condie, K.C. Geochemistry of Proterozoic Granitic Plutons from New Mexico, USA. *Chem. Geol.* **1978**, *21*, 131–149. [CrossRef]
34. Damon, P.E.; Livingston, D.E.; Mauger, R.L.; Gilletti, B.J.; Alor, P. Edad Del Precámbrico Anterior Y De Otras Rocas Del Zócalo De La Región De Caborca-Altar De La Parte Noroccidental Del Estado De Sonora: Boletín Instituto De Geología. *Univ. Nac. Autónoma México* **1962**, *64*, 11–64.
35. Ransome, F.L. *Geology of the Globe Copper District, Arizona*; US Government Printing Office: Washington, DC, USA, 1903.
36. Viveros-Martínez, A. Estudio Geológico De La Sierra De Cabullona, Municipio De Agua Prieta, Estado De Sonora. Bachelor's Thesis, Universidad Nacional Autónoma de México, Ciudad de México, Mexico, 1965.
37. CONAGUA. Actualizacion De La Disponibilidad Media Anual De Agua En El Acuifero Los Vidrios (2602), Estado De Sonora. Mexico, DF: Subdireccion General Tecnica, CONAGUA. 2018. Available online: https://sigagis.conagua.gob.mx/gas1/Edos_Acuiferos_18/sonora/DR_2602.pdf (accessed on 20 May 2020).
38. Clemons, R.E. *Geology of the Florida Mountains, Southwestern New Mexico*; New Mexico Bureau of Mines and Mineral Resources: Socorro, NM, USA, 1998.
39. Trauger, F.D.; Herrick, E.H. *Ground Water in Central Hachita Valley Northeast of the Big Hatchet Mountains, Hidalgo County, New Mexico*; New Mexico State Engineer Office: Albuquerque, NM, USA, 1962.
40. Richardson, G.B. *Report of a Reconnaissance in Trans-Pecos Texas: North of the Texas and Pacific Railway*; Von Boeckmann-Jones Company, State Printers, 1904. Available online: https://repositories.lib.utexas.edu/handle/2152/24408 (accessed on 20 April 2017).

41. Hanson, R.T.; McLean, J.S.; Miller, R.S. *Hydrogeologic Framework and Preliminary Simulation of Ground-Water Flow in the Mimbres Basin, Southwestern New Mexico*; US Geological Survey, US Department of the Interior: Washington, DC, USA, 1994; Volume 94. [CrossRef]
42. Díaz-Cabrera, P. Simulación Numérica Del Acuífero Superior Del Valle De Mexicali, Baja California, México. Master's Thesis, CICECE. División de Ciencias de la Tierra, Ensenada BC, Ciudad de México, Mexico, 2001.
43. Gordon, C.H. Art. Vi.—Mississippian (Lower Carboniferous) Formations in the Rio Grande Valley, New Mexico. *Am. J. Sci. (1880–1910)* **1907**, *24*, 58. Available online: https://ngmdb.usgs.gov/Prodesc/proddesc_88702.htm (accessed on 20 April 2017). [CrossRef]
44. Kottlowski, F.E.; Kottlowski, F.E. *Stratigraphic Studies of the San Andres Mountains, New Mexico*; New Mexico Institute of Mining, State Bureau of Mines and Mineral Resources: Socorro, NM, USA, 1956.
45. Gilluly, J.; Cooper, J.R.; Williams, J.S. *Late Paleozoic Stratigraphy of Central Cochise County, Arizona*; U.S. Geological Survey: Reston, VS, USA, 1954. [CrossRef]
46. CONAGUA. Actualizacion De La Disponibilidad Media Anual De Agua En El Acuifero Josefa Ortiz De Dominguez (0829), Estado De Chihuahua. Mexico, DF: Subdireccion General Tecnica, CONAGUA. 2018. Available online: https://www.gob.mx/cms/uploads/attachment/file/103590/DR_0829.pdf (accessed on 20 June 2020).
47. Nations, D.; Stump, E. *Geology of Arizona*; Kendall Hunt Publishing Company: Dubuque, IA, USA, 1996.
48. CONAGUA. Actualizacion De La Disponibilidad Media Anual De Agua En El Acuifero Tecate (0202), Estado De Baja California. Mexico, DF: Subdireccion General Tecnica, CONAGUA. 2018. Available online: https://sigagis.conagua.gob.mx/gas1/Edos_Acuiferos_18/BajaCalifornia/DR_0202.pdf (accessed on 20 June 2020).
49. Gilmont, N.L. Geology of the Puerto La Bandera Area. Master's Thesis, University of Northern Arizona, Flagstaff, AZ, USA, 1978.
50. Minjárez Sosa, I.; Saavedra, R.M.; Grijalva Noriega, F.J.; Ochoa Granillo, J.A.; Tapia Villaseñor, E.M.; Rangel Medina, M.; Montijo González, A. *Actividades Hidrogeológicas En El Acuífero Río Santa Cruz, Estado De Sonora, Primera Etapa*; CILA: Hermosillo, Mexico, 2011.
51. CONAGUA. Actualizacion De La Disponibilidad Media Anual De Agua En El Acuifero Rio Alisos (2613), Estado De Sonora. Mexico, DF: Subdireccion General Tecnica, CONAGUA. 2018. Available online: https://sigagis.conagua.gob.mx/gas1/Edos_Acuiferos_18/sonora/DR_2613.pdf (accessed on 25 September 2020).
52. Force, E.R. *The Bisbee Group of the Tombstone Hills, Southeastern Arizona: Stratigraphy, Structure, Metamorphism, and Mineralization*; US Government Printing Office: Washington, DC, USA, 1996. [CrossRef]
53. Bilodeau, W.L.; Kluth, C.F.; Vedder, L.K. Regional Stratigraphic, Sedimentologic, and Tectonic Relationships of the Glance Conglomerate in Southeastern Arizona. *Mesoz. Rocks South. Ariz. Adjac. Areas Ariz. Geol. Soc. Dig.* **1987**, *18*, 229–256.
54. Brown, T.E. *Stratigraphy of the Washita Group in Central Texas*; Department of Geology, Baylor University: Waco, TX, USA, 1971.
55. Hill, R.T. *Preliminary Notes on the Topography and Geology of Northern Mexico and Southwest Texas, and New Mexico*; American Geologist: Austin, Texas, USA, 1891; Volume 8, pp. 133–141.
56. CONAGUA. Actualizacion De La Disponibilidad Media Anual De Agua En El Acuifero Palomas-Guadalupe Victoria (0812), Estado De Chihuahua. Mexico, DF: Subdireccion General Tecnica, CONAGUA. 2018. Available online: https://sigagis.conagua.gob.mx/gas1/Edos_Acuiferos_18/chihuahua/DR_0812.pdf (accessed on 25 September 2020).
57. Zeller, R.A.; Molina, R. *Reconnaissance Geologic Map of Southern Animas Mountains*; New Mexico Bureau of Mines and Mineral Resources: Socorro, NM, USA, 1962.
58. CONAGUA. Actualizacion De La Disponibilidad Media Anual De Agua En El Acuifero Arroyo San Bernardino (2630), Estado De Sonora. Mexico, DF: Subdireccion General Tecnica, CONAGUA. 2018. Available online: https://sigagis.conagua.gob.mx/gas1/Edos_Acuiferos_18/sonora/DR_2630.pdf (accessed on 25 September 2020).
59. McDonald, C.C.; Loeltz, O.J. *Water Resources of Lower Colorado River-Salton Sea Area as of 1971: Summary Report*; US Government Printing Office: Washington, DC, USA, 1976; Volume 486.
60. Landis, E.R.; Dane, C.H.; Cobban, W.A. Stratigraphic Terminology of the Dakota Sandstone and Mancos Shale, West-Central New Mexico. US Government Printing Office: Washington, DC, USA, 1973.
61. Taliaferro, N.L. An Occurrence of Upper Cretaceous Sediments in Northern Sonora, Mexico. *J. Geol.* **1933**, *41*, 12–37. [CrossRef]
62. Lasky, S.G. Newly Discovered Section of Trinity Age in Southwestern New Mexico. *AAPG Bull.* **1938**, *22*, 524–540. [CrossRef]
63. Kennedy, M.P.; Moore, G.W. Stratigraphic Relations of Upper Cretaceous and Eocene Formations, San Diego Coastal Area, California. *AAPG Bull.* **1971**, *55*, 709–722. [CrossRef]
64. Haxel, G.B.; Jacobson, C.E.; Wittke, J.H. Mantle Peridotite in Newly Discovered Far-Inland Subduction Complex, Southwest Arizona: Initial Report. *Int. Geol. Rev.* **2015**, *57*, 871–892. [CrossRef]
65. Scholle, P.A. *Geologic Map of New Mexico*; New Mexico Bureau of Geology and Mineral Resources: Socorro, NM, USA, 2003.
66. Richard, S.M.; Reynolds, S.J.; Spencer, J.E.; Pearthree, P.A. Geologic Map of Arizona: Arizona Geological Survey Map 35, 1 Sheet, Scale 1:1,000,000. 2000. Available online: http://hdl.handle.net/10150/630741 (accessed on 15 October 2020).
67. Department of Planning County of San Diego. General Plan Update. Appendix D, Groundwater Study. In *Department of Planning and Land Use*; San diego County, Planning and Development Services; 2010. Available online: https://www.sandiegocounty.gov/content/dam/sdc/pds/gpupdate/docs/BOS_Aug2011/EIR/Appn_D_GW.pdf (accessed on 15 October 2020).
68. Darton, N.H. *Geology and Underground Water of Luna County, New Mexico*; Government Printing Office: Washington, DC, USA, 1916.

69. Flynn, C.J. Geology of the La Gloria-Presa Rodriguez Area, Baja California, Mexico. Master's Thesis, San Diego State University, San Diego, CA, USA, 1968.
70. Olmsted, F.H.; Loeltz, O.J.; Irelan, B. *Geohydrology of the Yuma Area, Arizona and California*; US Government Printing Office: Washington, DC, USA, 1973; Volume 486.
71. Trauger, F.D.; Lavery, N.G. Geohydrology of the Upper Pipe Line Draw Area, Grant County, New Mexico. In *Consultant report to Exxon Minerals*; Exxon Company: Houston, TX, USA, 1976.
72. O'Brien, K.M.; Stone, W.J. *A Two-Dimensional Hydrologic Model of the Animas Valley, Hidalgo County, New Mexico*; New Mexico Bureau of Mines and Mineral Resources: Socorro, NM, USA, 1983.
73. CONAGUA. Actualizacion De La Disponibilidad Media Anual De Agua En El Acuifero Rio Santa Cruz (2615), Estado De Sonora. Mexico, DF: Subdireccion General Tecnica, CONAGUA. 2018. Available online: https://sigagis.conagua.gob.mx/gas1/Edos_Acuiferos_18/sonora/DR_2615.pdf (accessed on 25 September 2020).
74. Morton, P.K. Geology and Mineral Resources of Imperial County, California. California Division of Mines and Geology. 1977. Available online: https://ngmdb.usgs.gov/Prodesc/proddesc_408.htm (accessed on 11 November 2020).
75. Mason, D.; Hipke, W. *Regional Groundwater Flow Model of the Tucson Active Management Area, Arizona. Model Update and Calibration*; Arizona Department of Water Resources: Phoenix, AZ, USA, 2013.
76. Connell, S.D. Stratigraphy of the Albuquerque Basin, Rio Grande Rift, Central New Mexico: A Progress Report. In *Stratigraphy and Tectonic Development of the Albuquerque Basin, Central Rio Grande Rift. Field-Trip Guidebook for the Geological Society of America, Mini-Papers*; New Mexico Bureau of Mines and Mineral Resources: Socorro, NM, USA, 2001; pp. A1–A27.
77. Myers, R.G.; Orr, B.R. *Geohydrology of the Aquifer in the Santa Fe Group, Northern West Mesa of the Mesilla Basin near Las Cruces, New Mexico*; US Geological Survey: Reston, VS, USA, 1985; Volume 84. [CrossRef]
78. Artim, E.R.; Pinckney, C.J. La Nacion Fault System, San Diego, California. *Geol. Soc. Am. Bull.* **1973**, *84*, 1075–1080. [CrossRef]
79. Walsh, S.L.; Demere, T.A. Age and Stratigraphy of the Sweetwater and Otay Formations, San Diego County, California. 1991. Available online: http://sntbberry.cityofsanteeca.gov/sites/FanitaRanch/Public/Remainder%20of%20the%20Record/(2)%20Reference%20Documents%20from%20EIR%20&%20Technical%20Reports/Tab%20083%20-%201991%20Walsh%20&%20Demere%20-%20Age%20and%20Stratigraphy.pdf (accessed on 14 May 2021).
80. Flint, L.E.; Flint, A.L.; Stolp, B.J.; Danskin, W.R. Water-Balance and Groundwater-Flow Estimation for an Arid Environment: San Diego Region, California. *Hydrol. Earth Syst. Sci. Discuss.* **2012**, *9*, 2717–2762.
81. Woodring, W.P. Distribution and Age of the Marine Tertiary Deposits of the Colorado Desert. *Carnegie Inst. Wash. Publ.* **1932**, *418*, 1–28. Available online: https://ngmdb.usgs.gov/Prodesc/proddesc_88957.htm (accessed on 24 May 2021).
82. Cohen, A.; Cassidy, C.; Crow, R.; Bright, J.; Crossey, L.; Dorsey, R.; Gootee, B.; House, K.; Howard, K.; Karlstrom, K. The Bouse Formation, a Controversial Neogene Archive of the Evolving Colorado River: A Scientific Drilling Workshop Report (28 February–3 March 2019–Bluewater Resort & Casino, Parker, Az, USA). *Sci. Drill.* **2019**, *26*, 59–67. [CrossRef]
83. Simons, F.S. Geologic Map and Sections of the Nogales and Lochiel Quadrangles, Santa Cruz County, Arizona. US Geological Survey: Reston, VS, USA, 1974. [CrossRef]
84. Page, W.R.; Gray, F.; Bultman, M.W.; Menges, C.M. *Hydrogeologic Investigations of the Miocene Nogales Formation in the Nogales Area, Upper Santa Cruz Basin, Arizona*; US Geological Survey: Reston, VS, USA, 2016. [CrossRef]
85. Thiros, S.A.; Bexfield, L.M.; Anning, D.W.; Huntington, J.M. *Conceptual Understanding and Groundwater Quality of Selected Basin-Fill Aquifers in the Southwestern United States*; US Geological Survey: Reston, VS, USA, 2010.
86. CONAGUA. Actualizacion De La Disponibilidad Media Anual De Agua En El Acuifero San Pedro (2616), Estado De Sonora. Mexico, DF: Subdireccion General Tecnica, CONAGUA. 2018. Available online: https://www.gob.mx/cms/uploads/attachment/file/104308/DR_2616.pdf (accessed on 20 April 2017).
87. Gilbert, G.K. Report on the Geology of Portions of California, Nevada, Utah, Colorado, New Mexico, and Arizona, Examined in the Years 1871, 1872, and 1873. *US Geol. Geog. Surv. West 100th Meridian (U. S.)* **1875**, *3*, 288.
88. Garcia, J. *New Mexico's Border with Mexico: Creating a Viable Agenda for Growth*; The Center for Latin American and Border Studies, New Mexico State University: Las Cruces, NM, USA, 2001.
89. Wilkins, D.W. *Summary of the Southwest Alluvial Basins Regional Aquifer-System Analysis in Parts of Colorado, New Mexico, and Texas*; US Geological Survey: Reston, VS, USA, 1998. [CrossRef]
90. Chamberlin, R.M.; Cather, S.M.; Anderson, O.J.; Jones, G.E. *Reconnaissance Geologic Map of the Quemado 30× 60-Minute Quadrangle, Catron County, New Mexico*; Open-File Report 406; New Mexico Bureau of Mines and Mineral Resources: Socorro, NM, USA, 1994.
91. Cleveland, G.B. *Geology of the Otay Bentonite Deposit, San Diego County, California*; Division of Mines, Department of Natural Resources, State of California: San Francisco, CA, USA, 1960; Volume 64.
92. Anderson, E.K. Hydraulic Conductivity of Pliocene & Pleistocene Sediments in the San Diego Bay Subsurface. 2009. Available online: https://www.carleton.edu/departments/geol/resources/comps/compspdffiles/2009/andersok_2009.pdf (accessed on 10 June 2021).
93. Dibblee, T.W., Jr. Geology of the Imperial Valley Region, California. *Geol. South. Calif. Calif. Div. Mines Bull.* **1954**, *170*, 21–28. Available online: https://ngmdb.usgs.gov/Prodesc/proddesc_73371.htm (accessed on 20 April 2017).
94. CONAGUA. Actualizacion De La Disponibilidad Media Anual De Agua En El Acuifero Arroyo Sahuaro (2604), Estado De Sonora. Mexico, DF: Subdireccion General Tecnica, CONAGUA. 2018. Available online: https://sigagis.conagua.gob.mx/gas1/Edos_Acuiferos_18/sonora/DR_2604.pdf (accessed on 17 February 2021).

95. CONAGUA. Actualizacion De La Disponibilidad Media Anual De Agua En El Acuifero Arroyo Seco (2607), Estado De Sonora. Mexico, DF: Subdireccion General Tecnica, CONAGUA. 2018. Available online: https://sigagis.conagua.gob.mx/gas1/Edos_Acuiferos_18/sonora/DR_2607.pdf (accessed on 17 February 2021).
96. CONAGUA. Actualizacion De La Disponibilidad Media Anual De Agua En El Acuifero Laguna Salada (0209), Estado De Baja California. Mexico, DF: Subdireccion General Tecnica, CONAGUA. 2018. Available online: https://www.gob.mx/cms/uploads/attachment/file/103406/DR_0209.pdf (accessed on 23 February 2021).
97. CONAGUA. Actualizacion De La Disponibilidad Media Anual De Agua En El Acuifero Sonoyta-Puerto Penasco (2603), Estado De Sonora. Mexico, DF: Subdireccion General Tecnica, CONAGUA. 2018. Available online: https://sigagis.conagua.gob.mx/gas1/Edos_Acuiferos_18/sonora/DR_2603.pdf (accessed on 19 February 2021). Estado De Sonora.
98. White, N.D.; Matlock, W.G.; Schwalen, H.C. An Appraisal of the Groundwater Resources of Avra and Altar Valleys, Pima County, Arizona. In *Arizona State Land Department Water-Resources Report*; Arizona State Land Department: Phoenix, AZ, USA, 1966.
99. Anderson, T.H.; Nourse, J.A.; McKee, J.W.; Steiner, M.B. Pull-Apart Basins at Releasing Bends of the Sinistral Late Jurassic Mojave-Sonora Fault System. *Spec. Pap.-Geol. Soc. Am.* **2005**, *393*, 97.
100. Longoria, J.F.; Jimenez, O.H. Sir-a Imagery in Geologic Studies of the Sierra Madre Oriental, Northeastern Mexico. Part 1 (Regional Stratigraphy): The Use of Morphostratigraphic Units in Remote Sensing Mapping. 1985. Available online: https://ntrs.nasa.gov/api/citations/19850011187/downloads/19850011187.pdf (accessed on 11 March 2021).
101. IGRAC. Transboundary Aquifers of the World. Scale 1:50.000.000. Available online: https://ggis.un-igrac.org/view/tba (accessed on 2 April 2021).
102. Magistrale, H.; Rockwell, T. The Central and Southern Elsinore Fault Zone, Southern California. *Bull. Seismol. Soc. Am.* **1996**, *86*, 1793–1803.
103. CONAGUA. Actualizacion De La Disponibilidad Media Anual De Agua En El Acuifero La Rumorosa—Tecate (0236), Estado De Baja California. Mexico, DF: Subdireccion General Tecnica, CONAGUA. 2018. Available online: https://sigagis.conagua.gob.mx/gas1/Edos_Acuiferos_18/BajaCalifornia/DR_0236.pdf (accessed on 11 March 2021).
104. Moyle, W.R., Jr. *Geohydrologic Maps of Southern California*; US Geological Survey: Reston, VS, USA, 1973. [CrossRef]
105. UNESCO/OEA/ISARM AMERICAS. *Sistemas Acuíferos Transfronterizos En Las Américas. Evaluación Preliminar*; UNESCO: Montevideo, Uruguay; Washington, DC, USA, 2015.
106. CONAGUA. Actualizacion De La Disponibilidad Media Anual De Agua En El Acuifero Valle De San Luis Rio Colorado (2601), Estado De Sonora. Mexico, DF: Subdireccion General Tecnica, CONAGUA. 2018. Available online: https://sigagis.conagua.gob.mx/gas1/Edos_Acuiferos_18/sonora/DR_2601.pdf (accessed on 9 March 2021).
107. Haxel, G.B.; Tosdal, R.M.; May, D.J.; Wright, J.E. Latest Cretaceous and Early Tertiary Orogenesis in South-Central Arizona: Thrust Faulting, Regional Metamorphism, and Granitic Plutonism. *Geol. Soc. Am. Bull.* **1984**, *95*, 631–653. [CrossRef]
108. Mason, D.; Bota, L. A Regional Groundwater Flow Model of the Tucson Active Management Area. 2006; Volume 159. Available online: https://new.azwater.gov/hydrology/model-data/regional-groundwater-flow-model-tucson-active-management-area-arizona-model (accessed on 29 March 2021).
109. Callegary, J.; Megdal, S.B.; Scott, C.A.; Vandervoet, P.L. *Five-Year Interim Report of the United States-Mexico Transboundary Aquifer Assessment Program: 2007–2012*; United States Geological Survey (USGS): Reston, VA, USA, 2013. [CrossRef]
110. Nassereddin, M.T. Hydrogeological Analysis of Groundwater Flow in Sonoita Creek Basin, Santa Cruz County, Arizona. 1967. Available online: https://citeseerx.ist.psu.edu/viewdoc/download?doi=10.1.1.882.9537&rep=rep1&type=pdf (accessed on 19 March 2021).
111. Bradbeer, G.E. Hydrogeologic Evaluation of the Sonoita Creek Aquifer. 1978. Available online: https://repository.arizona.edu/handle/10150/191669?show=full (accessed on 30 August 2021).
112. ADWR. *Upper San Pedro Basin*; Arizona Department of Water Resources: Phoenix, AZ, USA, 2009.
113. ADWR. *Santa Cruz Ama*; Arizona Department of Water Resources: Phoenix, AZ, USA, 2009.
114. Hawley, J.W.; Hibbs, B.J.; Kennedy, J.F.; Creel, B.J.; Remmenga, M.D.; Johnson, M.; Lee, M.M.; Dinterman, P. *Trans-International Boundary Aquifers in Southwestern New Mexico*; Rep. X-996350-01; New Mexico Water Resources Research Institute: Las Cruces, NM, USA, 2000; Volume 3.
115. Baldridge, W.S.; Olsen, K.H. The Rio Grande Rift: What Happens When the Earth's Lithosphere Is Pulled Apart? *Am. Sci.* **1989**, *77*, 240–247. Available online: http://www.jstor.org/stable/27855725 (accessed on 14 May 2021).
116. International Boundary Water Commission (Ed.) *Minute 242, Permanent and Definitive Solution to the International Problem of the Salinity of the Colorado River*; 1973. Available online: https://www.ibwc.gov/Files/Minutes/Min242.pdf (accessed on 25 April 2021).

water

MDPI

Article

Current Status and Future Directions in Modeling a Transboundary Aquifer: A Case Study of Hueco Bolson

Rocky Talchabhadel [1,*], Helene McMillan [1], Santosh S. Palmate [1], Rosario Sanchez [2], Zhuping Sheng [1,†] and Saurav Kumar [1,*]

[1] Texas A&M AgriLife Research, Texas A&M University, El Paso, TX 79927, USA; helene.mcmillan@ag.tamu.edu (H.M.); santosh.palmate@ag.tamu.edu (S.S.P.); zsheng@yahoo.com (Z.S.)
[2] Texas Water Resources Institute, Texas A&M University, 1500 Research Parkway, MS 2260, College Station, TX 77845, USA; Rosario@tamu.edu
* Correspondence: rocky.talchabhadel@ag.tamu.edu (R.T.); saurav@tamu.edu (S.K.)
† Retired.

Abstract: The Hueco Bolson aquifer is a binational aquifer shared by the United States of America (USA) and Mexico that is strongly interconnected with the transboundary river, Rio Grande/Rio Bravo. Limited recharge, increasing urbanization, and intensified agriculture have resulted in the over-drafting of groundwater resources and stressed the aquifer, threatening its sustainability if mitigation actions are not taken soon. Research indicates that the aquifer's hydraulic gradients and flow directions have changed due to the high groundwater withdrawal rates from the two major cities—El Paso (USA) and Ciudad Juarez (Mexico). This paper presents a comprehensive overview of the Hueco Bolson aquifer modeling history and makes a case for future modeling and binational engagement efforts. First, we discuss the evolution of groundwater modeling for Hueco Bolson from the past to recent times. Second, we discuss the main water management issues in the area, including water quality and quantity, stakeholders' participation, and climate change. To address the challenges of holistic water management, we propose developing a graphical quantitative modeling framework (e.g., system model and Bayesian belief network) to include experts' opinions and enhance stakeholders' participation in the model. Though the insights are based on a case study of Hueco Bolson, the approaches discussed in this study can provide new strategies to overcome the challenges of managing a transboundary aquifer.

Keywords: groundwater; Hueco Bolson; MODFLOW; Rio Grande; transboundary aquifer

Citation: Talchabhadel, R.; McMillan, H.; Palmate, S.S.; Sanchez, R.; Sheng, Z.; Kumar, S. Current Status and Future Directions in Modeling a Transboundary Aquifer: A Case Study of Hueco Bolson. *Water* **2021**, *13*, 3178. https://doi.org/10.3390/w13223178

Academic Editors: Sharon B. Megdal and Anne-Marie Matherne

Received: 1 July 2021
Accepted: 4 November 2021
Published: 10 November 2021

1. Introduction and History of Modeling Efforts

An aquifer is considered transboundary if it is part of a "system of surface water and groundwater situated in more than one nation" [1]. This applies when (i) an internal groundwater body is hydraulically connected to a transboundary river and (ii) a domestic river is interlinked to a transboundary aquifer [2]. For the Rio Grande–Hueco Bolson (Rio Bravo–Valle de Juarez) water system (Figure 1a), both are the case. The USA–Mexico international border follows the Rio Grande river course above the Hueco Bolson aquifer, and the internal aquifer stretches into the USA and Mexico. The Hueco Bolson lies within the state-triangle of New Mexico (USA), Texas (USA), and Chihuahua (Mexico) and is recognized as a transboundary aquifer by the USA and Mexico [3,4]. It has been declared a priority aquifer under the Transboundary Aquifer Assessment Program (TAAP) [5] of the USA and Mexico, highlighting the importance of Hueco Bolson for the transboundary Paso del Norte Region with its sister cities of El Paso (Texas) and Ciudad Juarez (Chihuahua).

The first hydrogeological studies of the Hueco Bolson were carried out by Slicher [6] and Richardson [7] in the early 1900s when significant pumping of the Hueco Bolson started. The first pump field, the Old Mesa field, was developed in 1904 [8]. Water withdrawal steadily increased until the 1950s, which triggered comprehensive studies about the geology, hydrogeology, and the groundwater resources of the El Paso area by the United States Geological Survey (USGS) [9,10], and has since accelerated due to the rapid population growth of the Paso del Norte Region [8]. In 2020, the combined sister cities had a population of approximately 2.2 million people, i.e., Ciudad Juarez with 1.51 million [11] and El Paso with 0.68 million [12]. Large water consumers of the region entirely rely on the Hueco Bolson as their drinking water supply, including the U.S. Army Air Defense Artillery Center Fort Bliss and several small communities [13]. Ciudad Juarez uses the Hueco Bolson as its main source for potable water [4], and El Paso Water (EPW, used to be called El Paso Water Utilities) relies on groundwater for approximately 40% of its total water supply in an average non-drought year—and this number is higher in drought years [14]. In addition, agriculture in the Paso del Norte Region depends on irrigation. In non-drought years, enough water is released from the Elephant Butte Reservoir (Figure 1a) into the downstream Rio Grande, and irrigation water demands can mainly be covered by surface water. However, groundwater must be pumped to meet the irrigation water demand in dry years [15].

White [8] estimated that approximately 40 million m^3 of water per year is recharged into the Rio Grande alluvium overlying the Hueco Bolson aquifer, which is far less than the withdrawal [16]. As the aquifer only receives such limited recharge in terms of rainfall, seepage or/and artificial deep-well injection and infiltration basins [13,17,18], the long-term mining of the Hueco Bolson resulted in a significant groundwater drawdown and decreased the water quality at some locations. The large water-level declines changed the groundwater flow directions since the 1960s [8,19] and caused flow intrusion from surrounding brackish water into the freshwater zone [13,20]. Proactive management strategies, mainly carried out by the EPW, include reducing per capita water use [21] and artificial aquifer recharge [18] to reduce freshwater pumping and slow down brackish water intrusion. In recent years, several projects (Table 1) have been carried out and contributed to a better understanding of the current situation of the Hueco Bolson.

Table 1. List of major groundwater models of the Hueco Bolson.

Key Findings	Study	Year Published
Saline water resources in Hueco Bolson using a two-dimensional electric-analog model for 1903–1963	Leggat and Davis [22]	1967
A two-layered digital model of the Hueco Bolson from 1903 to 1973 using a computer program developed by Bredehoeft and Pinder [23]	Meyer [24]	1976
Report 3—Hydrogeology of the Hueco Basin: Prepared for thePublic Services Board, City of El Paso, Texas	Lee Wilson and Associates [25]	1985
Summary of U.S. Geological Survey ground-water-flow models of basin-fill aquifers in the southwest alluvial basins' region, Colorado, New Mexico, and Texas	Kernodle [26]	1992
Simulation of groundwater and saline water in Hueco Bolson aquifer using the modular model developed by McDonald and Harbaugh [27] and solute transport three-dimensional flow model developed by Kipp [28]	Groschen [29]	1994
Groundwater model using a modified version of MODFLOW 96 developed by Harbaugh and McDonald [30]; the model was simulated from 1903 to 1996	Heywood and Yager [31]	2003
The MODFLOW model was updated to include input data from 1997 to 2002	Hutchison [32]	2004
Groundwater Flow for Administration and Management in the Lower Rio Grande Basin. Main Report; Technical Report prepared for the State of New Mexico	Papadopoulos and Associates [33]	2007
Updated model for Hueco Bolson aquifer using MODFLOW-2005 and MT3DMS solute transport code	Hutchison [34,35]	2016

Eastoe et al. [36,37] showed that surface water from the Rio Grande infiltrated far deeper into the groundwater body before the Elephant Butte Dam was constructed than after its construction in 1916. The groundwater chemistry and isotope study of Anderholm and Heywood [38] indicated that the infiltration of precipitation and the runoff from the Franklin Mountains is the main source of groundwater along the ridge, and dilute recharge water mixes with sodium chloride brine as groundwater moves away from the recharge area. The location of fresh and brackish groundwater seems to be controlled by stratigraphic and structural changes in the El Paso area of the Hueco Bolson [39]. In some parts of the basin, the deeper-lying saline groundwater [32] can upwell through fractures due to fault step-overs [39]. Effects of the surrounding land-use activities, i.e., agriculture and wastewater treatment systems, are reflected in the lower groundwater quality near the Rio Grande than farther away from the river [40].

The first study that conducted groundwater modeling across the Hueco Bolson was by Leggat and Davis [22] in 1967, using an electric-analog approach (refer to Table 1 for some major studies). Groundwater modeling became more user-friendly after USGS published MODFLOW-96 with several graphical interfaces. The latest model, to our knowledge, was commissioned by EPW [34], which is a groundwater model developed with MODFLOW-2005 which concentrated on developing a conjunctive use strategy for surface water and groundwater supplies and locating the production wells for the Kay Bailey Hutchinson Desalination Plant. The model also includes a chloride transport model using the MT3DMS (Modular 3-D Multi-Species Transport model) solute transport code [34].

The management of an aquifer, particularly a transboundary aquifer, is a social enterprise that needs cooperation among stakeholders [41,42]. The numerical models developed should enable such cooperation by providing robust unbiased results and promoting education and intuition building for all stakeholders whose partnership is sought. We propose that quantitative graphical models such as the Bayesian belief network (BBN) and system modeling may be necessary in addition to the numerically intensive physically based models.

2. Physical Setting, Data, and Numerical Modeling

The Hueco Bolson is located in the border triangle of Far West Texas, southern New Mexico, and northern Chihuahua (Figure 1a). It covers approximately 6500 km^2 with approximately 2/3 lying in the USA and 1/3 lying in Mexico, the Hueco Mountain range in the east, and the Sierra Juarez Mountain range in the south. In the North, the Hueco Bolson borders the Tularosa Basin (Figure 1a). This boundary between the two aquifers is not geological or hydrogeological, as they are hydraulically connected but have already been divided by Richardson [7]. This study only covers the heavily pumped Hueco Bolson. For modeling purposes, the hydraulic connection between Hueco Bolson and the Tularosa Basin will be represented as inflow at the northern boundary.

The climate in the Chihuahuan Desert is arid with an average annual precipitation of 253 mm and potential evapotranspiration of 1773 mm based on Climate Research Unit (CRU) gridded time series from the recent 30 years [43]. The Rio Grande flow depends on mountain snowpack runoff, upstream water diversions, reservoir releases, as well as agricultural and urban return flows. The river flows below Caballo Reservoir are managed by releasing the water from two reservoirs (i.e., Elephant Butte and Caballo). Most of the precipitation occurs as rainfall during the monsoon season from June to September. Generally, the sparse rainfall over the basin floor outside the Rio Grande Valley evaporates or transpires from the vadose zone before it can infiltrate water-table depths and recharge the aquifer system. Applied irrigation water has a better chance of infiltrating the water table because the water table is within several meters of the land surface in the Rio Grande Valley and flood irrigation is dominant in the region [31]. Pecans and cotton are the dominant crops, and gravel, fine sandy loam, and loamy fine sands are predominant soil types [34].

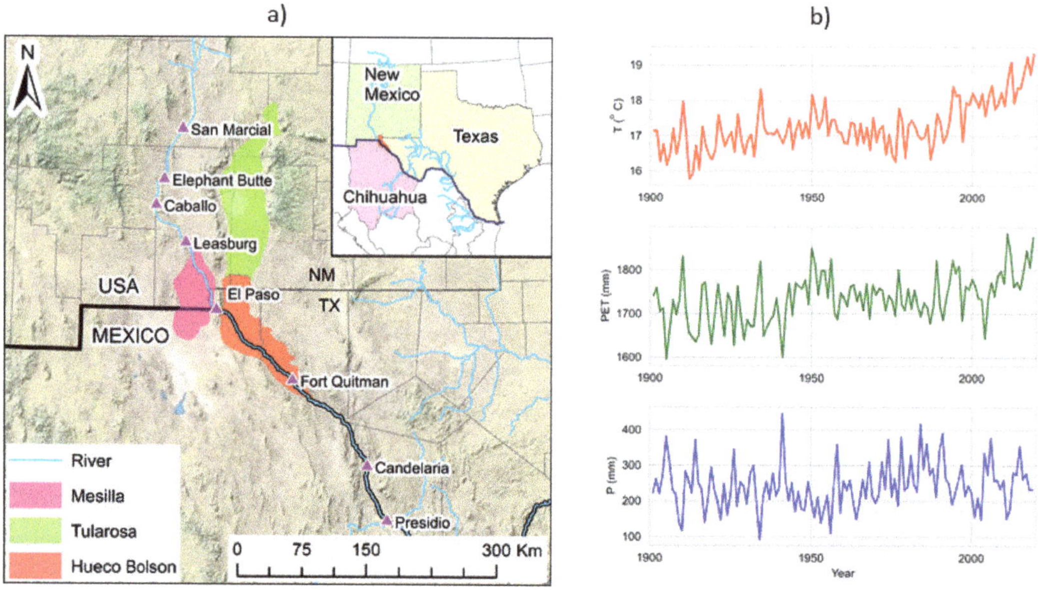

Figure 1. (**a**) Map showing Hueco Bolson and the adjacent Mesilla and Tularosa aquifers overlaid over administrative borders and the river channel. Triangles are streamflow gauging stations. Elephant Butte and Caballo are two reservoirs controlling the release of streamflow. Inset shows the Hueco Bolson aquifer (red), located in two states of the USA states (Texas and New Mexico) and the state of Chihuahua, Mexico; and (**b**) the inter-annual variation in average temperature (T), potential evapotranspiration (PET), and precipitation (P) from 1901 to 2019 across the Hueco Bolson [43].

2.1. Geology and Hydrogeology

The Hueco Bolson follows the structural depression associated with the Rio Grande Rift. The northern part is dominated by north striking faults, whereas the southern part, mainly located in Mexico, is characterized by northwest-striking faults [39]. The main aquifer is of unconfined and semiconfined nature and consists of basin-fill deposits up to 2500 m thickness [10]. It is commonly divided into hydrogeologic unit—alluvial facies, alluvial–fan facies, lacustrine–playa facies, and recent alluvial facies [31]. The deeper layers are Tertiary age deposits, whereas the recent alluvial facies are considered Quaternary and along the Rio Grande [31]. The freshwater body of Hueco Bolson is surrounded by naturally occurring brackish and saline groundwaters (>1000 TDS) [17]. Freshwater is mainly present on the west side of the Hueco Bolson at a depth of 320 m and more [44]. Towards the east, salinity levels increase, and the freshwater layer thins to less than 30 m [19]. Groundwater chemistry analysis by EPW showed a salinity increase with depth, and TDS values were measured as high as 35,000 mg/L [17].

Groundwater Observations

Figure 2a shows a monthly variation of groundwater pumping from the Hueco Bolson in the USA and Mexico regions from 1969 to 2013. The figure shows a substantial month-to-month variation. There is a clear indication of a sharp rise in groundwater pumping in Mexico until the year 2000 (from 2 million m^3 per month in 1969 to approximately 12 million m^3 per month in 2000). In contrast, the temporal evolution of groundwater pumping in the USA shows a less obvious trend. However, a mild rising trend of groundwater pumping in the USA from 1969 to 1990 (from approximately 7 million m^3 per month in 1969 to approximately 8.5 million m^3 per month in 1990) is observed, followed by a sharp declining trend until 2000. The non-parametric rank-based Mann–Kendall method [45,46], in conjunction with the Theil–Sen slope method [47,48], showed a statistically significant declining trend at a significance level of 99% for all months (January to December) in Mexico from 1969 to 2013. In contrast, there is a mixed pattern of increasing, decreasing, and no trends (statistically significant and non-significant) in the USA. We computed the trend magnitude and significance using the pyMannKendall package [49] following Yue and Wang's [50] pre-whitening method to limit the serial correlation contaminated in the observation time series [51]. We observe that the average monthly rise in groundwater pumping in Mexico is approximately 3 million m^3 per month per decade, which varies from approximately 2.2 million m^3 per month per decade in February to approximately 3.5 million m^3 per month per decade in July. In comparison, there is no clear tendency in the USA. Additionally, the temporal fluctuation is low compared to Mexico. However, with a noticeable oscillation, there has been an increasing trend of groundwater pumping since 2009 in the USA.

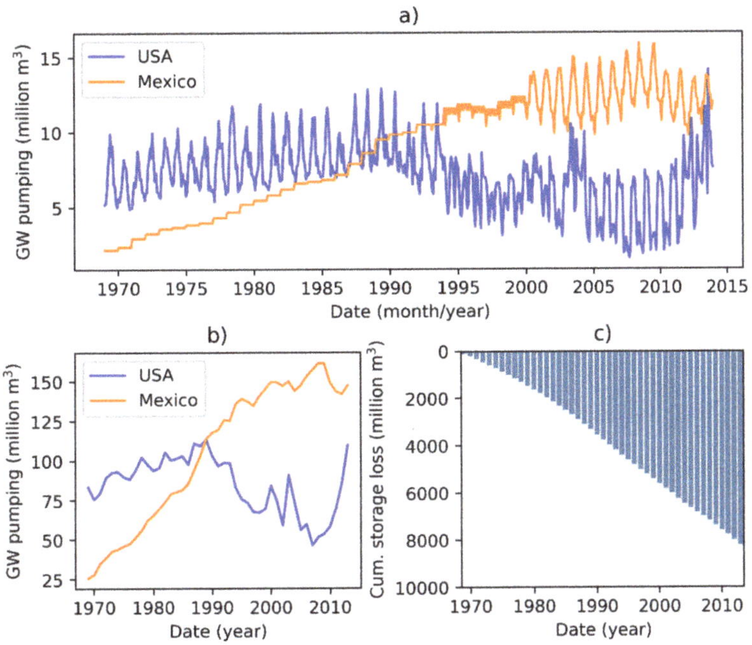

Figure 2. (**a**) Temporal variation in monthly total groundwater (GW) pumping in USA (average value of 346 wells) and Mexico (average value of 266 wells); (**b**) yearly variation of GW pumping in USA and Mexico. Daily data from these wells were used to compute monthly and annual total GW pumping; and (**c**) cumulative storage loss from the aquifer with respect to the initial year taken in this study (1969). Note: bars are inverted here to represent storage loss.

Overall, the rising groundwater withdrawal raises a question on the sustainability of water availability of the Hueco Bolson aquifer. Figure 2b shows the temporal variation in annual total groundwater pumping in the USA and Mexico. The annual groundwater pumping in Mexico steadily increased from approximately 25 million m^3 in 1969 to approximately 150 million m^3 in 2013. Figure 2c informs the aquifer's cumulative storage loss with respect to the initial year taken here as 1969. These observations provide an insight into the stress of the aquifer. However, numerical modeling would explain the local and regional groundwater dynamics and provide a spatiotemporal distribution of the hydraulic heads. Therefore, we aimed to numerically simulate the historical groundwater dynamics and explore the possible stress under climate change and different groundwater pumping scenarios.

Similarly, based on the ground-based observed data from nine stations from 1998 to 2013, we found that the TDS values of river water were higher downstream during both irrigation and non-irrigation periods. For example, the average TDS value in Fort Quitman, located 120 km south of El Paso, was approximately 2000 mg/L—almost double the average for the El Paso area. Additionally, the average TDS values in the dry year (2003, annual precipitation = 135 mm) were over 10% higher than during the wet year (2007, annual precipitation = 360 mm). In addition to hydroclimatic factors, the soil salinity levels of upland soils are directly associated with soil texture, permeability, and irrigation. The salts in the soils are brought in either through irrigation or have a geological origin. The soil textures across the selected study area show that sand and cobble dominate near the El Paso gauge, whereas loam and silty loam dominate Fort Quitman.

In the early 20th century, before the impact of major groundwater pumping on both sides of the border occurred, the general groundwater flow was south from the Texas–New Mexico line and east to southeast from the Sierra de Juarez toward the valley of Rio Grande [8]. As a result, underflows across the political boundaries were from Mexico to the USA, enhanced by increased groundwater withdrawal from El Paso and an even larger hydraulic gradient from south to North [13]. However, the net flow has reversed since the 1960s due to increased pumping in Ciudad Juarez [19]. Groundwater level declines are sharp in downtown areas of El Paso and Ciudad Juarez. Cones of depression are formed at these pumping centers, which affect the flow patterns. Heavy pumping decreases hydraulic heads and induces the movement of surrounding saline water into the freshwater zone. TDS values of pumped groundwater have increased for more than 70 years and even necessitate well abandonment in the El Paso area [31]. Ashworth [52] suggested that the annual increase in salinity was approximately 10–30 mg/L per year from the 1950s to 1990, and in parts of downtown El Paso and Ciudad Juarez, an annual increase of 40 to 100 mg/L per year was observed. For the years after 1979, Sheng [19] observed an annual rise of 80 to 120 mg/L per year at some locations. This indirect reduction in the fresh groundwater resource especially occurred in the wellfields near the El Paso Airport and northeast of El Paso [13,44].

2.2. Numerical Modeling

Numerical modeling for the Hueco Bolson Aquifer has been largely based on the MODFLOW. MODFLOW solves the groundwater flow equation using linear and nonlinear numerical solution methods [53]. The first MODFLOW model was initially developed by Heywood and Yager [31] in MODFLOW 96 over the larger Rio Grande area, including the Hueco Bolson aquifer (Table 1). The Heywood and Yager Hueco Bolson aquifer model [31] consists of ten layers of 165 rows and 100 columns in a variable grid, with the finer grid in El Paso and Juarez area [54]. The model was calibrated with data from 1903 to 1996. It was later upgraded to MODFLOW 2000 and then to MODFLOW 2005 [55]. The packages included in the MODFLOW model are: well packages, multi aquifer well packages stream package, drain package, evapotranspiration package, recharge package, horizontal flow barrier, and flow and head boundary package [56]. EPW then updated the model to include input data from 1997 to 2002 [32]. We used the EPW model and developed the current MODFLOW model for Hueco Bolson.

2.2.1. Current MODFLOW Model

The latest model developed for the area of interest has a ten vertical-layers grid. The top nine layers are 30 m thick and the bottom layer is 276 m thick (Figure 3c,d). The first 30 m, i.e., the top layer, is set as alluvial deposits and the other remain as fluvial facies. The horizontal domain is represented with rectangular grids (Figure 3a). We simulated monthly water stress from 1969 to 2013. Groundwater levels measured at wells and pumping data were used to calibrate the model. The inflow boundary condition was set in New Mexico (shown in green in Figure 3b). Similarly, the outflows from the USA and Mexico were placed at the southeast corner (shown in blue and brown, respectively, in Figure 3b).

Accurate boundaries of aquifers are hard to establish, especially for the connected aquifers (e.g., Hueco–Tularosa). Although no clear physical boundary separates the Hueco Bolson from the Tularosa basin, the recognized Hueco Bolson aquifer (Figures 1a and 3a) does extend to New Mexico [31]. In the USA, groundwater in a shared aquifer is managed by each state [19].

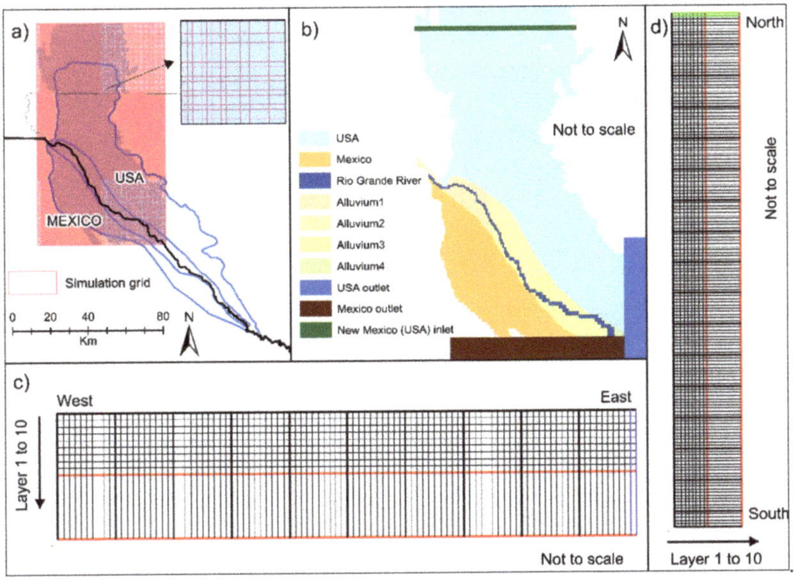

Figure 3. (**a**) Computational grid consisting of 165 rows and 100 columns. Inset shows a close location highlighting different grid sizes varying from 500 to 1170 m; (**b**) the location of inflow and outflow boundaries and other features; (**c**) West to East cross-section showing ten vertical layers; and (**d**) North to South cross-section showing ten vertical layers. Four alluvium areas were defined based on hydraulic characteristics along the river channel. The red color in (**c**,**d**) shows the computation grid's bottommost layer has a different vertical thickness compared to the other nine layers.

2.2.2. Updated MODFLOW Coupled with a Watershed Model

It is critical to analyze both the surface water and groundwater on either side of the border. We are developing a coupled watershed–groundwater model to simulate surface and subsurface hydrologic processes [57]. The Soil and Water Assessment Tool (SWAT) will be used as a watershed model that allows a better and fine resolution simulation of the surface water and land use activities. Figure 4a shows the subbasins (N = 69) of the SWAT model. The number of hydrologic response units (HRUs) in the SWAT model is 1243 (not shown in Figure). The water balance is maintained at the HRU level. The HRU is the smallest unit for the computation in the SWAT, which is delineated based on a unique combination of soil types, land use/crop type classes, and topography. Figure 4b shows the linking of HRUs with the MODFLOW grids and represents linking river cells and SWAT river reach.

Though the SWAT model has its groundwater components, the model itself is lumped [58]. Conversely, the MODFLOW model has challenges in computing the distributed groundwater recharge. The coupled model allows HRU-based groundwater recharge to the MODFLOW model, and grid-based outputs from MODFLOW are sent back to the SWAT model. Similarly, river–aquifer flow exchange and water transfer by pumping occur at the MODFLOW river and pumping cells with the SWAT river reach and HRUS. The exchange happens on a daily scale, and the coupled model is run simultaneously. Details on the coupling are available in Kim et al. [58]. The coupled model assesses flux exchange across the border and between different formations using the Zonebudget package [59] that computes the water budget [60]. The SWAT model was set up for the Middle Rio Grande River from San Marcial to Presidio. Detailed hydrologic analyses were performed for Rincon valley, and the details are available in Ahn et al. [61]. The coupled watershed–groundwater model is still under evaluation and will be made available as a separate technical publication.

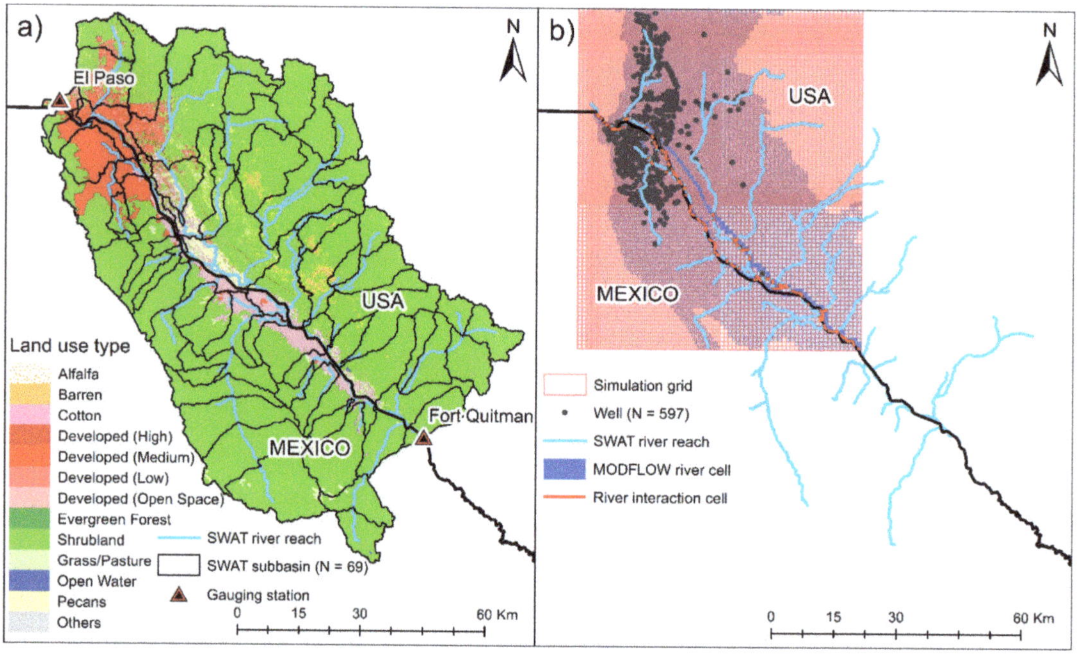

Figure 4. (**a**) Distribution of subbasins of Soil and Water Assessment Tool (SWAT) model superimposed over the land use type across the study area; (**b**) the location of wells, and river interactions between SWAT river reach and groundwater (MODFLOW) river cell used for the surface–subsurface linkage.

2.2.3. Simulation of Solute Transport (Salinity)

An elevated salinity level has significant environmental and economic impacts on nature as well as agricultural and urban water management. These include agricultural yield reduction, increased water use for salinity leaching, reduced life of water-using appliances and water delivery systems, and increased cost of water use, either by water softening system or dispensed water [62,63]. Under the current groundwater depletion scenario, the brackish groundwater will continue to intrude into fresh groundwater and will probably affect well locations more in the future. A few modeling works are available for the Hueco Bolson aquifer [34,35]. We plan to implement a solute transport model into the coupled watershed–groundwater model.

3. Water Management Efforts and Issues

In recognizing the continued depletion of fresh groundwater in the aquifer, El Paso and Ciudad Juarez have made efforts to address management issues of shared groundwater resources via cooperation [13]. EPW has reduced its pumping from Hueco Bolson since the mid-1990s, implemented a managed aquifer recharge program to inject reclaimed water, and constructed a desalination plant to utilize brackish groundwater [13,17,20]. Ciudad Juarez is considering an alternative source from La Mesilla Bolson/Conejos Medanos aquifer to reduce the stress on Hueco Bolson. The conjunctive management of Hueco Bolson and the Rio Grande is an important part of the water supply of El Paso as both contribute approximately 40% each to the potable water use of EPW in non-drought years [14]. In addition, EPW uses the Mesilla Bolson (17%) and desalination (5%) to supply the city with potable water [14]. To secure water supply, the EPW is planning increased use of reclaimed water, desalinated groundwater from more remote parts of the Hueco Bolson, and importing additional water sources from other basins [64]. On the other hand, Ciudad Juarez uses Hueco Bolson as its main source of portable water and uses the Mesilla Bolson/Conejos Medanos aquifer as a complementary source as water is imported via a 40 km aqueduct [4].

In this section, we highlight issues related to water quantity and quality, climate change impacts, governance and jurisdiction, and the prospect of integrated groundwater management (IGM).

3.1. Water Quantity and Quality

Based on the carbon dating of the groundwater samples, Anderholm and Heywood [38] found that groundwater is 12,000 years old near the New Mexico/Texas state line. The Hueco Bolson aquifer has been pumped for its ancient groundwater for over a century, resulting in groundwater decline [8]. The cross-border exchange of both the surface and subsurface water has complicated the understanding of water availability on both sides of the border. The groundwater flow between the USA and Mexico has changed over the past decades depending on which city caused stronger water withdrawal [13,19].

Water quality is a crucial factor that guides the usability of water. Due to the surrounding brackish water, groundwater quantity and quality issues of the Hueco Bolson are strongly interlinked. The rapid groundwater depletion due to heavy groundwater pumping is not only a concern by itself but has also resulted in brackish groundwater intrusion into freshwater sections of the aquifer. The water quality decrease in public wells was so low in the El Paso area that in some cases, well abandonment was required [31]. Due to the inter-basin flow of poor-quality groundwater, freshwater storage depletion will be even faster than that calculated from groundwater pumping only [19].

In 1979, the Texas Water Development Board (TWDB) estimated that the Hueco Bolson could not be used as a freshwater supply as the groundwater body will have slightly saline conditions from brackish groundwater intrusion by 2031 [65]. This scenario was fortuitously prevented as various measures were put in place, including water injection [65]. In addition, the EPW has used reclaimed wastewater to recharge the Hueco Bolson aquifer at its northeast wells since 1985 [18] and the infiltration basin after 2001 [19] to minimize the negative impacts of decreasing groundwater levels. However, to our knowledge, there are no similar operations on the Mexican side of the aquifer. Therefore, technical water management actions on both sides on the border must be implemented to reduce the negative effects of the scarce groundwater and create new quasi-stable conditions of the Hueco Bolson through a bilateral and participatory approach.

3.2. Impacts of Climate Change

It is ubiquitous that the entire globe is warming and is being more pronounced by the anthropogenic release of greenhouse gases (GHGs) [66]. Looking at the near-surface temperature of the CRU dataset from 1951 to 2019, we find that the study area already experienced an increase of 1 °C in annual average temperature between 1951–1990 and 1991–2019. The hydrologic and the water cycle are deeply affected by alterations in climate variables. Surface water and groundwater are affected in one way or another due to shifts in precipitation patterns, intensity, and temperature. Additionally, the Rio Grande receives snowmelt runoff in southern Colorado and northern New Mexico [19], which is anticipated to be affected by changing climate.

Generally, aquifers have high storage capacities and are less sensitive to climate change than surface water bodies. However, the decrease in mountain front recharge (currently very little in the Hueco Bolson) and the temperature rise impact the hydrologic cycle and water availability. In addition, prolonged drought may affect water quantity, deteriorate groundwater quality, and stop or reduce groundwater recharge. Therefore, understanding hydrologic processes in aquifer systems is critical to developing adaptive management strategies under changing climates.

3.3. Governance and Jurisdiction Considerations

The governance and jurisdiction of common pool groundwater resources are challenging [67]. The issues are compounded by the transboundary nature of the Hueco Bolson. In Texas, the Hueco Bolson is recognized as a major aquifer by the TWDB and one of the major water sources by the Far West Regional Water Planning Group (E), which makes water management recommendations to the TWDB. According to the New Mexico constitution, all water bodies belong to the public, and the State Engineer permits rights under the doctrine of prior appropriation (first in time, first in right) to use the water. Property rights and governance structures related to groundwater resources are principally different within New Mexico and Texas in the USA and the two countries. In Mexico, groundwater rights lie with the Mexican federal authorities. Groundwater governance is centralized under the federal water agency Comision Nacional del Agua (CONAGUA). For surface water and groundwater shared with the USA, there is a common assumption that the de facto authority is with the International Boundary and Water Commission—Mexico Section (CILA) [68].

Within the USA, groundwater rights are juridical issues of each state. In Texas, the landowner owns the groundwater, and the rule of capture, also known as the principle of "the biggest pump", governs the groundwater use [4].The dichotomy in regulations is highlighted by the Texas surface water and groundwater jurisdictions. Surface water is considered to be owned by the people of Texas and allocated for beneficial use according to the doctrine of prior appropriation. Groundwater, however, is governed by the "rule of capture", giving the landowner authority to use/pump as much water as needed without liability to neighbors for drying up their wells. The separate management of groundwater and surface water systems described as *hydroschizophrenia* is often a key hurdle in solving complicated common pool water system sharing problems [69]. *Hydroschizophrenia* is compounded by the transboundary nature of the water system [70]. In the case of Hueco, as other authors have suggested (e.g., Hargrove et al. [71]), it can be extended to include the myriad of different regulations that govern the connected surface water and groundwater in three jurisdictions (Texas, New Mexico, and Chihuahua).

There have been water-sharing agreements and treaties to address transboundary water sharing in the region. The Rio Grande Compact [72] is an interstate compact in the USA between the states of Colorado, New Mexico, and Texas, and approved by the United States Congress to equitably apportion the waters of the Rio Grande Basin. Similarly, the international treaties between the USA and Mexico (e.g., 1944 treaty [73]) deal with issues related to boundary water sharing. The 1944 treaty *"Utilization of Waters the Colorado and Tijuana Rivers and of the Rio Grande"* allocated waters in the international segment of the Rio Grande from Fort Quitman, Texas, to the Gulf of Mexico. This treaty also authorized the two countries to construct, operate and maintain dams on the main channel of the Rio Grande. A recent analysis of transboundary water delivery for compliance with the treaty has demonstrated a delivery regime that is not similar to what was anticipated at the time of signing, largely attributable to changing water use [74]. Uncertainties in the water deliveries, particularly large deficits in some years with low water availability, are a significant problem in water management.

According to Sanchez and Eckstein [42], small-scale, informal, and non-binding approaches seem more viable for future transboundary groundwater management between Mexico and Texas. A holistic approach that captures the opinions of growers, stakeholders, and experts is necessary to address associated socioecological issues. The transboundary water rights can be modeled in the system model by generating different water allocations scenarios. However, binational interests are difficult to define without experts' interpretation of casual chains [70,75].

3.4. Integrated Groundwater Management (IGM) for Hueco Bolson

IGM may be defined as the "structured process that promotes the coordinated management of groundwater and related resources (including conjunctive management with surface water), taking into account non-groundwater policy interactions, in order to achieve balanced economic, social welfare, and ecosystem outcomes over space and time" [76]. Given the transboundary nature, the problem space (*referring to issues/domains that must be engaged to solve the problem*) for an IGM strategy for Hueco includes several components shown in Figure 5. Any decision support tool and integrated modeling derived for IGM support connections between elements in the problem space. The issues related to water allocation, water scarcity, irrigation, water quality, water supply, diverse jurisdictions, and climate change have been discussed earlier in the paper. JEDI represents *justice, equity, diversity*, and *inclusion* in the IGM process [77]. The four components of JEDI applied to water, or environmental systems may be thought of as:

1. **Justice**—the right to an equitable, safe, healthy, productive, and sustainable environment for all community members;
2. **Equity**—impartiality and fairness in the procedures, processes, and allocation of resources;
3. **Diversity**—including a broad demographic mix (including race, age, gender, ethnicity, cultural background, geography) within a group or organization, which reflects the makeup of the community;
4. **Inclusion**—ability of diverse individuals to participate fully in all aspects, including the decision-making processes.

Given the stakeholder buy-in needed to manage the transboundary common pool resource, it is anticipated that aspects of JEDI, as defined above, become important to develop and successfully apply the management plan.

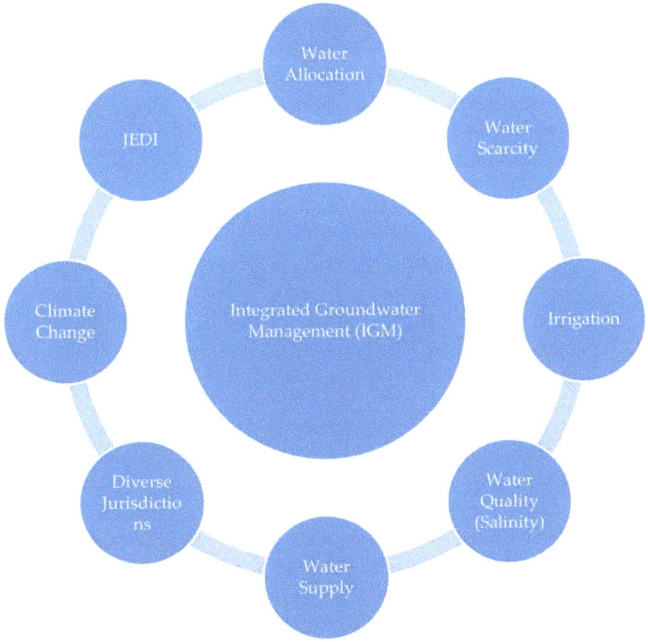

Figure 5. Problem space for integrated groundwater management (IGM). Modeling tools and other decision support tools should address the whole problem space. JEDI is used to represent justice, equity, diversity, and inclusion.

For Hueco Bolson modeling and planning, given the *hydroschizophrenia* (discussed above) and the inequities among the communities on two sides of the international border and within each jurisdiction (see McDonald and Grinesiki [78], Grineski and Collins [79], Grineski et al. [80], Moya et al. [81]), JEDI becomes an important consideration. Although environmental justice and equity are key policy objectives in the region [62] and worldwide [82–84], they are seldom considered in knowledge production using models. However, any sustainable and acceptable management solution will need all stakeholders' buy-in and knowledge co-production [85]. Furthermore, centering our science and communication framing around JEDI [77,86] can provide an essential point of access for all communities to engage with scientific communication, preventing critical gaps in stakeholder representation. However, such engagement at scale in different countries remains a challenge.

Nevertheless, the tools developed for IGM should be designed to allow JEDI engagement. From a model development and integration perspective, this design may include more intuitive and straightforward modeling tools that are easier to understand and allow interaction without any computing limitation. Different models in conjunction with existing numerical models, such as the system model, can provide a participatory modeling environment to bring together binational stakeholders and build a collaborative Hueco Bolson model for the transboundary region. Based on the stakeholders' opinions, the Hueco Bolson model can be developed by conceptualizing and quantitatively analyzing alternative management scenarios. The proposed approach aims to bring the scientific community, stakeholders, and decision makers together in developing equitable and inclusive IGM following open science, coordinated, and networked efforts [87,88]. We present the future direction in groundwater modeling in the subsequent section.

4. Future Outlook

Future work will require a continued collaboration of the two countries and researchers to tackle the various management issues of the transboundary aquifer. To develop an IGM supporting tool for Hueco Bolson, we are trying to develop a system-of-systems (SoS) models supported by the system model and BBN. These models, as planned, will utilize the existing calibrated model (SWAT and MODFLOW) as an "expert" and integrate opinions from other expert stakeholders. Additionally, new insights through additional data and groundwater simulations can be integrated or used as input parameters in the existing numerical model. In this section, we highlight future outlooks based on water data and decision modeling.

4.1. Protocol for Transboundary Data Sharing and Collection

The International Boundary & Water Commission (IBWC) oversees the application of USA–Mexico treaties related to boundary demarcation and the national ownership of water resources [5]. Transboundary Aquifer Assessment Program (TAAP) is a binational scientific effort that enables data sharing, harmonization, and knowledge improvement of the transboundary aquifers. Federal agencies, academic institutions, and research institutes analyze the common data to better understand the groundwater dynamics of transboundary aquifers. The TAAP project plans to develop an interactive and open-data portal, including numerically simulated outcomes. The distribution of input data currently used in existing models is not consistent with the JEDI principles. Due to the ease of water, land use, and other data access in the USA, the USA data dominate and bias the models. In the models, water system processes on the Mexican side are assumed to follow a similar trend as on the USA side (where data are available to calibrate the models), which is an untested and potentially incorrect assumption. Similarly, this data imbalance is also valid across other divides (e.g., urban-rural).

Under the TAAP program, a memorandum of understanding between CILA for Mexico and the Texas Water Research Institute (TWRI) for the USA has been brought on to exchange data between countries and fill data gaps in urban and rural areas in both countries. Additionally, we will estimate agricultural withdrawal by compiling and analyzing land use, land cover data, and crop acreage data to help overcome data gaps in the rural areas where data are sometimes scarce. We may also consider using some data for validation instead to balance the data distribution.

4.2. System Model

A system model explicitly represents the diverse set of connections within a system, the sensitivity of each link, dynamic mechanism, and feedback loops. The concept of system dynamics was first introduced by Forrester [89] to determine problems and evolve possible solutions by boosting our system thinking capacity to extrapolate and interpolate in a broader sense. The system model formalizes the main causal chains and complex mechanisms in a meaningful manner [90]. The benefit of the system model is that the relationship between the "cause" and "effect" can be easily inferred and simulated by stocks and flows processes. Additionally, it can deal with a high degree of nonlinear problems, which are commonly present in managed environmental systems such as an aquifer [91].

An aquifer is one of the most complex systems consisting of interacting dynamic variables and balancing feedback loops. Many studies in the literature have demonstrated the importance of a system model in managing aquifer systems and their connected subsystems [92]. Recently, Afshar et al. [93] integrated surface water and groundwater into a single cyclic storage system model to simulate a long-term outlook. Barati et al. [94] developed a smart groundwater governance system model introducing an index and four indicators, namely equitability, efficiency, sustainability, and democracy. Balali and Viaggi [95] employed a system model to identify limits to growth and additional risks of aquifer development. Niazi et al. [96] determined the recharge and discharge dynamics within the aquifer system based on the long-term aquifer responses to hydrological variability.

The Hueco Bolson aquifer system model (Figure 6) can act as a mental model that allows stakeholders from Mexico and the USA to agree on a common representation, which creates new insights and unifies the knowledge of different stakeholders, water practitioners, and researchers. Figure 6 shows the connectivity of the surface water system with the groundwater system in the Hueco Bolson. From both Mexico and the USA sides, surface water delivery and groundwater pumping are taking place to meet the regional water demand. Here, water availability depends on the surface water storage in the reservoirs upstream and the amount of water recharged into the Hueco Bolson aquifer. Additionally, lateral flows from the Tularosa basin of the USA add a significant amount of inflows to the Hueco Bolson aquifer system. Therefore, the interaction between the surface water recharge into the aquifer system and managed aquifer recharge by EPW [18,19] and the aquifer water pumping back to the surface for agricultural irrigation [18,19] as per the water demand indicates an important water-balancing feedback loop of the model. This loop could help understand the dynamic nature of the Hueco Bolson aquifer and the change in water fluxes over time.

The collaborative aquifer system model can further simulate the anticipated effects of land-use change, climate change, and human activities on regional water supply and demand and understand their dynamic relationships [90]. Additionally, the system model will help provide insights and alternative opportunities for allocating the aquifer water, policy support, and participatory strategic planning to mitigate future impacts [97–99]. Policy and decision-makers can utilize scenario-based analysis to obtain a set of optimal solutions and trade-offs between transboundary aquifers [100]. The benefits of the system model include the management of the binational aquifer, exploitation of uncontaminated groundwater, control of groundwater fluctuations, conjunctive use of groundwater [101–103], and securing the future water [104]. Although only the water-system model is shown in Figure 6, one can conceive a system-of-systems model that informs decision-makers on the interaction among environment (water and land use), social (population growth rate and anthropogenic activities), economic (water demand, water productivity, and wastewater discharge), and political (policy, decision, and management framework) system.

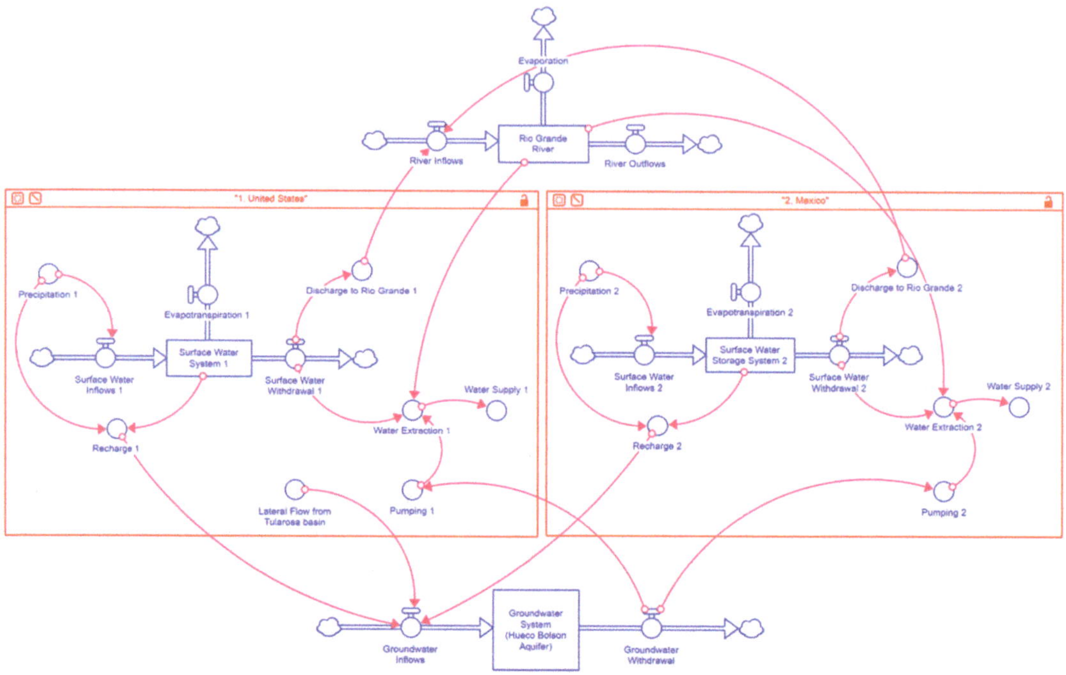

Figure 6. System model concentrating on the water system for a transboundary aquifer—Hueco Bolson.

4.3. Bayesian Belief Network (BBN)

The BBN is a graphical model that shows probabilistic relationships between different quantitative and qualitative variables. The BBN can predict and analyze associational relationships, i.e., cause and effect, even if some data entries are missing or/and the precise relationship between variables is unknown [105]. The formalism also allows using a wide range of algorithms to learn the network from data or/and use experts' knowledge for some prior associations. Ground-based and remotely sensed data, numerically simulated outputs, and stakeholders' opinions can be used to generate and learn the model. Integrating knowledge from these different domains makes BBNs a useful tool for problems with a high level of uncertainty and complexity in water management issues [106]. It has been shown that BBNs are well suited as planning tools to incorporate the system view of various stakeholder groups next to numeric data [107]. Furthermore, involving the different expert groups can help overcome the problem of incomplete data and to link variables [108]. Finally, the BBN allows assessing the alteration in conditional probabilities upon the perturbations on any factors.

We aimed to develop a BBN model to understand the surface and sub-surface water dynamics in the Hueco Bolson transboundary aquifer system. Figure 7 shows ongoing efforts to develop the BBN for Hueco Bolson. Different nodes are from various thematic groups, including legislative framework, agriculture, weather and climate, demography, and the economy are acyclically connected. Conditional probabilities for different combinations will be input into the model based on earth observations, ground-based observations, modeled results, experts' knowledge, stakeholders' participation, and legal frameworks. For instance, the land-use transition can be quantified using satellite images and annual crop data layers at high spatial resolution. The legal framework will be based on a county to the state level and will be different for the USA and Mexico. Similarly, demographic data could range from the household, county, etc., to other administrative boundaries. Farmers' input can be provided on a farm field level. In BBN modeling, data from several sources will first be transformed into conditional probability based on observations.

Overall, the BBN model will likely consist of a range of variables, including hydroclimate, groundwater, crop types, profit outlook, land use transition, grower risk tolerance, consumption per capita, industrial performance, and population growth. Expert knowledge and data-driven approaches, i.e., different statistical algorithms, are used to learn the relationship between the variables. Like the system model discussed earlier, this model will allow intuition building among stakeholders. However, the key difference is the deeper and more direct integration of expert opinion with BBNs. Thus, these SoS modeling approaches will be helpful to capture the opinion and knowledge from different stakeholders, including growers, academicians, planners, and decision-makers. Incorporating experts' knowledge will help build trust in the modeling system. Furthermore, from a technical perspective, it will help overcome some of the anticipated incomplete data problems to drive more physically based models [108].

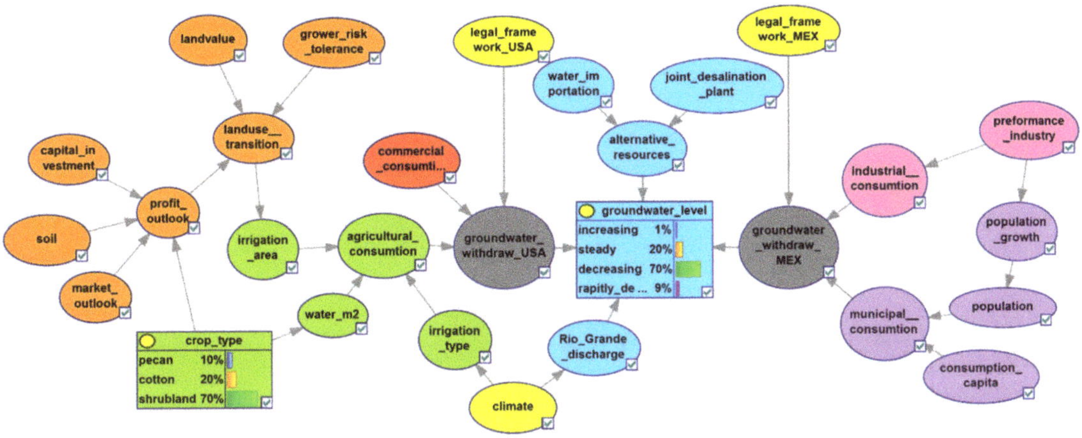

Figure 7. Simplified Bayesian belief network for a transboundary aquifer. Each variable is represented by one node and has an accompanying conditional probability table.

5. Conclusions

This paper highlights the current status and future works in the water system management for the transboundary Hueco Bolson aquifer between the USA and Mexico. The challenges involved in managing a transboundary aquifer are of a technical, social, political, and juridical nature, including groundwater withdrawal, brackish groundwater intrusion, multi-discipline multi-stakeholders participation, centralized or/and decentralized governance structures, different jurisdictions. Therefore, around the core of groundwater science, it is important to integrate other disciplines. The first step for Hueco Bolson has been to establish a recognition of both countries and all counties that share the groundwater body. The next step (ongoing) is to cooperate for transboundary data sharing and use the data to modify and improve the models. Finally, it is critical that models are developed in support of IGM with JEDI principles. Data for modeling groundwater not only means the traditionally recorded observations but also includes remotely sensed data, stakeholder engagement, and expert knowledge. These data can help solve the problems and challenges related to the aquifer using an integrated SoS approach along with physically based numerical simulations.

Author Contributions: Conceptualization, R.T. and S.K.; methodology, R.T.; software, R.T., H.M. and S.S.P.; validation, R.T., S.K., R.S. and Z.S.; formal analysis, R.T., H.M. and S.S.P.; investigation, R.T. and H.M.; resources, R.S. and S.K.; data curation, R.T. and H.M.; writing—original draft preparation, R.T. and H.M.; writing—review and editing, R.T., H.M., S.S.P., R.S. and Z.S.; visualization, R.T.; supervision, S.K. and R.S.; project administration, R.S. and S.K.; funding acquisition, R.S. and S.K. All authors have read and agreed to the published version of the manuscript.

Funding: This research was funded by the United States Geological Survey (USGS) under the Transboundary Aquifer Assessment Project (TAAP): USGS Project No. G17AC00440 and United States Department of Agriculture (USDA) NIFA Hatch project 1022633.

Institutional Review Board Statement: Not applicable.

Informed Consent Statement: Not applicable.

Data Availability Statement: Data used in the study will be provided on request.

Acknowledgments: The authors are grateful to the United States Geological Survey (USGS) and the United States Department of Agriculture (USDA) for financial support.

Conflicts of Interest: The authors declare no conflict of interest.

References

1. Fitzmaurice, M. Convention on Law of the Non-Navigational Uses of International Watercourses. *Leiden J. Int. Law* **1997**, *10*, 501–508. [CrossRef]
2. Eckstein, G. Implications of the UN Watercourses Convention for Groundwater Resources. International Water Law Project Blog. 2014. Available online: https://www.internationalwaterlaw.org/blog/2014/08/05/professor-gabriel-eckstein-implications-of-the-un-watercourses-convention-for-groundwater-resources/ (accessed on 13 April 2021).
3. Sanchez, R.; Lopez, V.; Eckstein, G. Identifying and Characterizing Transboundary Aquifers along the Mexico–US Border: An Initial Assessment. *J. Hydrol.* **2016**, *535*, 101–119. [CrossRef]
4. Sanchez, R.; Eckstein, G. Aquifers Shared Between Mexico and the United States: Management Perspectives and Their Transboundary Nature. *Groundwater* **2017**, *55*, 495–505. [CrossRef] [PubMed]
5. Tapia-Villaseñor, E.M.; Megdal, S.B. The US-Mexico Transboundary Aquifer Assessment Program as a Model for Transborder Groundwater Collaboration. *Water* **2021**, *13*, 530. [CrossRef]
6. Slichter, C.S. *Observations on the Ground Waters of the Rio Grande Valley*; Water-Supply Paper 141; US Geological Suvey: Reston, VA, USA, 1905; p. 93.
7. Richardson, G.B. *El Paso Folio, Texas*; Atlas of the United States No. 166; US Geological Survey: Reston, VA, USA, 1909.
8. White, D.E. *Summary of Hydrologic Information in the El Paso, Texas, Area, with Emphasis on Ground-Water Studies, 1908–1980*; Texas Water Development Board: Austin, TX, USA, 1987; p. 75.
9. Sayre, A.N.; Livingston, P.P. *Ground-Water Resources of the El Paso Area, Texas*; Water Supply Paper; US Government Printing Office: Washington, DC, USA, 1945; Volume 919.
10. Knowles, D.B.; Kennedy, R.A. *Ground-Water Resources of the Hueco Bolson, Northeast of El Paso, Texas*; US Government Printing Office: Washington, DC, USA, 1958.
11. INEGI Censo Población y Vivienda. 2020. Available online: https://www.inegi.org.mx/programas/ccpv/2020/default.html#Tabulados (accessed on 19 April 2021).
12. United States Census Bureau. US Census Bureau QuickFacts: El Paso City, Texas. Available online: https://www.census.gov/quickfacts/elpasocitytexas (accessed on 19 April 2021).
13. Sheng, Z.; Devere, J. Understanding and Managing the Stressed Mexico-USA Transboundary Hueco Bolson Aquifer in the El Paso Del Norte Region as a Complex System. *Hydrogeol. J.* **2005**, *13*, 813–825. [CrossRef]
14. El Paso Water Utilities Water Resources. Available online: https://www.epwater.org/our_water/water_resources (accessed on 21 April 2021).
15. Sheng, Z.; Michelsen, A.; Fahy, M.P.; Ashworth, J. Conflicts and Cooperation: Water Resources Planning in Far West Texas. In Proceedings of the World Water & Environmental Resources Congress 2003, Philadelphia, PA, USA, 23–26 June 2003; American Society of Civil Engineers: Philadelphia, PA, USA; pp. 1–10.
16. Zhuping, S.; Mace, R.E.; Fahy, M.P. The Hueco Bolson: An Aquifer at the Crossroads. In *Aquifers of West Texas*; Texas Water Development Board: Austin, TX, USA, 2001.
17. Hibbs, B.J.; Boghici, R.N.; Hayes, M.E.; Ashworth, J.B.; Hanson, A.T.; Samani, Z.A.; Kennedy, J.F.; Creel, B.J. *Transboundary Aquifers of the El Paso/Ciudad Juarez/Las Cruces Region—Chapter 4 Southeastern Hueco Aquifer*; Texas Water Development Board and New Mexico Water Resources Research Institute Report; US Environmental Protection Agency: Dallas, TX, USA, 1997.
18. Sheng, Z. An Aquifer Storage and Recovery System with Reclaimed Wastewater to Preserve Native Groundwater Resources in El Paso, Texas. *J. Environ. Manag.* **2005**, *75*, 367–377. [CrossRef]

19. Sheng, Z. Impacts of Groundwater Pumping and Climate Variability on Groundwater Availability in the Rio Grande Basin. *Ecosphere* **2013**, *4*, art5. [CrossRef]
20. Eastoe, C.J.; Olivas, A.G.; Hibbs, B.J. Tracers of Groundwater Mixing in the Hueco Bolson Aquifer, Ciudad Juárez, Mexico. *J. Environ. Eng. Geosci.* **2016**, *22*, 195–207. [CrossRef]
21. Capt, T.; Mirchi, A.; Kumar, S.; Walker, W.S. Urban Water Demand: Statistical Optimization Approach to Modeling Daily Demand. *J. Water Resour. Plan. Manag.* **2021**, *147*, 04020105. [CrossRef]
22. Davis, M.E.; Leggat, E.R. *Preliminary Results of the Investigation of the Saline-Water Resources in the Hueco Bolson near El Paso, Texas*; US Geological Survey: Reston, VA, USA, 1967.
23. Bredehoeft, J.D.; Pinder, G.F. Digital Analysis of Areal Flow in Multiaquifer Groundwater Systems: A Quasi Three-Dimensional Model. *Water Resour. Res.* **1970**, *6*, 883–888. [CrossRef]
24. Meyer, W.R. *Digital Model for Simulated Effects of Ground-Water Pumping in the Hueco Bolson, EL Paso Area, Texas, New Mexico and Mexico*; US Geological Survey: Reston, VA, USA, 1976; p. 36.
25. Wilson, L. *Report 3—Hydrogeology of the Hueco Basin: Prepared for the Public Services Board*; City of El Paso, Texas; Lee Wilson & Associates Inc.: Santa Fe, NM, USA, 1985.
26. Kernodle, J.M. *Summary of US Geological Survey Ground-Water-Flow Models of Basin-Fill Aquifers in the Southwest Alluvial Basins Region, Colorado, New Mexico, and Texas*; Open-File Report; US Geological Survey: Reston, VA, USA, 1992.
27. McDonald, M.G.; Harbaugh, A.W. *A Modular Three-Dimensional Finite-Difference Ground-Water Flow Model*; US Geological Survey: Reston, VA, USA, 1988.
28. Kipp, K.L. *HST3D: A Computer Code for Simulation of Heat and Solute Transport in Three-Dimensional Ground-Water Flow Systems*; US Geological Survey: Reston, VA, USA, 1987.
29. Groschen, G.E. *Sumulation of Ground-Water Flow and the Movement of Saline Water in the Hueco Bolson Aquifer, El Paso, Texas, and Adjacent Areas*; Open-File Report; US Geological Survey: Reston, VA, USA, 1994.
30. Harbaugh, A.W.; McDonald, M.G. *User's Documentation for MODFLOW-96, an Update to the US Geological Survey Modular Finite-Difference Ground-Water Flow Model*; Open-File Report; US Geological Survey: Reston, VA, USA, 1996.
31. Heywood, C.E.; Yager, R.M. *Simulated Ground-Water Flow in the Hueco Bolson, an Alluvial-Basin Aquifer System near El Paso, Texas*; US Department of the Interior, US Geological Survey: Reston, VA, USA, 2003.
32. Hutchison, W.R. *Hueco Bolson Groundwater Conditions and Management in the El Paso Area*; EPWU Hydrogeology Report 04-01; El Paso Water—Publich Service Board: El Paso, TX, USA, 2004.
33. Papadopoulos, S.S. *Groundwater Flow for Administration and Management in the Lower Rio Grande Basin*; Main Report; Technical Report Prepared for the State of New Mexico; New Mexico State Engineer: Santa Fe, NM, USA, 2007.
34. Hutchison, W.R.; Montgomery & Associates. *Groundwater Flow and Transport Model for Hueco Bolson Aquifer*; Report; El Paso Water—Publich Service Board: El Paso, TX, USA, 2016; Volume 1.
35. Hutchison, W.R.; Montgomery & Associates. *Groundwater Flow and Transport Model for Hueco Bolson Aquifer*; Appendices; El Paso Water—Publich Service Board: El Paso, TX, USA, 2016; Volume 2.
36. Eastoe, C.J.; Hutchison, W.R.; Hibbs, B.J.; Hawley, J.; Hogan, J.F. Interaction of a River with an Alluvial Basin Aquifer: Stable Isotopes, Salinity and Water Budgets. *J. Hydrol.* **2010**, *395*, 67–78. [CrossRef]
37. Eastoe, C.J.; Hibbs, B.J.; Olivas, A.G.; Hogan, J.F.; Hawley, J.; Hutchison, W.R. Isotopes in the Hueco Bolson Aquifer, Texas (USA) and Chihuahua (Mexico): Local and General Implications for Recharge Sources in Alluvial Basins. *Hydrogeol. J.* **2008**, *16*, 737–747. [CrossRef]
38. Anderholm, S.K.; Heywood, C.E. *Chemistry and Age of Ground Water in the Southwestern Hueco Bolson, New Mexico and Texas*; US Geological Survey: Reston, VA, USA, 2003.
39. Budhathoki, P.; Doser, D.I.; Thapalia, A.; Langford, R.P.; Avila, V.M. Geological and Geophysical Studies of the Structure and Stratigraphy of the Northwestern Hueco Bolson Aquifer, El Paso, Texas. *Geosphere* **2018**, *14*, 731–748. [CrossRef]
40. Ging, P.B.; Humberson, D.G.; Ikard, S.J. *Geochemical Assessment of the Hueco Bolson, New Mexico and Texas, 2016–2017*; US Geological Survey: Reston, VA, USA, 2020.
41. Sanchez, R.; Rodriguez, L.; Tortajada, C. Effective Transboundary Aquifer Areas: An Approach for Transboundary Groundwater Management. *J. Am. Water Resour. Assoc.* **2020**, *56*, 360–378. [CrossRef]
42. Sanchez, R.; Eckstein, G. Groundwater Management in the Borderlands of Mexico and Texas: The Beauty of the Unknown, the Negligence of the Present, and the Way Forward. *Water Resour. Res.* **2020**, *56*, 1–20. [CrossRef]
43. Harris, I.; Osborn, T.J.; Jones, P.; Lister, D. Version 4 of the CRU TS Monthly High-Resolution Gridded Multivariate Climate Dataset. *Sci. Data* **2020**, *7*, 109. [CrossRef]
44. Druhan, J.L.; Hogan, J.F.; Eastoe, C.J.; Hibbs, B.J.; Hutchison, W.R. Hydrogeologic Controls on Groundwater Recharge and Salinization: A Geochemical Analysis of the Northern Hueco Bolson Aquifer, Texas, USA. *Hydrogeol. J.* **2008**, *16*, 281–296. [CrossRef]
45. Mann, H.B. Nonparametric Tests Against Trend. *Econometrica* **1945**, *13*, 245. [CrossRef]
46. Kendall, M. *Rank Correlation Methods*; Charles Griffin: London, UK, 1975; p. 202.
47. Theil, H. A Rank-Invariant Method of Linear and Polynomial Regression Analysis. In *Henri Theil's Contributions to Economics and Econometrics*; Raj, B., Koerts, J., Eds.; Advanced Studies in Theoretical and Applied Econometrics; Springer: Dordrecht, The Netherlands, 1992; Volume 23, pp. 345–381, ISBN 978-94-010-5124-8.

48. Sen, P.K. Estimates of the Regression Coefficient Based on Kendall's Tau. *J. Am. Stat. Assoc.* **1968**, *63*, 1379–1389. [CrossRef]
49. Hussain, M.; Mahmud, I. PyMannKendall: A Python Package for Non Parametric Mann Kendall Family of Trend Tests. *JOSS J. Open Source Softw.* **2019**, *4*, 1556. [CrossRef]
50. Yue, S.; Wang, C.Y. Applicability of Prewhitening to Eliminate the Influence of Serial Correlation on the Mann-Kendall Test: Technical Note. *Water Resour. Res.* **2002**, *38*, 4-1. [CrossRef]
51. Talchabhadel, R.; Karki, R.; Thapa, B.R.; Maharjan, M.; Parajuli, B. Spatio-Temporal Variability of Extreme Precipitation in Nepal. *Int. J. Climatol.* **2018**, *38*, 4296–4313. [CrossRef]
52. Ashworth, J.B. *Evaluation of Ground-Water Resources in EI Paso County, Texas*; Texas Water Development Board: Austin, TX, USA, 1990; p. 33.
53. Harbaugh, A.W. *MODFLOW-2005, the US Geological Survey Modular Groundwater Model—The Groundwater Flow Process*; Techniques and Methods 6-A16; US Geological Survey: Reston, VA, USA, 2005.
54. Hutchison, W.R.; Hibbs, B.J. Ground Water Budget Analysis and Cross-Formational Leakage in an Arid Basin. *Groundwater* **2008**, *46*, 384–395. [CrossRef] [PubMed]
55. Harbaugh, A.W. *MF2KtoMF05UC, a Program to Convert MODFLOW-2000 Files to MODFLOW-2005 and UCODE_2005 Files*; US Geological Survey: Reston, VA, USA, 2007.
56. Harbaugh, A.W.; Banta, E.R.; Hill, M.C.; McDonald, M.G. *MODFLOW-2000, The US Geological Survey Modular Ground-Water Model—User Guide to Modularization Concepts and the Ground-Water Flow Process*; Open-File Report; US Geological Survey: Reston, VA, USA, 2000.
57. Ahn, S.; Sheng, Z.; Jung, C.; Park, S.; Bailey, R.; Granados-Olivas, A.; Mirchi, A.; Samimi, M.; Hargrove, W.L. Coupled SWAT-MODFLOW Modeling for Determining Groundwater Sustainability Under Climate and Pumping Scenarios in a Semi-Arid Agricultural Watershed. In Proceedings of the AGU Fall Meeting, San Francisco, CA, USA, 9–13 December 2019.
58. Kim, N.W.; Chung, I.M.; Won, Y.S.; Arnold, J.G. Development and Application of the Integrated SWAT–MODFLOW Model. *J. Hydrol.* **2008**, *356*, 1–16. [CrossRef]
59. Harbaugh, A.W. *A Computer Program for Calculating Subregional Water Budgets Using Results from the US Geological Survey Modular Three-Dimensional Finite-Difference Ground-Water Flow Model*; Open-File Report; US Geological Survey: Reston, VA, USA, 1990.
60. Sheng, Z.; Jung, C.; Ahn, S.; Sanchez, R.; Tracy, J. Cross-Border Groundwater Movement Affecting Aquifer Storage Depletion in the Hueco Bolson. In Proceedings of the AGU Fall Meeting, San Francisco, CA, USA, 9–13 December 2019.
61. Ahn, S.; Abudu, S.; Sheng, Z.; Mirchi, A. Hydrologic Impacts of Drought-Adaptive Agricultural Water Management in a Semi-Arid River Basin: Case of Rincon Valley, New Mexico. *Agric. Water Manag.* **2018**, *209*, 206–218. [CrossRef]
62. Pitman, M.G.; Läuchli, A. Global Impact of Salinity and Agricultural Ecosystems. In *Salinity: Environment-Plants-Molecules*; Läuchli, A., Lüttge, U., Eds.; Springer: Dordrecht, The Netherlands, 2002; pp. 3–20, ISBN 978-0-306-48155-0.
63. Cooley, H.; Phurisamban, R.; Gleick, P. The Cost of Alternative Urban Water Supply and Efficiency Options in California. *Environ. Res. Commun.* **2019**, *1*, 042001. [CrossRef]
64. Texas Water Development Board (TWDB). *2007 State Water Plan—Chapter 2—Region E*; Texas Water Development Board: Austin, TX, USA, 2007.
65. Muller, D.A.; Price, R.D. *Ground-Water Availability in Texas, Estimates and Projections through 2030*; Report 238; Texas Department of Water Resources: Austin, TX, USA, 1979; p. 73.
66. Xu, Y.; Ramanathan, V.; Victor, D.G. Global Warming Will Happen Faster than We Think. *Nature* **2018**, *564*, 30–32. [CrossRef] [PubMed]
67. Petersen-Perlman, J.; Megdal, S.; Gerlak, A.; Wireman, M.; Zuniga-Teran, A.; Varady, R. Critical Issues Affecting Groundwater Quality Governance and Management in the United States. *Water* **2018**, *10*, 735. [CrossRef]
68. Sanchez, R.; Eckstein, G. The Path towards Groundwater Management in the Borderlands of Mexico and Texas. *Wiley Interdiscip. Rev. Water* **2020**, *7*, e1399. [CrossRef]
69. Llamas, M.R.; Martínez-Santos, P. Intensive Groundwater Use: Silent Revolution and Potential Source of Social Conflicts. *J. Water Resour. Plan. Manag.* **2005**, *131*, 5. [CrossRef]
70. Jarvis, T.; Giordano, M.; Puri, S.; Matsumoto, K.; Wolf, A. International Borders, Ground Water Flow, and Hydroschizophrenia. *Groundwater* **2005**, *43*, 764–770. [CrossRef]
71. Hargrove, W.L.; Sheng, Z.; Granados, A.; Heyman, J.M.; Mubako, S.T. Impacts of Urbanization and Intensification of Agriculture on Transboundary Aquifers: A Case Study. *J. Am. Water Resour. Assoc.* **2021**, *57*, 170–185. [CrossRef]
72. Rio Grande Compact Commission Report. Available online: Https://Www.Usbr.Gov/Uc/Albuq/Water/RioGrande/Pdf/Rio_Grande_Compact.Pdf (accessed on 15 September 2021).
73. Utilization of Waters of the Colorado and Tijuana Rivers and of the Rio Grande, Treaty between the United States of America and Mexico. Available online: https://Www.Ibwc.Gov/Files/1944Treaty.Pdf (accessed on 15 September 2021).
74. Sandoval-Solis, S.; McKinney, D.C. Risk Analysis of the 1944 Treaty between the United States and Mexico for the Rio Grande/Bravo Basin. In Proceedings of the World Environmental and Water Resources Congress 2011, Palm Springs, CA, USA, 22–26 May 2011; American Society of Civil Engineers: Palm Springs, CA, USA, 2011; pp. 1924–1933.
75. Eckstein, G.; Eckstein, Y. A Hydrogeological Approach to Transboundary Ground Water Resources and International Law. *Am. Univ. J. Int. Law Rev.* **2003**, *19*, 201–258.

76. Jakeman, A.J.; Barreteau, O.; Hunt, R.J.; Rinaudo, J.-D.; Ross, A.; Arshad, M.; Hamilton, S. Integrated Groundwater Management: An Overview of Concepts and Challenges. In *Integrated Groundwater Management: Concepts, Approaches and Challenges*; Jakeman, A.J., Barreteau, O., Hunt, R.J., Rinaudo, J.-D., Ross, A., Eds.; Springer International Publishing: Cham, Switzerland, 2016; pp. 3–20, ISBN 978-3-319-23576-9.
77. The JEDI Collaborative. The Path Forward to Systematic Change Justice. Equity. Diversity. Inclusion. Available online: https://jedicollaborative.com/ (accessed on 1 October 2021).
78. McDonald, Y.J.; Grineski, S.E. Disparities in Access to Residential Plumbing: A Binational Comparison of Environmental Injustice in El Paso and Ciudad Juárez. *Popul. Environ.* **2012**, *34*, 194–216. [CrossRef]
79. Grineski, S.E.; Collins, T.W. Environmental Injustices in Transnational Context: Urbanization and Industrial Hazards in El Paso/Ciudad Juárez. *Environ. Plan. A* **2010**, *42*, 1308–1327. [CrossRef]
80. Grineski, S.E.; Collins, T.W.; Ford, P.; Fitzgerald, R.; Aldouri, R.; Velázquez-Angulo, G.; de Lourdes Romo Aguilar, M.; Lu, D. Climate Change and Environmental Injustice in a Bi-National Context. *Appl. Geogr.* **2012**, *33*, 25–35. [CrossRef]
81. Moya, E.M.; Chávez-Baray, S.M.; Wood, W.W.; Martinez, O. Nuestra Casa: An Advocacy Initiative to Reduce Inequalities and Tuberculosis along the US-Mexico Border. *Int. Public Health J.* **2016**, *8*, 107–119.
82. Brulle, R.J.; Pellow, D.N. Environmental Justice: Human Health and Environmental Inequalities. *Annu. Rev. Public Health* **2006**, *27*, 103–124. [CrossRef]
83. Teron, L. A Language of (In)Justice: Expanding the Sustainability Planning Lexicon. *Environ. Justice* **2015**, *8*, 221–226. [CrossRef]
84. Martinez-Alier, J.; Temper, L.; Del Bene, D.; Scheidel, A. Is There a Global Environmental Justice Movement? *J. Peasant Stud.* **2016**, *43*, 731–755. [CrossRef]
85. Aaltonen, K. Project Stakeholder Analysis as an Environmental Interpretation Process. *Int. J. Proj. Manag.* **2011**, *29*, 165–183. [CrossRef]
86. Taylor, G. Justice, Equity, Diversity, Inclusion (JEDI): Part 1 (The What and Why of JEDI). Available online: https://www.americantrails.org/resources/justice-equity-diversity-inclusion-jedi-part-1-the-what-and-why-of-jedi (accessed on 5 July 2021).
87. Goldman, A.; Emani, S.; Perez-Angel, L.; Rodriguez-Ramos, J.; Stegen, J.; Fox, P. Special Collection on Open Collaboration Across Geosciences. *Eos*, 6 January 2021; p. 102. [CrossRef]
88. Sharma, S.; Dahal, K.; Nava, L.; Gouli, M.R.; Talchabhadel, R.; Panthi, J.; Roy, T.; Ghimire, G.R. Natural Hazards Perspectives on Integrated, Coordinated, Open, Networked (ICON) Science. Available online: https://www.essoar.org/doi/10.1002/essoar.10508384.1 (accessed on 25 October 2021).
89. Forrester, J.W. *Urban Dynamics*; MIT Press: Cambridge, MA, USA, 1969.
90. Valderrama, J.M.; Ibáñez, J.; Alcalá, F.J.; Dominguez, A.; Yassin, M.; Puigdefábregas, J. The Use of a Hydrological-Economic Model to Assess Sustainability in Groundwater-Dependent Agriculture in Drylands. *J. Hydrol.* **2011**, *402*, 80–91. [CrossRef]
91. Ringleb, J.; Sallwey, J.; Stefan, C. Assessment of Managed Aquifer Recharge through Modeling—A Review. *Water* **2016**, *8*, 579. [CrossRef]
92. Phan, T.D.; Bertone, E.; Stewart, R.A. Critical Review of System Dynamics Modelling Applications for Water Resources Planning and Management. *Clean. Environ. Syst.* **2021**, *2*, 100031. [CrossRef]
93. Afshar, A.; Khosravi, M.; Ostadrahimi, L.; Afshar, A. Reliability-Based Multi-Objective Optimum Design of Nonlinear Conjunctive Use Problem; Cyclic Storage System Approach. *J. Hydrol.* **2020**, *588*, 125109. [CrossRef]
94. Barati, A.A.; Azadi, H.; Scheffran, J. A System Dynamics Model of Smart Groundwater Governance. *Agric. Water Manag.* **2019**, *221*, 502–518. [CrossRef]
95. Balali, H.; Viaggi, D. Applying a System Dynamics Approach for Modeling Groundwater Dynamics to Depletion under Different Economical and Climate Change Scenarios. *Water* **2015**, *7*, 5258–5271. [CrossRef]
96. Niazi, A.; Prasher, S.; Adamowski, J.; Gleeson, T. A System Dynamics Model to Conserve Arid Region Water Resources through Aquifer Storage and Recovery in Conjunction with a Dam. *Water* **2014**, *6*, 2300–2321. [CrossRef]
97. Ryu, J.H.; Contor, B.; Johnson, G.; Allen, R.; Tracy, J. System Dynamics to Sustainable Water Resources Management in the Eastern Snake Plain Aquifer Under Water Supply Uncertainty. *JAWRA J. Am. Water Resour. Assoc.* **2012**, *48*, 1204–1220. [CrossRef]
98. Mashaly, A.F.; Fernald, A.G. Identifying Capabilities and Potentials of System Dynamics in Hydrology and Water Resources as a Promising Modeling Approach for Water Management. *Water* **2020**, *12*, 1432. [CrossRef]
99. Nozari, H.; Moradi, P.; Godarzi, E. Simulation and Optimization of Control System Operation and Surface Water Allocation Based on System Dynamics Modeling. *J. Hydroinform.* **2021**, *23*, 211–230. [CrossRef]
100. Andricevic, R.; Kitanidis, P.K. Optimization of the Pumping Schedule in Aquifer Remediation under Uncertainty. *Water Resour. Res.* **1990**, *26*, 875–885. [CrossRef]
101. Safavi, H.R.; Enteshari, S. Conjunctive Use of Surface and Ground Water Resources Using the Ant System Optimization. *Agric. Water Manag.* **2016**, *173*, 23–34. [CrossRef]
102. Ghasemi, A.; Saghafian, B.; Golian, S. System Dynamics Approach for Simulating Water Resources of an Urban Water System with Emphasis on Sustainability of Groundwater. *Environ. Earth Sci* **2017**, *76*, 637. [CrossRef]
103. Madioune, D.H.; Diaw, M.; Mall, I.; Orban, P.; Faye, S.; Dassargues, A. Hydrogeological Characterization and Hydrodynamic Behaviour of the Overexploited Diass Aquifer System (Senegal) Inferred from Long Term Groundwater Level Monitoring. *Am. J. Water Resour.* **2020**, *8*, 104–117.

104. Ratna, D.; Puji, A. Using System Dynamic Modeling for Improving Water Security in the Coastal Area: A Literature Review. *Open Access Maced. J. Med. Sci.* **2020**, *8*, 143–154. [CrossRef]
105. Heckerman, D. A Tutorial on Learning with Bayesian Networks. In *Innovations in Bayesian Networks*; Holmes, D.E., Jain, L.C., Eds.; Studies in Computational Intelligence; Springer: Berlin/Heidelberg, Germany, 2008; Volume 156, pp. 33–82, ISBN 978-3-540-85065-6.
106. Phan, T.D.; Smart, J.C.R.; Capon, S.J.; Hadwen, W.L.; Sahin, O. Applications of Bayesian Belief Networks in Water Resource Management: A Systematic Review. *Environ. Model. Softw.* **2016**, *85*, 98–111. [CrossRef]
107. Singto, C.; Fleskens, L.; Vos, J.; Quinn, C. Applying Bayesian Belief Networks (BBNs) with Stakeholders to Explore and Codesign Options for Water Resource Interventions. *Sustain. Water Resour. Manag.* **2020**, *6*, 23. [CrossRef]
108. Bertone, E.; Sahin, O.; Richards, R.; Roiko, A. Extreme Events, Water Quality and Health: A Participatory Bayesian Risk Assessment Tool for Managers of Reservoirs. *J. Clean. Prod.* **2016**, *135*, 657–667. [CrossRef]

water

MDPI

Article

Investigating Management of Transboundary Waters through Cooperation: A Serious Games Case Study of the Hueco Bolson Aquifer in Chihuahua, Mexico and Texas, United States

Alex Mayer [1,*], Josiah Heyman [2], Alfredo Granados-Olivas [3], William Hargrove [4], Mathew Sanderson [5], Erica Martinez [6], Adrian Vazquez-Galvez [3] and Luis Carlos Alatorre-Cejudo [7]

1 Department of Civil Engineering, Center for Environmental Resource Management, University of Texas at El Paso, El Paso, TX 79968, USA
2 Department of Sociology and Anthropology, Center for Inter-American and Border Studies, University of Texas at El Paso, El Paso, TX 79968, USA; jmheyman@utep.edu
3 Departamento de Ingeniería Civil y Ambiental, Universidad Autónoma de Ciudad Juárez, Cd Juárez 32315, Mexico; agranados@uacj.mx (A.G.-O.); fvazquez@uacj.mx (A.V.-G.)
4 Center for Environmental Resource Management, University of Texas at El Paso, El Paso, TX 79968, USA; wlhargrove@utep.edu
5 Department of Sociology, Anthropology, and Social Work, Kansas State University, Manhattan, KS 66506, USA; mattrs@ksu.edu
6 Environmental Science and Engineering PhD Program, University of Texas at El Paso, El Paso, TX 79968, USA; emartinez15@miners.utep.edu
7 Departamento de Arquitectura, Universidad Autónoma de Ciudad Juárez, Cd Juárez 32315, Mexico; luis.alatorre@uacj.mx
* Correspondence: amayer2@utep.edu

Citation: Mayer, A.; Heyman, J.; Granados-Olivas, A.; Hargrove, W.; Sanderson, M.; Martinez, E.; Vazquez-Galvez, A.; Alatorre-Cejudo, L.C. Investigating Management of Transboundary Waters through Cooperation: A Serious Games Case Study of the Hueco Bolson Aquifer in Chihuahua, Mexico and Texas, United States. *Water* **2021**, *13*, 2001. https://doi.org/10.3390/w13152001

Academic Editor: Sharon B. Megdal

Received: 25 May 2021
Accepted: 18 July 2021
Published: 21 July 2021

Abstract: Management of transboundary aquifers is a vexing water resources challenge, especially when the aquifers are overexploited. The Hueco Bolson aquifer, which is bisected by the United States–Mexico border and where pumping far exceeds recharge, is an apt example. We conducted a binational, multisector, serious games workshop to explore collaborative solutions for extending the life of the shared aquifer. The value of the serious game workshop was building knowledge, interest, understanding, and constituency among critical stakeholders from both sides of the border. Participants also learned about negotiations and group decision-making while building mutual respect and trust. We did not achieve consensus, but a number of major outcomes emerged, including: (1) participants agreed that action is called for and that completely depleting the freshwater in the shared aquifer could be catastrophic to the region; (2) addressing depletion and prolonging the life of the aquifer will require binational action, because actions on only one side of the border is not enough; and (3) informal binational cooperation will be required to be successful. Agreeing that binational action is called for, the serious games intervention was an important next step toward improving management of this crucial binational resource.

Keywords: groundwater depletion; transboundary aquifers; binational resource management; serious games; stakeholder cooperation

1. Introduction

Excessive groundwater pumping is leading to rapid depletion of aquifers around the world, in the context of climate change, dwindling supplies, and increasing demand [1–6]. Aquifers have long been used as a means to buffer annual variation in meeting water demands in many regions, at the cost of long-term depletion. Aquifers thus represent the source of long-term adaptive capacity, but many are at risk. Moving forward, it will be crucial to establish trajectories of remaining freshwater over time, and to identify means of extending or prolonging the life of these aquifers. Transboundary aquifers have had sufficient storage in the past to support autonomous, uncoordinated extraction, but the need for transboundary coordination is emerging rapidly.

Groundwater aquifers are common pool resources. Ostrom's [7] fundamental insight was that common pool resources without institutional structure governing use are vulnerable to over-extraction, as each user accesses them without consideration of overall impact. On the other hand, imposition of a single overarching authority, she demonstrated, tends to be less effective than shared collaboration, and we add that overarching regulation is particularly hard to achieve at international borders. Rather, she suggested that increased knowledge of future risk, shared information, and mutual trust in knowledge encourage the emergence of voluntary shared governance of such resources. Furthermore, while envisioning a watershed approach to water governance, all parties would expect some benefits and advantages from a transboundary approach to groundwater management.

Due to the characteristics described above, transboundary groundwater is an important common pool resource that is particularly at risk [8]. The international border means that conventional institutionalized governance by the jurisdictions involved does not cover all access sites and users. At borders, the risk of the open access situation is heightened, and sharing is made more difficult. The literature on transboundary groundwater cooperation/governance has two characteristics. It either (1) consists of cataloging cases where aquifers are transected by borders (e.g., [9] for the region and [10] globally), but noting the absence of transboundary governance institutions in most cases [11,12] or (2) addresses legal and political issues in envisioning such hypothetical transboundary institutions, [13–18]). Important work has been done on the hydrology of transboundary aquifers, cataloging knowledge and gaps around transboundary aquifers as physical entities (e.g., [19]). The central challenge is that there are few actual cases of transboundary governance to study, and little has been done in terms of researching efficacy of transboundary management in those cases.

There are just seven cases around the world of such aquifers with agreements of any kind (mostly data sharing) and only one transboundary aquifer (the Franco–Swiss Genevan aquifer) is effectively managed [20,21]. Notably, the Genevan aquifer has shown sustainable water levels over the last 30 years as compared to levels before the signing of the agreement between Switzerland and France in 1970s, and the governing agreement was recently renewed. The one scholarly case study [22] of the Genevan aquifer shows that cooperation emerged as a response to serious decline, via mutual recognition of a common resource and involvement of local (subnational) actors. Other transboundary aquifer governance cases are either too new, with limited empirical research, or so far have only been studied in terms of the legal/administrative framework, separate from the hydrological and climate dynamics [20–25]. An important need, among others, is research that addresses the key goals of increased knowledge of future risk, shared information, and building mutual trust across bounded jurisdictions.

We address the groundwater in the Hueco Bolson/Valle de Juárez aquifer (names used in the United States (US) and Mexico (MX) respectively, see Figure 1). The aquifer, hereafter referred to as the HB is bisected by the Rio Grande/Rio Bravo del Norte, which delineates the border between the Mexican state of Chihuahua (CH) and the US state of Texas (TX), and thus also separates the two largest users of groundwater in the region, El Paso Water, EPW (similar but not identical to the city of El Paso) and the Junta Municipal de Agua y Saneamiento, JMAS, conterminous with the municipality of Ciudad Juárez. There are many other smaller users on both sides of the border, including small rural utilities and agricultural users.

Surface water is governed at the transboundary scale in the region by the 1906 Treaty in this river particular segment [26]. Definite volumes are allocated to MX and the US, and within the US separate compacts divide that share between Colorado, New Mexico, and TX. Specific binationally-coordinated institutions, the International Boundary and Water Commission (IBWC, US) and Comisión Internacional de Límites y Aguas (CILA, MX), govern surface water. The IBWC and CILA, in Minute 242 (Minutes reflect decisions of the

IBWC and CILA that are binding obligations of the US and MX, once signed by the two governments), committed MX and the US to developing "a comprehensive transboundary solution to the extant and emerging groundwater disputes along the border" [27], but has never been fulfilled, although the joint US–MX Transboundary Aquifer Assessment Program (TAAP) has been successful in expanding knowledge around US–MX shared aquifers [12,28].

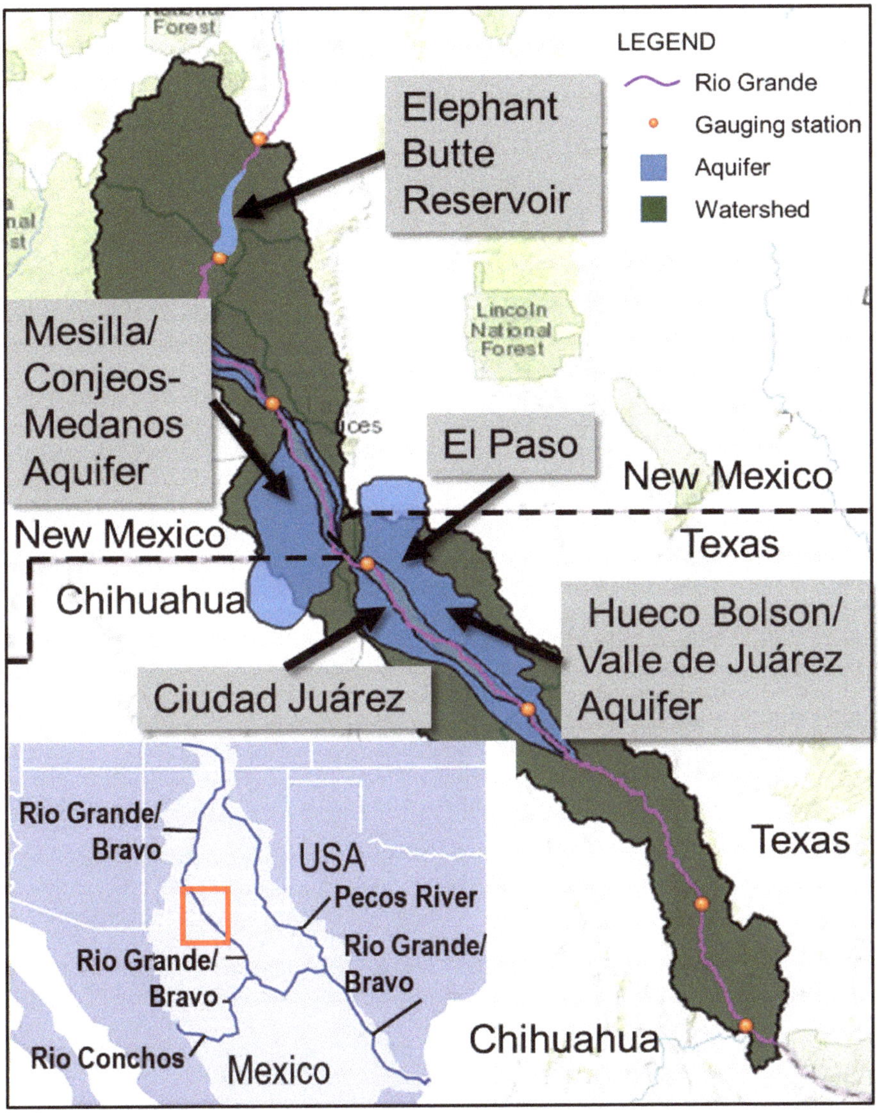

Figure 1. Study area.

Thus, at present, there is minimal shared governance over the groundwater in the HB. Hydrologically, the groundwater constitutes a transboundary common pool in the sense that the water is efficiently mobile across the border, and any one actor on one side, affects quantity and quality of the whole. However, this common pool resource is governed by the rules and regulations of the individual states and/or countries who share the resource, chiefly TX in the US and the Federal Government of MX. In MX, access to groundwater is governed by the federal agency CONAGUA, and in TX access to groundwater is delegated to private surface owners [29]. These access governance regimes are uncoordinated binationally. Furthermore, the hydrological fact that surface and subsurface water are connected and that, as discussed below, subsurface water is used when surface water is insufficient, does not enter into this strictly delineated governance system.

The specific institutional gaps seen for US–MX transboundary groundwater are common to international borders [12,30]. The bureaucratic machinery of territorial nation states are effective for organized societal action inside borders, but less well-designed for transboundary action [31]. While some cooperative actions extend across borders [32], in most cases, actors' influence terminates at their national border. This causes a notable institutional disparity and sometimes incongruity at such sites [33], including the US–MX border. One example among many differences in approach among institutions is that water capital funding for JMAS in Juárez is mainly derived from federal sources, while EPW is able to set its own capital investment priorities, according to availability of resources from the state and national level.

Another barrier to governance at international borders is methodological nationalism, the ways that knowledge is enclosed inside of and limited by nation-state units (and replicated with smaller units like states) [34,35]. For example, the main planning document for TX shaping EPW's investments, the Texas Water Development Board Region E Plan [36], does not account for groundwater extraction in the HB by MX, let alone capital and policy measures south of the border. Yet the groundwater moves efficiently under the border. EPW likely does account for activities on the MX side, but there is no explicit shared modeling of the commons.

An important aspect of our serious game was an attempt to transcend the knowledge gaps implied in methodological nationalism by creating a shared water budget for the HB and bringing it for discussion and consensus to a binational group of water stakeholders. By focusing on the unified groundwater budget as a knowledge object, we emphasized discussion of joint groundwater stocks and volumes extracted (see [21]), rather than rules for which parties within nations are allowed physically to access aquifers, consistent with the admonitions in [37] to distinguish groundwater from aquifers. Methodological nationalism is part of a wider range of bridges and gaps, such as cultural and linguistic differences, and ambivalent attitudes toward the other country: beneficent feelings based on shared relationships, but also deep-seated prejudices of nationalism and superiority/inferiority [38]. While unequal power over water is widespread, perhaps unavoidable, at borders, we follow the findings of Zeitoun and Warner [39] that there is opportunity within hydrohegemony for the construction of more equitable and cooperative relations.

The need for a binational forum, or space for interaction, where stakeholders could evaluate the most recent data and scenarios regarding groundwater evolution along the binational HB has been a continuous challenge for local and regional water users. Furthermore, water agencies responsible for water management on either side of the border struggle to communicate to local water users about management the HB. We implemented a serious games approach to address these issues in the HB with the specific aim of exploring binational cooperation as an approach to prolonging the life of the transboundary aquifer. To our knowledge, we report insights from the first such effort in the Paso del Norte region of the US–MX border.

2. Materials and Methods

2.1. Serious Games Background

Preparing for and managing water security risks is complicated by a wide variety of challenges, including scientific uncertainty and complexity, limited resources, competing priorities, and differences in risk perception. To move towards effective mitigation and adaptation strategies, stakeholders need to develop a collective sense of the risks that they face, how they could prepare for and manage water security risks, and the decision-making approaches that will allow them to respond collaboratively and adaptively to emerging threats. Achieving these goals requires that stakeholders learn together and from each other to create a collective intelligence and shared understanding [40–43]. Serious games have gained attention as a way to advance actions to mitigate or adapt to risks associated with the environment and natural resources [44–48]. A serious game is an exercise that directly engages participants in working to solve a realistic but hypothetical challenge with the intent that they learn new material or approaches. Serious games can provide an opportunity for participants to experiment with solutions in an environment where they can freely express opinions.

With increased concern over the suitability of current management arrangements to handle future water security issues, water resource management has been at the forefront of many serious games efforts (Madani et al. [45] estimated that about one-third of environmental management-serious game topics had a water resources management theme) Applications have included watershed planning, drought management, drinking water access and safety, and conflict resolution [49–57]. In these applications, researchers have evaluated the effect of the games on the participants has been a focus, including assessing the degree of social learning [44,53–57], change in beliefs and values [58], and success in conveying the complex interconnectedness of water resources problems [59,60]. Researchers have investigated whether serious games can inform modeling; for example, Aubert et al. [61] argue that serious games can be used to elicit preference weights in the context of multi-criteria decision analysis and Addamatti et al. [62] use serious games to develop agent-based models of water users. The apparently few serious games applications to groundwater resources [59,63,64] have emphasized the notion of groundwater as a common pool resource and the necessity for users to collaborate to sustainably manage the resource.

The approach taken in the present work to the game objectives fits best with the "Design and Recommend" model in the typology described by Bots and Van Daalen [65]; or, in other words, using the game as "design studio". Since the objective of the present work is to identify better solutions through cooperation, our efforts also fit with the description of "games as interventions" by Rodela et al. [44]. In the games as interventions model, it is important that the management scenarios be as realistic as possible, and that participants play their real roles to the greatest extent possible. Social learning is also emphasized in the games as interventions model. As far as the authors know, the current effort is the first to apply serious games to binational management of a transboundary groundwater resource.

2.2. Modeling Methodology

We rely on past, substantial work on the hydrogeology of the HB to develop our groundwater balance model (e.g., [66–71]). We embedded the following key concepts into the groundwater model used in the games: (a) a simple model of a single aquifer compartment with binational pumping is sufficient for exploring the sustainability of the aquifer; (b) accordingly, the lifetime of the freshwater portion of the aquifer can be quantified by a depletion time which could be lengthened or shortened by changing pumping rates in either country; (c) current situation and projected business as usual scenarios and the associated depletion times are useful for exploring alternative, binational management strategies; (d) extending the lifetime a meaningful amount beyond the business-as-usual projection requires reducing pumping by substantial amounts; (e) the basis for assigning shares of reduced pumping to each city (e.g., equal percentage vs. equal volume) significantly impacts the relative burden of pumping reductions for each city; and (f) an array of potential water supply and demand reduction options exist for offsetting pumping reductions, each with a different costs.

The single compartment groundwater model is stated as an aquifer water balance: $D = \sum Q - \sum R$, where D is the depletion rate or change in storage, $\sum Q$ is the sum of pumping over both countries and all water use sectors and $\sum R$ is the sum over all sources of aquifer recharge. We implicitly assume that there is no groundwater outflow except via pumping. The associated depletion time is then $T = V_{fw}/D$, where V_{fw} is the volume of recoverable freshwater in the aquifer. We first estimate depletion rates and times according to present day circumstances, which we call the current situation, summarized in Tables 1 and 2. For demand from the HB, we use the average pumping rates from the most recent five years of available pumping data to reflect the most recent patterns of use by El Paso, Ciudad Juárez, and other users. Other users include agricultural wells in the Valle de Juárez irrigation district in CH, the El Paso County Water Improvement District #1 in TX, industrial users not served by EPW in TX, and small rural water utilities in both TX and CH. Recharge estimates for the HB are highly uncertain, but estimates for potential sources of recharge have been derived from groundwater elevation mapping, geochemical surveys, and groundwater flow modeling (see Table 2). In addition to values for pumping from the HB, we note in Table 1 the other water supply sources to which each city has access.

Table 1. Current situation: annual water demand from El Paso and Ciudad Juárez (kAF).

User	Total Demand	Hueco Bolson Aquifer	Rio Grande	Mesilla-Conejo-Medanos Aquifer	Desalination
Ciudad Juárez	151	121	NA	30	NA
El Paso	118	53	30	27	8
Other	NA	14	NA	NA	NA
Total	269	188	NA	NA	NA

Table 2. Annual recharge rates for the Hueco Bolson (kAF).

Recharge Component	Recharge
Mountain front	9
Lateral inflow from Tularosa basin	0
Engineered artificial recharge	6
Seepage from Rio Grande channel	1
Leakage from irrigation & return flow canals	17
Total	33

The amount of freshwater remaining in the HB is calculated using historical estimates of freshwater volumes in the aquifer and estimates of pumping that have occurred from the timing of the freshwater volume estimates to the present. Estimates of recoverable volumes of freshwater and brackish water in the HB range from 7.5 MAF to 10 MAF and up to 20 MAF, respectively [71–75]. We use the Heywood and Yager [72] estimate of recoverable freshwater volume of 9 MAF as of 2003 because this estimate is within the range of other freshwater volume estimates and the conceptual basis for the estimate is consistent with other hydrogeologic models proposed for the HB. We estimate that approximately 2.5 MAF of groundwater have been depleted from the HB since 2003, leaving about 6.5 MAF of recoverable fresh groundwater. Using the current rate of depletion of 155 kAF/yr and the recoverable freshwater volume estimate of 6.5 MAF, the recoverable freshwater will be completely depleted in approximately 42 years.

The Business-as-Usual (BAU) scenario (see Table 3) is meant to set the stage for the discussions by envisioning a future that assumes that urban populations and thus water demands will increase and there will be no significant change in policies or human behavior that would slow depletion of the HB. The BAU scenario spans a 50-year period (2020–2070) and pumping from the HB over the period is based on assumptions regarding (a) increases in population and corresponding water demand for the two cities; (b) climate-change-induced reduction in surface water available to EPW; (c) proportional increases in pumping from the HB in response to increased demand overall for the cities and reductions in surface water availability for EPW, the only utility that also uses surface water; (d) pumping by users other than EPW and JMAS would remain the same as in the current situation; (e) recharge would remain constant; and (f) per capita use rates would remain constant. Table 3 summarizes the basis for the projected demand for the two cities and sources of the associated information. Given the BAU depletion rate of 209 kAF/yr and the recoverable freshwater volume estimate of 6.5 MAF, the recoverable freshwater could be completely depleted in 31 years.

Table 3. Summary of business-as-usual scenario.

	Ciudad Juárez	El Paso	Other	Total	Units
Population increase	66%	33%	NA	NA	
Reduction in Rio Grande supply	NA	40%	NA	NA	
Average demand	204	141	NA	NA	kAF/yr
Average HB pumping	164	63	14	242	kAF/yr
Recharge	NA	NA	NA	33	kAF/yr
Depletion rate	NA	NA	NA	209	kAF/yr

Reductions in HB pumping, if any, would result in mismatches between future supply and demand. Based on experience in prior stakeholder meetings and informal interactions with the two city water utilities, we identified alternatives for offsetting pumping reductions in Table 4. We used values from state reports (TWDB) for estimates of unit costs for each option (also in Table 4, unit costs include amortized capital and operating costs.) and used a simple calculation of volume multiplied by unit costs to determine total costs associated with implementing each option. Upper limits for each of the options also were established.

Table 4. Options for offsetting pumping reductions.

Option	Description	Cost (US$/kAF)	Maximum Amount (kAF/yr)
Desalination	A desalination plant is constructed and operated to jointly serve Ciudad Juárez and El Paso and would draw from brackish portions of the HB.	518	CJ: total demand EP: total demand
Aquifer recharge with treated wastewater	Treated tertiary effluent is applied to recharge basins overlying the HB to recharge the freshwater aquifer and reduce brackish water intrusion.	1000	CJ: 133 EP: 71
Direct potable reuse	Treated tertiary effluent is piped to water treatment plants and blended with current water supplies.	850	CJ: 133 EP: 71
Imported water	Groundwater is secured in remote aquifers and pipelines and pumping plants are constructed.	2400	CJ: total demand EP: total demand
Incentivized household water conservation	Educational and financial incentive campaigns are implemented to reduce household and commercial water use.	367	CJ: 15 EP: 30
Reduce infrastructure leaks	The cities repair leaking water distribution systems and continue leak detection and replacement campaigns.	2295	CJ: 27 EP: 4

2.3. Workshop Implementation

Stakeholders from both sides of the border were invited to participate, with the intention to have roughly equal participation from US and MX stakeholders. We also intended to recruit roughly equal numbers of participants from the municipal and industrial (M&I) sector and the non-M&I sector. We did not include agricultural users, both in order to simplify the framework of the discussion and because that the agricultural sector has a significantly lower impact currently on the long-term trajectory of the HB. We compiled a list of 30 potential participants and sent invitations by email. Where necessary, we followed up with phone calls or text messages. In several cases, contacts from our initial list of invitees recommended additional or alternative potential participants. We received 20 positive responses and we communicated the final details of the workshop to them. To prepare, participants were provided with and asked to read two documents one week before the first session: one describing the intention and schedule of sessions and key questions to be addressed in each session, and another with a more detailed description of the activities in the first two sessions.

Six sessions, each 60–80 min, were held over a four-month period. The first session was held in November 2020 and the remaining five sessions were held in consecutive weeks in January and February 2021. Due to restrictions imposed by the COVID-19 pandemic, the sessions were held on Zoom with simultaneous Spanish and English translation. Agendas were sent out ahead of each session and, in some cases, participants were asked to read documents and use spreadsheets calculations to support their decisions in the upcoming sessions. All documents used in the workshops were provided in both Spanish and English and numerical values were presented in both metric and English units. Each session began with a recap of the preceding session and ended with a description of goals for the next session and, in some cases, assignments to complete ahead of the next sessions.

In Session 1, the participants were introduced to each other, the format of the sessions, and the key questions to be addressed in each session. The workshop organizers presented the concepts behind the single compartment groundwater model, the calculation of depletion rates and times, and the information supporting the calculations. The basis for the current situation depletion rate and times were explained, followed by a similar presentation on the projected, BAU scenario. The presentations were followed by moderated small group discussions of the following questions. (a) Are the current situation and BAU scenarios reasonable? (b) What can be done to mitigate or adapt to the BAU scenario? (c) What would be the potential impacts of depletion of the HB?

Session 2 focused first on the exploration by participants of setting targets for reducing depletion rates and extending aquifer depletion times. The results of pre-session participant polling on acceptable reductions and binational sharing in the reduction of the depletion rate were used as foundation for a discussion of factors motivating the selection of targets. A second pre-session poll on options for technologies, policies, and broader approaches for meeting the target reductions was used to motivate preliminary discussions of advantages and disadvantages and potential binational approaches for implementing the options.

In Session 3, participants worked with the first version of a spreadsheet that provided estimates of recoverable freshwater depletion time based on potential reductions in HB pumping rates for the two cities. Two general schemes were offered for apportioning the reductions in pumping rates for the two cities: equal volumetric reductions and equal fractional reductions. For either scheme, no reduction corresponds to the BAU scenario. The spreadsheet was used to motivate discussion of how to share the reduction in depletion and, correspondingly pumping rate between the two cities.

In Session 4, participants explored strategies for offsetting pumping reductions using a second spreadsheet that provided unit cost estimates (US$/kAF see Table 4) for the options in the spreadsheet. Participants were asked to identify annual volumes for each of the options, based on their individual preferences, to offset the deficit between supply and demand. The spreadsheet calculates the cost of each option and total costs, given the annual volume of water to be used by Ciudad Juárez and El Paso. Several combinations of options were presented, to give the participants an idea of the range of possible volumes and associated costs.

In Session 5, the workshop organizers presented estimates of the cost of doing nothing (BAU scenario); that is, what would happen if fresh groundwater in the HB were depleted in 31 years. The basis for these costs was as follows: (a) 57 kAF/yr of HB pumping would have to be replaced for El Paso, (b) the current cost of groundwater for El Paso is US$150/AF, (c) HB pumping for El Paso is replaced by a 50/50 mix of imported water (US$2,400/AF) and desalination (US$518/AF), and (d) an estimated cost of replacing El Paso HB pumping of US$74 million. The cost of replacing 147 kAF/yr of HB pumping for Ciudad Juárez was not estimated precisely, given that unit costs of water replacement for Ciudad Juárez were unreliable, but a coarse estimate of a US$100 million to US$200 million was deemed reasonable. Participants were again asked to choose options and how much water supply would be gained or demand would be reduced for offsetting pumping reductions.

In the last session (Session 6), a poll was given for participants to choose their top three choices from seven options for meeting a target reduction in depletion times and associated reductions in pumping. The options are described in Table 5 and the costs and freshwater aquifer lifetimes are shown in Figure 2. After the selection of the options, the workshop organizers presented two options for sharing the costs between the two cities: (a) each city pays for implementing their options alone and (b) the total cost for the two cities is shared 50/50 between the two cities. Finally, a summary of what was learned over the entire workshop was presented, followed by a discussion of next steps.

Table 5. Primary water supply gain or demand reduction alternatives for each option and city.

Option	CJ Portfolio	EP Portfolio
1. Reduce each city's pumping by 15%	• aquifer recharge w/treated wastewater • imported water	• desalination from local aquifers • aquifer recharge w/treated wastewater
2. Reduce each city's pumping by 20 kAF/yr	• aquifer recharge w/treated wastewater • imported water	• desalination from local aquifers • aquifer recharge w/treated wastewater
3. Reduce each city's pumping by 35%	• aquifer recharge w/treated wastewater • imported water	• desalination from local aquifers • aquifer recharge w/treated wastewater • imported water
4. Reduce each city's pumping by 40 kAF/yr	• aquifer recharge w/treated wastewater • imported water	• desalination from local aquifers • aquifer recharge w/treated wastewater • imported water
5. Reduce each city's pumping by 35% + reduce demand by 13%	• aquifer recharge w/treated wastewater • imported water • household conservation • leak reduction	• desalination from local aquifers • aquifer recharge w/treated wastewater • imported water • household conservation
6. Reduce each city's pumping by 40 kAF/yr + reduce demand 36 kAF/yr	• aquifer recharge w/treated wastewater • imported water • household conservation • leak reduction	• desalination from local aquifers • aquifer recharge w/treated wastewater • imported water • household conservation
7. Do nothing (business as usual)	• not applicable	• not applicable

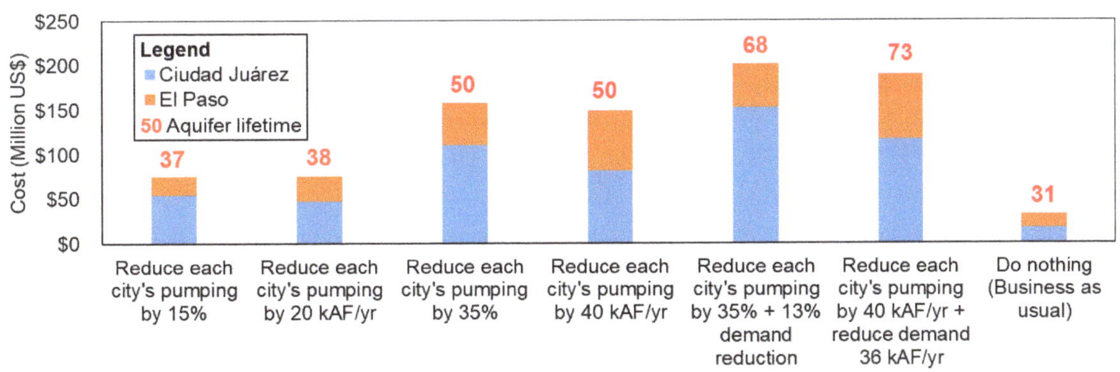

Figure 2. Costs of options for meeting a target reduction in pumping and associated depletion times (numerical values in red bold).

2.4. Data Collection

Note takers kept detailed notes of discussions, and comments in the Zoom chat function were saved. Several survey or polling instruments were used to collect information regarding the participants' beliefs and attitudes and their choices for increasing the aquifer lifetimes, including:

- a short, 29-item survey (available upon request) administered at the beginning of Session 1 and again at the end of Session 6, which was designed to learn about participants' knowledge about water use and conservation in the HB, and their beliefs about groundwater use responsibility and cooperation for management. Surveys were administered online, in English and Spanish, and followed IRB protocols for human subjects.
- polling on acceptable reductions and binational sharing in the reduction of the depletion rate and on options for technologies, policies, and broader approaches for meeting the target reductions (prior to Session 2)
- polling on options for meeting target reductions in depletion times and associated reductions in pumping (Session 6)
- a survey regarding participants' opinions on the workshop salience and relevance, format of the workshop, and overall satisfaction with the workshop. In addition, bilingual students took notes during all sessions, and the notes were analyzed to develop common themes that arose during the discussions.

Due to a low response rate from participants in Mexico on the post-workshop survey ($n = 1$), it was not possible to conduct a pre-post analysis of individual perspectives as we had anticipated. However, the survey responses to the preworkshop were sufficient for descriptive analysis, and for an aggregated analysis that allows group-level comparisons of perspectives before and after the workshop, both of which we discuss below, along with a synthesis of session observations.

3. Results

3.1. Participation in Sessions

Table 6 shows the participation in sessions. While more participants came from the US, substantial participation came from both countries. Organizations with participants included JMAS Juárez; the Mexican Society of Engineers—Chihuahua; Junta Central de Agua y Saneamiento-Chihuahua; Proteccion Civil Juárez; El Paso Water; the Bureau of Reclamation; El Paso Electric; TCEQ; EPA; Ysleta del Sur Pueblo; Friends of the Rio Bosque; Fort Bliss; The Frontera Land Alliance; the Lower Valley Water District; and the Hunt Companies (a large business and residential real estate developer and manager).

Table 6. Summary of participation in the sessions.

Participation	Mexico	US	Total
any session	7	13	20
more than 1 session	6	12	18
more than 2 sessions	5	10	15
more than 3 sessions	5	9	14
more than 4 sessions	2	4	6
all 6 sessions	1	2	3

3.2. Synthesis of Session Observations

The most important outcome to emerge was that, although discussion moved progressively through issues as described above, in the end, a consensus list of pumping reduction measures could not be achieved. The technical demand of numerous, detailed, interlocking decisions at that level were beyond a short group discussion. However, we did effectively discuss major policy parameters. In all, the sessions generated a foundation of relationships that, if pursued and developed further, could allow for even more movement toward a consensus on pumping reductions.

Participants found the basic model that produced scenarios for the HB to be reasonable and credible. The knowledge and beliefs survey revealed that participants from MX and US both were aware that pumping rates far exceed recharge rates, and that groundwater depletion is a serious problem. Notably, participants from the US were more likely than participants from Mexico to disagree with the statement that "decreases in groundwater elevation are greater than 1 foot per year". Conversely, participants from Mexico were more likely than participants from the US to agree with the statement "El Paso and Ciudad Juárez contribute equally to groundwater depletion", to agree that "freshwater in the HB will be completely depleted in a few decades", and to believe that the lack of water in the future will limit future economic growth.

When examining the BAU scenario of depletion of HB freshwater in approximately 31 years, all participants were motivated to consider ways to extend it or avoid it altogether. Likewise, they agreed that HB depletion is a shared issue. Groundwater was understood to be mobile, a common pool good: that which is consumed on one side is lost from both sides; that saved on one side is saved for both. Moreover, the social-economic fate of each side (especially Juárez, the more groundwater dependent side currently) was understood to matter to all. A sense of mutual engagement and commitment was palpable.

Lurking in the background was a model result that participants felt was revelatory. In 2051, if pumping from the HB has to be completely replaced because it is depleted, the one-time, sudden replacement costs would be US$100s of millions. Participants wrestled—in an engaged and serious fashion—with three considerations. A shorter freshwater depletion timeline would require less expensive measures to replace supplies or reduce demand, but it would mean quicker introduction of costly alternative supplies and/or drastic conservation measures. A longer timeline puts off the expensive full transition and gives more time to adjust. However, it is costlier to accomplish a long depletion timeline as more/larger alternatives to pumping need investment.

In the first round, pumping reductions were proposed, ranging from 3–97% (translating to a 46-year lifetime to an indefinite time). The most commonly chosen reduction was about 38%, which gives a 50-year lifetime (instead of 31-year lifetime with BAU). Fifty is an easy number to envision, a typical planning horizon, and falls approximately in the middle range of the discussion amongst all participants. The group then wrestled with the question of how to partition the responsibility of reduction between the two countries. An equal volume reduction has a large percentage impact on El Paso, as it does not currently pump as much volume, but an equal percentage reduction would have a huge impact on Ciudad Juárez, as an equal percentage of a large pumped volume is a larger amount to replace. Initially, the most common assigned responsibility was a reduction of 70% in MX (ranging from 50% to 70%).

The conversation turned when participants from JMAS made it clear that Ciudad Juárez would have trouble reducing pumping by more than 15% without significant financial help from, for example, the Mexican federal government or the US. A 15% reduction on both sides would extend the lifetime from 31 to 37 years. It is notable that realizing this constraint did not keep the participants from discussing deeper reductions in pumping, but with a recognition that these larger reductions would need to be funded and overcome in a binational sense. The sessions that followed worked their way through hard considerations: the difficulty but desirability of affecting the timeline; impact and fairness of reductions on two sides; and the kinds of measures (volume saved, cost, applicability to the US, Mexico, or both) needed to reduce pumping to the chosen goal.

The seven-option poll (see Table 5 and Figure 2 for description of options) captured many of the elements of the discussion. The results of the poll are shown in Figure 3, including a simple weighted sum of the choices, where the first choice is given three points, the second two points, and the third one point. Option 4 scores highest; Options 1, 2, 3, and 5 cluster about half the score for Option 4; Option 6 was least preferred of the reduction alternatives, and no one chose BAU (Option 7). Option 4's notable qualities are its moderate timeline (50 years to depletion, weakly favored throughout the workshop) and its choice to reduce pumping by an equal volume, not percentage, which is relatively favorable to Juárez in a situation that would generally be very stressful.

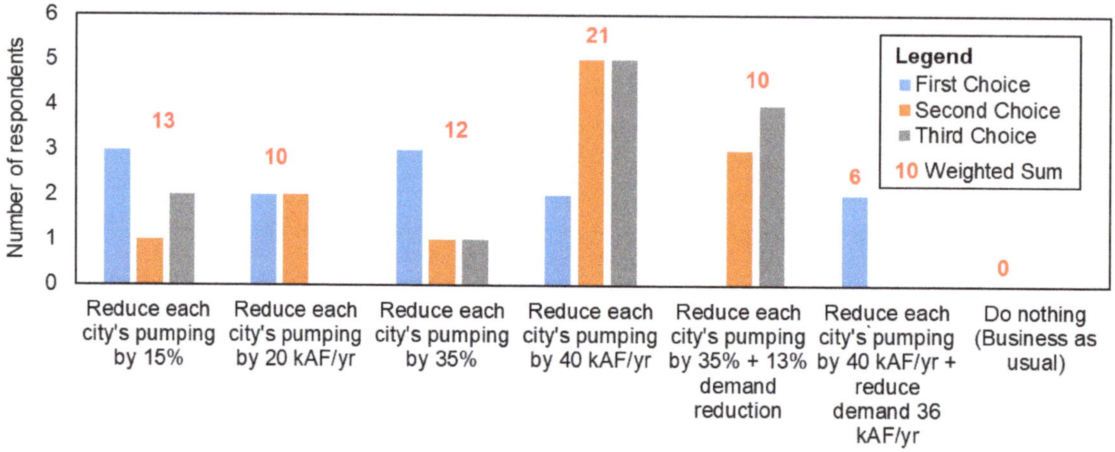

Figure 3. Results of polling on options for reducing depletion times (numerical values in red bold are weighted sums of choices).

We then moved to discuss possible policy measures. This conversation occurred in the frame of the achieved recognition that the burden and benefit of managing a common pool resource affected both sides—the US cannot go it alone—and acknowledgement that for Ciudad Juárez to reach reductions of pumping above 15%, external funding would be necessary. Conservation was identified as important for both cities. The most effective actions to achieve significant conservation included incentivized reduction in outdoor water use in EP and reduction of infrastructure leaks in Ciudad Juárez. For EP, incentivizing reductions in outdoor water use is more expensive than continuing to use local groundwater, but it is a relatively inexpensive way to reduce pumping. For Juárez, reducing infrastructure leaks is relatively expensive and it would require additional funding to reconstruct old water infrastructure. The participants also found desalination to be an attractive option for both sides: more expensive than local freshwater but less expensive than other interventions like long distance importation and direct potable reuse, but no consensus or clear direction emerged on the level and mix of specific measures.

Although we cannot say that the serious game resulted in a single, clear resolution, it certainly did constitute an effective common dialogue about a common concern and put many important perspectives and considerations on the table. Furthermore, it was shown that stakeholders were capable of interacting in an open forum to discuss sensitive issues of the common problem, binational water management, and to envision common solutions.

4. Discussion

The evolution of the conversation was notable and seemed to demonstrate a degree of social learning. In the end, no one proposed accepting the business-as-usual timeline of 31 years to freshwater depletion. However, target reductions in pumping varied widely (33%–99%). Over the course of the workshops, participants were capable of understanding the relevance, or importance, of joining a collaborative effort. That is, the serious games provided an opportunity to take a more holistic view, and to appreciate how we are all together in the "same boat" in facing aquifer depletion.

The discussions meaningfully brought out value-based issues in a format that was otherwise not available to participants. For example, when the discussion identified the difficulty that Ciudad Juárez would have in reducing pumping by much more than 15% without significant outside help, the dialogue in all sessions that followed took that concern seriously. Some voted for a 15% reduction or 20 kAF (its equivalent) in the final vote among options, and even those who voted for longer timelines did so with accompanying discussion of how to fund added help for Ciudad Juárez. The serious games format proved helpful in creating an open but thoughtful dialogue about important variables and values among actors of differing viewpoints, and in turn, in creating the shared community needed for common pool governance of the Hueco Bolson.

The small number of participants and the irregularity of participation were shortcomings for this activity. Ideally, it should be done in a format where the activity could be completed in one to two days and conducted in person, face-to-face. The participation would have been more consistent, the use of the model for determining impacts could have been more efficient, and the decision points could likely have been made with more consensus/agreement. In addition, the consensus building process would likely have been improved by having professional facilitators who were not part of the project and who were seen as more "neutral". The lack of consensus or agreement does not nullify our results but points to the difficulty of the decision making and to the need for thorough consideration, negotiation, and consensus building that would have been better accomplished at an in-person event.

The workshop sessions occurred between 9 and 12 months into the COVID-19 pandemic and the accompanying formal restrictions against travel across the border and holding in-person meetings in general. While we did not ormally question the participants about potential impacts of holding the workshop by teleconference, our impression is that most participants had experienced other meetings on these types of platforms, both before and after the conditions imposed by the COVID-19 pandemic. They felt comfortable expressing their opinions and broadly participated in the different tasks during the sessions. Nevertheless, some participants expressed that an in-person setting may have generated a friendlier environment because they would have mingled with other participants during the process. In addition, participants could not easily have one on one discussions among themselves as they might have at an in person event. For the session moderators, it was not easy to follow who was present at the sessions, and thus to ensure that everyone had a chance to speak and provide input. In addition, the session moderators noted that facilitators could have helped participants use the spreadsheet models at an in person event and perhaps this would have motivated the participant to experiment with the spreadsheet between sessions.

The participants generally agreed that binational cooperation and solutions are needed, but the dialogue was still partly limited by institutional (methodological) nationalism, evident in for example, the tendency to allocate quantitative responsibilities and costs to major utilities of each respective country. However, considerable progress was made in discussing problems and solutions as a shared problem, and this is not a trivial outcome. Considering the long and tricky history of tension and suspicion—as well as cooperation—between the two countries, and especially considering the political context of the serious games during the time period of late 2020, the tenor of the serious games should be viewed positively. Furthermore, trust was a very important moral value between participant stakeholders since most of them were very interested as well as knowledgeable of the current situation on binational water resources. Some of them had a long history of local interest on the issue and direct involvement on addressing water problems along both sides of the border and the serious games approach facilitated a common ground approach to a complex, binational water management problem.

Finally, the serious games intervention was but a moment in the long-term trajectory of the management of transnational water. Such an approach had not been applied in this region before, despite the recognized history of binational collaboration in regard to transboundary water resources at the Paso del Norte. The history of past US-Mexico border water agreements is very incremental and multiactor, but such an approach is a first step into potential informal agreements to extend the life of the most important water resource in the region, one which if completely depleted would result in catastrophic consequences for society on both sides of the border. Notably, our approach supports the Texas–Mexico stakeholder survey findings in [29] suggesting starting with incremental regional arrangements. Such platforms keep stakeholders talking and informed about the main water issues that could affect future sustainable development along a critical section of US–Mexico border.

Author Contributions: Conceptualization, A.M., J.H., A.G.-O., W.H., M.S.; methodology, A.M., J.H., A.G.-O., W.H., M.S.; analysis, A.M., J.H., A.G.-O., W.H., M.S.; writing—original draft preparation, A.M., J.H., A.G.-O., W.H., M.S.; writing—review and editing, A.M., J.H., A.G.-O., W.H., M.S., E.M., L.C.A.-C., A.V.-G.; project administration, W.H., J.H.; funding acquisition, W.H., J.H. All authors have read and agreed to the published version of the manuscript.

Funding: This material is based upon work supported by the National Institute of Food and Agriculture, U.S. Department of Agriculture, under award number 2015-68007-23130.

Institutional Review Board Statement: This study was approved by the Institutional Review Board (IRB) of the University of Texas at El Paso, Protocol 714550-1, approved 5 February 2015.

Informed Consent Statement: Informed consent was obtained from all subjects involved in the surveys.

Acknowledgments: We gratefully acknowledge the time and energy of the workshop participants. We acknowledge technical support from Patricia Esparza, Nuria Andreu-Garcia, and Janette Rodriguez, Carlo Ivan Romero-Gameros, and Carolina Salazar-Gonzalez.

Conflicts of Interest: The authors declare no conflict of interest.

References

1. Gleeson, T.; Wada, Y.; Bierkens, M.F.; Van Beek, L.P. Water balance of global aquifers revealed by groundwater footprint. *Nature* **2012**, *488*, 197–200. [CrossRef] [PubMed]
2. Wada, Y.; Van Beek, L.P.; Van Kempen, C.M.; Reckman, J.W.; Vasak, S.; Bierkens, M.F. Global depletion of groundwater resources. *Geophys. Res. Lett.* **2010**, *37*. [CrossRef]
3. Dalin, C.; Wada, Y.; Kastner, T.; Puma, M.J. Groundwater depletion embedded in international food trade. *Nature* **2017**, *543*, 700–704. [CrossRef] [PubMed]
4. Konikow, L.F.; Kendy, E. Groundwater depletion: A global problem. *Hydrogeol. J.* **2005**, *13*, 317–320. [CrossRef]
5. Aeschbach-Hertig, W.; Gleeson, T. Regional strategies for the accelerating global problem of groundwater depletion. *Nat. Geosci.* **2012**, *5*, 853–861. [CrossRef]
6. Jasechko, S.; Perrone, D. Global groundwater wells at risk of running dry. *Science* **2021**, *372*, 418–421. [CrossRef]
7. Ostrom, E. *Governing the Commons: The Evolution of Institutions for Collective Action*; Cambridge University Press: Cambridge, UK, 1990.
8. Albrecht, T.R.; Varady, R.G.; Zuniga-Teran, A.; Gerlak, A.K.; De Grenade, R.R.; Lutz-Ley, A.; Martín, F.; Megdal, S.B.; Meza, F.; Melgar, D.O.; et al. Unraveling transboundary water security in the arid Americas. *Water Int.* **2018**, *43*, 1075–1113. [CrossRef]
9. Hathaway, D.L. Transboundary Groundwater Policy: Developing Approaches in the Western and Southwestern United States 1. *J. Am. Water Resour. Assoc.* **2011**, *47*, 103–113. [CrossRef]
10. Rivera, A.; Candela, L. Fifteen-year experiences of the internationally shared aquifer resources management initiative (ISARM) of UNESCO at the global scale. *J. Hydrol. Reg. Stud.* **2018**, *20*, 5–14. [CrossRef]
11. Eckstein, G.E. Managing buried treasure across frontiers: The international Law of Transboundary Aquifers. *Water Int.* **2011**, *36*, 573–583. [CrossRef]
12. Megdal, S.B.; Scott, C.A. The importance of institutional asymmetries to the development of binational aquifer assessment programs: The Arizona-Sonora experience. *Water* **2011**, *3*, 949–963. [CrossRef]
13. Hayton, R.D.; Utton, A.E. Transboundary groundwaters: The Bellagio draft treaty. *Nat. Resour. J.* **1989**, *29*, 663.
14. Eckstein, G.; Hardberger, A. State practice in the management and allocation of transboundary ground water resources in North America. *Yearb. Int. Environ. Law* **2007**, *18*, 96. [CrossRef]
15. Aureli, A.; Eckstein, G. Strengthening cooperation on transboundary groundwater resources. *Water Int.* **2011**, *36*, 549–556. [CrossRef]
16. Puri, S.; Appelgren, B.; Arnold, G.; Aureli, A.; Burchi, S.; Burke, S.; Margat, J.; Pallas, P. *Internationally Shared (Transboundary) Aquifer Resources Management: Their Significance and Sustainable Management*; United Nations Educational, Scientific and Cultural Organization (UNESCO): Paris, France, 2001.
17. Puri, S.; Aureli, A. Transboundary aquifers: A global program to assess, evaluate, and develop policy. *Groundwater* **2005**, *43*, 661–668. [CrossRef] [PubMed]
18. Varady, R.G.; Scott, C.A.; Megdal, S.B.; McEvoy, J.P. *Transboundary Aquifer Institutions, Policies, and Governance: A Preliminary Inquiry*; Citeseer: Princeton, NJ, USA, 2010.
19. Szocs, T.; Rman, N.; Rotár-Szalkai, Á.; Tóth, G.; Lapanje, A.; Černák, R.; Nádor, A. The upper pannonian thermal aquifer: Cross border cooperation as an essential step to transboundary groundwater management. *J. Hydrol. Reg. Stud.* **2018**, *20*, 128–144. [CrossRef]

20. Eckstein, G. *The International Law of Transboundary Groundwater Resources*, 1st ed.; Earthscan Water Text; Routledge: Abingdon, UK; New York, NY, USA, 2017. [CrossRef]
21. Sindico, F. *International Law and Transboundary Aquifers*; Edward Elgar Publishing: Cheltenham, UK; Northampton, MA, USA, 2020.
22. de los Cobos, G. The Genevese transboundary aquifer (Switzerland-France): The secret of 40 years of successful management. *J. Hydrol. Reg. Stud.* **2018**, *20*, 116–127. [CrossRef]
23. Eckstein, G. The Newest Transboundary Aquifer Agreement: Jordan and Saudi Arabia Cooperate Over the Al-Sag/Al-Disi Aquifer. International Water Law Project Blog. 2015. Available online: https://www.internationalwaterlaw.org/blog/2015/08/31/the-newest-transboundary-aquifer-agreement-jordan-and-saudi-arabia-cooperate-over-the-al-sag-al-disi-aquifer/ (accessed on 13 November 2020).
24. Hussein, H. The Guarani Aquifer System, highly present but not high profile: A hydropolitical analysis of transboundary groundwater governance. *Environ. Sci. Policy* **2018**, *83*, 54–62. [CrossRef]
25. Sindico, F.; Hirata, R.; Manganelli, A. The Guarani Aquifer System: From a Beacon of hope to a question mark in the governance of transboundary aquifers. *J. Hydrol. Reg. Stud.* **2018**, *20*, 49–59. [CrossRef]
26. International Boundary and Water Commission (USIBWC). Convention between the United States and Mexico: Equitable Distribution of the Waters of the Rio Grande. International Boundary and Water Commission; 1906. Available online: https://www.ibwc.gov/Files/1906Conv.pdf (accessed on 30 June 2021).
27. Mumme, S.P. Minute 242 and beyond: Challenges and opportunities for managing transboundary groundwater on the Mexico–US border. *Nat. Resour. J.* **2000**, *40*, 341–378.
28. Tapia-Villaseñor, E.M.; Megdal, S.B. The US-Mexico Transboundary Aquifer Assessment Program as a Model for Transborder Groundwater Collaboration. *Water* **2021**, *13*, 530. [CrossRef]
29. Sanchez, R.; Eckstein, G. Groundwater Management in the Borderlands of Mexico and Texas: The Beauty of the Unknown, the Negligence of the Present, and the Way Forward. *Water Resour. Res.* **2020**, *56*. [CrossRef]
30. Callegary, J.; Megdal, S.; Villaseñor, E.T.; Petersen-Perlman, J.; Sosa, I.M.; Monreal, R.; Gray, F.; Noriega, F.G. Findings and lessons learned from the assessment of the Mexico-United States transboundary San Pedro and Santa Cruz aquifers: The utility of social science in applied hydrologic research. *J. Hydrol. Reg. Stud.* **2018**, *20*, 60–73. [CrossRef]
31. Payan, T.; Cruz, P.L. Introduction. In Binational Commons: Institutional Development and Governance on the U.S.-Mexico Border. 2020. Available online: https://www.jstor.org/stable/10.2307/j.ctv16b77c6 (accessed on 15 July 2021).
32. Alvarez, R. Borders and Bridges: Exploring a New Conceptual Architecture for (U.S.-Mexico) Border Studies: Borders and Bridges. *J. Lat. Am. Caribb. Anthropol.* **2012**, *17*, 24–40. [CrossRef]
33. Payan, T.; Cruz, P.L.; Villarreal, C.P. Human Mobility at the US-Mexico Border. In *Binational Commons: Institutional Development and Governance on the US-Mexico Border*; University of Arizona Press: Tucson, AZ, USA, 2020; p. 318.
34. Jeffery, C.; Schakel, A.H. Insights: Methods and Data beyond Methodological Nationalism. *Reg. Stud.* **2013**, *47*, 402–404. [CrossRef]
35. Moore, S.A.; Rosenfeld, H.; Nost, E.; Vincent, K.; Roth, R.E. Undermining methodological nationalism: Cosmopolitan analysis and visualization of the North American hazardous waste trade. *Environ. Plan. A Econ. Space* **2018**, *50*, 1558–1579. [CrossRef]
36. Far West Texas Water Planning Group. *Far West Texas Water Plan*; Texas Water Development Board: Austin, TX, USA, 2021.
37. Rivera, A. What Should We Manage: Aquifers or Groundwater? *Groundwater* **2021**. [CrossRef] [PubMed]
38. Davidow, J. *The US and Mexico: The Bear and the Porcupine*; Markus Wiener Pub: Princeton, NJ, USA, 2004.
39. Zeitoun, M.; Warner, J. Hydro-hegemony—A framework for analysis of trans-boundary water conflicts. *Water Policy* **2006**, *8*, 435–460. [CrossRef]
40. Pahl-Wostl, C. Transitions towards adaptive management of water facing climate and global change. *Water Resour. Manag.* **2007**, *21*, 49–62. [CrossRef]
41. Pahl-Wostl, C.; Craps, M.; Dewulf, A.; Mostert, E.; Tabara, D.; Taillieu, T. Social learning and water resources management. *Ecol. Soc.* **2007**, *12*. [CrossRef]
42. Pahl-Wostl, C.; Sendzimir, J.; Jeffrey, P.; Aerts, J.; Berkamp, G.; Cross, K. Managing change toward adaptive water management through social learning. *Ecol. Soc.* **2007**, *12*. [CrossRef]
43. Huitema, D.; Mostert, E.; Egas, W.; Moellenkamp, S.; Pahl-Wostl, C.; Yalcin, R. Adaptive water governance: Assessing the institutional prescriptions of adaptive (co-) management from a governance perspective and defining a research agenda. *Ecol. Soc.* **2009**, *14*. [CrossRef]
44. Rodela, R.; Ligtenberg, A.; Bosma, R. Conceptualizing serious games as a learning-based intervention in the context of natural resources and environmental governance. *Water* **2019**, *11*, 245. [CrossRef]
45. Madani, K.; Pierce, T.W.; Mirchi, A. Serious games on environmental management. *Sustain. Cities Soc.* **2017**, *29*, 1–11. [CrossRef]
46. Den Haan, R.-J.; Van der Voort, M.C. On evaluating social learning outcomes of serious games to collaboratively address sustainability problems: A literature review. *Sustainability* **2018**, *10*, 4529. [CrossRef]
47. Hallinger, P.; Wang, R.; Chatpinyakoop, C.; Nguyen, V.-T.; Nguyen, U.-P. A bibliometric review of research on simulations and serious games used in educating for sustainability, 1997–2019. *J. Clean. Prod.* **2020**, *256*, 120358. [CrossRef]
48. Liarakou, G.; Sakka, E.; Gavrilakis, C.; Tsolakidis, C. Evaluation of serious games, as a tool for education for sustainable development. *Eur. J. Open Distance E Learn.* **2012**, *15*, 96.

49. Wang, K.; Davies, E.G.R. A water resources simulation gaming model for the Invitational Drought Tournament. *J. Environ. Manag.* **2015**, *160*, 167–183. [CrossRef] [PubMed]
50. Jarvis, W.T. Scientific Mediation through Serious Gaming Facilitates Transboundary Groundwater Cooperation. *Water Resour. Impact* **2018**, *20*, 3.
51. Bolte, J.P.; Hulse, D.W.; Gregory, S.V.; Smith, C. Modeling biocomplexity–actors, landscapes and alternative futures. *Environ. Model. Softw.* **2007**, *22*, 570–579. [CrossRef]
52. Hill, H.; Hadarits, M.; Rieger, R.; Strickert, G.; Davies, E.G.; Strobbe, K.M. The invitational drought tournament: What is it and why is it a useful tool for drought preparedness and adaptation? *Weather. Clim. Extrem.* **2014**, *3*, 107–116. [CrossRef]
53. Bathke, D.J.; Haigh, T.; Bernadt, T.; Wall, N.; Hill, H.; Carson, A. Using Serious Games to Facilitate Collaborative Water Management Planning Under Climate Extremes. *J. Contemp. Water Res. Educ.* **2019**, *167*, 50–67. [CrossRef]
54. Gomes, S.; Hermans, L.; Islam, K.; Huda, S.; Hossain, A.; Thissen, W. Capacity Building for Water Management in Peri-Urban Communities, Bangladesh: A Simulation-Gaming Approach. *Water* **2018**, *10*, 1704. [CrossRef]
55. Ferrero, G.; Bichai, F.; Rusca, M. Experiential Learning through Role-Playing: Enhancing Stakeholder Collaboration in Water Safety Plans. *Water* **2018**, *10*, 227. [CrossRef]
56. Medema, W.; Mayer, I.; Adamowski, J.; Wals, A.E.J.; Chew, C. The Potential of Serious Games to Solve Water Problems: Editorial to the Special Issue on Game-Based Approaches to Sustainable Water Governance. *Water* **2019**, *11*, 2562. [CrossRef]
57. Medema, W.; Furber, A.; Adamowski, J.; Zhou, Q.; Mayer, I. Exploring the Potential Impact of Serious Games on Social Learning and Stakeholder Collaborations for Transboundary Watershed Management of the St. Lawrence River Basin. *Water* **2016**, *8*, 175. [CrossRef]
58. Marini, D.; Medema, W.; Adamowski, J.; Veissière, S.; Mayer, I.; Wals, A. Socio-Psychological Perspectives on the Potential for Serious Games to Promote Transcendental Values in IWRM Decision-Making. *Water* **2018**, *10*, 1097. [CrossRef]
59. Meinzen-Dick, R.; Janssen, M.A.; Kandikuppa, S.; Chaturvedi, R.; Rao, K.; Theis, S. Playing games to save water: Collective action games for groundwater management in Andhra Pradesh, India. *World Dev.* **2018**, *107*, 40–53. [CrossRef]
60. Goodspeed, R.; Babbitt, C.; Briones, A.L.G.; Pfleiderer, E.; Lizundia, C.; Seifert, C.M. Learning to Manage Common Resources: Stakeholders Playing a Serious Game See Increased Interdependence in Groundwater Basin Management. *Water* **2020**, *12*, 1966. [CrossRef]
61. Aubert, A.H.; Bauer, R.; Lienert, J. A review of water-related serious games to specify use in environmental Multi-Criteria Decision Analysis. *Environ. Model. Softw.* **2018**, *105*, 64–78. [CrossRef]
62. Adamatti, D.F.; Sichman, J.S.; Rabak, C.; Bommel, P.; Ducrot, R. JogoMan: A Prototype Using Multi-Agent-Based Simulation and Role-Playing Games in Water Management. In Proceedings of the Conference on Multi-Agent Modelling for Environmental Management, Les Arcs, Bourg Saint Maurice, France, 21–25 March 2005.
63. Kocher, M.; Martin-Niedecken, A.L.; Li, Y.; Kinzelbach, W.; Wang, H.; Bauer, R.; Lunin, L. Save the Water—A China water management game project. In Proceedings of the 3rd International GamiFIN Conference (GamiFIN 2019), Levi, Finland, 8–10 April 2019. [CrossRef]
64. Dray, A.; Perez, P.; Le Page, C.; D'Aquino, P.; White, I. Who wants to terminate the game? The role of vested interests and metaplayers in the ATOLLGAME experience. *Simul. Gaming* **2007**, *38*, 494–511. [CrossRef]
65. Bots, P.W.; van Daalen, C.E. Participatory model construction and model use in natural resource management: A framework for reflection. *Syst. Pract. Action Res.* **2008**, *21*, 389. [CrossRef]
66. Sanchez, R.; Rodriguez, L.; Tortajada, C. Transboundary aquifers between Chihuahua, Coahuila, Nuevo Leon and Tamaulipas, Mexico, and Texas, USA: Identification and categorization. *J. Hydrol. Reg. Stud.* **2018**, *20*, 74–102. [CrossRef]
67. Eastoe, C.J.; Hibbs, B.J.; Olivas, A.G.; Hogan, J.F.; Hawley, J.; Hutchison, W.R. Isotopes in the Hueco Bolson aquifer, Texas (USA) and Chihuahua (Mexico): Local and general implications for recharge sources in alluvial basins. *Hydrogeol. J.* **2008**, *16*, 737–747. [CrossRef]
68. Hibbs, B.; Eastoe, C.; Hawley, J.; Granados-Olivas, A. Multiyear study of the binational Hueco Bolson Aquifer reformulates key conceptual models of groundwater flow. In Proceedings of the 2015 UCOWWR/NIWR/CUAHSI Annual Conference, Water Is Not for Gambling: Utilizing Science to Reduce Uncertainty, Carbondale, IL, USA, 16–18 June 2015; pp. 81–86.
69. Granados-Olivas, A.; Creel, B.; Sanchez-Flores, E.; Chavez, J.; Hawley, J. Thirty years of groundwater evolution: Challenges and opportunities for binational planning and sustainable management of the transboundary Paso del Norte watersheds. In *The US-Mexican Border Environment: Progress and Challenges for Sustainability*; Lee, E., Ganster, P., Eds.; San Diego State University Press: San Diego, CA, USA, 2012; pp. 201–217.
70. Creel, B.J.; Hawley, J.W.; Kennedy, J.F.; Granados-Olivas, A. Groundwater resources of the New Mexico-Texas-Chihuahua border region. *New Mex. J. Sci.* **2006**, *44*, 11–29.
71. Hawley, J.W.; Kennedy, J.F.; Granados-Olivas, A.; Ortiz, M.A.; Hibbs, B.J. Hydrogeologic Framework of the Binational Western Hueco Bolsón-Paso del Norte Area, Texas, New Mexico, and Chihuahua: Overview and Progress Report on Digital-Model Development. In *WRRI Technical Completion Report*; The New Mexico Water Resources Research Institute: Las Cruces, NM, USA, 2009.
72. Heywood, C.E.; Yager, R.M. *Simulated Ground-Water Flow in the Hueco Bolson, an Alluvial-Basin Aquifer System Near El Paso, Texas*; US Department of the Interior, US Geological Survey: Reston, VA, USA, 2003; Volume 2, Number 4108.

73. Sheng, Z.; Mace, R.E.; Fahy, M.P. The Hueco Bolson: An aquifer at the crossroads. In *Aquifers of West Texas*; Texas Water Development Board: Austin, TX, USA, 2001; pp. 66–75.
74. Sheng, Z. Impacts of groundwater pumping and climate variability on groundwater availability in the Rio Grande Basin. *Ecosphere* **2013**, *4*, 1–25. [CrossRef]
75. Groschen, G.E. *Simulation of Ground-Water Flow and the Movement of Saline Water in the Hueco Bolson Aquifer, El Paso, Texas, and Adjacent Areas*; US Department of the Interior, US Geological Survey: Reston, VA, USA, 1994.

water

MDPI

Article

Modeling as a Tool for Transboundary Aquifer Assessment Prioritization

Ashley E. P. Atkins [1,*], Saeed P. Langarudi [2] and Alexander G. Fernald [1,2]

1 New Mexico Water Resources Research Institute, Las Cruces, NM 88003, USA; afernald@nmsu.edu
2 College of Agricultural, Consumer and Environmental Sciences, New Mexico State University, Las Cruces, NM 88003, USA; lang@nmsu.edu
* Correspondence: apage@nmsu.edu

Abstract: Transboundary aquifers are critical global water supplies facing unprecedented threats of depletion; existing efforts to assess these resources do not adequately account for the complexities of transboundary human and physical system interactions to the determinant of the impact of assessment outcomes. This study developed a system dynamics model with natural, human, and technical system components for a section of the transboundary Mesilla Basin/Conejos-Médanos aquifer to evaluate the following dynamic hypothesis: how and when information from a transboundary aquifer assessment is reported and perceived, in scenarios where two countries follow identical and different timeframes, dynamically impacts the behaviors of the shared aquifer. Simulation experiments were conducted to quantitatively assess the dynamics of transboundary aquifer assessment information reporting and perception delays. These critical feedbacks have not previously been incorporated practically in simulation and analysis. Simulation results showed that the timing and content of reporting can change the dynamic behavior of natural, human, and technical components of transboundary aquifer systems. This study demonstrates the potential for modeling to assist with prioritization efforts during the data collection and exchange phases to ensure that transboundary aquifer assessments achieve their intended outcomes.

Keywords: transboundary aquifers; human and natural systems; assessment; system dynamics modeling

Citation: Atkins, A.E.P.; Langarudi, S.P.; Fernald, A.G. Modeling as a Tool for Transboundary Aquifer Assessment Prioritization. *Water* **2021**, *13*, 2685. https://doi.org/10.3390/w13192685

Academic Editors: Sharon B. Megdal and Anne-Marie Matherne

Received: 1 July 2021
Accepted: 22 September 2021
Published: 28 September 2021

Publisher's Note: MDPI stays neutral with regard to jurisdictional claims in published maps and institutional affiliations.

1. Introduction

International groundwater depletion jeopardizes the well-being of groundwater-dependent natural and human systems, as well as the global populations that rely on agriculture and other critical exports from regions impacted by these water scarcity trends [1–4]. The challenges associated with groundwater depletion are exacerbated for the world's 592 identified transboundary aquifers and groundwater bodies [5,6]. Transboundary groundwater systems, which exist in nearly every country, serve as critical water supplies for populations with distinctive characteristics, histories, and priorities [7,8]. Successful transboundary groundwater management necessitates data and information produced through efforts such as assessments [9]. Collecting and exchanging data to increase understanding regarding these shared resources has emerged as a foundational component of assessments [10].

While recognition has grown regarding the importance and structure of transboundary aquifer assessments, constraints exist that impede their success. Assessments are funded with finite resources and conducted by a limited number of professionals with capacity limits related to the amount of information and analysis they can produce. Logistical challenges, resulting from things such as a lack of data and meta-data standards and conflicting binational priorities and structures, further complicate exchange and coordination. Transboundary aquifers are part of complex, interconnected natural and human systems. Determining what steps an assessment can take to achieve its objectives requires consideration of not just what information it produces, but how and when this information is communicated and perceived within these complex systems. Given these realities, where should assessment resources be directed to produce the most impactful results? This study investigated modeling as a tool to assist with the prioritization of efforts within the data collection and exchange phase of transboundary aquifer assessments, hypothesizing that for scenarios in which two countries follow either identical or different timeframes, how and when information from a transboundary aquifer assessment is reported and perceived can dynamically impact behaviors of the shared water system. It should be noted that, while this study looks at a scenario for an aquifer shared between two countries, many transboundary aquifers are shared by more than two countries.

This research utilized system dynamics modeling [11] because the foundational structure of system dynamics maintains inherent similarities to hydrologic structures and the non-linear feedback characteristic of human and natural systems [12]. Similar to all models, the model developed in this study is only an abstract simplification of the problem [13]. The model in this study represents a simplification of interconnected natural and human components of a transboundary groundwater system to help make sense of its complexities. The model development process for this research was guided by the acknowledgment of the dominant influence of human behavior, executed through human decision-making, on hydrologic systems. This acknowledgment is the central driver of this study, which attempts to progress understanding of the complexities of human decision-making in innovative ways to understand hydrologic trends within the Anthropocene for transboundary systems.

The model explored the potential role of reporting and perception delays of water availability information in a transboundary groundwater system, positing that researchers can use modeling to understand interconnected human and natural processes to analyze the systemic impact of potential transboundary aquifer assessment efforts [14]. The model investigated the potential role of reporting and perception delays of water availability information in a transboundary groundwater system. This research addressed the following questions: Should assessments focus solely on the type of information they aim to produce and exchange? Or does how and when that information is reported and perceived necessitate prioritization? How do similarities and differences between nations regarding information reporting and perception delays manifest within a transboundary system? This study examined the dynamic hypothesis that how and when information from a transboundary aquifer assessment is reported and perceived, in scenarios where two countries follow identical and different timeframes, impacts the behaviors of the shared water system. A dynamic hypothesis is a logical explanation that relates the feedback structure of a complex system to its dynamic behavior [15].

Simulation experiments were conducted that investigate different reporting and perception delay realities in scenarios where two countries that share an aquifer system pursue identical and different transboundary aquifer assessment timeframes. Results that display oscillations indicate instability in the system. As an example, throughout the COVID-19 pandemic, oscillatory trends have persisted as decision-makers attempted to react to rapidly reported and perceived information and find a balance between heightening or loosening social-distancing related restrictions [16]. These oscillatory behaviors are not unique to the COVID-19 pandemic or to transboundary groundwater decision-making. The presence of these oscillatory behaviors indicates the need for policy optimization to achieve more stable outcomes.

2. Materials and Methods

The study site for this model encompasses the internationally neighboring communities of Sunland Park and Santa Teresa in New Mexico, United States (U.S.) and Anapra and San Jerónimo in Chihuahua, Mexico. These populations utilize a portion of the Mesilla Basin/Conejos-Médanos, which is one of four priority transboundary aquifers along the U.S.–Mexico border designated through the Transboundary Aquifer Assessment

Program [17]. The U.S. and Mexico manage the Mesilla Basin/Conejos-Médanos separately. The aquifer supports the populations that live within the area of the Mesilla Basin/Conejos-Médanos; additionally, these resources supply water for industrial operations on both sides of the border. In Mexico, water from the Mesilla Basin/Conejos-Médanos is pumped to meet the growing demands of neighboring Ciudad Juárez. A map of the study site is available in the Supplementary Materials. See [18] for further background about this study site. The model developed for this research depicts simplified natural, human, and technical components to better understand system behaviors and outcomes (Figure 1). Core behaviors and interconnections for this transboundary region have been modeled previously [18,19]. This study expands on the assumptions from those past efforts, which is explained in the subsections below, to facilitate the quantitative analysis of reporting and perception delays within a transboundary aquifer assessment process. While the model is based specifically on a section of the Mesilla Basin/Conejos-Médanos, the applicability of core behaviors to other transboundary systems makes the findings from the dynamic results insightful for arid and semi-arid regions with transboundary aquifers.

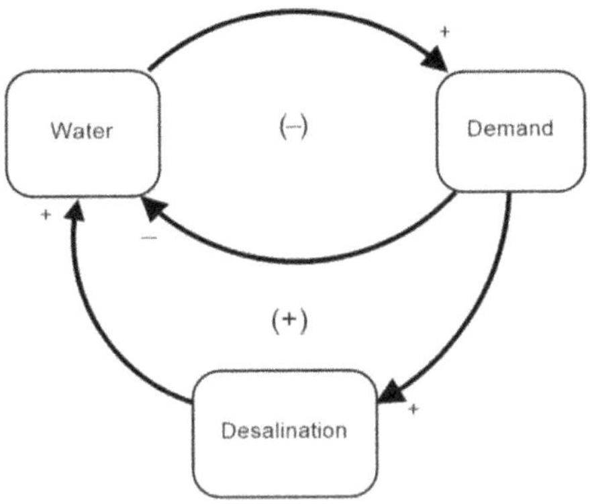

Figure 1. Showcases a simplified representation of the relationships between the model's three primary components.

The model developed for this study has three primary components, water, demand, and desalination (Figure 1), that are detailed in their correspondingly titled subsections. In Figure 1, the (+) indicates a positive or reinforcing loop, and the (−) indicates a negative or balancing loop; this structure was validated in water resources research modeling [20,21]. The water module contains hydrologic dynamics, the demand module contains components relating to the dynamics of water demand, and the desalination module contains components relating to the dynamics of desalination infrastructure and operation. Each module's subsection has a figure, referred to as a stock and flow diagram, that depicts its key components. These stock and flow diagrams are simplifications and do not include all the information needed to recreate the model. The stock and flow diagrams were developed in Stella Architect. Full model documentation is available in the Supplementary Materials section. Information about how to access an online version of the model that allows users to view and experiment with the model is included in the Supplementary Materials. The following standard system dynamics confidence building tests were utilized in the development of this model: boundary adequacy, structure assessment, dimensional consistency, parameter assessment, extreme conditions, integration error, behavior anomaly, surprise behavior, and sensitivity analysis [15]. From these tests, we confirmed that the feedbacks (see [18] for exogenous, endogenous, and excluded parameters) were necessary and aligned with the established hydrologic and decision-making theory. The dimensions were determined to be consistent, and the model underwent rigorous tests under extreme scenarios to ensure that it responded appropriately and logically.

2.1. Water

The quantity and quality-related water dynamics for this study were all rigorously validated through the standard system dynamics confidence building tests discussed in Section 2 [18]. They align with key hydrologic research findings for the Mesilla Basin/Conejos-Médanos [22–28]. This region of the Mesilla Basin/Conejos-Médanos relies almost solely on groundwater for their drinking water supply. As such, the model only investigates groundwater dynamics. Many transboundary aquifers around the world have intrinsic though not fully understood connections to surface water [29]. While these realities do not apply in the study area for this research, the dynamics of surface water and groundwater connectivity should be accounted for when trying to understand the impact of transboundary aquifer assessments on behaviors for regions dependent on both supplies.

The stock and flow diagrams for each module use standard system dynamics modeling representation. Stocks, depicted in Figure 2 as rectangles, are a fundamental part of system dynamics modeling. They are measurable quantities, such as the brackish water stock in Figure 2. System dynamics models allow users to pursue an analysis that accounts for ranges of stock quantities, which reflects the uncertainty that oftentimes exists regarding quantities of freshwater in aquifers. Model stocks change based on model inflows and model outflows. In the model, freshwater withdrawal is a model outflow from freshwater and a model inflow to withdrawn water. The symbology utilized for freshwater withdrawal represents a flow, and it is used throughout the model to depict flows. As an example, a well that pumps freshwater from an aquifer represents a model outflow that decreases the aquifer's freshwater stock, but a model inflow increases the stock of water withdrawn from the aquifer. Converters are components of the system that indirectly affect stocks by directly impacting other converters or flows that are connected to stocks. Circles are used to show converters in this study; stored water availability serves as an example of a converter (Figure 2). Stocks and converters outlined with dotted lines differentiate components that come from a different module, such as the demand or desalination module in the context of Figure 2. Solid black arrows portray connections from stocks to converters, flows to flows, converters to converters, and converters to flows.

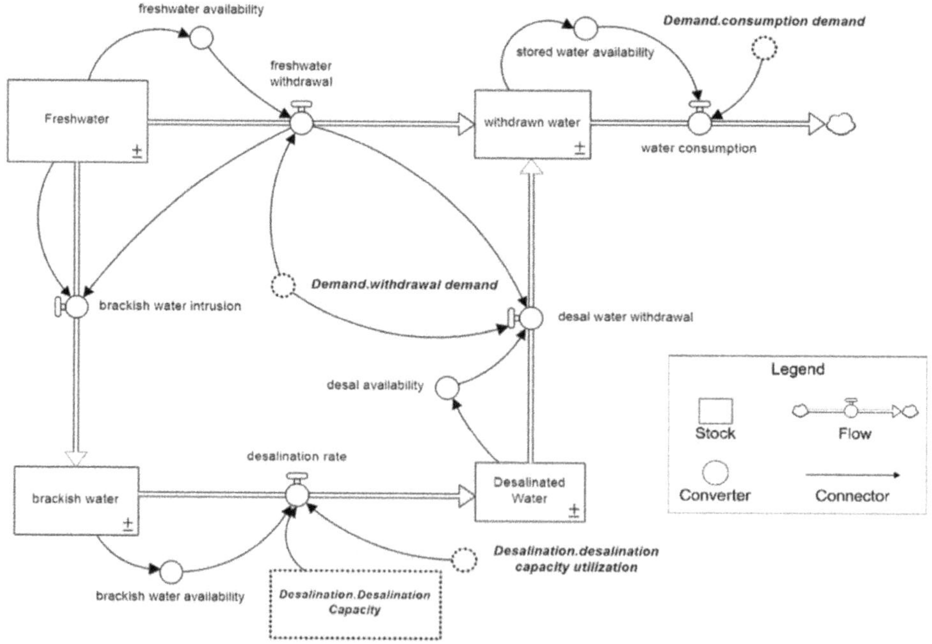

Figure 2. A simplification of the model's water component.

2.2. Demand

The demand module (Figure 3) explores human decision-making dynamics within the context of water demand. It centers around the understanding that perceptions of water availability and the water demand gap are key drivers of water demand [18]. This module was developed based on assumptions rooted in historical water demand trends and system interconnections. In this model, reported demand gap influences are perceived as a water demand gap [18]. While it is commonplace for water models that incorporate demand to calculate demand based primarily on population growth, data from the U.S. Geological Survey (USGS) shows that population increases to do not mean water usage will increase [30]. Our model does not calculate water as a function solely tied to population growth; in this model, the reported demand gap influences the perceived water demand gap [18]. Despite population growth, water use in the United States by 2010 was less than it was in 1970 [31]. This model relies on the assumption that perceptions of water availability have a critical influence on water demand. For example, perceptions of abundant groundwater availability led Albuquerque, New Mexico residents to use a peak of 272 gallons per person per day in 1989 [32,33]. Data instead revealed trends of groundwater depletion from a finite aquifer. In response, the city reduced its per capita water use to approximately 121 gallons per day by 2019, to reflect its updated perceptions of water availability [34–37]. Similar examples on different scales are abundant throughout history.

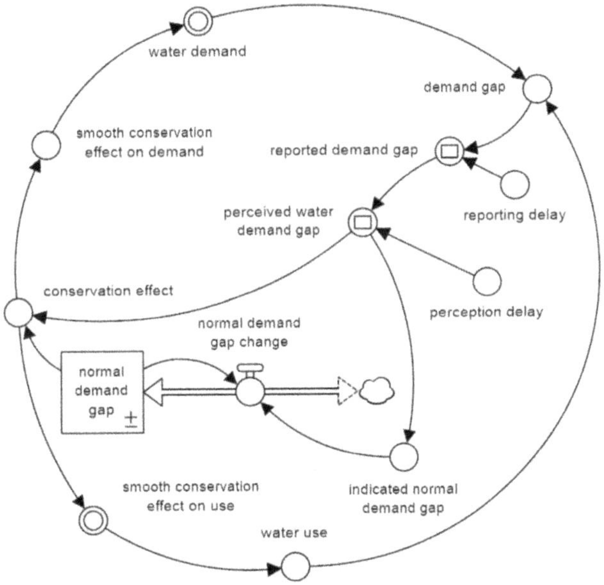

Figure 3. Key components of the demand module.

The reported demand gap is not instantaneously produced, and in this model, variations in reporting delay realities impact this timeline. The Results and Discussion section describes the context of reporting delay realities in further detail. Reported information does not become perceived by the system or understood in a way that dynamically impacts perceptions of water demand and availability immediately. The study also evaluates varying perception delays timelines.

Both Country A and Country B in the model have identical demand modules. Figure 3 demonstrates the demand module for Country A. The generic names Country A and County B were chosen to reflect the transferability of this study outside the Mesilla Basin/Conejos-Médanos. The demand divide between both countries, however, is not identical. This model explores a scenario where Country A is a majority water user. Country B uses 20% of the water that Country A does. The water use divide reflects approximate water use distributions in the studied section of the Mesilla Basin/Conejos-Médanos [18]. Uneven divides between water use or the spatial distribution of transboundary aquifers across borders is a common reality. For example, approximately 90% of the Genevese Aquifer is in Switzerland, while approximately 10% is in neighboring France [38].

The model assumes that, to minimize a demand gap, demand must be decreased, or supply must be increased. Decreasing demand and increasing supply can occur simultaneously in this model. Increasing the water supply was investigated in this model through the implementation of inland desalination, which is discussed further in the desalination subsection below. The conservation effect in this study is an aggregate decision rule that acts based on water supply and demand. When the demand gap increases, the conservation effect accounts for scenarios where there is a collective response to reducing water demand. In Figure 3, the collective response comes from Country A. We assume that demand gap does not immediately impact the conservation effect decision rule. An anchoring and adjustment process takes place to produce a normal demand gap [39]. This module considers bounded rationality by taking this anchoring and adjustment heuristic (rule of thumb for decision-making) into account [40].

2.3. Desalination

The desalination module (Figure 4) reflects the reality that water decision-makers implement policies in the present to meet future needs. Policies that involve changes to built and natural environments, such as the implementation of desalination, have binding characteristics and cannot be easily adjusted. Desalination represents an alternative water supply option that can be pursued to increase freshwater supply in this study site. Inland desalination has specifically received attention as a potential policy for this region, as well as other arid and semi-arid inland regions. Pursuing desalination will impact the built and natural environments; including it as a policy option in the model provides insight into its dynamics for this and other inland regions considering desalination. The model simplifies the options decision-makers have available to decrease demand or increase supply to lessen the demand gap. The reported demand gap from both countries impacts the total demand gap. The policy perception delay represents the time between when reported information was perceived in a way that impacts society's perceptions of water availability and policies that reflect those perceptions were implemented. In the simulation experiments conducted in this study, the policy perception delay remains set at a constant 2 years for all runs. Differing political structures between countries that share transboundary aquifers likely means differing policy implementation timelines. Variations in this delay for transboundary aquifers need further investigation. The desalination component of the model was rigorously assessed [18].

2.4. Simulation Experiments

Both countries maintain equivalent reporting and perception delays in Runs 1–3 (Table 1). In Run 1, there is a 1-year reporting delay and a 2-year perception delay. In this scenario, water availability information was collected, analyzed, and reported within the span of 1 year. The reported information is understood and contributes to the public perception of water availability in the system, 2 years into the entire process. A 2-year policy perception delay exists in every run in this study. This delay accounts for the time between the information being perceived and implemented as policy. In Run 2, both countries exhibit a 5-year reporting delay and a 6-year perception delay. Similar to Run 1, there is a 1-year delay between the reporting delay and the perception delay. This 1-year delay remains consistent across all runs in the study. Run 3 represents the lengthiest cumulative delay, with a 10-year reporting delay and an 11-year perception delay.

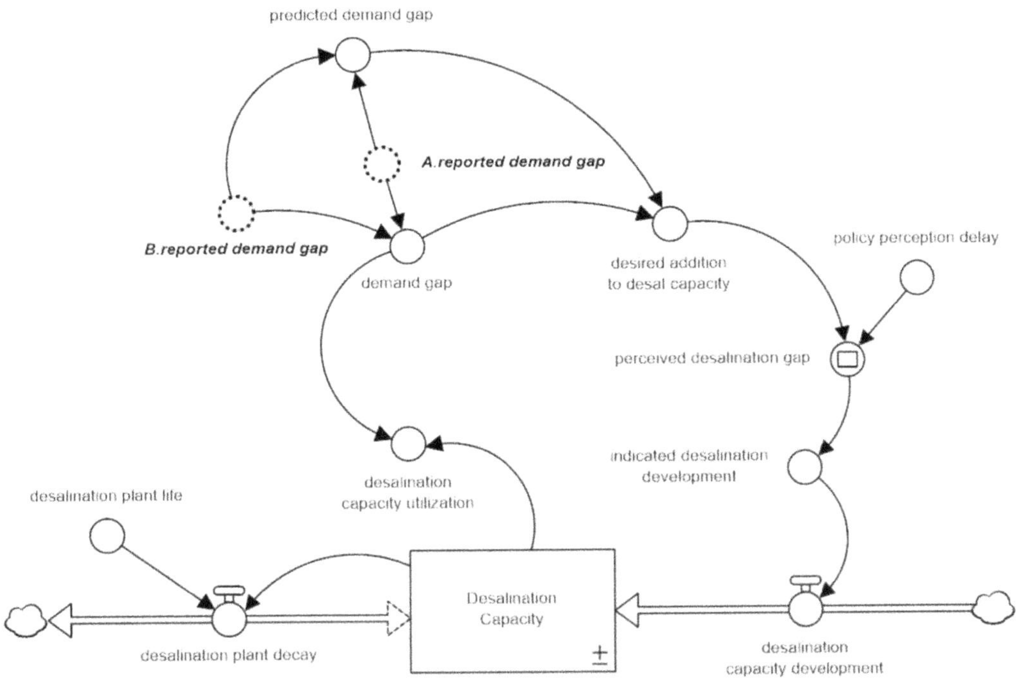

Figure 4. A depiction of the desalination module's core components.

As shown in Table 1, both countries maintain differing reporting and perception delays in Run 4 and Run 5. In each of these runs, one country has the reporting and perception delay from Run 1, while the other country has the reporting and perception delay from Run 2. In Run 4, Country A has a 1-year reporting delay and a 2-year perception delay; Country B has a 5-year reporting delay and a 6-year perception delay. In Run 5, Country A has a 5-year reporting delay and a 6-year perception delay; Country B has a 1-year reporting delay and a 2-year perception delay. In the absence of assessments or agreements, countries that share transboundary aquifers act on their own to develop, process, and implement information. However, in an ideal scenario, scenarios can pursue agreements or assessments together and develop, process, and implement information on the same timeline. This study investigates the impacts of both options: when two countries that share a transboundary groundwater system follow the same and different timelines.

Transboundary groundwater resources can be referred to as common pool resources [41,42]. Rather than homogenously examining the impact of transboundary aquifer assessments through the lens of common-pool resource theory, this study design recognizes that a plethora of dynamics within the system can and have in practice, as witnessed through TAAP, result in delays. The tests selected for this study are a way to empirically evaluate the impact of these heterogeneous delays on the system. A plethora of dynamics can contribute to delays; the delays in this study were chosen based on the assumptions below. The purpose of this study is not to recreate the entire system and examine every possible influencing factor for a delay. Rather, it is to provide dynamic insight into the impact that delays themselves might have on the system and the effectiveness of transboundary aquifer assessments.

The perception and reporting delays explored in Runs 1–5 are compared against different conservation parameters and maximum conservation in Runs 6–18. These parameters control people's response to water shortage in the model as shown by Equation (1), where f(x) is the effect of water shortage on water demand, x is normalized water shortage, m is maximum conservation, and p is conservation parameter. The conservation parameter (p) represents people's responsiveness in conservatory reaction to water shortage. The maximum conservation (m) places a limit on the quantity of water that the system can conserve. Equation (1) implies that water demand reacts to water shortage in the opposite direction, but the significance of this reaction depends on m and p.

$$f(x) = \max(m, 1 - p\,(x - 1)) \tag{1}$$

Runs 1–15 explore the same runs with different conservation parameters and a constant maximum conservation parameter. The runs with a 0.1 conservation parameter are the least sensitive, and the runs with a 0.9 conservation parameter are the most sensitive. The maximum conservation limit in Runs 1–15 means that water usage can be reduced by up to 50%. In Runs 16–18, the maximum conservation limit is set to reflect an extreme scenario of up to a 90% possible reduction in water usage.

Table 1. Details for each of the runs conducted in this study.

Run	Country	Perception Delay	Reporting Delay	Conservation Parameter	Maximum Conservation Parameter
1	A	1	2	0.5	0.5
1	B	1	2	0.5	0.5
2	A	5	6	0.5	0.5
2	B	5	6	0.5	0.5
3	A	10	11	0.5	0.5
3	B	10	11	0.5	0.5
4	A	1	2	0.5	0.5
4	B	5	6	0.5	0.5
5	A	5	6	0.5	0.5
5	B	1	2	0.5	0.5
6	A	1	2	0.1	0.5
6	B	1	2	0.1	0.5
7	A	5	6	0.1	0.5
7	B	5	6	0.1	0.5
8	A	10	11	0.1	0.5
8	B	10	11	0.1	0.5
9	A	1	2	0.1	0.5
9	B	5	6	0.1	0.5
10	A	5	6	0.1	0.5
10	B	1	2	0.1	0.5
11	A	1	1	0.9	0.5
11	B	1	2	0.9	0.5
12	A	5	6	0.9	0.5
12	B	5	6	0.9	0.5
13	A	10	10	0.9	0.5
13	B	11	11	0.9	0.5
14	A	1	2	0.9	0.5
14	B	5	6	0.9	0.5
15	A	5	6	0.9	0.5
15	B	1	2	0.9	0.5
16	A	1	2	0.1	0.9
16	B	1	2	0.1	0.9
17	A	1	2	0.5	0.9
17	B	1	2	0.5	0.9
18	A	1	2	0.9	0.9
18	B	1	2	0.9	0.9

The run periods were chosen to show differing cumulative reporting and perception delay realities. Beginning a study and producing reporting results within the span of 1 year, as exhibited in Run 1, is arguably an expedited timeline. The delay between submission and publication of peer-reviewed research alone can span a year or longer [43]. These delays, however, were shortened in the face of extreme circumstances such as the COVID-19 pandemic, which has expedited medical research and publication timelines [44,45]. In the context of extreme water-related local circumstances, such as elevated lead levels in Flint, Michigan and water shortages in Cape Town, South Africa, traditional timelines and procedures have also adapted [46–48]. The National Science Foundation's grants are generally awarded for no more than 5 years; Run 2 showcases behaviors associated with a 5-year reporting delay. The time between identifying an area of research that needs data collection, securing funding and resources, collecting data, analyzing data, and ultimately reporting that data and analysis can take much longer than 5 years. Run 3 shows these realities with a 10-year reporting delay. It should also be noted that reporting and perception delays can also extend well beyond the selected times from these runs, particularly for transboundary regions that face additional coordination challenges.

All simulation experiments in this study are conducted over a 50-year period. This period was selected to reflect a realistic planning horizon for the region. New Mexico and Texas, the two states on the U.S. side of the border that the Mesilla Basin/Conejos-Médanos falls within, either developed or are developing 50-year water plans. Regional and local planning horizons around the world vary; most fall within increments at or under a 50-year period. The 50-year planning horizons were utilized previously in system dynamics water modeling and simulations. While occasionally 100-year horizons are pursued by water managers, plans that extend beyond this are rare.

3. Results and Discussion

3.1. *Users with Identical Perception and Reporting Delays*

The perception and reporting delays associated with assessment information impact the dynamics of the shared system and lead to different freshwater quantity outcomes (Figure 5). Runs 1–3 showcase scenarios where both countries follow identical timelines; assessment information regarding water availability is reported, perceived, and implemented through policies at the same time for each country. These runs show that even a few years of difference in delays can impact the overall effectiveness of assessments. While these simulation experiments serve as a helpful baseline for understanding transboundary aquifer assessment behaviors and impacts, their level of synced coordination may be difficult to attain in practice, even in regions with established cooperation mechanisms. Therefore, it remains important to investigate the possibility of each county following a different timeline.

3.2. *Users with Different Perception and Reporting Delays*

The simulation experiments for Runs 4 and 5 show that differences between countries in how and when information from an assessment is reported and perceived ultimately affects outcomes for freshwater resources and system components connected with freshwater resources. Runs 4 and 5 have cumulative delay differences of 4 years between each country; there is a 4-year difference between the perception delays and the reporting delays in Runs 4 and 5 (Figure 5). These simulation experiments are intended to explore the potential impacts of coordinated assessments that progress on divergent timelines. The 4-year difference, while seemingly small compared with the 50-year reporting period, notably impacts the behavior of the system. These simulation experiments show that differences between countries in how and when information from a transboundary aquifer assessment is reported and perceived ultimately affects outcomes for freshwater resources. The reporting and perception delay maintained by Country A, the majority water user, dictates the overall withdrawal behaviors and freshwater behavior of the system as exhibited in the comparison between Run 1 and Run 4 (Figure 5). However, the reporting and perception delays of Country B have an influence on the overall behaviors. Comparing the behavior of the freshwater supply between Run 2 and Run 5 showcases an example of this minority water user influence (Figure 5).

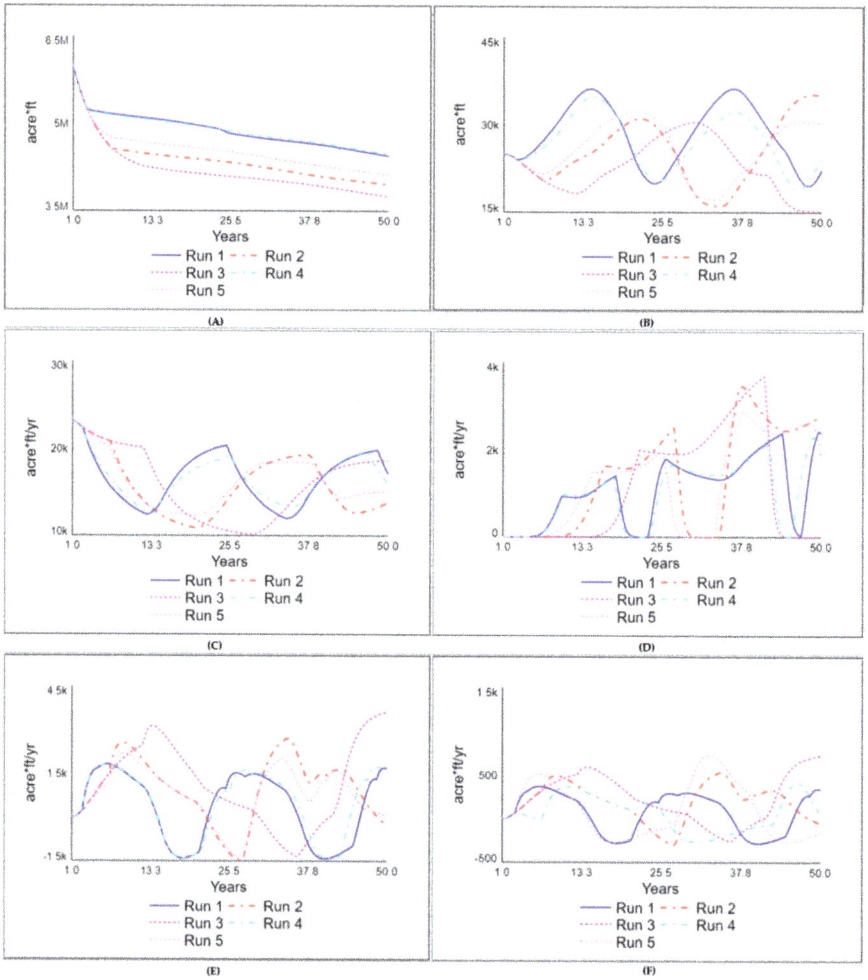

Figure 5. Results of Runs 1–5 for (**A**) freshwater; (**B**) withdrawn water; (**C**) freshwater withdrawal; (**D**) desalinated water withdrawal; (**E**) Country A demand gap; and (**F**) Country B demand gap. The name for each graph corresponds with the name of the selected stock in the model. For example, (**A**) showcases the dynamic quantity of freshwater in the system over the run period of 50 years.

3.3. Practical Implications

An important finding is that all simulation experiments in this study exhibit undesirable behaviors due to the oscillatory behavior of the results, which reflects instability in the system. The decision-makers in Run 1 react most rapidly to changes in freshwater availability and most predictably with almost identical amplitudes and periods for withdrawn water (Table 2). All runs start out on the same freshwater depletion trajectory. Run 1 reacts the most quickly and aggressively to the depletion trend and achieves the best result for the freshwater supply at the end of the 50-year simulation. Run 2 and Run 3 continue the same freshwater depletion trajectory as each other until they begin their staggered responses. The simulations showcase a tradeoff; scenarios with increased delays result in reduced oscillatory behaviors but lead to a greater use of water resources. Optimizing the runs to reduce volatility by maximizing the period and minimizing the amplitude of an oscillation can result in more stable outcomes. In the context of this study, an optimization of transboundary aquifer assessments would mean that both countries simultaneously receive and perceive water availability information from assessments quickly but react less aggressively and with more foresight for long-term, systemic trends.

Table 2. A description first and second amplitudes and periods for Runs 1–5 for withdrawn water, which is also displayed in Figure 5B.

Run	Amplitude 1	Amplitude 2	Period 1	Period 2
1	16.1	16.2	10	12
2	10.5	15	14	12
3	12.3	15.3	18	18
4	14.3	11.7	12	10
5	15.1	13.3	11	24

3.4. System Sensitivity

Runs 6–15 showcase behaviors for scenarios with a different conservation parameter (p) and a constant maximum conservation (m) (Figure 6). Note that p controls the strength of the negative (balancing) feedback loop in Figure 1, which goes through demand gap, water demand, and withdrawal. We know that negative feedback loops when coupled with significant delays can generate oscillatory behaviors [49]. Higher values of p strengthen this feedback loop and potentially generate greater oscillations. When p is set to 0.5, the model produces relatively large oscillatory behavior (Figure 5). Setting p to 0.9, such as it is in Runs 11–15, strengthens the feedback loop and produces similar dynamic behaviors (Figure 6). When p is set at 0.1 in Runs 6–10, the negative feedback loop is weakened and almost knocked off, and the system becomes insensitive in this case (Figure 6). In the context of this study, an assessment that increases sensitivity above the threshold has a negligible impact on the behaviors of the system. Assessment outputs produced below the threshold will likely not meet their intended outcomes, given the effects of the system's insensitivity.

3.5. Extreme Scenarios

Runs 16–18 test each of these three different p values in an extreme scenario where m is set to 0.9, meaning that the system can reduce water usage up to 90% (Figure 7). Even in this extreme scenario, the behaviors for p at 0.5 and 0.9 are similar, and the behavior for p at 0.1 results in less sensitive and less oscillatory behaviors. These results strengthen confidence in the model as they corroborate our previous knowledge of the systems. They also reveal that a threshold for p exists somewhere between 0.1 and 0.5. Making investments that increase the sensitivity of the system for conservation between the threshold and 0.9 have little impact on the overall behaviors of the system. In the context of a transboundary aquifer assessment, assessment outputs intended to increase system sensitivity can likely have a negligible impact on overall behaviors.

Assessments producing outputs in a system with sensitivity below the threshold likely cannot meet their intended outcomes given the absence of the critical feedback loop between availability and demand that drives behaviors in this scenario. Future research needs to identify the threshold for a system and for a transboundary system, the implications of differing thresholds. These initial results showcase the critical impact that conservation parameters can have on the outcomes of assessments.

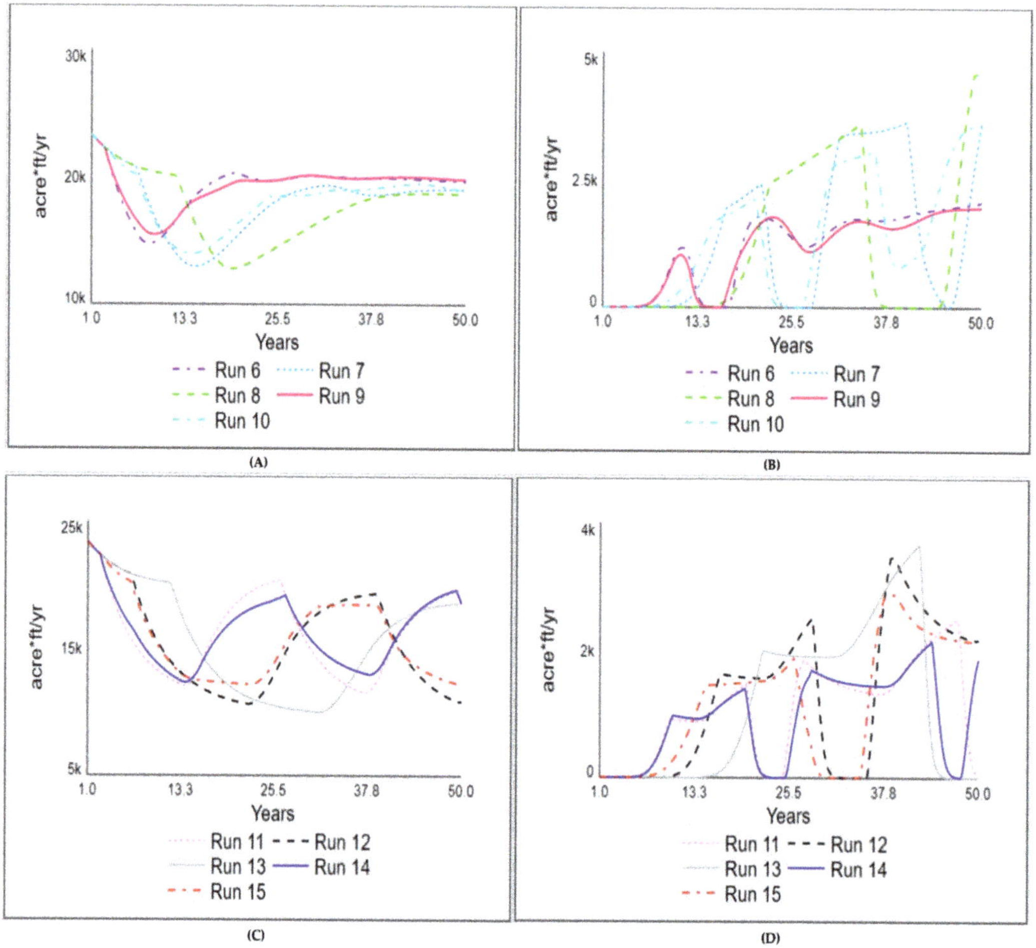

Figure 6. Results of runs 6–10 for (**A**) freshwater withdrawal and (**B**) desalinated water withdrawal. Results of runs 11–15 for (**C**) freshwater withdrawal and (**D**) desalinated water withdrawal.

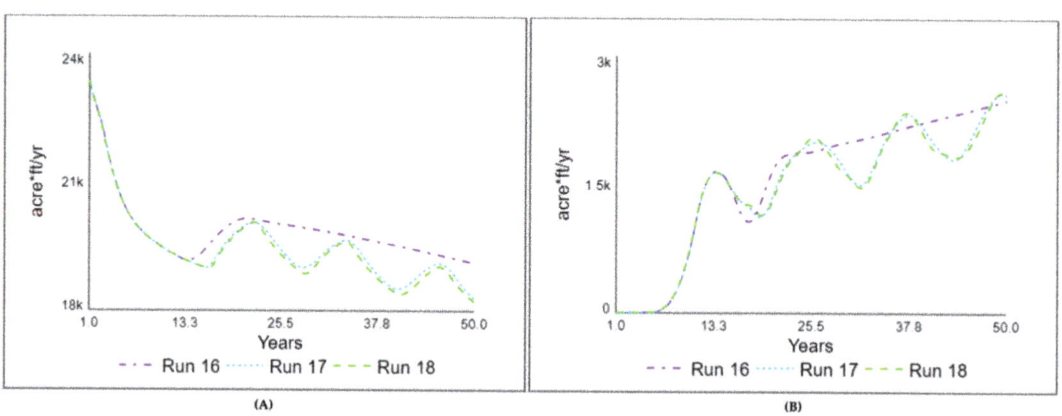

Figure 7. Results of Runs 16–18 for (**A**) freshwater withdrawal and (**B**) desalinated water withdrawal.

3.6. Future Work

Future research needs to identify the conservation parameter threshold for a system and, for transboundary resources, the implications of differing thresholds between countries. The findings regarding the threshold are also widely applicable beyond transboundary systems and should be incorporated more broadly in research investigating the potential impacts of water policies to ensure that decisions can meet their intended outcomes. Additional future work requires exploration into the complex relationships between information reporting, perception, and policy implementation. Langarudi et al. (2021) provides insight into breaking down the information perception process so that its individual components can be analyzed within a system [50]. To carry out our initial investigation, each run in this study had a 1-year gap between the reporting delay and the perception delay and a 2-year gap between the perception delay and the policy delay. These timeframes—and the complex feedbacks that drive them—are not always so straightforward. For example, the science that identified climate change is not new and has been increasing for decades [51,52].

Despite the depth of reported climate change information, decision-making behaviors have not reversed the global warming trends identified in scientific literature. How does the accumulation of reported information affect perceptions? How are policy implementation timeframes impacted by the formulation of divergent perceptions, either within a nation or between nations? These are just some of the questions that may need to be accounted for when pursuing future research. The intricacies of interactions between human decision-making and the hydrologic system might benefit from the innovative application of hybrid modeling that combines system dynamics and agent-based modeling methodologies. While system dynamics models are well-suited for hydrologic structures, agent-based models can allow for a more complex analysis of the rules that govern decision-making behaviors, particularly when studying questions with spatial components [53].

4. Conclusions

This study explored the dynamic hypothesis that how and when information from a transboundary aquifer assessment is reported and perceived impacts the behaviors of the shared water system. The simulation experiments showed a tradeoff; scenarios with reduced oscillatory behavior resulted in greater water use. Based on the evidence of the simulations, we conclude that the explored perception dynamics change the behavior of the transboundary water system. The simulations conducted in this study produced oscillatory behaviors that reflect instability in the system. Optimizing the runs to produce more stable results would mean that both countries receive and perceive water availability information from assessments on the same timeline and react to that information less aggressively and with long-term planning foresight. An optimization that accounts for the sensitivity of a system related to conservation parameters is also a key component in ensuring that the goals of an assessment are met and that investments are made efficiently. Determining how to accomplish this optimization within the complexities of human decision-making behaviors requires further investigation. Transboundary groundwater assessments have been recognized as key components of effective transboundary groundwater management. Understanding what impedes the success of assessments and how assessment characteristics impact the overall system are important areas of exploration to ensure that assessments achieve their intended outcomes. Modeling, as exhibited in this study, serves as a useful tool with the potential to assist with the prioritization efforts within the data collection and exchange phase of transboundary aquifer assessments.

Supplementary Materials: The Supplementary Materials are attached separately as part of the submission. The following are available online at https://www.mdpi.com/article/10.3390/w13192685/s1, 1: Model Access, 2: Map of Study Site, A map of the study site, as it appeared in Page et al., "A Dynamic Hydro-Socio-Technical Policy Analysis of Transboundary Desalination Development," Journal of Environmental Accounting and Management 7, no. 1 (2019). 3: Model Documentation.

Author Contributions: Conceptualization, A.E.P.A.; methodology, A.E.P.A., S.P.L.; software, A.E.P.A., S.P.L.; validation, A.E.P.A., S.P.L.; formal analysis, A.E.P.A., S.P.L.; investigation, A.E.P.A.; resources, A.G.F.; data curation, A.E.P.A., S.P.L.; writing—original draft preparation, A.E.P.A.; writing—review and editing, A.E.P.A., A.G.F., S.P.L.; visualization, A.E.P.A., A.G.F., S.P.L.; supervision, A.G.F., S.P.L.; project administration, A.G.F.; funding acquisition, A.G.F. All authors have read and agreed to the published version of the manuscript.

Funding: Funding for this research was provided by the U.S. Bureau of Reclamation/New Mexico State University Cooperative Agreement R16AC00002 and the State of New Mexico Legislature NMWRRI2021, administered by the New Mexico Water Resources Research Institute.

Institutional Review Board Statement: Not applicable.

Informed Consent Statement: Not applicable.

Data Availability Statement: Please refer to the Supplementary Materials for access to the model and the model equations.

Acknowledgments: Thank you to Carolina Mijares at the New Mexico Water Resources Research Institute for support managing the projects under which this research was carried out.

Conflicts of Interest: The authors declare no conflict of interest.

References

1. Famiglietti, J.S. The global groundwater crisis. *Nat. Clim. Chang.* **2014**, *4*, 945–948. [CrossRef]
2. Konikow, L.F.; Kendy, E. Groundwater depletion: A global problem. *Hydrogeol. J.* **2005**, *13*, 317–320. [CrossRef]
3. Siebert, S.; Burke, J.; Faures, J.M.; Frenken, K.; Hoogeveen, J.; Döll, P.; Portmann, F.T. Groundwater use for irrigation—A global inventory. *Hydrol. Earth Syst. Sci.* **2010**, *14*, 1863–1880. [CrossRef]
4. Wada, Y.; Van Beek, L.P.H.; Van Kempen, C.M.; Reckman, J.W.T.M.; Vasak, S.; Bierkens, M.F.P. Global depletion of groundwater resources. *Geophys. Res. Lett.* **2010**, *37*, 37. [CrossRef]
5. IGRAC (International Groundwater Resources Assessment Centre). *Transboundary Aquifers of the World*; UNESCO International Hydrologic Programme; IGRAC: Delft, The Netherlands, 2015.
6. Wada, Y.; Heinrich, L. Assessment of transboundary aquifers of the world—vulnerability arising from human water use. *Environ. Res. Lett.* **2013**, *8*, 024003. [CrossRef]
7. Eckstein, G.; Eckstein, Y. A Hydrogeological Approach to Transboundary Ground Water Resources and International Law. *Am. Univ. Int. Law Rev.* **2003**, *19*, 201–258.
8. Ganoulis, J.; Fried, J. *Transboundary Hydro-Governance: From Conflict to Shared Management*; Springer: Cham, Switzerland, 2018. Available online: https://libezp.nmsu.edu:2176/lib/nmsu/reader.action?docID=5419660&ppg=10 (accessed on 1 February 2021).
9. Albrecht, T.R.; Varady, R.G.; Zuniga-Teran, A.; Gerlak, A.K.; Staddon, C. Governing a shared hidden resource: A review of governance mechanisms for transboundary groundwater security. *Water Secur.* **2017**, *2*, 43–56. [CrossRef]
10. Tapia-Villaseñor, E.; Megdal, S. The U.S.-Mexico Transboundary Aquifer Assessment Program as a Model for Transborder Groundwater Collaboration. *Water* **2021**, *13*, 530. [CrossRef]
11. Forrester, J.W. *Industrial Dynamics*; Productivity Press: Cambridge, MA, USA, 1961.
12. Winz, I.; Brierley, G.; Trowsdale, S. The Use of System Dynamics Simulation in Water Resources Management. *Water Resour. Manag.* **2008**, *23*, 1301–1323. [CrossRef]
13. Sterman, J.D. All models are wrong: Reflections on becoming a systems scientist. *Syst. Dyn. Rev.* **2002**, *18*, 501–531. [CrossRef]
14. Blair, P.; Buytaert, W. Socio-hydrological modelling: A review asking "why, what and how?". *Hydrol. Earth Syst. Sci.* **2016**, *20*, 443–478. [CrossRef]
15. Sterman, J. *Business Dynamics: Systems Thinking and Modeling for a Complex World*; Irwin/McGraw-Hill: New York, NY, USA, 2000. Available online: https://books.google.com/books?id=CCKCQgAACAAJ (accessed on 1 February 2021).
16. Glaubitz, A.; Fu, F. Oscillatory dynamics in the dilemma of social distancing. *Proc. R. Soc. A Math. Phys. Eng. Sci.* **2020**, *476*, 20200686. [CrossRef]
17. Bingaman, J. United States-Mexico Transboundary Aquifer Assessment Act. In *International Groundwater Law and the US-Mexico Border Region*; Brill Nijhoff: Leiden, The Netherlands, 2006.
18. Page, A.; Langarudi, S.; Forster-Cox, S.; Fernald, A. A Dynamic Hydro-Socio-Technical Policy Analysis of Transboundary Desalination Development. *J. Environ. Account. Manag.* **2019**, *7*, 91–118. [CrossRef]
19. Page, A. A Public Policy Modeling Analysis of the Sustainability of Transboundary Desalination within the Sunland Park-Santa Teresa-San Jerónimo-Anapra Hydrologic-Social System. Master's Thesis, New Mexico State University, Las Cruces, NM, USA, 2018.
20. Di Baldassarre, G.; Wanders, N.; AghaKouchak, A.; Kuil, L.; Rangecroft, S.; Veldkamp, T.I.E.; Garcia, M.; van Oel, P.; Breinl, K.; Van Loon, A.F. Water shortages worsened by reservoir effects. *Nat. Sustain.* **2018**, *1*, 617–622. [CrossRef]

21. Di Baldassarre, G.; Sivapalan, M.; Rusca, M.; Cudennec, C.; Garcia, M.; Kreibich, H.; Konar, M.; Mondino, E.; Mård, J.; Pande, S.; et al. Sociohydrology: Scientific Challenges in Addressing the Sustainable Development Goals. *Water Resour. Res.* **2019**, *55*, 6327–6355. [CrossRef]
22. Gates, J.S.; Stanley, W.; Ackermann, H. Availability of fresh and slightly saline ground water in the basins of westernmost Texas. In *Open-File Report*; US Geological Survey: Reston, VA, USA, 1978; pp. 1258–2331.
23. Hawley, J. *Challenges and Opportunities for Brackish Groundwater Resource Development in New Mexico—Prediction Hydro-Science from an Octogenarian Hydrogeologist's Perspective*; White Paper; Hawley Geomatters: Albuquerque, NM, USA, April 2016.
24. Hawley, J.; Kennedy, J.; Creel, B. The Mesilla Basin Aquifer System of New Mexico, West Texas, and Chihuahua—An Overview of Its Hydrogeologic Framework and Related Aspects of Groundwater Flow and Chemistry. In *Aquifers of West Texas*; Angle, E.S., Mace, R.E., Mullican, W.F., III, Eds.; Texas Water Development Board: Austin, TX, USA, 2001; pp. 76–99.
25. Ojeda, G. Aquifer recharge estimation at the Mesilla Bolson and Guaymas aquifer systems, Mexico. In *Isotope Based Assessment of Groundwater Renewal in Water Scarce Regions*; Edmunds, W.M., Ed.; International Atomic Energy Agency: Vienna, Austria, 2001; pp. 23–44.
26. Sheng, Z. Impacts of groundwater pumping and climate variability on groundwater availability in the Rio Grande Basin. *Ecosphere* **2013**, *4*, 1–25. [CrossRef]
27. Texas Water Development Board and New Mexico Water Resources Research Institute. *Transboundary Aquifers of the El Paso/Ciudad Juarez/Las Cruces Region*; US Environmental Protection Agency: Washington, DC, USA, October 1997.
28. Witcher, J.; King, J.P.; Hawley, J.; Williams, J.; Cleary, M.; Bothern, L. *Sources of Salinity in the Rio Grande and Mesilla Basin Groundwater*; WRRI Technical Completion Report 330; New Mexico Water Resources Research Institute: Las Cruces, NM, USA, February 2004.
29. Rivera, A. Transboundary aquifers along the Canada–USA border: Science, policy and social issues. *J. Hydrol. Reg. Stud.* **2015**, *4*, 623–643. [CrossRef]
30. Maupin, M.A.; Kenny, J.F.; Hutson, S.S.; Lovelace, J.K.; Barber, N.L.; Linsey, K.S. Estimated use of water in the United States in 2010. In *Circular 1405*; US Geological Survey: Reston, VA, USA, 2014. [CrossRef]
31. Cooley, H.; Donnelly, K. *Proposed Seawater Desalination Facilities in California*; The Pacific Institute: Oakland, CA, USA, 2012. Available online: http://pacinst.org/app/uploads/2014/04/desalination-facilities.pdf (accessed on 27 January 2017).
32. Selcraig. Albuquerque learns it really is a desert town. *High Country News*, 26 December 1994.
33. Indicators Progress Commission. *City of Albuquerque Progress Report 2004*; City of Albuquerque: Albuquerque, NM, USA, 2004.
34. Indicators Progress Commission. *City of Albuquerque Progress Report 2020*; City of Albuquerque: Albuquerque, NM, USA, 2020.
35. Albuquerque Bernalillo County Water Utility Authority. *2015 Annual Report: Groundwater on the Rebound*; Albuquerque Bernalillo County Water Utility Authority: Albuquerque, NM, USA, 2015.
36. Fleck, J. Aquifer rising despite drought. *Albuquerque Journal*, 6 October 2013.
37. Davis, T. Aquifer on the rebound. *Albuquerque Journal*, 23 August 2019.
38. Cobos, G.D.L. The Genevese transboundary aquifer (Switzerland-France): The secret of 40 years of successful management. *J. Hydrol. Reg. Stud.* **2018**, *20*, 116–127. [CrossRef]
39. Tversky, A.; Kahneman, D. Judgment under uncertainty: Heuristics and biases. *Science* **1974**, *185*, 1124–1131. [CrossRef]
40. Simon, H.A. Bounded Rationality. In *Utility and Probability*; Eatwell, J., Milgate, M., Newman, P., Eds.; Palgrave Macmillan: London, UK, 1990; pp. 15–18. [CrossRef]
41. Roumasset, J.A.; Wada, C.A. Chapter 5—Integrated Groundwater Resource Management. In *Sustainable Economic Development*; Balisacan, A.M., Chakravorty, U., Ravago, M.-L.V., Eds.; Academic Press: San Diego, CA, USA, 2015; pp. 77–89. [CrossRef]
42. Petersen-Perlman, J.D.; Megdal, S.B.; Gerlak, A.K.; Wireman, M.; Zuniga-Teran, A.A.; Varady, R.G. Critical Issues Affecting Groundwater Quality Governance and Management in the United States. *Water* **2018**, *10*, 735. [CrossRef]
43. Björk, B.-C.; Solomon, D. The publishing delay in scholarly peer-reviewed journals. *J. Inf.* **2013**, *7*, 914–923. [CrossRef]
44. Horbach, S.P.J.M. Pandemic publishing: Medical journals strongly speed up their publication process for COVID-19. *Quant. Sci. Stud.* **2020**, *1*, 1056–1067. [CrossRef]
45. Lipworth, W.; Gentgall, M.; Kerridge, I.; Stewart, C. Science at Warp Speed: Medical Research, Publication, and Translation During the COVID-19 Pandemic. *J. Bioethical Inq.* **2020**, *17*, 555–561. [CrossRef]
46. Hanna-Attisha, M.; Lachance, J.; Sadler, R.C.; Schnepp, A.C. Elevated Blood Lead Levels in Children Associated With the Flint Drinking Water Crisis: A Spatial Analysis of Risk and Public Health Response. *Am. J. Public Health* **2016**, *106*, 283–290. [CrossRef] [PubMed]
47. Simpson, N.; Shearing, C.; Dupont, B. Climate gating: A case study of emerging responses to Anthropocene Risks. *Clim. Risk Manag.* **2019**, *26*, 100196. [CrossRef]
48. Visser, W.P. A perfect storm: The hydropolitics of Cape Town's water crisis. *J. Transdiscipl. Res. S. Afr.* **2018**, *14*, 10. [CrossRef]
49. Bradley, J. Morrison and Rogelio Oliva, Integration of Behavioral and Operational Elements Through System Dynamics. In *The Handbook of Behavioral Operations*; Wiley: New York, NY, USA, 2017. Available online: https://papers.ssrn.com/abstract=2997235 (accessed on 1 August 2021).
50. Langarudi, S.P.; Silva, C.G.; Fernald, A.G. Measure more or report faster? Effect of information perception on management of the commons. *Syst. Dyn. Rev.* **2021**, *37*, 72–92. [CrossRef]

51. Arrhenius, S. XXXI. On the influence of carbonic acid in the air upon the temperature of the ground. *Lond. Edinb. Dublin Philos. Mag. J. Sci.* **1896**, *41*, 237–276. [CrossRef]
52. Marx, W.; Haunschild, R.; Thor, A.; Bornmann, L. Which early works are cited most frequently in climate change research literature? A bibliometric approach based on Reference Publication Year Spectroscopy. *Scientometrics* **2017**, *110*, 335–353. [CrossRef]
53. Langarudi, S.; Sabie, R.; Bahaddin, B.; Fernald, A. A Literature Review of Hybrid System Dynamics and Agent-Based Modeling in a Produced Water Management Context. *Modelling* **2021**, *2*, 12. [CrossRef]

MDPI
St. Alban-Anlage 66
4052 Basel
Switzerland
Tel. +41 61 683 77 34
Fax +41 61 302 89 18
www.mdpi.com

Water Editorial Office
E-mail: water@mdpi.com
www.mdpi.com/journal/water

www.ingramcontent.com/pod-product-compliance
Lightning Source LLC
Chambersburg PA
CBHW061607220926
51850CB00005B/444

* 9 7 8 3 0 3 6 5 8 4 5 1 5 *